Chemical Engineering at Supercritical Fluid Conditions

Chemical Engineering at Supercritical Fluid Conditions

Edited by

Michael E. Paulaitis
Johan M. L. Penninger
Ralph D. Gray, Jr.
Phillip Davidson

Copyright © 1983 by Ann Arbor Science Publishers
230 Collingwood, P.O. Box 1425, Ann Arbor, Michigan 48106

Library of Congress Catalog Card Number 82-71529
ISBN 0-250-40564-4

Manufactured in the United States of America
All Rights Reserved

Butterworths, Ltd., Borough Green, Sevenoaks
Kent TN15 8PH, England

PREFACE

Although the concept of a dense gas or supercritical fluid as a solvent has been established for nearly a century, there is renewed interest in supercritical fluids resulting from potential applications in synfuels processes, improved separation methods, and as non-toxic, non-carcinogenic solvents. In this respect, the term "supercritical fluids" is a rather restricted description of these research interests, as well as of the subject matter presented in this book. Much of this material is concerned with investigations of phase equilibria at elevated pressures. The important difference, however, between this latter more general area of research interest and that involving "supercritical fluids" is in emphasis and hopefully application. Whereas a comprehensive knowledge of phase equilibria at elevated pressures is undoubtedly invaluable for a large number of chemical engineering processes, interest in supercritical fluids implies an extension of this knowledge to ultimately using high pressure phase equilibria phenomena—such as solvents that are highly compressible at certain operating conditions—to accomplish desired separations or other process objectives. This approach is chemical engineering oriented, but also requires an understanding of phase equilibrium behavior at elevated pressures, which is just now beginning to evolve within the field of chemical engineering. It is hoped that this book will contribute to the present understanding of phase equilibria at elevated pressures, and provide worthwhile directions to explore new chemical engineering applications for supercritical fluids.

Michael E. Paulaitis
Johan M. L. Penninger
Ralph D. Gray, Jr.
Phillip Davidson

ACKNOWLEDGMENTS

The work presented in this book represents the collective effort of a large number of people to whom we are greatly indebted. We especially express our appreciation to the authors for their contributions, including preparation of the manuscripts. We would also like to thank the following people for reviewing various parts of the book: G. Alexander, T. Bergstresser, G. Brunner, W. H. Corcoran, J. DiAndreth, W. O. Eisenbach, K. E. Gubbins, R. Kander, M. B. King, M. T. Klein, M. A. McHugh, S. Murad, J. P. O'Connell, M. Radosz, S. I. Sandler, E. Slocum, M. Thies, C. Torres, H. J. Trappeniers, and R. Weber. Finally, we are indebted to John O'Connell, chairman of the Thermodynamics and Transport Properties Committee of the AIChE, and Bori Franko, chairman of the High Pressure Committee of the AIChE, for their support in organizing the symposium which eventually led to this book.

Michael E. Paulaitis received his PhD in Chemical Engineering from the University of Illinois before joining the faculty of the University of Delaware, where he is currently Associate Professor of Chemical Engineering. He is also the Associate Editor of *IEC Fundamentals* and a consultant to Dow Chemical Company, Gulf Research and Development Company, and the Procter and Gamble Company in the area of supercritical fluid separations. His research interests include phase equilibria at elevated pressures and solvent extraction with supercritical fluid solvents.

Johan M. L. Penninger obtained his education in his native country, the Netherlands. There he attended the Eindhoven University of Technology and obtained a doctorate in Chemical Technology. His career swung between industrial and educational positions in different countries, and covered a wide range of the engineering profession (Middle East Technical University, Ankara, Turkey; International Business Development with Akzo Zout Chemie, the Netherlands; University of Cincinnati, Ohio; Occidental Research Corporation, Irvine, California). After almost five years in the U.S., Dr. Penninger returned to the Netherlands to assume his current position with Akzo Zout Chemie as Head Scouting Research. He has published 25 papers in areas of thermal hydrocracking, hydroformulation catalysis, in situ infrared spectroscopy of reacting species at high temperature and pressure, and design of specific high-pressure equipment. He is a member of the Royal Institute of Engineers (the Netherlands) and the American Chemical Society, and is an officer of the High Pressure Committee of the American Institute of Chemical Engineers.

Ralph D. Gray, Jr. is currently a Senior Staff Engineer with Exxon Research & Engineering Company in Florham Park, New Jersey, and an Adjunct Professor of Chemical Engineering at Manhattan College. He received his PhD in Chemical Engineering from the University of Delaware in 1965 before joining Exxon as a member of the applied thermodynamics group. His professional interests include applications of thermodynamics to process calculations and phase equilibria at elevated pressures.

Phillip Davidson is Sales Manager of Superpressure, Inc., a firm specializing in the manufacture of high-pressure equipment, industrial diaphragm-type compressors and materials test equipment. He is also Sales Manager of Hygrodynamics, Inc., a sister company specializing in the manufacture of precision humidity measurement and control apparatus. Mr. Davidson is a long-standing member of the American Chemical Society (ACS) and the American Institute of Chemical Engineers (AIChE), and recently became a member of the American Society for Testing and Materials (ASTM). He is currently serving on the AIChE High Pressure Committee and the ASTM Committee G-4 on the Compatibility and Sensitivity of Materials in Oxygen-Enriched Atmospheres. Mr. Davidson holds a BS in Chemical Engineering from Northeastern University, Boston, Massachusetts, and has more than 30 years of experience in the engineering, application and sale of high-pressure apparatus.

This book is dedicated to Dick Greiger-Block.

CONTENTS

Part I
Experimental Data on Phase Equilibrium Behavior

Overview of Part I 1
M. E. Paulaitis

1. Phase Equilibria In Fluid and Solid Mixtures At High Pressure 3
   W. B. Streett

2. Some Vapour/Liquid and Vapour/Solid Equilibrium
   Measurements of Relevance for Supercritical Extraction
   Operations, and Their Correlation 31
   *M. B. King, D. A. Alderson, F. H. Fallah, D. M. Kassim,
   K. M. Kassim, J. R. Sheldon, and R. S. Mahmud*

3. The Rapid Depressurization of Hot, High Pressure Liquids
   or Supercritical Fluids 81
   M. E. Kim-E and R. C. Reid

4. Solubility of Oxygenated Hydrocarbons In Supercritical
   Carbon Dioxide 101
   M. S. Kuk and J. C. Montagna

5. High Pressure Fluid Phase Equilibria of
   Alcohol-Water-Supercritical Solvent Mixtures 113
   M. A. McHugh, M. W. Mallett, and J. P. Kohn

6. Solid Solubilities In Supercritical Fluids at Elevated Pressures 139
   M. E. Paulaitis, M. A. McHugh, and C. P. Chai

7. An Experimental Method for Measuring Solubilities of
   Heavy Fossil-Fuel Fractions In Compressed Gases to
   100 Bar and 300°C 159
   A. Monge and J. M. Prausnitz

8. Measuring the Properties of Petroleum Reservoir Fluids up to 20,000 psia (138 MPa) and 400°F (200°C) 173
   R. Simon

Part II
Thermodynamic Theories and Equations of State

Overview of Part II 183
   R. D. Gray, Jr.

9. Three-Phase Equilibrium and the Tricritical Point 185
   B. Widom

10. Thermodynamic Models for Fluid Mixtures Near Critical Conditions 199
    J. C. Rainwater and M. R. Moldover

11. Molecular Thermodynamics of Dilute Solutes In Supercritical Solvents 221
    D. A. Jonah, K. S. Shing, V. Venkatasubramanian, and K. E. Gubbins

12. Mean-Field Lattice-Gas Description of Fluid-Phase Equilibria 245
    L. A. Kleintjens and R. Koningsveld

13. Binary Phase Diagrams from a Cubic Equation of State ... 263
    G. T. Hong and M. Modell

14. Phase Equilibria of High-Boiling Organic Solutes In Compressed Supercritical Fluids—Equation of State with New Mixing Rule 323
    K. W. Won

15. The Correlation and Prediction of Critical States of Mixtures Using a Corresponding States Principle 341
    A. S. Teja and R. L. Smith

16. Corresponding States Theories for Chain Molecules 359
    C. K. Hall and B. A. Hacker

Part III
Applications

Overview of Part III 375
   J. M. L. Penninger

17. Experimental Observations on a Systematic Approach to Supercritical Extraction of Coal 377
   W. S. Fong, P. C. F. Chan, P. Pichaichanarong, W. H. Corcoran, and D. D. Lawson

18. Liquefaction of Lignite Using Low Cost Supercritical Solvents 395
   W. P. Scarrah

19. The Supercritical Gas Extraction of Lignites and Wood . 409
   A. Olcay, T. Tugrul, and A. Calimli

20. Supercritical Fluid Extraction of Oil Sands and Residues from Oil and Coal Hydrogenation 419
   W. O. Eisenbach, K. Niemann, and P. J. Göttsch

21. Separation of Finely Dispersed Solids from Low-Volatile Viscous Media by Gas Extraction 435
   D. Stützer, G. Brunner, and S. Peter

22. The Preparation of Acid-Catalyzed Silica Aerogel 445
   W. J. Schmitt, R. A. Greiger-Block, and T. W. Chapman

23. The Adsorption of Phenol from Dense Carbon Dioxide onto Activated Carbon 461
   R. G. Kander and M. E. Paulaitis

24. Analysis of Dense (Supercritical) Gas Systems 477
   L. G. Randall

25. Liquefaction of Biomass with Supercritical Fluids In a High Pressure/High Temperature Flow Reactor 499
   P. Köll, B. Brönstrup, and J. O. Metzger

26. Thermal Organic Reactions In Supercritical Fluids 515
   J. O. Metzger, J. Hartmanns, D. Malwitz, and P. Köll

Index 535

OVERVIEW

PART I. EXPERIMENTAL DATA ON
PHASE EQUILIBRIUM BEHAVIOR

Michael E. Paulaitis
  Department of Chemical Engineering
  University of Delaware
  Newark, DE 19711

Part I covers phase equilibrium behavior for mixtures at elevated pressures with emphasis on experimental work. Chapter 1 reviews several different classes of phase diagrams for binary mixtures, and discusses qualitative features of these diagrams relevant to supercritical-fluid (SCF) solvent extraction processes. An analysis of solubility enhancements for solids in supercritical fluids illustrates how a comprehensive understanding of the high-pressure phase equilibrium behavior for binary mixtures can provide useful concepts for practical applications of SCF solvents. Another such illustration, given in Chapter 6, deals specifically with phase equilibria at elevated pressures for solid-fluid, binary mixtures. An experimental study of biphenyl-carbon dioxide mixtures coupled with a thermodynamic analysis based on the Peng-Robinson equation of state suggests a novel method for separating solid solutes from SCF solvents.

Experimental studies of the phase equilibrium behavior for alcohol-water-SCF solvent systems are described in Chapters 4 and 5, and the results are used to evaluate the feasibility of extracting alcohols from aqueous solutions using SCF solvents. These extractions represent a large class of important separations which involve the recovery of organic chemicals from aqueous solutions. Measured SCF solvent loadings and alcohol/water selectivities as a function of solvent density for carbon dioxide (Chapter 4) and for ethane (Chapter 5) are discussed in relationship to optimizing the SCF solvent extraction and to developing energy-efficient techniques for separating SCF solvents and extracts.

A thermodynamic study of rapid depressurization of compressed superheated liquids and supercritical fluids is described in Chapter 3. An analysis based upon the Peng Robinson equation of state is developed to identify temperatures and pressures at which compressed carbon dioxide will reach a state where explosive depressurization can occur. Experiments with compressed carbon dioxide did not produce a depressurization explosion at the predicted conditions, but examples of explosive depressurizations reported previously in the literature are shown that suggest this phenomenon can be a potential hazard in handling compressed gases.

The remaining chapters in Part I emphasize experimental methods for measuring phase equilibria at elevated pressures relevant to SCF solvent extraction processes. In Chapter 2, various experimental techniques are described and new experimental data are reported for a number of binary and ternary mixtures. In Chapters 7 and 8, new experimental techniques are presented. A new method for measuring solubilities in compressed gases and supercritical fluids is described in Chapter 7 which has the advantage that solubilities of complex solutes, such as coal tar fractions, can be determined. Experimental solubilities in compressed methane and methane/water mixtures are reported for well-defined hydrocarbon mixtures and for two narrow-boiling heavy fossil-fuel fractions. In Chapter 8, a new visual cell system is described for measuring properties of reservoir fluids. Measurements include quantities, compositions, densities, viscosities, and interfacial tensions of vapor and liquid phases in equilibrium at pressures up to 20,000 PSIA and temperatures up to 200°C. Special features of the experimental apparatus and the methods used for measurements are discussed.

In general, these first eight chapters demonstrate an important relationship between understanding phase equilibria for simple binary and ternary mixtures at elevated pressures, and the development of SCF solvent extraction processes. The results from these experimental studies and the new experimental techniques provide important information for further work on phase equilibria at elevated pressures and ultimately for future advances in processes utilizing SCF solvents.

CHAPTER 1

PHASE EQUILIBRIA IN FLUID AND
SOLID MIXTURES AT HIGH PRESSURE

W.B. Streett
School of Chemical Engineering
Cornell University

ABSTRACT

Following a brief discussion of the application of phase rule principles to one- and two-component phase diagrams, the three-dimensional features of several important classes of pressure-temperature-composition phase diagrams for binary mixtures are described in detail. Two- and three-phase equilibria between gas, liquid and solid phases are included in the discussion, and the emphasis is on understanding the qualitative features of three-dimensional PTX diagrams. The importance of certain features in supercritical fluid extraction is pointed out.

1. INTRODUCTION

Supercritical extraction depends upon the ability of supercritical fluids to dissolve selectively varying amounts of relatively nonvolatile substances. The solvent properties of the fluid can be varied within wide limits by changing its pressure and temperature. Although supercritical extraction involves multicomponent systems, the limiting case of a binary system consisting of a supercritical solvent and a single solute provides a convenient approach to understanding the important phase equilibrium principles upon which the process depends. We consider here some important features of two-component phase diagrams in pressure-temperature-composition (PTX) space.

Early studies of fluid mixtures revealed new phenomena that are not present in pure fluids. The most important of these are the additional types of phase equilibria that arise from the extra degrees of freedom introduced by varying the number and proportions of the components. The variety and complexity of phase behavior observed in early experiments, even in two-component mixtures of relatively simple molecules, seemed, at first, chaotic. The discovery of the phase rule by Gibbs, in 1875, brought a measure of order by providing a framework for the interpretation and classification of phase diagrams, and led to a period of intensive experimental study, lasting until about 1915.

During this period most of the known types of fluid phase diagrams were discovered; however, much of what was learned in the early years about phase equilibria and critical phenomena in fluid mixtures gradually disappeared from the literature of physics and chemistry.

A revival of interest in experimental thermodynamics of fluids was brought about in the second quarter of this century by the growth of chemical engineering technology, especially in the natural gas and petroleum industries. Many properties of multicomponent phase diagrams were rediscovered and further explored by chemical engineers, and the number of systems subjected to study was greatly expanded. Concurrently, the development of high pressure technology led to systematic studies, by physicists and engineers, of the effects of high pressure on phase equilibria in fluid systems. The systematic application of pressure in the study of fluid phase equilibria, and in particular the effects of pressure on critical properties, brought new insights into the relations between the three types of two-phase equilibria (liquid-liquid, gas-liquid, and gas-gas) that occur in fluid systems. Experiments of the last two decades, especially those of G.M. Schneider, have shown that there are continuous transitions between these three types [1-3]. A picture of the unity and continuity of critical phenomena in fluid mixtures has gradually emerged, bringing a greater order to the classification and interpretation of many types of phase behavior that previously seemed unrelated. The understanding of these phenomena is only slowly being assimilated into the current literature of chemistry and chemical engineering, mainly because of the difficulties of interpreting and explaining multi-dimensional phase diagrams. The phase behavior that is utilized in modern supercritical extraction processes has long been understood, but only after the discovery in the 1960's of the ability of supercritical gases to absorb nonvolatile substances selectively, were the commercial applications realized.

The independent variables most convenient for the measurement and study of phase equilibria in fluid systems are pressure and temperature. In fluid systems changes in both temperature and pressure produce dramatic changes in phase behavior, and a three-dimensional diagram in pressure, temperature and a third variable is required for complete description of a two-component system. For qualitative descriptions and comparisons of phase diagrams the most convenient choice for this third variable is composition. PTX diagrams provide a basis for the design of separation processes such as distillation and extraction.

The following sections are devoted to the interpretation of several types of three-dimensional PTX diagrams for binary mixtures. The study of these diagrams is largely an exercise in descriptive geometry, requiring the visualization of lines and surfaces in three dimensions. Schematic three-dimensional drawings are used, together with two-dimensional diagrams cut by planes of constant T or P, or formed by projections of lines and points on the PT coordinate planes. The emphasis here is on the qualitative features of PTX diagrams. Discussions of the thermodynamics of fluid phase equilibria can be found in the books by Rowlinson [4], Prausnitz [5], King [6] and in similar works.

2. THE PHASE RULE AND THE INTERPRETATION OF PHASE DIAGRAMS

In the form applicable to nonreacting systems the phase rule is expressed by the simple relation

$$\underline{f} = \underline{c} + 2 - \underline{p}, \qquad (1)$$

where $\underline{f}$ is the number of independent variables (sometimes called "degrees of freedom"), $\underline{c}$ is the number of components and $\underline{p}$ the number of phases. Since the maximum number of independent variables is two for a one-component system and three for a two-component system, the phase behavior of these systems can be completely described by volumes, surfaces, lines and points in three-dimensional space. For systems of three or more components, complete description of the phase behavior requires diagrams of higher dimensionality, although two- and three-dimensional diagrams can effectively describe certain limiting features of the more complex multi-dimensional diagrams.

In the construction and interpretation of phase diagrams the phase rule serves as an important guide; it imposes definite constraints on the geometry of the features that describe the existence or coexistence of fixed numbers of phases. Another useful concept in the analysis of phase diagrams—one that is always used but seldom explained in textbooks of thermodynamics and physical chemistry—is the distinction between "field" and "generalized density" variables. Field variables are defined to be those that are the same in all phases at equilibrium, such as pressure, temperature and chemical potential; generalized density variables are those that are different for different phases, such as density, composition, internal energy, index of refraction, etc. The distribution between field and density variables can impose additional geometrical constraints on the resulting phase diagram. The geometrical constraints imposed by the phase rule on the equilibrium between a fixed number of phases, in one- and two-component systems, are summarized in the following table:

Number of Phases in Equilibrium

One Component	Two Components	Degrees of Freedom	Geometrical Features
3	4	0	points
2	3	1	lines
1	2	2	surfaces
-	1	3	volumes

A system of n phases at equilibrium is described by n geometrical features of the appropriate type: a one-component system of three phases is described by three points, a two-component system of three phases is described by three lines, etc. In the most general case no further conclusions can be drawn about the mutual relations in space of the points, lines, surfaces and volumes that make up the phase diagram. For each phase rule variable that is a field variable, there is a degeneracy that reduces the total number of variables by $p - 1$, where p is the number of phases in a multiphase system (i.e., $p > 1$); this results in further geometrical constraints that greatly simplify the structure and interpretation of many of the fluid phase diagrams commonly encountered in chemistry and chemical engineering. In experimental studies of phase equilibria, the two intensive properties most commonly measured are P and T, both field variables.

The geometry of the points, lines and surfaces in the PVT diagram of a one-component system, consisting of solid, liquid and gas phases, is shown in figure 1 for the "normal" case where the liquid contracts on freezing. The gas-liquid critical point, C, is defined as the point at which the gas and liquid phases become identical. The equality of properties reduces the number of degrees of freedom to zero, hence the critical point is a unique point in PVT space. The pairs of lines that represent two coexisting phases are: (1) A'D' and A"D", solid and liquid phases; (2) A"C and A'"C, liquid and gas phases; and (3) A'B' and A'"B'", solid and gas phases. Because P and T are field variables, each pair of lines lies in a ruled surface perpendicular to the PT plane, and has a common projection on that plane, as shown in the PT projection at the right. The three projections AC, AD and AB are the vapor pressure, melting and sublimation curves. The three points that represent three coexisting phases, A', A" and A'", project as the single point A in the PT plane. The projection of the two-phase lines on the PV plane is shown on the left. The

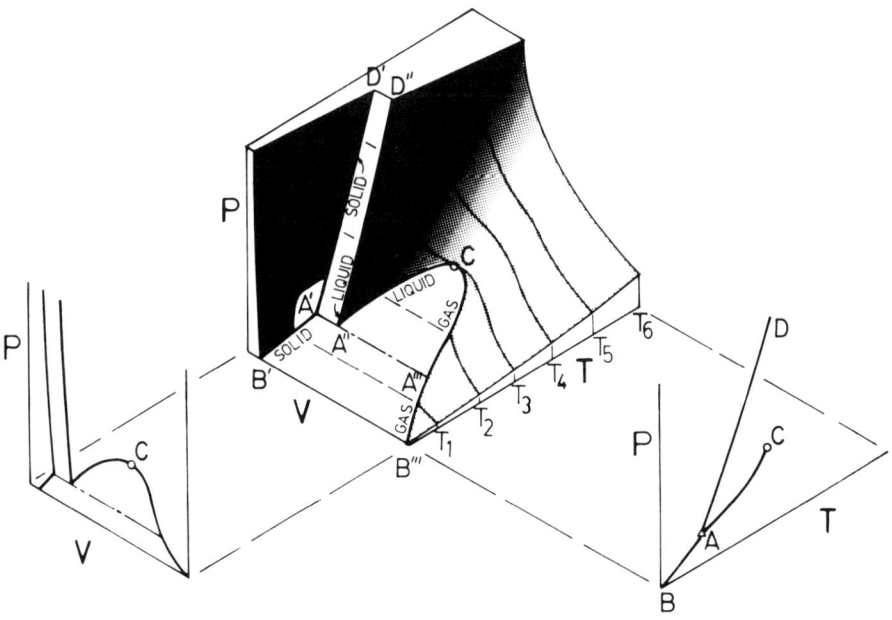

Figure 1. PVT diagram for a pure substance.

six curved lines extending across the PVT surface in the central figure are isotherms cut by planes of constant temperature $T_1 \ldots T_6$.

In experimental studies of two-component systems, the properties most commonly measured are P, T, and X, a combination of two field variables (P and T) and one generalized density (X). If X_1 is the mole fraction of component 1, the mole fraction of the second is $X_2 = 1 - X_1$, hence the composition of a phase is defined by a single variable. Regions of two-, three- and four-phase equilibrium are described in PTX space by pairs of surfaces, triplets of lines and quadruplets of points, respectively. The geometrical constraints imposed by the use of two field variables require that these features have common PT projections; i.e. two surfaces representing two coexisting phases project as a single surface, three lines representing three coexisting phases project as a single line, etc.

The simplest type of PTX diagram for a fluid system is one that describes the gas-liquid equilibrium of a system in which the liquids are miscible in all proportions. An example is shown in figure 2a. Lines $A_\alpha C_\alpha$ and $A_\beta C_\beta$ are the vapor pressure curves of the pure components, α and

7

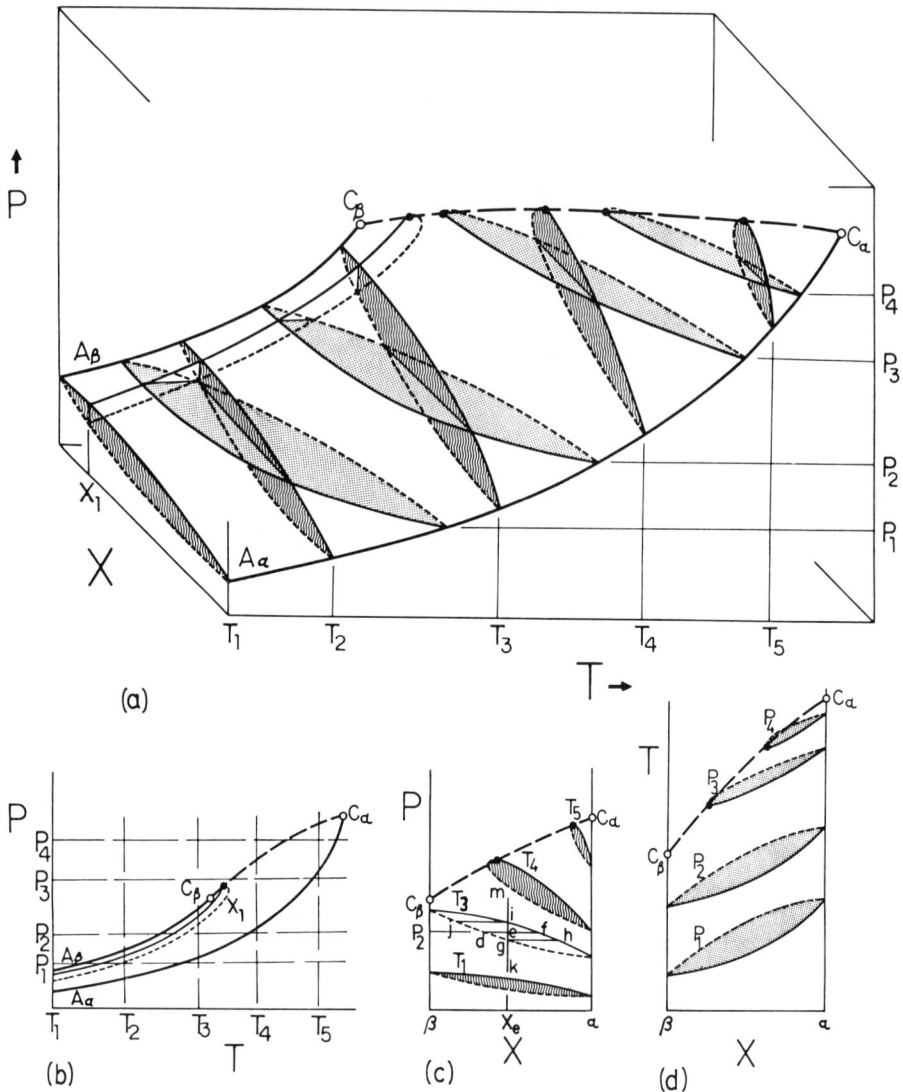

Figure 2. PTX diagram for a class I system.

β, in the two PT planes of the diagram, ending in critical
points C_α and C_β. The mixture critical line $C_\alpha C_\beta$
is continuous in PTX space. The two surfaces representing
saturated gas and liquid phases extend across the diagram,
and are bounded in space by the lines $A_\alpha C_\alpha$, $A_\beta C_\beta$
and $C_\alpha C_\beta$. With the P-axis vertical, the lower surface
represents the gas phase and the upper surface the liquid

phase. It is difficult to use shading of these surfaces to show their structure, because one would be obscured by the other. The convention used in figure 2, and elsewhere in this paper, is to use sections cut by planes of constant P, T, and X, commonly called isobars, isotherms and isopleths, and to shade portions of these sections lying <u>within the two-phase region</u>; the boundary lines of these shaded planar sections lie in the surfaces representing two coexisting phases--the surfaces of principal interest. Lines in liquid surfaces are solid lines, lines in gas surfaces are short-dash lines, and critical lines are dark long-dash lines. Pure component critical points are open circles, and mixture critical points (on isotherms, isobars and isopleths) are dark circles. Figure 2b shows projections of the boundary lines of the two-phase region on a PT coordinate plane, with vertical and horizontal lines to mark the isotherms and isobars shown as shaded sections in figure 2a (T_1-T_5 and P_1-P_4). Figures 2c and d show projections of selected isotherms and isobars on PX and TX planes. A horizontal line through one of the isotherms in figure 2c (e.g. def) is a tie line, a line that connects the points representing coexisting gas and liquid phases at fixed P and T. In section 3 the classification of fluid phase diagrams on the basis of critical and three-phase lines is discussed, and several representative diagrams are described in detail. Section 4 is devoted to a discussion of phase diagrams that include solid phases.

3. CLASSIFICATION AND DESCRIPTION OF TWO-COMPONENT FLUID PHASE DIAGRAMS

Recent experimental and theoretical studies have shown that there are continuous transitions between phase diagrams that exhibit gas-liquid, liquid-liquid and gas-gas phase separations. In many systems critical lines change continuously from one type of phase separation to another.

The most important experimental contributions in this field in recent years are those of G.M. Schneider (for reviews see Schneider [1-3], and other references therein). Two features of his work are particularly important: (1) the application of high pressures (up to about 7000 atm) to the study of critical behavior in gas-liquid and liquid-liquid mixtures, and (2) the study of families of two-component systems in which one component remains the same and the other is systematically changed (e.g. CO_2 mixed with families of hydrocarbons of increasing molecular weight). In most of his experiments Schneider concentrated on measurements of the PT traces of critical lines and three-phase lines of the type liquid-liquid-gas. These lines, together with pure-component vapor pressure curves (and in some cases azeotropic lines), form the principal boundaries in PT space of the surfaces representing equilibrium between two fluid phases. Other recent experiments by W.B. Streett (see Streett [7-8], Tsang, et al. [9], and references therein) on binary mixtures in which helium, hydrogen or neon is the light component, have explored gas-gas equilibria and pressure-induced solidification in mixtures of simple molecules at pressures as high as 10,000 atm. In these systems critical lines and three-phase lines of the type gas-liquid-solid have been measured. Plots of these boundary lines on PT diagrams fall naturally into several different types, providing a convenient basis for the classification of fluid phase equilibria.

In the classification of phase diagrams, the most important analytical work is that of Scott and Van Konynenberg [10,11], who demonstrated that most of the experimentally observed binary fluid phase diagrams can be described qualitatively by the van der Waals equation of state. Using as a basis the PT projections of critical and three phase lines resulting from their calculations, Scott and Van Konynenberg grouped fluid phase diagrams into five major classes, designated I to V. They recognized a sixth class that occurs in some aqueous systems, but was not among those predicted by the van der Waals equation. The classification scheme is outlined in figure 3, which illustrates the principal lines (one degree of freedom) and points (zero degrees of freedom) that form the boundaries in PT space of the pairs of surfaces that describe equilibrium between two phases. The types of boundary lines are: (1) pure component vapor pressure curves (solid lines); (2) critical lines (dashed lines); (3) three-phase lines (dash-dot lines). The types of points are: (1) pure component critical points (open circles); and (2) critical end points (triangles), formed by the intersection of a critical line with a three-phase region liquid-liquid-gas. These lines and points provide only a skeletal outline of the phase diagram; to describe fully the different types of phase behavior they represent, it is necessary to consider the three-dimensional structure of the full PTX diagrams. Several examples are considered in the remainder of this section. The diagrams illustrated are purely qualitative, and do not represent specific real systems. In some instances, certain characteristic features of real systems have been exaggerated or disorted, to better illustrate three-dimensional features. In earlier reviews by Schneider [2,3] more emphasis is placed on specific diagrams for real systems.

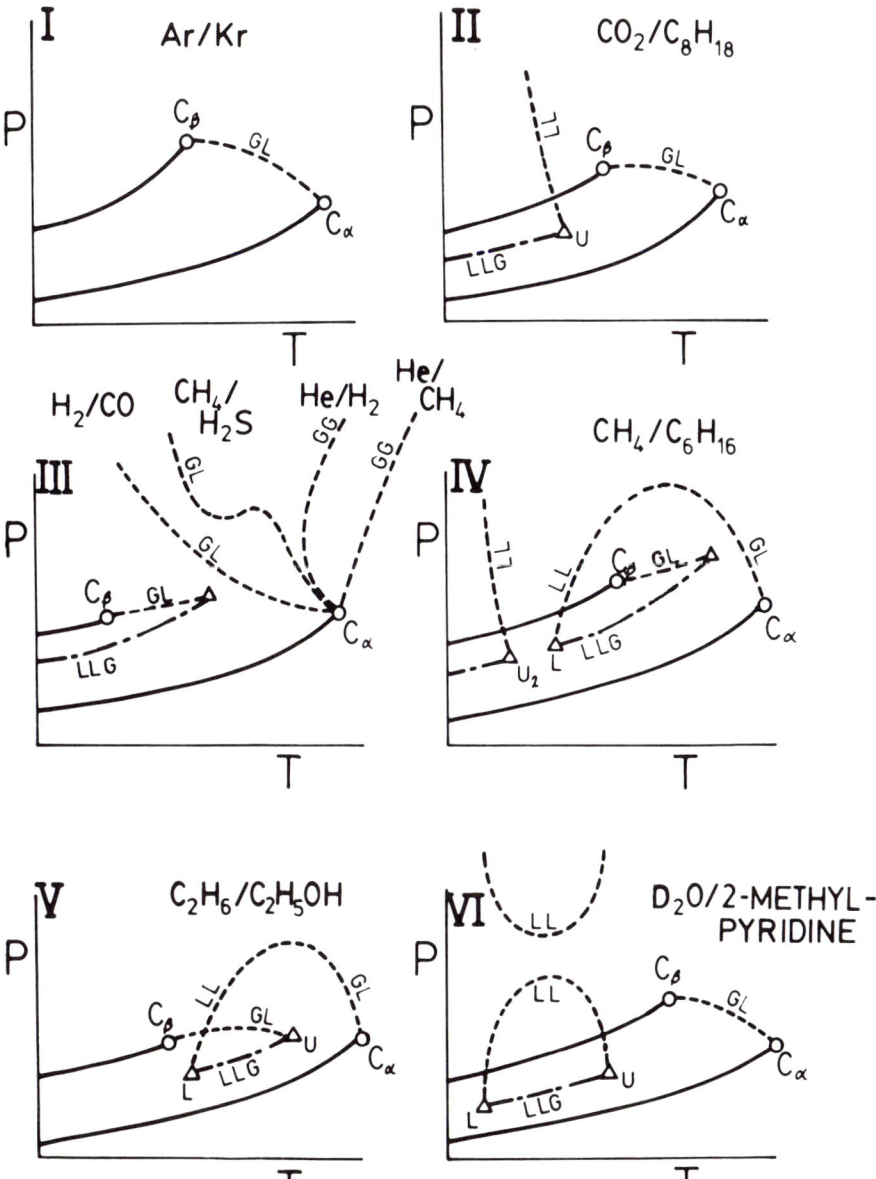

Figure 3. The six classes of fluid phase diagrams, following van Konynenberg and Scott (1980).

Before proceeding it is useful to summarize the main PT features of the six classes and the distinctions between them. Distinctions between classes in figure 3 are drawn mainly on the basis of critical lines. In classes I, II

and VI the gas-liquid critical line is continuous between the critical points of the pure components C_α and C_β.
In class I there is no liquid-liquid phase separation; in class II there is a liquid-liquid phase separation, with a single liquid-liquid critical line bounded at low pressures by an upper critical end point, U, where it intersects the three-phase region liquid-liquid-gas (LLG); in class VI, one or two liquid-liquid critical lines are bounded at low pressures by upper and lower critical end points, U and L.
In classes IV and V, the branch of the gas-liquid critical line originating in C_α has a maximum in temperature and passes continuously into a liquid-liquid critical line, terminating in a lower critical end point. In class IV there is a second liquid-liquid phase separation at lower temperature, with a critical line ending in a second upper critical end point. In the remaining class, III, the branch of the gas-liquid critical line originating in C_α rises to very high pressures, sometimes after passing through maxima and minima in pressure and/or a minimum in temperature. It is likely that these critical lines end at high pressures in a critical end point formed by an intersection with a three-phase region gas-liquid-solid (Streett, 1974), although such points lie at pressures in excess of 15,000 atm in many systems. In some cases the critical line rises to supercritical temperatures at high pressures leading to the so-called gas-gas equilibrium.

In interpreting fluid phase diagrams, it is useful to keep in mind geometrical constraints on the shape of the PTX surface in the vicinity of critical lines. A critical point is defined as a limiting point at which the differences in properties between two coexisting phases vanish. In isothermal and isobaric sections, as in figures 2c and d, lines connecting two coexisting phases (tie lines) are parallel to the X axis, and it follows from the definition of a critical point that the limiting tie line connecting identical phases is parallel to the X axis. Hence critical points on isotherms and isobars are either maxima or minima, where the PX loop is tangent to a line parallel to the X axis. This geometric constraint results from the use of two field variables (P and T) in the representation of the phase diagram. It establishes the additional constraint that in the PT projection the critical line is an envelope, tangent to all sections cut by planes of constant composition. Maxima in P and T in these sections (isopleths) have no particular significance. These geometric constraints can also be deduced from the condition of material stability and the laws of thermodynamics [4].

3.1 Class I Systems

The upper left diagram of figure 3, class I, illustrates the simplest type of system: the liquids are miscible in all proportions, there are no azeotropes, and the critical lines are continuous between the critical points of the pure components (C_α and C_β). The critical line may have a maximum or minimum in P or T. A PTX diagram for class I is shown in figure 2; the curvature of the critical line is slight, and mixtures of this type often show small departures from Raoult's law. At temperatures and pressures below those of the two critical points C_α and C_β, the isotherms and isobars extend smoothly across the diagram between the pure-component vapor-pressure curves (isotherms T_1, T_2; isobars P_1, P_2). At temperatures and pressures between C_α and C_β the isotherms pull away from the β side of the diagram, forming rounded loops with critical points (dark circles) that are maxima on isotherms and minima on isobars. The critical point on the isopleth, X_1, is neither a maximum nor a minimum in T or P, as explained above, but a point of tangency between the isopleth and the critical envelope on a PT projection (figure 2b). Typical examples include argon + krypton, nitrogen + oxygen, methane + krypton, and ethane + n-heptane [4]. Variations of class I behavior include critical lines with maxima and/or minima in P and T, and azeotropes [2,4].

3.2 Class II and Class III Systems

Before considering class II and class III systems a further examination of the geometry of three-phase equilibrium in PTX diagrams is in order. It was pointed out in section 2 that the three lines representing three coexisting phases lie in a ruled surface, and have a common projection on the PT plane. This is illustrated in figure 4, where the three-phase region is shown as a shaded surface, containing the three lines labeled "liquid α", "liquid β" and "gas". (The convention used here is to show lines representing three coexisting phases as dash-dot lines.) The line LLG, in the near PT plane, is the common projection of these lines. The parallel lines used as shading in the three-phase surface are tie lines; the intersections of these lines with the dash-dot lines form triplets of points that represent the PTX properties of three coexisting phases at fixed P and T. In this example the two liquid lines in the three-phase region end in an upper critical end point which is also an end point of a liquid-liquid critical line. Critical end points can be described as limiting points at which two of three coexisting phases become identical; the equality of properties reduces the

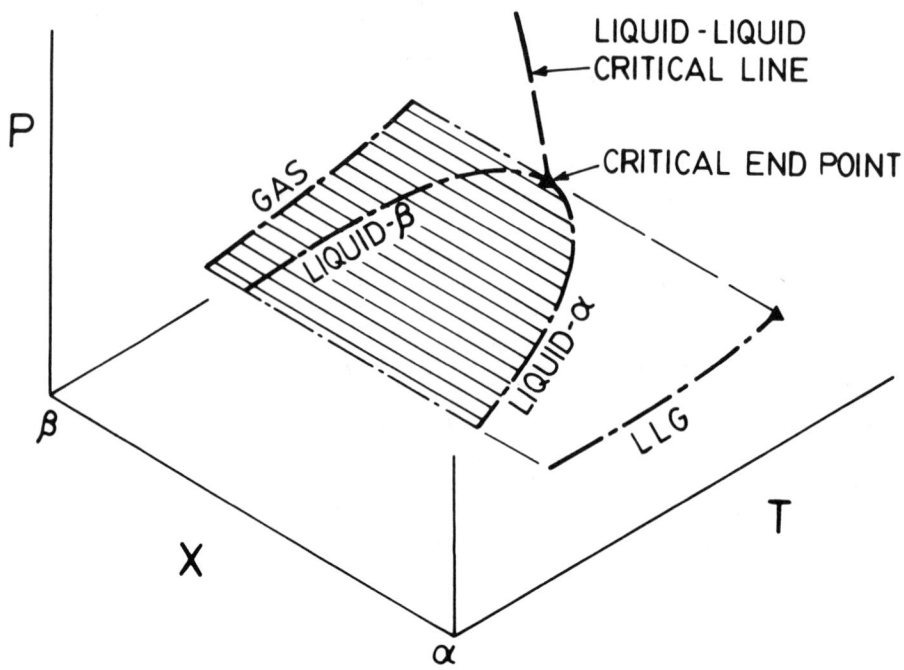

Figure 4. The PTX representation of a three-phase region liquid-liquid-gas. The dash-dot lines within the diagram represent the three coexisting phases, and the triangle denotes a critical end point.

number of degrees of freedom to zero. Three-phase regions can have an upper or lower critical end point, or both. Lower critical end points are usually type liquid-liquid, while upper critical end points can be either liquid-liquid or gas-liquid, although the distinction between the two becomes obscure in many cases. The phase behavior of class II systems combines the features shown in figures 2 and 4. In the class II system illustrated in figure 3, the critical line is continuous between the pure component critical points, and the region of coexistence between two liquid phases is confined to temperatures below the liquid-liquid critical line. The slope of this line can be either negative or positive in the PT projection. Examples include mixtures of carbon dioxide with octane, 2-hexanol, and 2-octanol [2]. Variations of class II include systems with positive and negative azeotropes.

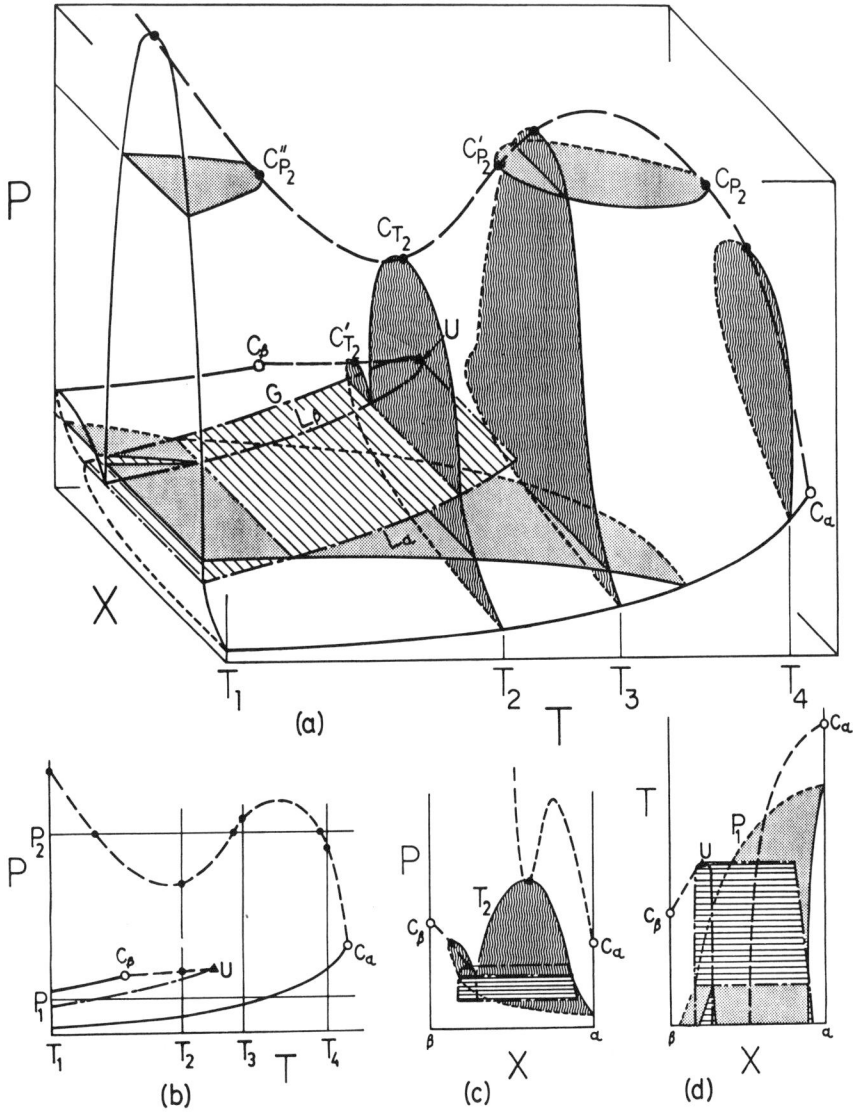

Figure 5. PTX phase diagram of a class III system.

An example of a class III system is shown in figure 5. The gas-liquid critical line is no longer continuous from C_α to C_β, but consists of two branches. One branch emerges from C_β and ends in the upper critical end point U, where it intersects the three-phase region, while the second branch, emerging from C_α, passes through maximum and minimum pressures, and then rises to very high pressures. At temperatures below that of U (e.g., T_1 and T_2) each isotherm includes three distinct two-phase regions: $L_\alpha + G$, lying below the three-phase pressure, and $L_\alpha + L_\beta$ and $L_\beta + G$ lying above. Between the temperature of C_β and U, the $L_\beta + G$ phase separation forms a closed loop ending in a gas-liquid critical point on the critical line $C_\beta - U$. With increasing temperature this loop decreases in size and vanishes at the critical end point U. Further details of the structure in the region of U are shown in figure 6, where the isotherm T_3 passes through a similar upper critical end point. At higher temperatures (T_3, T_4 in figure 5a) each isotherm consists of a single gas-liquid phase separation.

At temperatures T_3 and T_4 in figure 5 the critical points are unambiguously gas-liquid critical points. At T_2 there are two critical points, C_{T_2} and C'_{T_2}; the latter is a gas-liquid critical point, but the former is a liquid-liquid critical point, even though it lies on the same critical line as the gas-liquid critical points of isotherms T_3 and T_4. The upper critical line passes continuously from gas-liquid to liquid-liquid between T_3 and T_2, and there are continuous transitions between the gas phase and the liquid phase L_β in that region.

The isobar P_2 in figure 5a lies between the maximum and minimum pressure on the upper critical line, and has three critical points, C_{P_2}, C'_{P_2} and C''_{P_2}. At pressures beyond the range of the diagram the upper critical line probably ends in an upper critical end point where it intersects a three-phase region gas-liquid-solid (this is discussed in section 4). Examples of class III behavior similar to that shown in figure 5 include ethane + methanol [12] and carbon dioxide + water [13]. Several variations of critical behavior in class III systems are illustrated by the critical lines in figure 3. Other variations include systems in which the three-phase region LLG lies at pressures above the vapor pressures of the two pure components [4].

3.3 Class IV and Class V Systems

Referring to figure 3, it can be seen that in class IV systems the gas-liquid critical line emerging from C_α reaches a maximum in pressure, and then drops down to cut the three phase line at upper and lower critical end points U_2 and L. The qualitative features of this critical line are not unlike those of the line labeled CH_4/H_2S in the class III diagrams. Class V is schematically the same as class IV, except that the liquids are completely miscible at all temperatures below the critical end point L. In real systems, class V behavior has been found in which the liquid-liquid phase separaton is confined to a narrow range of temperatures lying between the pure component critical points C_α and C_β. This is illustrated in figure 6.

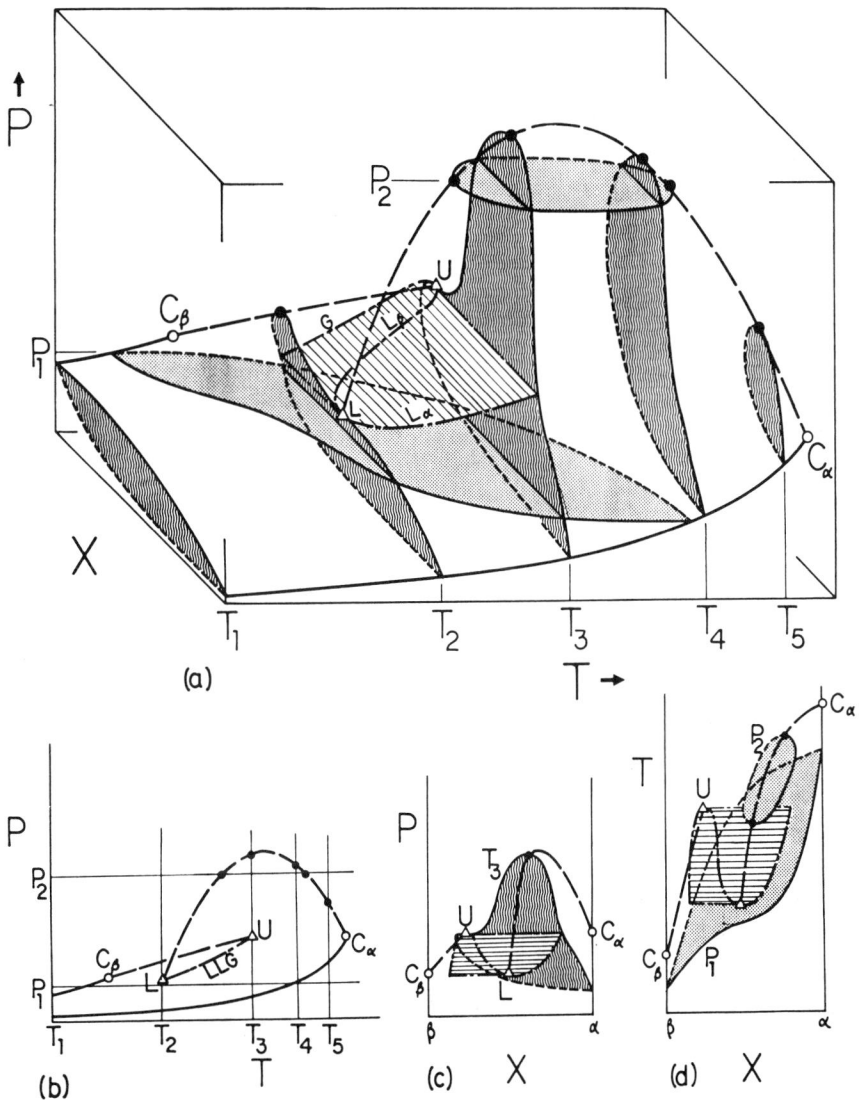

Figure 6. PTX phase diagram for a class V system.

In this example the isotherms T_2 and T_3 pass through the critical end points L and U, while the isobar P_1 lies below the pressure of L and does not pass through the three-phase region. An isotherm between T_2 and T_3 has two critical points, both pressure maxima (cf. isotherm T_2 in figure 5). Examples of systems that exhibit class V behavior similar to that illustrated in figure 6 include mixtures of ethane with ethanol, n-propanol and n-butanol. The class IV behavior illustrated in figure 3 has been found in mixtures of carbon dioxide with nitrobenzene and in several other carbon dioxide + hydrocarbon mixtures (see Schneider [2] for references).

Several systems have been found in which the temperature difference between the critical end points U and L is less than one degree C. The coincidence of these two points would result in a tricritical point--a limiting point at which three coexisting phases become identical. According to the phase rule a tricritical point in a binary mixture would have a negative degree of freedom, which means that the probability of finding such a point--which would be an "accident of nature"--is vanishingly small. However, by adding small amounts of a third component to these systems, pseudo-binary mixtures that exhibit a close approach to tricritical behavior can be formed [14].

3.4 Class VI Systems.

Class VI systems have thus far been discovered only in aqueous mixtures. In these systems the LLG three-phase region is bounded by upper and lower critical end points of the liquid-liquid type, and the liquid-liquid critical lines emerging from these points do not extend into the gas-liquid critical region (figure 3). PTX diagrams for a class VI system are shown in figure 7. In this figure the liquid-liquid critical line is continuous between the critical end points U and L, with a pressure maximum between. In some cases the liquid-liquid phase separation reappears at higher pressures, as in P_2 in figure 7, and is bounded by a critical line having a minimum in pressure. Examples have also been found in which the upper branch of the liquid-liquid phase separation is present but the lower branch is absent. In other cases the miscibility gap is absent, and the two branches are continuous. In these systems isobars at pressures above that of U form closed loops where they intersect the liquid-liquid region, with maximum and minimum critical temperatures (P_2 in figure 7).

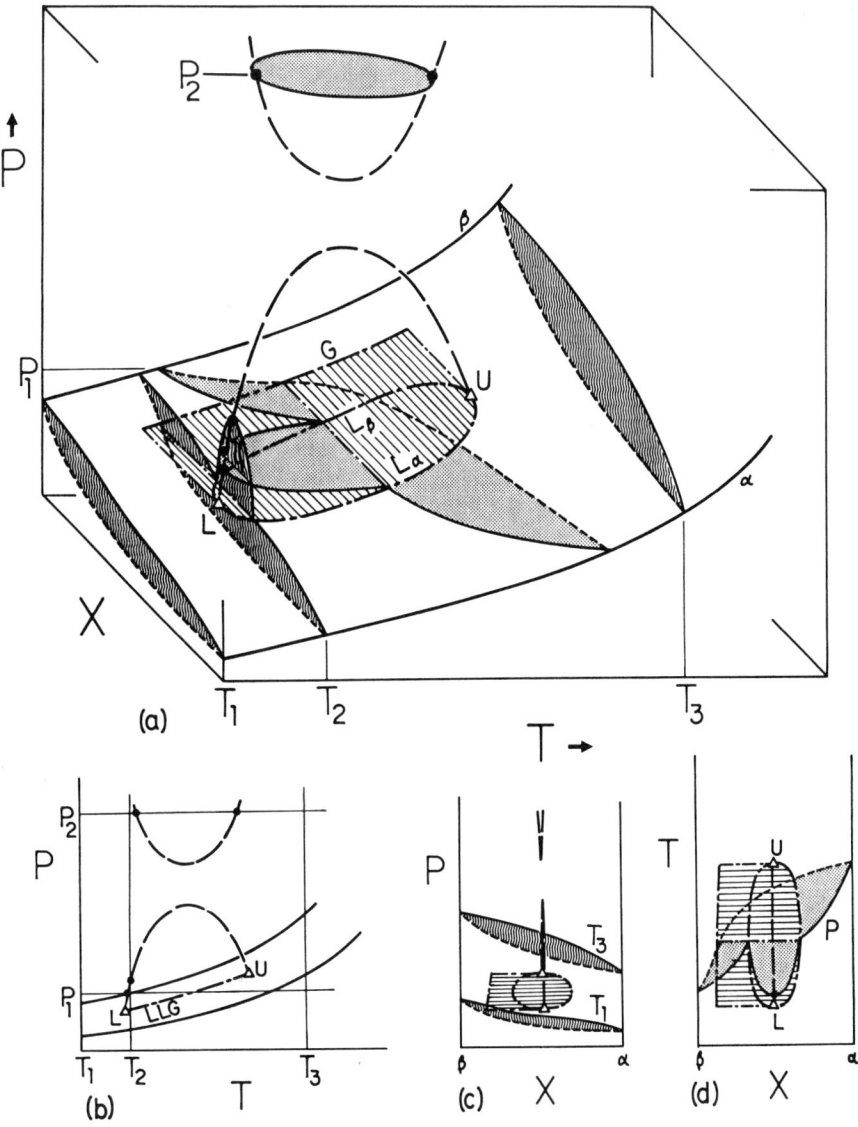

Figure 7. PTX phase diagram for a class VI system.

The diagrams shown here illustrate only a few of the known types of fluid phase equilibria; in particular, there are numerous variations of azeotropic behavior that are not shown. These include azeotropes bounded above or below (or both) by intersections of azeotropic lines with pure-component vapor pressure curves, and the transformation of heterogeneous azeotropes into homogeneous azeotropes, as a result of the equality of composition between the two liquid phases. There are also systems that exhibit double azeotropes, one positive and one negative [4,6].

4. PHASE DIAGRAMS FOR BINARY MIXTURES WITH WIDELY SEPARATED CRITICAL TEMPERATURES

Supercritical extraction usually involves the separation of relatively nonvolatile components, often in solid phases, through selective solubility in supercritical gases. A binary mixture consisting of a nonvolatile solute and a near-critical or supercritical solvent is likely to be a system in which the critical temperatures of the pure components are far apart. In many such systems the critical temperature of the solvent lies below the triple point of the solute, and there is no common range of temperature in which both pure components are liquid. In this section the phase diagrams of several systms of this type are described, and these are classified, as in the previous section, mainly on the basis of critical lines and three-phase lines. In this case the three-phase lines are type solid-liquid-gas, instead of type liquid-liquid-gas as in the phase diagrams of the preceeding section.

The phase diagrams of interest here are a subset of the class III systems illustrated in figure 3. In general they differ from the diagram of figure 3 in that the temperature of C_β lies below the triple point temperature of component α. The transition between these systems and those illustrated in the class III diagram of figure 3 seems to occur as a result of the sequence of changes shown in figure 8. Figure 8a is a PT projection of the principal boundary lines of a system in which two completely miscible liquids freeze to form partially miscible solid phases S_1 and S_2. The full lines are the sublimation, melting and vapor pressure curves of the pure components, α and β. The critical line is class I. Point Q is a quadruple point, where four phases, S_1, S_2, L and G, are in equilibrium (it consists of 4 points in PTX space, having a common projection on the PT plane). Radiating from this point are four lines (in PTX space four triplets of lines) representing equilibria between the four possible combinations of three phases drawn from this group. If the temperature of C_β lies far below that of A_α (the triple point of α) the three-phase region $S_2 + L + G$ (line QA_α in figure 8a) curves upward, and intersects the gas-liquid critical line to form critical end points U_1 and U_2, as shown in figure 8b. No liquid phase exists in the temperature range between U_1 and U_2.

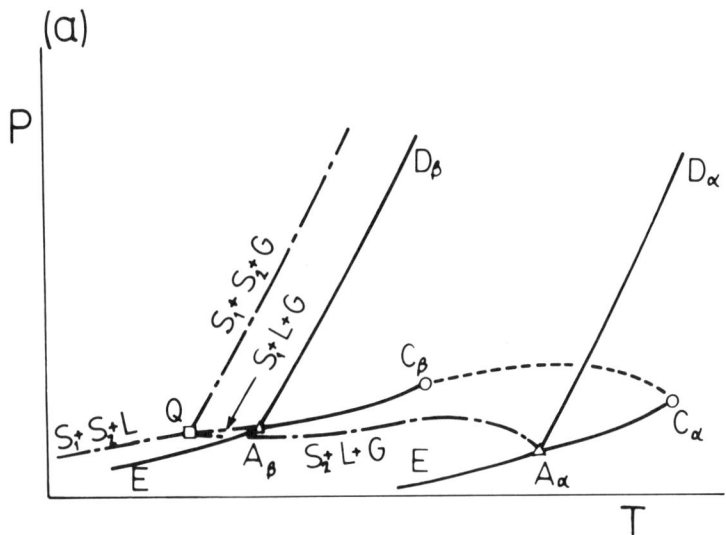

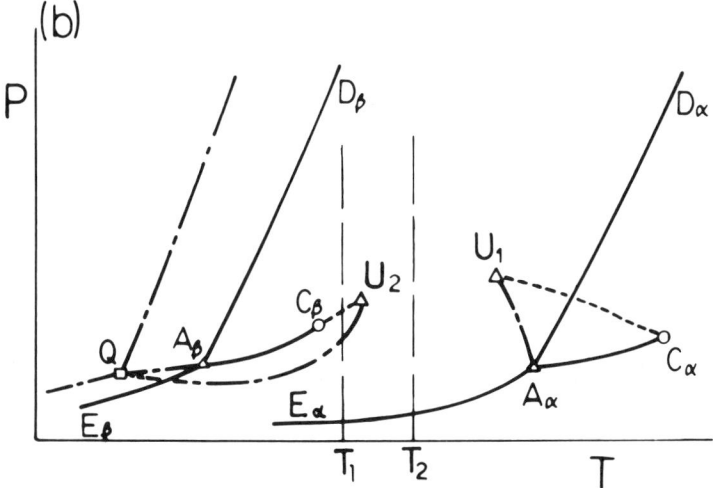

Figure 8. Transition from class I systems (figure 8a) to class III systems (8b) as the difference between the critical temperatures of the pure components increases (see text for discussion).

The three-sided region $A_\alpha C_\alpha U_1$, bounded by the vapor pressure curve $A_\alpha C_\alpha$, the critical line $C_\alpha U_1$, and the three-phase boundary $A_\alpha U_1$, is a region of gas-liquid phase separation. At temperatures between this region and U_2, the solid phase of component α is in equilibrium with a β-rich gas phase. If the vapor pressure of α is small at the temperature of C_β and below, the solubility of the α solid phase in the β liquid may be very low, with the result that U_2 and Q are virtually coincident with C_β and A_β, respectively.

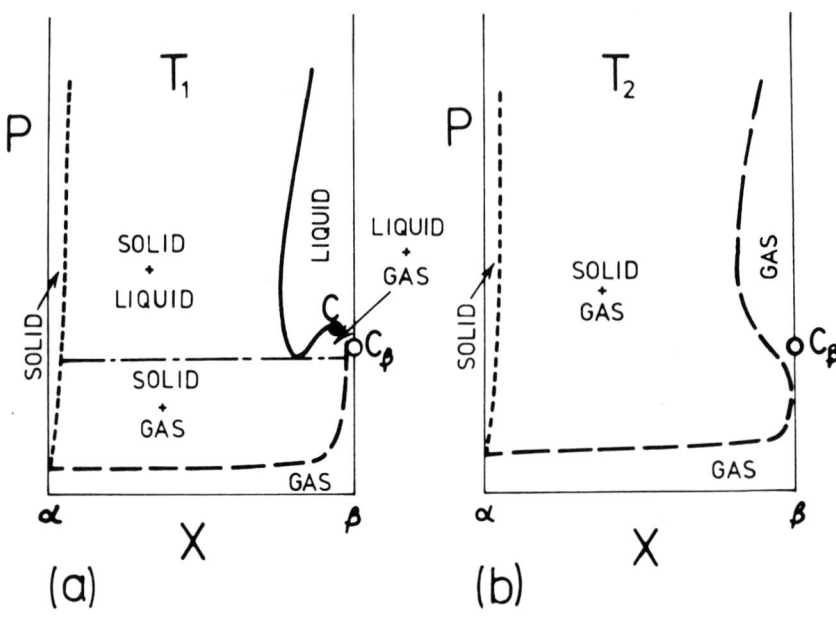

Figure 9. PX diagrams for isotherms T_1 and T_2 from figure 8b.

Figure 9 shows two PX isotherms corresponding to the temperatures T_1 and T_2 in figure 8. T_1 lies between C_β and U_2 and cuts both the three-phase line QU_2 and the critical line $C_\beta U_2$. In figure 9a (temperature T_1) the horizontal line connects the three coexisting phases (solid, liquid and gas) and point C is the gas-liquid critical point. As the temperature increases from T_1 to U_2, the liquid+gas loop on the right of 9a decreases in size, and vanishes at U_2 in a horizontal inflection point, similar to the one shown at point U on isotherm T_3 of figure 6. An isotherm at T_2 in figure 8b can be expected to have the shape shown in figure 9b. The region just above the nose on the right of this diagram, where the slope of the gas-phase boundary is negative, illustrates a type of behavior sometimes associated with enhanced solubility of a nonvolatile solute (α) in a supercritical solvent (β). With increasing pressure there is a dramatic increase in the solubility of α in the gas phase—shown by the negative slope of the gas boundary in this region. Both experimental and theoretical studies suggest that at much higher pressures the trend toward enhanced solubility is usually reversed, and the gas phase boundary again takes on a positive slope as shown in figure 9b.

In figure 9, and in figures 11 and 12 below, the more volatile component is assumed to have a finite solubility in the solid phase, increasing with pressure. This assumption is made mainly in the interest of describing the qualitative features of systems that include solid phases. In many systems of practical interest the solubility of gases in solids is small, and can be neglected. In the limiting case of zero solubility, the solid phase boundaries in figure 9 (the dotted lines) would coincide with the left-hand vertical axis--in other words the solid phase would be pure α.

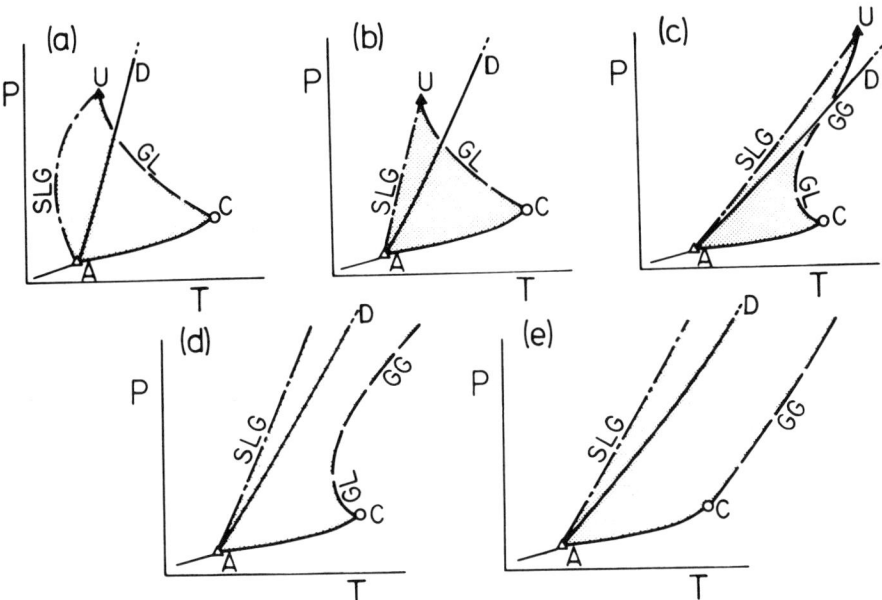

Figure 10. Sequence of changes in critical lines and solid-liquid-gas three-phase lines as the difference in pure-component critical temperatures increases. PTX diagrams corresponding to figures 10b and 10d are shown in figures 11 and 12, respectively.

The right-hand portion of figure 9b is relevant to the phase behavior of systems of interest in supercritical extraction when the less volatile component is a liquid.

With increasing difference between the critical temperatures of the pure components, critical lines and solid-liquid-gas three-phase lines can be arranged in the general sequence shown in figures 10a to 10e. These diagrams belong to class III in the classification scheme of figure 3. The three-phase boundary $A_\alpha U_1$ in figure 8b has a negative slope. Examples include carbon dioxide + diphenylamine, ethane + napthalene, ethylene + hexachlorethane, etc. (for references see Rowlinson and Richardson [15]). In figure 10a the three-phase boundary AU has a temperature minimum and regions of both negative and positive slope; examples include hydrogen + carbon monoxide and hydrogen + methane [9,16]. In figures 10b and c the critical and three-phase lines intersect at a critical end point U, which in the case of 10c lies at a temperature above the critical point C. An example of this behavior has been found in the system helium + argon, where the end point U lies at about 11,000 atm and 205 K, about 55 K above the argon critical temperature [17]. In figures 10d and e the critical and three-phase lines do not intersect, and the region of equilibrium between two fluid phases is opened to supercritical temperatures at high pressures, leading to the so-called gas-gas equilibrium. This term is somewhat misleading, since these phase separations occur in fluid mixtures that have been compressed to liquid-like densities at supercritical temperatures (above the critical temperatures of both pure components). The work of Schneider [1-3] clearly shows that there is a logical continuity between phase diagrams that exhibit gas-gas equilibrium and more conventional diagrams. The evidence suggests that gas-gas phase separation is a kind of liquid-liquid phase separation displaced to high temperatures at high pressures. These separations occur in mixtures that would probably form partially miscible liquids, except for the accident of nature that causes one to freeze at temperatures above the critical temperature of the other. The tendency to form partially miscible liquids asserts itself at high pressures, where the fluids have been compressed to liquid-like densities. The fact that these phase separations often extend to temperatures above the pure-component critical points has little to do with the distinctions between gas and liquid phases, in pure fluids or in gas-liquid mixtures, at low pressures. The term fluid-fluid phase separation is perhaps more appropriate for this phenomenon.

Figure 11a shows a PTX diagram for a system that has a PT projection similar to figure 10b. Examples include neon + argon [18] and hydrogen + carbon dioxide [19].

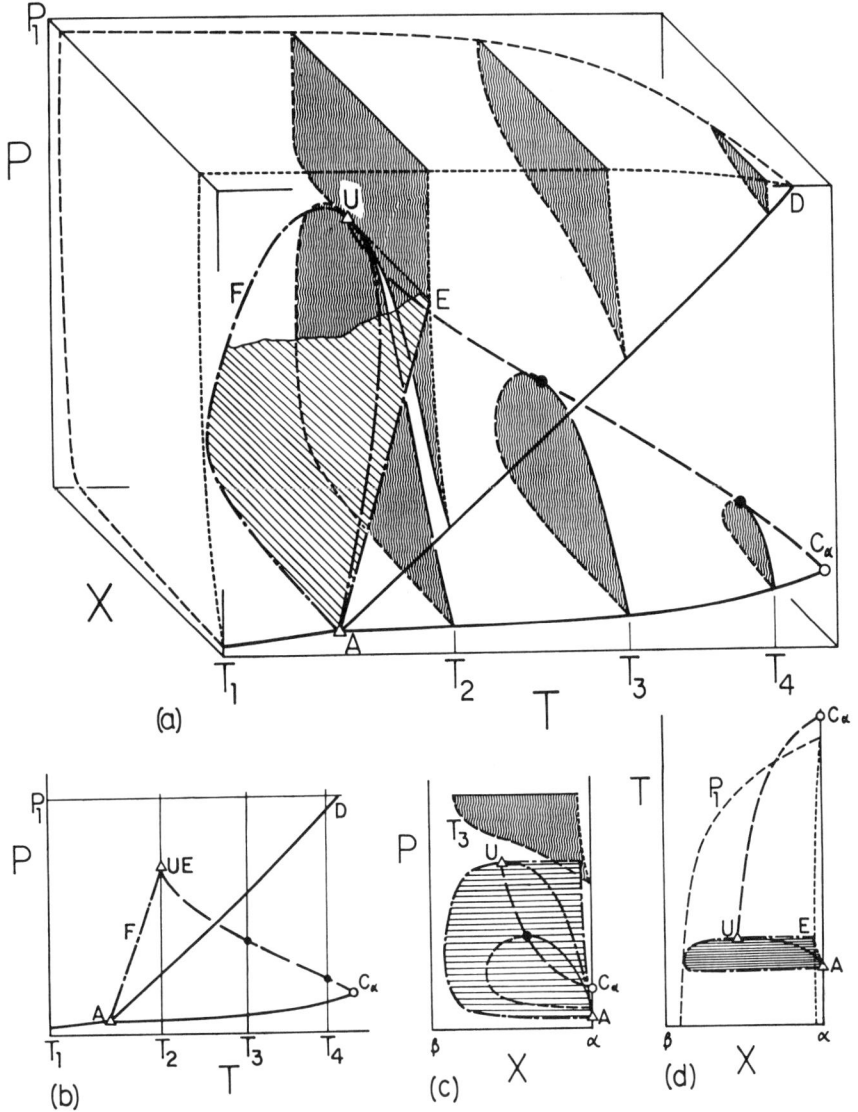

Figure 11. PTX diagram corresponding to figure 10b. In the system $H_2 + CO_2$ [19] the critical end point U lies at a pressure of 2000 atm.

In figure 11 the three-phase region is shaded by parallel lines as in figures 4-6; a portion is left unshaded to show details of isotherm T_2, where it passes through the critical end point U. This three-phase region forms a ruled surface containing the three lines, AFU, AU and AE, that represent gas, liquid and solid phases, respectively. The surface is perpendicular to the PT plane (because P and T are field variables that are the same in each of the three equilibrium phases), hence these three lines have a common projection AFUE in the PT diagram of figure 11b. The three-phase surface emerges from the triple point A of the less volatile component and ends at high pressures in the boundary UE, where the solid phase E is in equilibrium with gas and liquid phases that become identical at the critical end point U. Isotherm T_1 lies below the triple point A, and consists of a single region of gas-solid phase separation. (It has been left unshaded in figure 11a in the interest of clarity.) The isotherm T_2 passes through the end point U, which is a point of tangency between the gas-liquid region and the solid-liquid and solid-gas regions. The distinction between the latter two regions no longer exists at temperatures above U (e.g. the upper branch of isotherm T_3). At temperatures above that of U, increasing pressure increases the mutual solubilities of the gas and liquid phases, which eventually become identical at a point on the critical line $C_\alpha U$.

Figure 12 shows a phase diagram for a system that has a PT projection similar to figure 10d. Here the critical line extending into the diagram from C_α passes through a temperature minimum at M and rises to higher temperatures with increasing pressures. The isotherm T_3 in figures 12a and c consists of two branches that are tangent at M. As the temperature approaches M from below, the isotherms take on a characteristic "waisted" shape, shown here in isotherm T_2, not unlike that exhibited by the solid + gas isotherm in figure 9b. The systems nitrogen + ammonia [20] and helium + hydrogen [21] exhibit this behavior. These systems have been studied at pressures as high as 10,000 atm. The critical and three-phase lines in these systems may eventually intersect at high pressures, as in figure 10c; however, the available data suggest that critical end points, if they do exist, lie at pressures in excess of 20,000 to 30,000 atm.

The behavior shown in figure 10e has been found in many systems in which helium is the light component, including helium + methane [22], helium + carbon dioxide, and helium + ethylene [23,24].

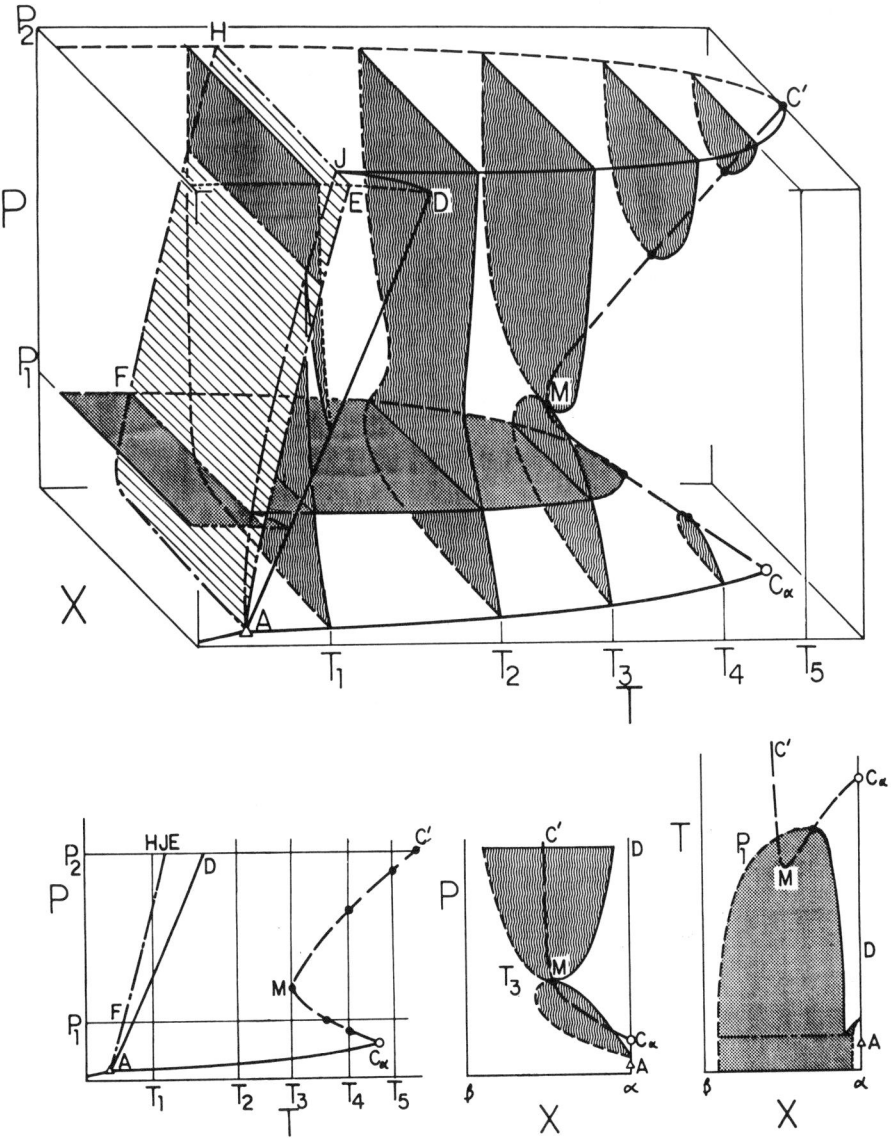

Figure 12. PTX diagram corresponding to figure 10d. The two-fluid region extends to supercritical temperatures at high pressures (gas-gas equilibrium).

5. CONCLUDING REMARKS

In the preceeding sections the broad qualitative features of several types of fluid and solid phase diagrams for binary mixtures were described. From this point of view, the phenomena that are important in supercritical fluid extraction are well understood. It is the ability of supercritical gases to dissolve nonvolatile substances selectively, coupled with the strong pressure dependence of solubility, that leads to useful applications of these phenomena in commercial processes. Although further experiments are likely to lead to new applications, there is a need to develop a more comprehensive understanding, based on the underlying intermolecular forces. A molecular approach based on perturbation theory and statistical mechanics (Streett and Gubbins [25], and references therein) appears to offer considerable promise of success.

ACKNOWLEDGEMENTS

The author's research on which this review is based has been supported, in part, by grants from the National Science Foundation, the Department of Energy, the Petroleum Research Fund, and NATO.

REFERENCES

1. Schneider, G.M., Adv. Chem. Phys., 17, 1 (1970).

2. Schneider, G.M., in Chemical Thermodynamics, Vol. 2, Specialist Periodical Reports, (Chemical Society, London, Chap. 4, 1978).

3. Schneider, G.M., Angew. Chem. Int. Ed. Engl., 17, 716 (1978).

4. Rowlinson, J.S., Liquids and Liquid Mixtures, 2nd Ed., (Butterworths, London, 1969).

5. Prausnitz, J.M., Molecular Thermodynamics of Fluid Phase Equilibria, (Prentice Hall, Englewood Cliffs, 1969).

6. King, M.B., Phase Equilibrium in Mixtures, (Pergamon Press, Oxford, 1969).

7. Streett, W.B., Can. J. Chem. Eng., 52, 92 (1974).

8. Streett, W.B., in High Pressure Technology, I.L. Spain and J. Pauuwe, Eds., (Marcel Dekker, New York, 1977).

9. Tsang, C.Y., P. Clancy, J.C.G. Calado and W.B. Streett, Chem. Eng. Commun., 6, 365 (1980).

10. Van Konynenberg, P.H., Ph.D. Thesis, University of California at Los Angeles, 1968.

11. Van Konynenberg, P.H. and R.L. Scott, Phil. Trans. Roy. Soc., 298, 495 (1980).

12. Kuenen, J.P., Phil. Mag. 6, 637 (1903).

13. Takenouchi, S. and G.C. Kennedy, Amer. J. Sci., 262, 1055 (1964).

14. Knobler, C.M., 1981, (private communication).

15. Rowlinson, J.S. and J.M. Richardson, Adv. Chem. Phys., 2, 85 (1959).

16. Tsang, C.Y. and W.B. Streett, Fluid Phase Equil., 6, 261 (1981).

17. Streett, W.B., and A.L. Erickson, Phys. Earth Planet Interiors, 5, 357 (1972).

18. Streett, W.B. and J.L.E. Hill, J. Chem. Phys., 54, 5088 (1971).

19. Tsang, C.Y. and W.B. Streett, Chem. Eng. Sci., 36, 993 (1981).

20. Krichevskii, I.R., Acta Phys-Chim. USSR, 12, 480 (1940).

21. Streett, W.B., Astrophys. J., 186, 1077 (1973).

22. Streett, W.B., A.L. Erickson and J.L.E. Hill, Phys. Earth Planet Interiors, 6, 69 (1972).

23. Tsiklis, D.S., Dokl. Akad. Nauk. USSR, 86, 1159 (1952).

24. Tsiklis, D.S., Dokl. Akad. Nauk. USSR, 91, 1361 (1953).

25. Streett, W.B. and K.E. Gubbins, paper presented at 8th Symposium on Thermophysical Properties, NBS, Gaithersburg, MD, June 1981 (to appear in the Proceedings of the Symposium, publ. by ASME).

CHAPTER 2

SOME VAPOUR/LIQUID AND VAPOUR/
SOLID EQUILIBRIUM MEASUREMENTS
OF RELEVANCE FOR SUPERCRITICAL
EXTRACTION OPERATIONS, AND
THEIR CORRELATION

M.B. King*
D.A. Alderson, Lever Bros, Port Sunlight, Cheshire, England
F.H. Fallah, Dupont Ltd, Maydown Works, Londonderry, N.Ireland
D.M. Kassim, Petroleum Research Centre, Baghdad, Iraq
K.M. Kassim*
J.R. Sheldon, Computer Aided Design Centre, Cambridge, England
R.S. Mahmud, Bechtel Petroleum Inc, San Francisco, CA94119, USA

*Chemical Engineering Department
University of Birmingham
England

INTRODUCTION

The potential scope for extractions carried out above the critical temperature and pressure of the solvent and the problems involved in developing such processes are discussed in another paper (1), where the need for both equilibrium and rate data is stressed. The present paper is concerned mainly with the determination and correlation of the equilibrium data, though the results of some simple extraction tests are also discussed against this background. Equipment is described which is suitable both for phase equilibrium determinations and for these simple extraction tests.

EQUIPMENT USED

The Static Cell (2)

For the initial work the stirred stainless steel equilibrium cell (total volume 500 ml) shown in Fig.1 was used. The cell has previously been described by Harris (3) and by Ellis and Valteris (4), though the mercury injection and sampling systems used in conjunction with the cell are not the same (2). This equipment is suitable for phase equilibrium studies in systems in which both phases are fluid.

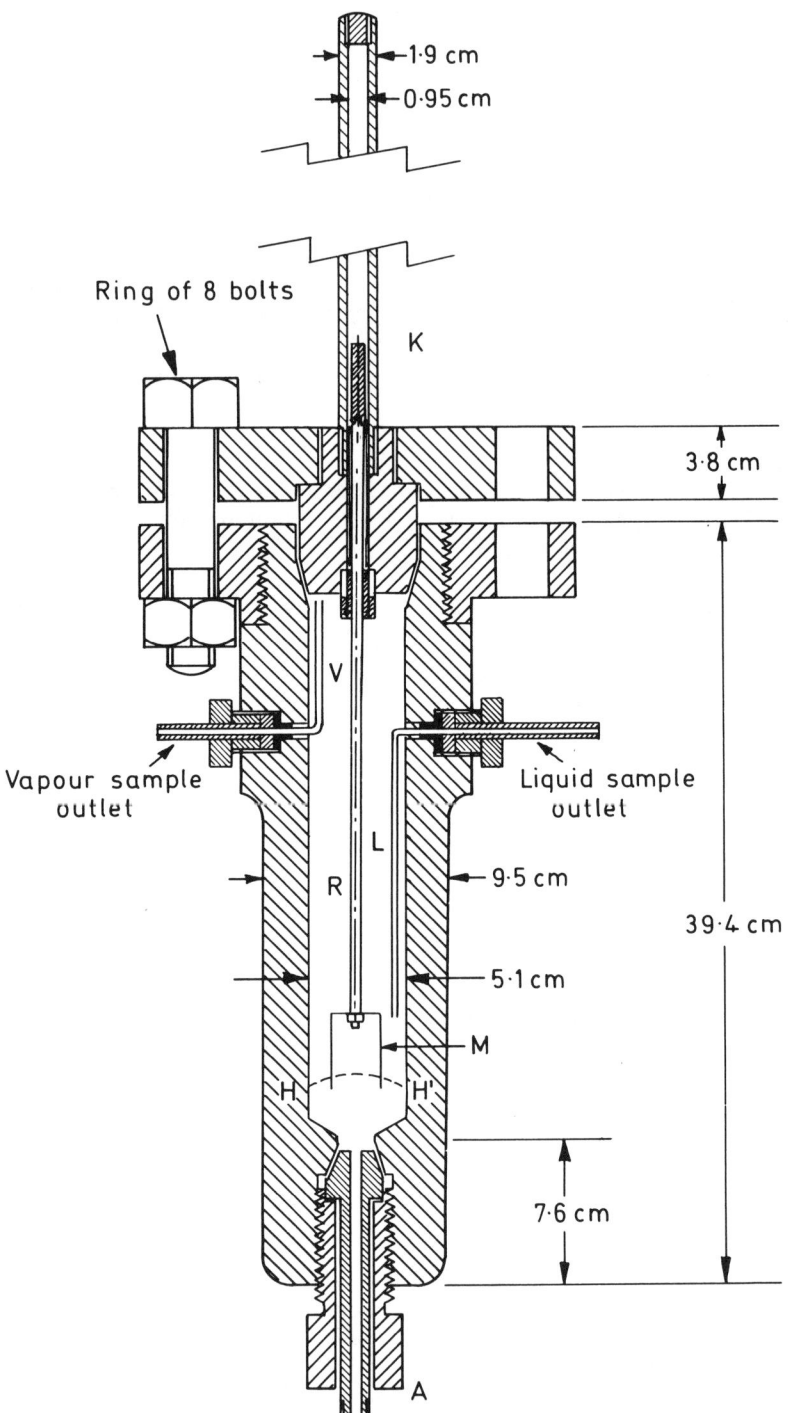

Figure 1. The Static stirred cell.

The systems studied with this cell consisted of a moderately heavy component (benzene, toluene and/or iso-octane) equilibrated with a light component (ethane or ethylene) at a temperature (73.5°C) which is above the critical temperature of the light component but below the critical temperature(s) of the heavy component(s).

The system under study is enclosed between the top of the cell and the interface (HH^1) with a mercury column which extends into the base of the cell. Stirring is provided by the perforated cup M which moves up and down continuously. The procedure is to leave the components under study to equilibrate in the stirred region under controlled conditions of temperature and pressure and then to extract smaples of the vapour and liquid phases for analysis. In order to keep the pressure constant during the sampling process, mercury is injected into the base of the cell, through port A. A line diagram showing the layout of the cell and ancilliary equipment is given in Fig.2.

At the start of each set of experiments, the equilibrium cell was evacuated, purged, evacuated again, charged with the heavy component or components and then filled with ethane or ethylene to the desired pressure. Pressures up to about 70 bar could be achieved by connecting the cell directly to the ethane or ethylene supply cylinder (C in Fig.2). Higher pressures were achieved using the "thermal compressor" T. This consisted of a non-magnetic stainless steel cylinder 30.5cm long, 7.6cm OD and 2.5cm ID. This was connected to the supply cylinder and cooled in liquid nitrogen until no further material would condense in it. It was then isolated from the supply cylinder and allowed to warm up until the compressor pressure exceeded the system pressure. Valve 3 was then opened to allow the system pressure to achieve the desired level. The rectangular region within the broken lines in Fig.2 was enclosed in an air bath maintained at 73.5°C. The air temperature was measured with mercury-in-glass thermometers inserted through the thick "Merionite" walls of the bath and the cell temperature was determined using a thermistor embedded in a thermowell which was immersed in the mercury at the foot of the cell. This thermistor was calibrated by measuring the vapour pressure on n-butane in the cell for a series of thermistor readings and obtaining the corresponding temperatures from vapour-pressure/temperature data reported in the literature(6).

During the sampling process, mercury is driven into the cell using compressed oil, the mercury/oil interface being located in the right hand arm of the stainless steel U tube (Q in Fig.2). The compressed oil is supplied either from the hand pump connected to the dead weight tester D or from the

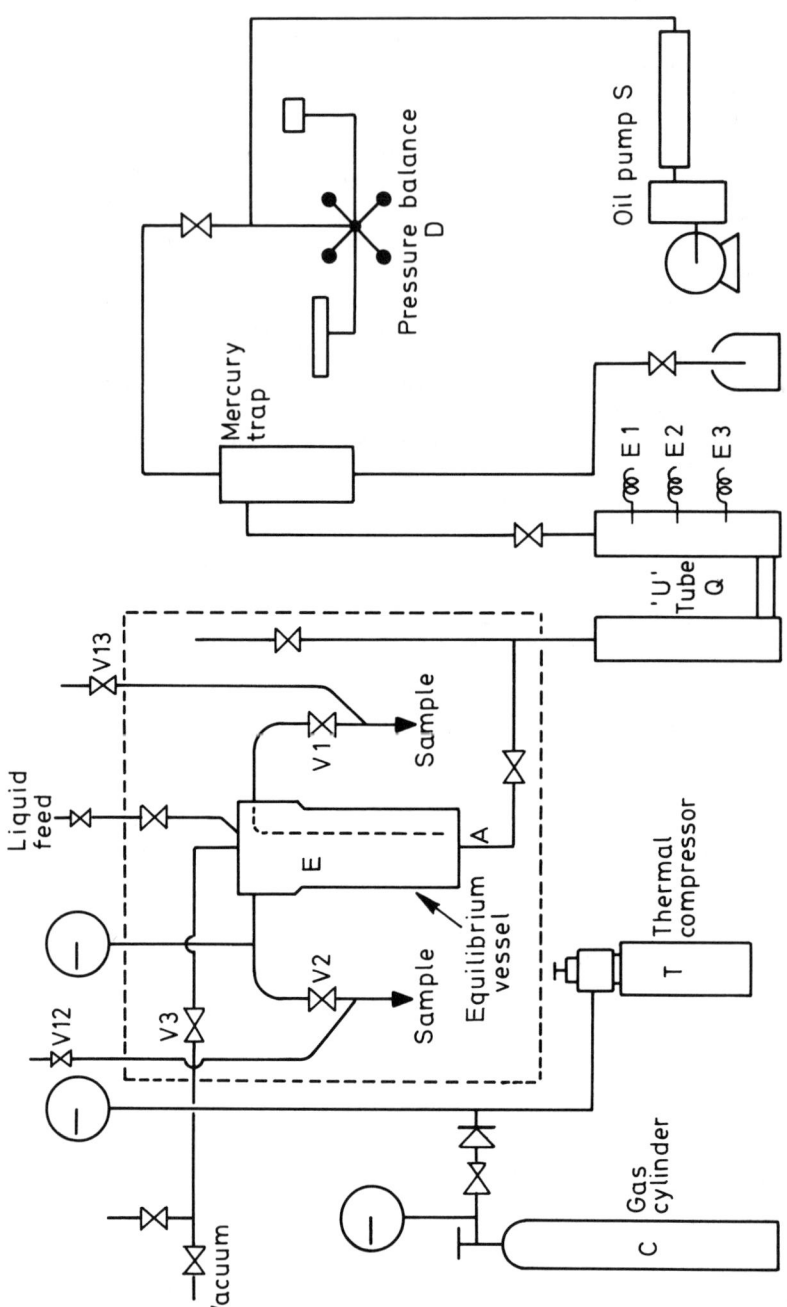

Figure 2. Line diagram of the static stirred cell and ancillary equipment.

electrically operated pump S. The pressure of the system is measured using the dead weight tester which also acts as a barostat. The position of the mercury meniscus in the right hand arm of the U tube Q (and hence of the mercury meniscus in the cell) is determined by three electrical level probes (E_1, E_2 and E_3 in Fig.2) which give an open circuit except when immersed in mercury. These probes are made to actuate three flash bulbs. During the equilibration period, the probes may also be made to actuate the electrically operated oil pump. If the mercury level in Q rises during the period due to loss of oil from the dead weight tester, fresh oil is automatically injected to compensate for the loss.

The system is stirred by the magnetically operated mechanism shown in Fig.1. This consists of the soft iron rod K which is moved up and down continuously by a tubular magnet and which is attached to the stainless steel shaft R and the perforated cup M. The time required for equilibration was found to depend on the volume of liquid initially introduced into the cell, being 8/12 hours for a 90 ml sample. After the equilibration period the stirrer is turned off for about 15 minutes to allow mist to settle from the vapour phase and also for any bubbles to rise from the liquid phase. The vapour and liquid samples are extracted via the capillary sampling tubes V and L respectively (Fig.1). They then pass via the valves V2 and V1 (Fig.2) to the sample collection equipment shown in Fig.3. Indentical sample collection equipment was used for the vapour and liquid samples. In each case the heavy condensible components are removed in condensers C1 and C2. These are maintained at a temperature of about $-75^\circ C$ using a mixture of acetone and solid carbon dioxide. The light component passes through these condensers into the expansion vessel F (Fig.3) which has a volume of about 36 dm^3. At the end of the expansion process the condensers are allowed to warm to room temperature and the temperature T and the pressure rise ΔP in the expansion vessel are noted. The amount of light component present in the sample is calculated from ΔP, T and the volume of the expansion system. The condensers are then immediately re-immersed in the acetone/solid carbon dioxide mixture (or in the case of comparatively volatile condensates, liquid air) and a gentle stream of nitrogen (about 100/150 ml/minute) is passed for about 15 minutes through the tubing connecting the expansion valve (V1 or V2) to the condensers. This nitrogen enters via valve V13 or V12. Droplets of the condensible components present in this tubing are evaporated into, or swept along by, the nitrogen carrier gas and may be seen to collect in the condensers. These are then warmed and the amount of condensible components present in the sample determined from the rise in weight, ΔM. In the case of a binary system, this completes the analysis process.

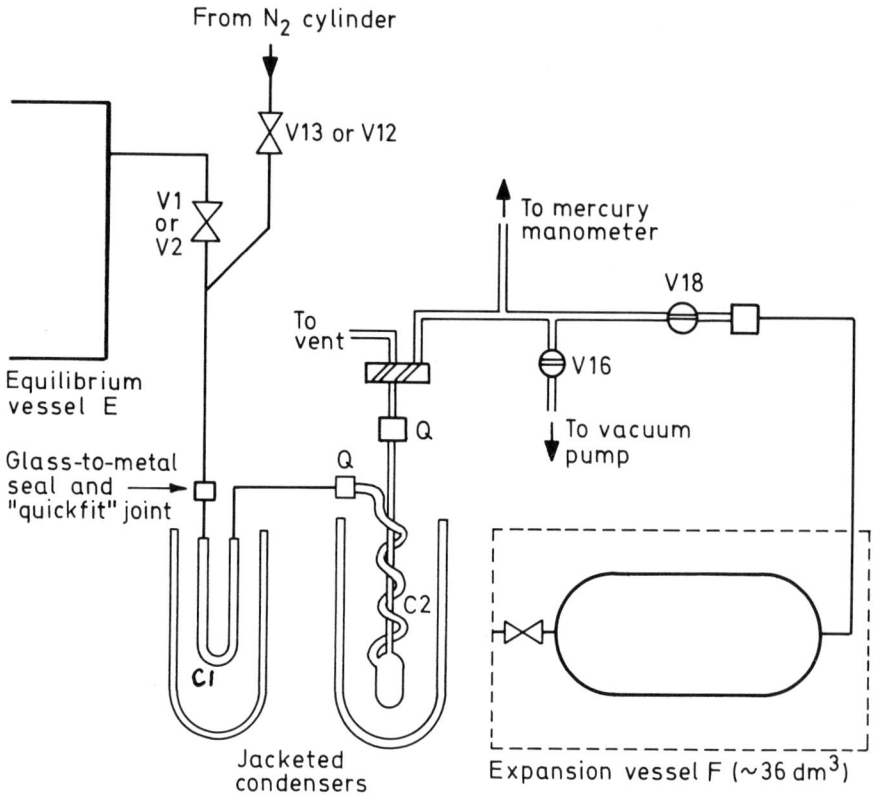

Figure 3. Sample collection equipment used with static stirred cell (not to scale) Q = "Quickfit" glass/glass joint.

In the case of a ternary system containing two condensible components a representative sample of the material in the condensers is extracted and analysed by G.L.C. The weight of each component in the sample is calculated from its weight fraction and ΔM. Further details of the sampling and analysis procedure are available (2).

Recirculation equipment used for equilibrium measurements on

fluid systems

This equipment was charged with "heavy" components via the liquid sample bomb and valves V18 and V8 (Fig.4) and the extracting gas was delivered from a mechanical compressor.

It has been used for studies on the systems Methane/ n-Hexane at 104.4°C, Ethylene/Benzene at 75°C, Carbon dioxide/ n-Octane at 40°C and 110°C, Carbon dioxide/n-Butanol at 40°C and 110°C, Carbon dioxide/n-Hexadecane at 60°C and 110°C and some other systems (1). In the above cases the temperatures investigated are above the critical temperature of the "light" component and below the critical temperature(s) of the "heavy" component(s). A few tests have also been carried out using sub-critical carbon dioxide as extractant.

A line diagram showing the layout of the equipment is given in Fig.4. The rectangular region within the broken lines is enclosed in an air bath. All units exposed to elevated pressure are constructed of stainless steel. The system under test is enclosed in the equilibrium vessel at controlled temperature. (This vessel is shown in detail in Fig.6. It has a capacity of about 300 cm^3, is provided with a solenoid-operated internal stirrer and is designed for use at temperatures and pressures up to 500°C and 500 bar. The temperature within the vessel is measured with the calibrated thermistors shown and remains constant to well within 0.1°C during operation). The vapour phase is recircuated via the vapour sampling bomb (volume 11.5 cm^3) and back to the equilibrium vessel where it bubbles up through the liquid. Periodically the liquid phase is also recirculated via the liquid sample bomb which has a volume of 34.7 cm^3. When it is considered that equilibrium has been reached the pressure is noted and representative samples of the vapour and liquid phases are sealed off in the sample bombs, the contents of which are analysed using the equipment shown in Fig.5. The analysis is carried out by allowing the contents of the sample bombs to expand very gently into previously evacuated glass condensers (in which the "heavy" components are removed) and thence into metal expansion vessels of known volume (34.34 dm^3 in the case of vapour samples and 6.46 dm^3 in the case of liquid samples).

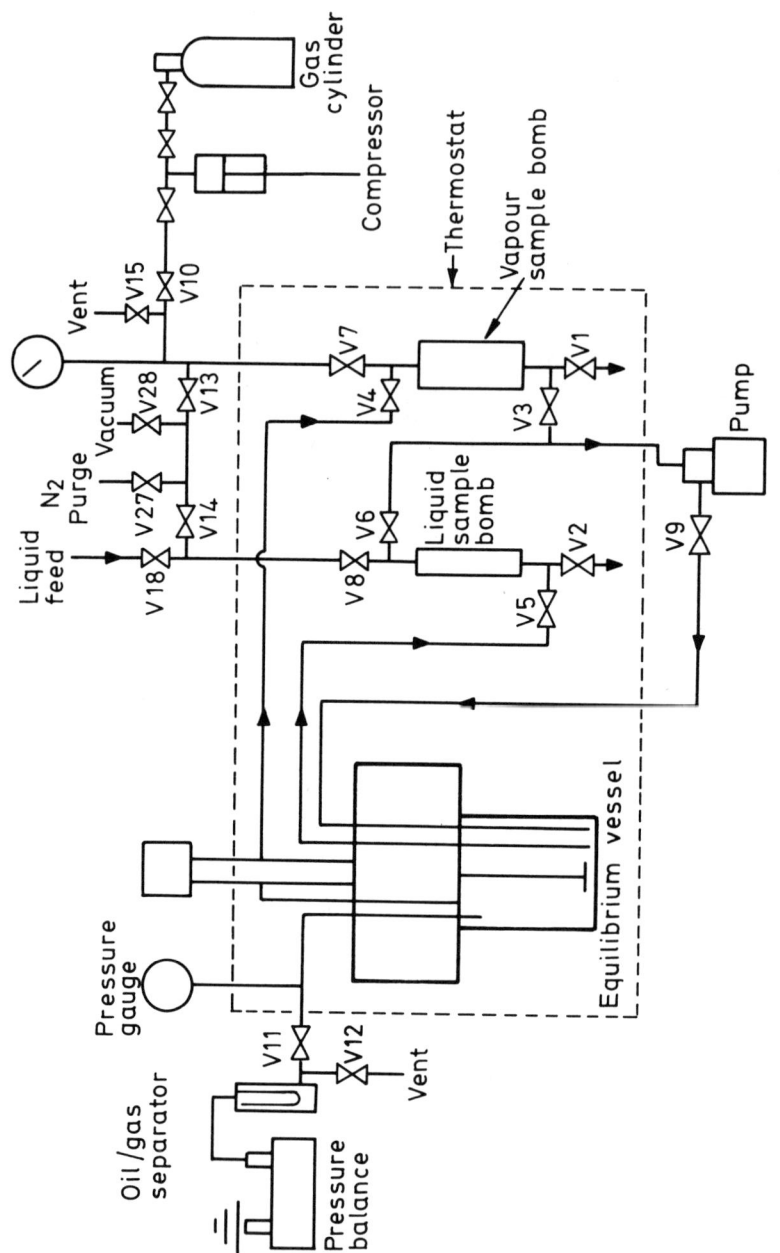

Figure 4. Flow diagram of the high pressure vapour/liquid recirculation apparatus.

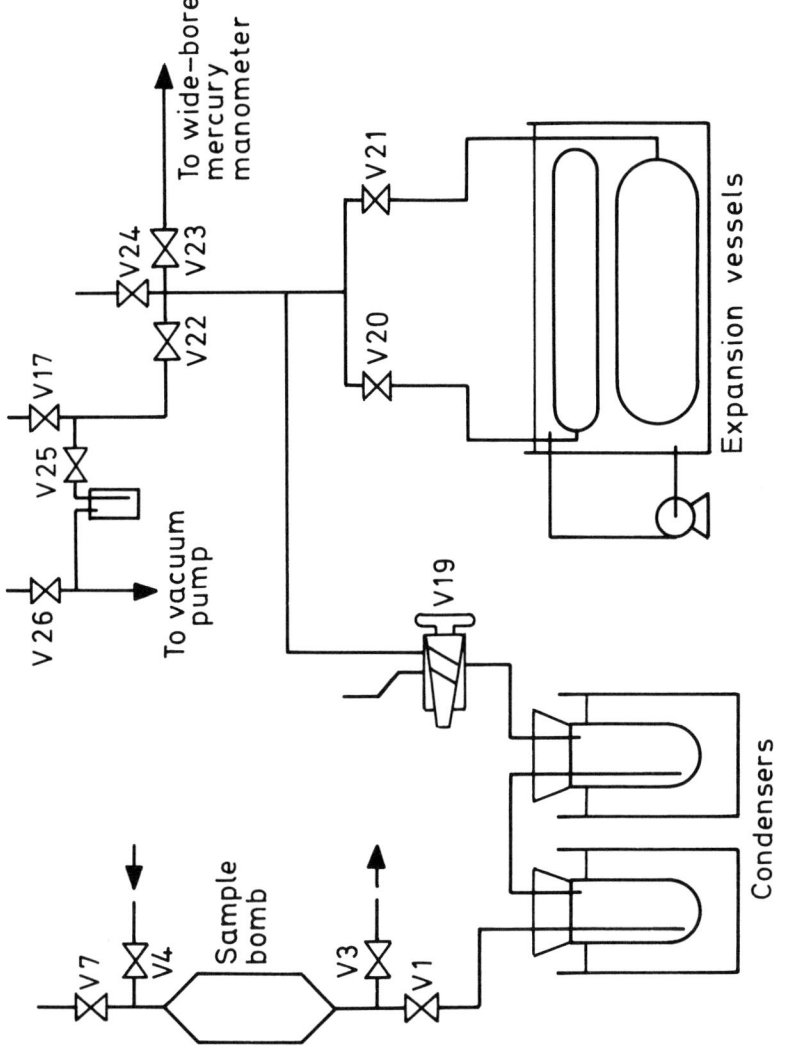

Figure 5. Sample collection system used with the recirculation equipment.

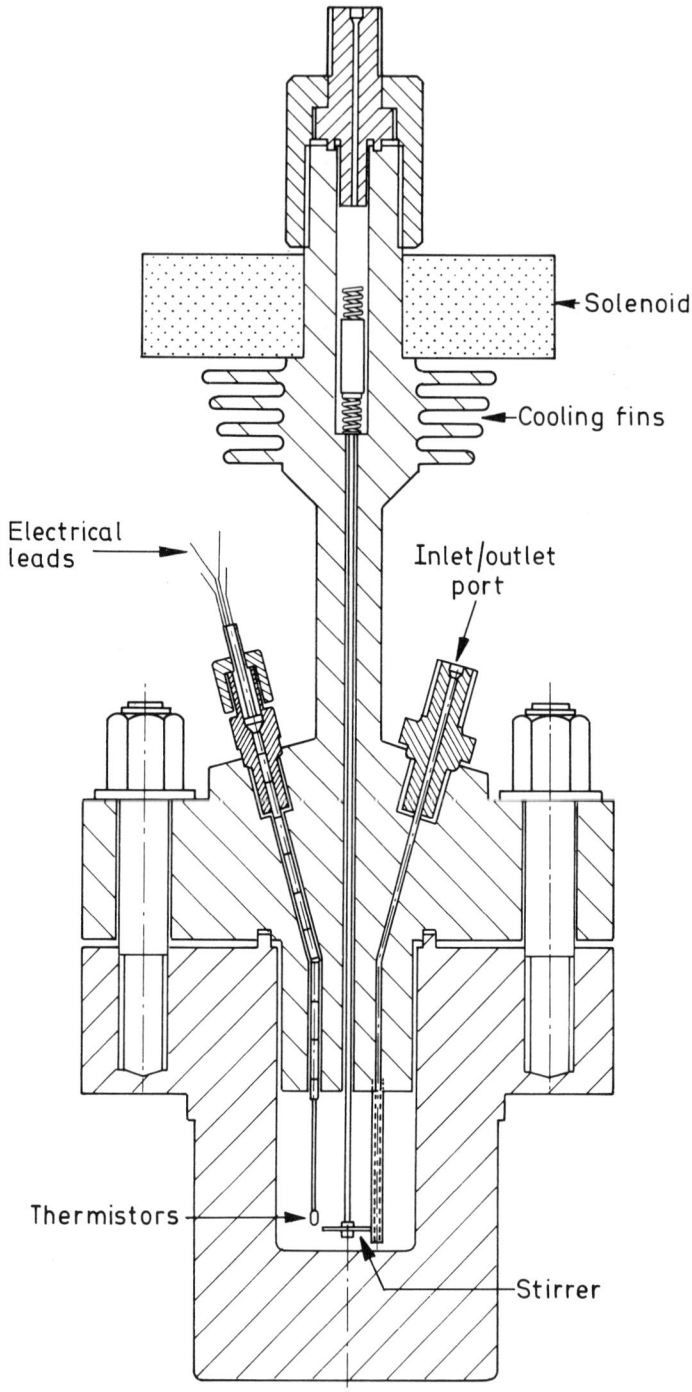

Figure 6. Equilibrium vessel used with vapour/liquid recirculation equipment.

These vessels are immersed in a water bath at temperature T. The increase in pressure (ΔP) in the expansion system is noted on a wide-bore manometer. At this stage some droplets of the heavy component (or components) remain in the sample bomb and the piping leading from this to the condensers. If the "heavy" material is comparatively volatile (as in the case of the systems containing hexane, benzene, octane and/or butanol) we find that a gentle slow stream of nitrogen introduced immediately above the sampling bomb (at V7 on Fig.5) and vented on the outlet line from the cooled condensers efficiently transfers this material. If the "heavy" material is comparatively involatile it is washed down into the condensers using a suitable volatile liquid solvent (1). (Both pentane and acetone were found to be suitable for the work on n-hexadecane). The solvent is then evaporated off and the weight of heavy components present in the sample is determined by weighing the condensers. The amount of the light component is calcuated from the volume of the expansion system, the temperature T and the pressure rise ΔP. In the case of a binary system, this completes the analysis. In the case of a ternary system, a representative sample of the condensate is removed and analysed by G.L.C.

The tubes carrying the phases out of the equilibrium vessel were packed with stainless steel mesh to inhibit entrainment of the wrong phase.

The vapour and liquid phases were recirculated using a small air-driven pump manufactured by Haskel Engineering. Although primarily designed for use with liquids the pump operates quite satisfactorily with compressed gases, provided the pressure generated in the cylinder per stroke is sufficient to operate the check valves. Whether or not this is the case depends of the pump geometry and spring loadings, on pressure and temperature and the nature of the gas. For example, the pump we used operated satisfactorily for ethylene at 73.5°C at pressures above 30 bar. Although used on this equipment for recirculation only, the pump will generate pressures upto 1000 bar in liquids. This facility was used in the apparatus described below.

Futher details of the equipment and analysis system are available (5).

Equipment used for equilibrium determinations of the loading of extracting phase in systems containing solid or liquid components, and also for simple extraction tests.

A line diagram of the equipment is shown in Fig.7. The stainless steel equilibrium cell and sampling bombs have capacities of 500 cm^3 and 37.49 cm^3 respectively. When a solid is to be extracted it is placed between layers of "knitmesh" in a stainless steel gauze "basket". This "basket" is lowered into the equilibrium cell and the lid bolted down. (Further details of this cell are available (21)). The cylinder was designed to make good contact with the walls of the cell so that the extracting fluid passes up through it rather than by-passing. When a liquid is to be extracted, a stainless steel base with a pipe 2cm long and 0.3cm in diameter is screwed into the bottom of the gauze cylinder (in place of the gauze bottom previously used) and the glass vapour/liquid contactor shown in Fig.8 is lowered onto this base with the stainless pipe projecting up into the central glass tube of the contactor. The region between the top of the contactor and the top of the "basket" is packed with knitmesh and the basket lowered into the equilibrium cell as before.

When used for equilibrium determinations, the system is charged to the desired pressure and the outlet valves, A, B, C, D and E are sealed. The extracting fluid (carbon dioxide) is then pumped round the closed circuit from the equilibrium cell to the sampling bomb and back until equilibrium has been sufficiently closely achieved (tests indicate that this requires about 20 minutes of pumping). The sampling bomb is then sealed off and the contents are analysed in exactly the manner described previously.

When used for extraction tests with liquid carbon dioxide, a carbon dioxide cylinder with dip tube is attached at F and the carbon dioxide is allowed to pass through valve A, up through the equilibrium cell and sampling bomb and out through valve B where the pressure is reduced to atmospheric. Material deposited from the stream is collected and the flow of effluent gas measured on a gas meter.

When used for extraction tests with supercritical carbon dioxide, compressed carbon dioxide at a pressure of about 100 bar is fed through valve C to the air driven pump which further compresses the stream to a suitable operating pressure. The stream is brought to the extraction temperature, passed up through the equilibrium cell and sampling bomb and out through valve B as in the liquid extraction case.

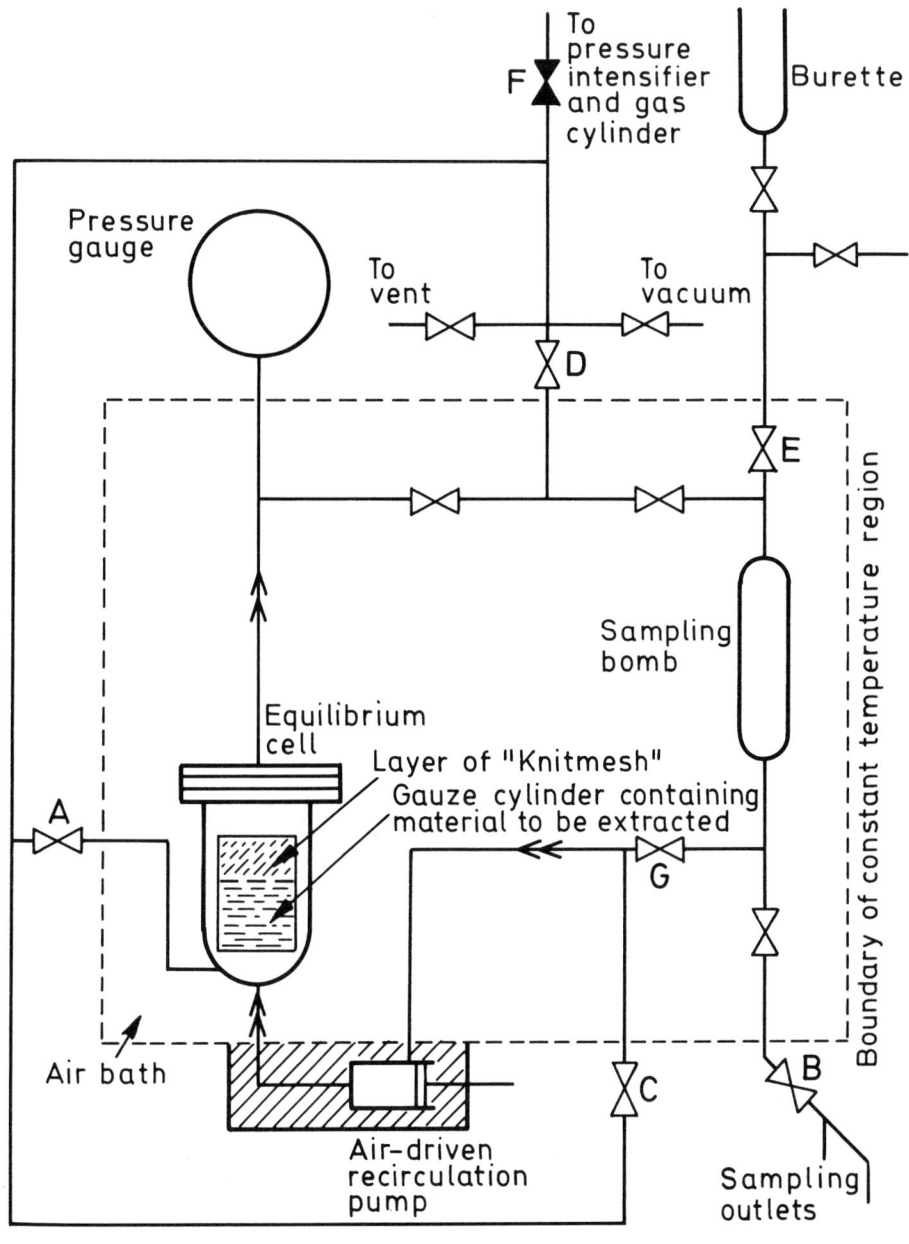

Figure 7. Layout of equipment for equilibrium vapour composition measurements over solid or liquid phases and for simple extraction tests.

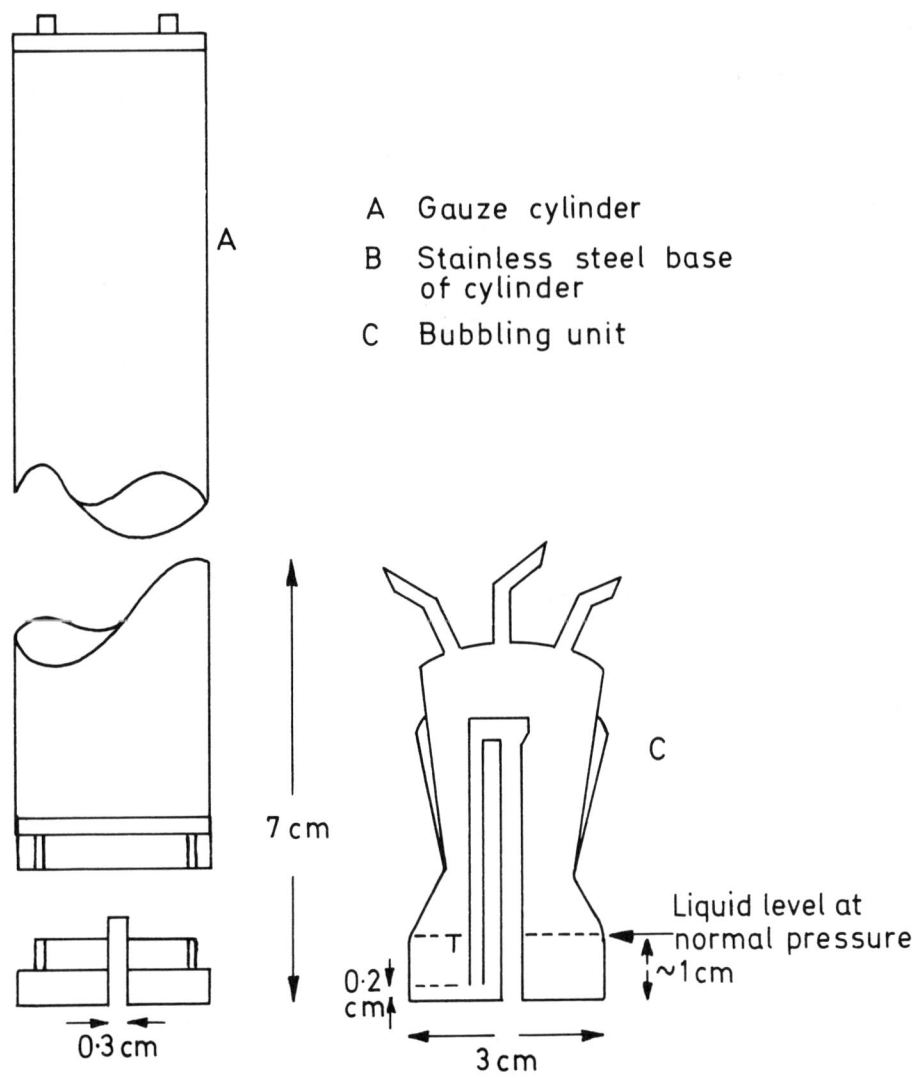

Figure 8. Solid-contacting and liquid-contacting devices.

Extractions of this type have been carried out on wood chips, rolled rape seed, crude palm oil and tobacco. The results of the tobacco extraction are described in an accompanying publication (1).

EQUILIBRIUM RESULTS AND THEIR CORRELATION USING THE REDLICH-KWONG EQUATION

Binary Vapour/Liquid Equilibria

Data obtained for the binary systems;

Ethylene/Benzene (at 73.5 and 75.0°C), Ethylene/Toluene (at 73.5°C), Ethylene/iso-Octane (at 73.5°C), Ethane/iso-Octane (at 73.5°C), Methane/n-Hexane (at 104.4°C), Carbon dioxide/n-Octane (at 40°C and 110°C), Carbon dioxide/n-Butanol (at 40°C and 110°C), Carbon dioxide/Oleic acid (at 35.0 and 60°C), and Carbon dioxide/mono-Olein (at 35 and 60°C) are summarised in Table 1 and are shown in the form of pressure/composition diagrams in Figs. 9 to 15. The data at 73.5°C were obtained with the static stirred cell and those at 40, 60 and 110°C with the vapour/liquid recirculation equipment described previously. Some of the data for the systems Carbon dioxide/Oleic acid and Carbon dioxide/mono-Olein were obtained with the vapour recirculation equipment described previously.

In each case the curve drawn between the experimental points is that predicted by the Redlich-Kwong equation of state using the parameters listed in Table 2 and 3. According to this well known simple equation;

$$P = \frac{RT}{(V-b_m)} - \frac{a_m}{T^{\frac{1}{2}} V(V+b_m)} \quad (1)$$

The use of this equation of state for phase equilibrium calculations is discussed briefly in an accompanying paper (1). For the present calculations the parameters a_m and b_m for the mixtures were related to the corresponding parameters for the pure components by the mixing rules given there. For binary systems these reduce to

$$a_m = x_1^2 a_{11} + x_2^2 a_{22} + 2x_1 x_2 (a_{11} a_{22})^{\frac{1}{2}} (1-\ell_{12}) \quad (2)$$

and
$$b_m = x_1 b_1 + x_2 b_2 \quad (3)$$

where x_1 and x_2 are the mole fractions of components 1 and 2.

Table 1

Vapour/Liquid Equilibrium Measurements in Binary Systems

(x_1 and y_1 are the mole fractions of the more volatile component in the liquid and vapour phases: y_1 (calc) is the vapour mole fraction predicted by the Redlich-Kwong equation (see text) symbol s indicates that system was single phase).

(A) Data obtained with static stirred cell

P(bar)	x_1	y_1	y_1 (calc)	P(bar)	x_1	y_1	y_1 (calc)
Ethane/iso-Octane at 73°C							
66.8	0.842	0.941	0.958	41.0	0.540	0.970	0.976
64.7	0.816	0.952	0.963	32.8	0.446	0.972	0.976
62.6	0.794	0.958	0.966	24.0	0.346	0.968	0.971
51.0	-	0.970	0.975	9.0	0.170	0.927	0.940
49.0	0.626	0.971	0.975	6.4	-	0.887	0.914
Ethylene/Benzene at 73.5°C							
85.5	0.672	-	0.947	50.6	0.367	-	0.965
84.7	-	0.937	0.950	38.5	0.275	0.951	0.963
75.2	-	0.955	0.958	29.2	0.221	-	0.958
66.2	-	0.972	0.963	21.3	-	0.941	0.950
-	-	-	-	8.3	-	0.907	0.898
Ethylene/iso-Octane at 73.5°C							
106.1	s	s	-	54.7	-	0.972	0.974
90.0	-	0.889	-	54.1	-	0.963	0.974
71.0	-	0.977	0.967	44.5	0.409	-	0.975
70.0	0.602	-	0.968	42.1	-	0.977	0.975
66.5	0.603	-	0.969	29.4	0.300	-	0.973
57.7	0.521	-	0.973	27.7	-	0.972	0.972
57.6	0.534	-	0.973	6.2	-	0.919	0.919
Ethylene/Toluene at 73.5°C							
98.9	-	0.960	0.958	58.5	0.471	0.982	0.983
87.9	0.696	0.967	0.971	53.3	0.429	0.988	0.984
75.6	0.571	0.968	0.978	39.6	0.322	-	0.984
75.2	0.592	0.970	0.978	30.3	0.245	0.978	0.983
70.1	0.550	0.976	0.980	10.1	0.077	0.968	0.967
64.1	0.511	0.982	0.982	7.6	0.055	0.948	0.957

Table 1 (continued)

Vapour/Liquid Equilibrium Measurements in Binary Systems

($\bar{P}$ is the average of the pressures at which the liquid and vapour samples were taken and $2\Delta P$ is the difference between them. The pressures for the liquid and vapour samples are thus $(\bar{P} + \Delta P)$ and $(\bar{P} - \Delta P)$ respectively. V^G and V^L are the molar volumes of the vapour and liquid phases. Other symbols as in Table 1A).

(B) Data obtained with vapour/liquid recirculation equipment

$\bar{P}$ (bar)	ΔP (bar)	x_1	y_1	y_1 (calc)	V^G (cm³/mol)	V^G (cm³/mol)
\multicolumn{7}{c}{Ethylene/Benzene at 75.0°C}
90.75	0.10	0.701	0.927	0.940	205.0	85.1
68.20	0.35	0.494	0.971	0.962	305.1	85.4
55.15	–	–	0.970	0.965	416.4	–
32.18	–0.13	0.250	0.959	0.960	726.8	90.1
\multicolumn{7}{c}{Methane/n-Hexane at 37.8°C (100.0°F)}
117.6	–2.9	0.426	0.970	0.976	181.6	98.4
77.5	–0.5	0.306	0.984	0.984	282.1	108.0
41.87	0.03	0.186	0.982	0.983	562.0	117.1
18.65	0.01	0.086	0.975	0.975	1337.1	126.0
\multicolumn{7}{c}{Methane/n-Hexane at 104.4°C (220°F) (Results with asterisk at 104.0°C)}
100.1	–0.3	0.350	0.930	0.918	271.2	125.9
80.25*	–0.25	0.284	0.922	0.918	360.9	129.3
50.75*	–0.25	0.182	0.906	0.905	569.2	137.2
31.40	0.10	0.117	0.865	0.876	938.1	141.2
\multicolumn{7}{c}{Carbon dioxide/n-Butanol at 40.0°C (Results with asterisk probably for 2nd liquid phase – see text and Fig.12a)}
89.26	–3.05	0.827	0.967*	–	63.6*	62.5
88.90	0.90	0.819	0.968*	–	65.7*	60.9
80.25	0.05	0.687	0.988	0.995	115.2	65.2
78.80	0.20	0.682	0.988	0.995	119.4	67.3
69.60	–0.10	0.579	0.994	0.997	198.7	72.1
55.15	–0.05	0.426	0.997	0.998	312.2	75.8
42.10	0.10	0.311	0.999	0.998	477.4	83.7
28.35	0.15	0.218	0.997	0.998	768.8	105.1

cont/.....

Table 1 (B) continued
Data obtained with vapour/liquid recirculation equipment

$\bar{P}$ (bar)	ΔP (bar)	x_1	y_1	y_1 (calc)	V^G (cm³/mol)	V^G (cm³/mol)
Carbon dioxide/n-Butanol at 110.0°C							
165.0	−0.5	0.773	0.832	0.900	85.5	78.7
153.95	−0.15	0.719	0.912	0.918	102.3	79.3
153.6	0.7	0.641	0.935	0.934	111.5	80.3
142.0	−2.2	0.573	0.948	0.946	127.4	81.4
120.4	−0.6	0.455	0.949	0.963	156.0	82.7
99.65	−1.15	0.359	0.948	0.970	215.5	83.2
90.65	0.35	0.350	0.946	0.972	250.9	87.1
70.15	0.05	0.294	0.942	0.974	342.0	90.2
50.20	0.00	0.213	0.937	0.972	502.7	97.7
32.45	2.25	0.147	0.925	0.962	899.2	102.0
20.25	0.05	0.085	0.905	0.952	1302.5	103.4
Carbon dioxide/n-Octane at 40.0°C							
73.80	−0.90	0.900	0.993	0.995	165.7	76.0
59.25	−0.75	0.674	0.995	0.996	275.9	100.5
42.05	−1.05	0.498	0.996	0.997	478.2	149.8
39.75	−0.25	0.520	0.996	0.997	500.0	155.2
27.45	−2.35	0.334	0.994	0.997	742.0	169.7
Carbon dioxide/n-Octane at 110.0°C							
140.0	−3.2	0.774	0.928	s	121.0	97.8
131.3	−1.3	0.744	0.953	0.933	136.8	102.0
119.7	0.2	0.696	0.958	0.953	164.4	109.1
99.3	−2.1	0.589	0.961	0.965	211.9	119.7
81.4	0.2	0.518	0.961	0.971	284.5	124.1
74.0	−0.5	0.492	0.960	0.972	309.3	−
60.2	0.1	0.436	0.961	0.973	409.4	133.4
40.9	0.0	−	0.939	0.972	656.6	−
29.0	0.0	0.211	0.938	0.965	−	−

Table 1 (continued)

Vapour/Liquid Equilibrium Measurements in Binary Systems

(C) Data obtained with vapour recirculation equipment

(v^G for the systems below was virtually identical with that for pure CO_2 under the given conditions. Liquid compositions (not obtained with this equipment) and calculated vapour compositions are discussed in text and shown in Figs.14 and 15. y_2 is the vapour mole fraction of the heavy component).

P (bar)	$y_2 \times 10^2$	P (bar)	$y_2 \times 10^2$
\multicolumn{4}{c}{Carbon dioxide/Oleic acid* at 35.0°C}			
191.3	(0.083)	130.6	0.079
180.9	0.106	114.8	0.052
176.8	0.092	96.1	0.027
150.6	0.076	85.1	0.015
140.3	0.085		
\multicolumn{4}{c}{Carbon dioxide/Oleic acid* at 60.0°C}			
181.6	0.0509	144.1	0.0282
177.5	0.0496	139.6	0.0147
162.3	0.0496	129.9	0.0139
\multicolumn{4}{c}{Carbon dioxide/mono-Olein+ at 35.0°C}			
178.9	0.0235	127.2	0.0092
167.8	0.0171	114.1	0.0087
152.0	0.0147	103.7	0.0073
146.8	0.0108		
\multicolumn{4}{c}{Carbon dioxide/mono-Olein+ at 60.0°C}			
189.2	0.0108	143.0	0.0024
174.0	0.0087	132.7	0.0023
156.8	0.0025		

* The oleic acid extracts and residues were better then 90% purity.

+ The initial sample and residues were of about 97% purity. They contained 1.5% of free acid and traces of di and tri glycerides. 97% of the acid groups present were oleic, with about 1% each of Margaric and Stearic acids. The extracts were of about 72% purity and contained 18.5% free acid and 9.5% triclycerides. 80% of the acid groups present were oleic with about 4% stearic acid, 3% each of margaric and palmitic acids and 1% godoleic acid (see text).

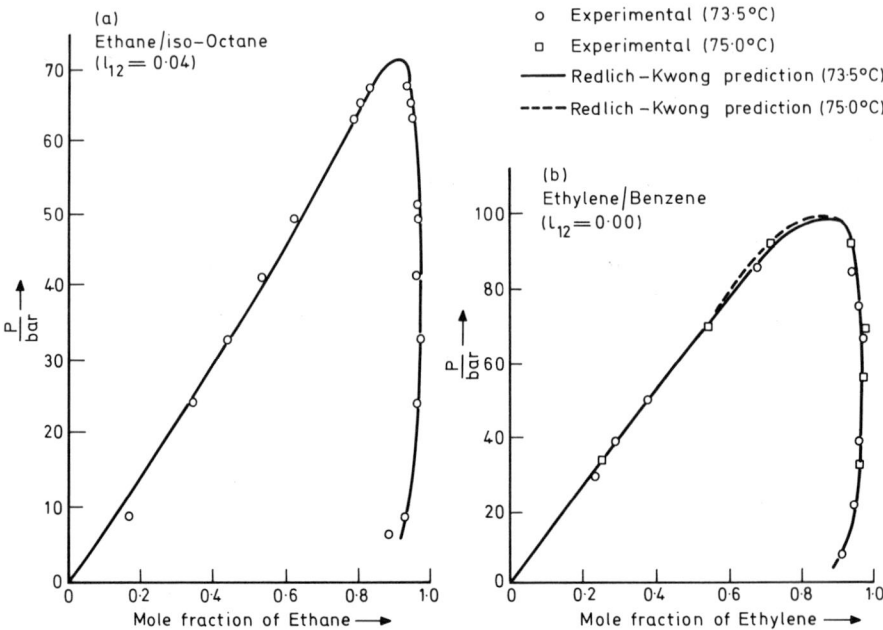

Figure 9. Pressure/composition diagrams for systems Ethane/iso-Octane (at 73.5°C) and Ethylene/Benzene (at 73.5 and 75.0°C).

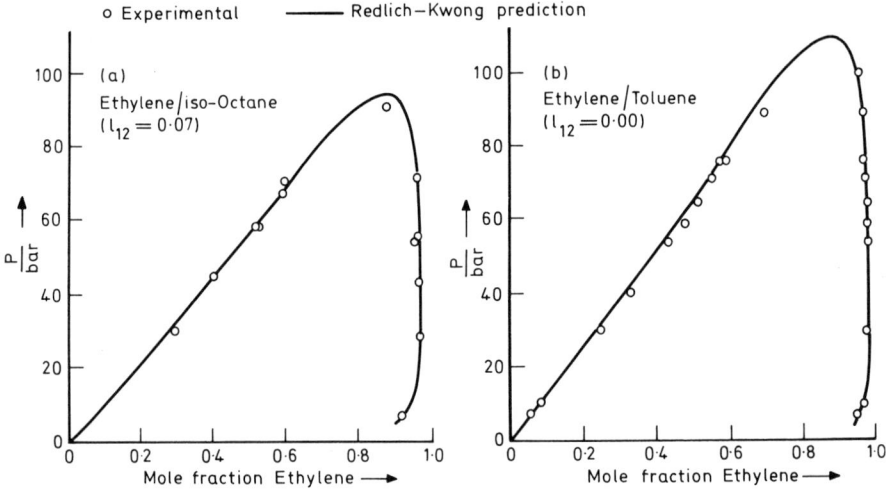

Figure 10. Pressure/composition diagrams at 73.5°C for the systems Ethylene/iso-Octane and Ethylene/Toluene.

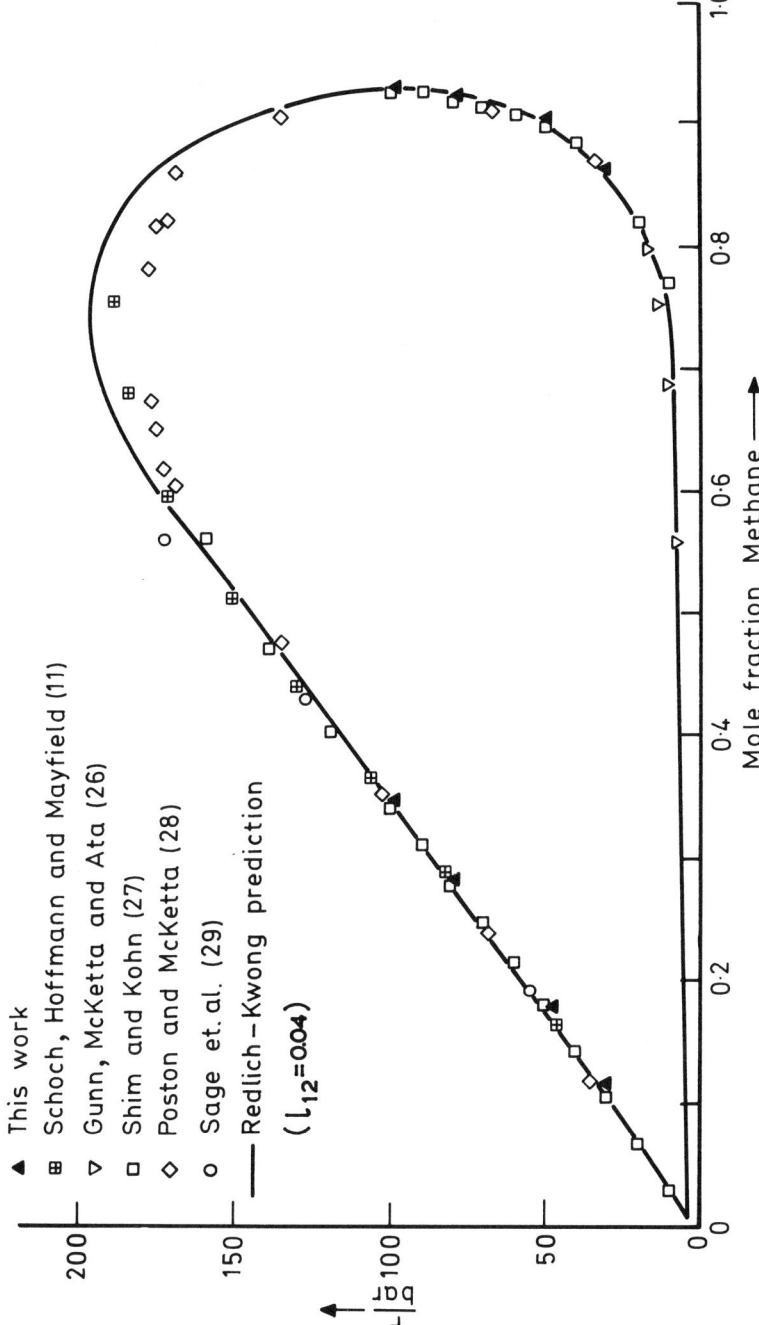

Figure 11. Pressure/composition diagram for system Methane/Hexane at 104.4°C.

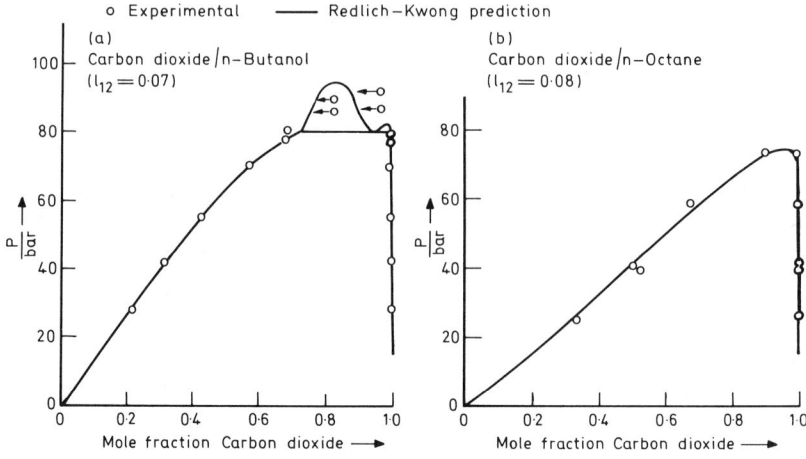

Figure 12. Pressure/composition diagrams for systems Carbon dioxide/n-Butanol and Carbon Dioxide/n-Octane at 40.0°C.

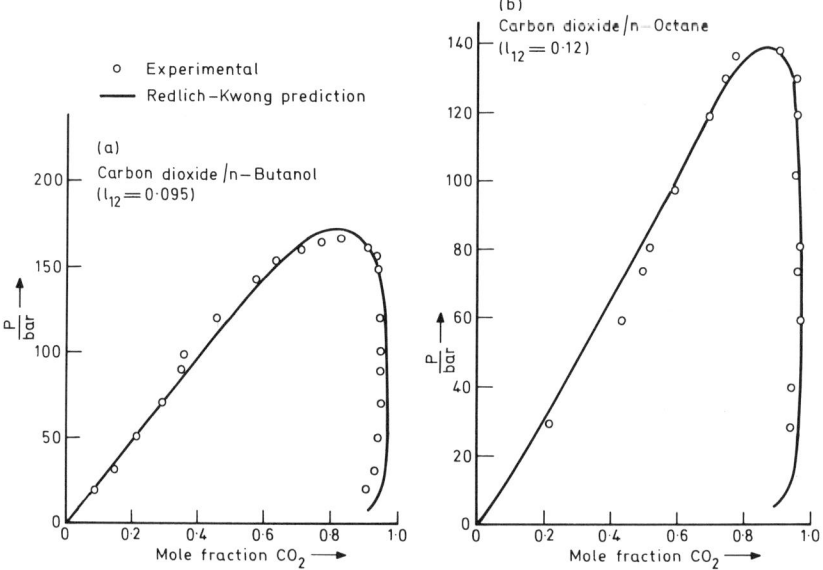

Figure 13. Pressure/composition diagrams for systems Carbon dioxide/n-Butanol and Carbon dioxide/n-Octane at 110.0°C.

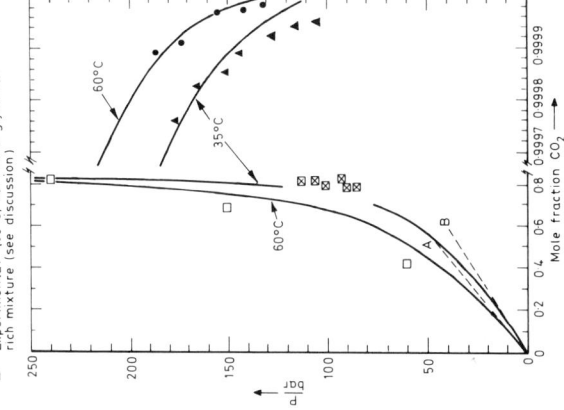

Figure 15. Pressure/composition diagrams for system Carbon dioxide/mono-Olein at 35 and 60°C (lines OA and OB give the terminal slopes of the bubble-point curves as obtained from gas solubility measurements at normal pressure).

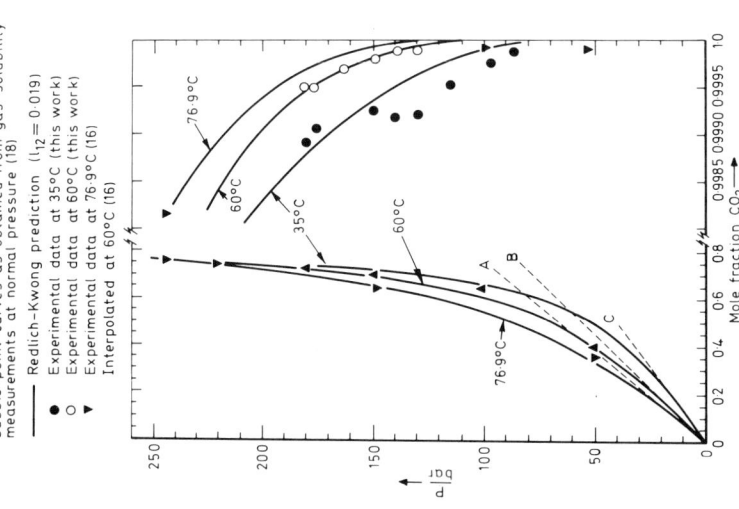

Figure 14. Pressure/composition diagrams for system Carbon dioxide/Oleic acid at 35.0, 60.0 and 76.9°C.

The pure component parameters a_{11} and b_{11} for the supercritical component were calculated from the critical constants listed by Ambrose (8) using the expressions

$$a_{11} = 0.42748 \; R^2 \; T_c^{2.5} \; P_c^{-1} \tag{4}$$

$$b_1 = 0.08664 \; RT^c \; P_c^{-1} \tag{5}$$

The parameters for the subcritical components apart from oleic acid and mono-olein were calculated from the liquid molar volume and vapour pressure using Zudkevitch's second method (9). The parameter ℓ_{12} (which allows for deviations from the geometric mean mixing rule $a_{ij} = (a_{ii}a_{jj})^{\frac{1}{2}}$ was adjusted to fit the data at each temperature (table 3). For hydrocarbon/hydrocarbon systems this parameter is normally close to zero while for carbon dioxide/hydrocarbon systems it is normally close to 0.1. It would appear from Table 3 that there is a tendency for ℓ_{12} to rise slightly with temperature. This may arise from the use of "universal" values of a and b for the supercritical component.

It is seen from the pressure/composition diagrams (Figs. 9 to 16) that, by selecting a suitable value for the deviation parameter ℓ_{12}, it is possible to obtain a surprisingly good fit for the compositions of the equilibrium phases in each of the binary systems shown. The fit obtained for the systems containing carbon dioxide is not quite as close as for those containing ethylene, the predicted amount of heavy component in the vapour phase tending to be a little low at pressures below about 50 bar. However, outside this range, the fit obtained for these systems also is very good. It is of interest that the Redlich-Kwong equation predicts a very small region of liquid/liquid partial miscibility close to the critical region in the system Carbon dioxide/n-Butanol at 40°C (similar behaviour has been reported for the system Ethylene/n-Propanol (13)). The experimental points in this region are consistent with this type of behaviour, though their quantitative prediction is not good.

Because of the scale used, the extent to which the vapour phase compositions are predicted by the Redlich-Kwong equation may not be clear from some of the figures. For this reason predicted compositions are given in addition to the experimental ones in Table 1.

Phasecompositions generated by the Redlich-Kwong equation are thermodynamically consistent and data which can be fitted to this equation over a substantial range must likewise be judged to be thermodynamically consistent. Although data

Table 2

Parameters a and b in the Redlich-Kwong equation (Eq.1) for the pure components used in the phase equilibrium calculations

Component	Temperature (°C)	a (bar)(dm^3/mol)^{2}K$^{\frac{1}{2}}$	b (dm^3/mol)
Ethane	>32.27	98.716	0.04508
Ethylene	> 9.19	78.551	0.04036
Carbon dioxide	>30.98	64.615	0.02970
iso-Octane	73.5	836.752	0.1407
	100.0	829.428	0.1407
Toluene	73.5	618.560	0.09273
	100.0	618.024	0.09295
Benzene	73.5	453.583	0.07590
	75.0	453.597	0.07592
n-Octane	40.0	968.769	0.1416
	110.0	953.923	0.1436
n-Butanol	40.0	619.724	0.08123
	110.0	572.205	0.08110
Oleic acid	35.0	6840.0	0.3019
	60.0	6600.0	0.3019
	76.9	6426.8	0.3019
mono-Olein	35.0	11090.0	0.3620
	60.0	10500.0	0.3620
n-Hexadecane	60.0	3391.7	0.2767
Naphthalene*	35.0	1281.3	0.1331
n-Dotriacontane**	35.0	18925.6	0.7788
n-Hexane	37.8	582.347	0.1081
	104.4	571.401	0.1084
Methane	>-82.6	31.496	0.02919

* From Soave's correlation (20)

** Mid-values from the correlations of Soave (20) and Brunner (17)

Table 3

Deviation parameters ℓ_{ij} used in conjunction with the Redlich-Kwong equation (Eq.1) and the mixing rule;

$$a_m = \sum_{i=1}^{n} \sum_{j=1}^{n} x_i x_j (a_{ii} a_{jj})^{\frac{1}{2}} (1 - \ell_{ij})$$

Component i	Component j	ℓ_{ij}	t (°C)
Ethane	iso-Octane	0.040	73.5
Ethane	Toluene	(0.000)*	73.5
iso-Octane	Toluene	-0.003	100.0
Ethylene	iso-Octane	0.070	73.5
Ethylene	Toluene	0.000	73.5
Ethylene	Benzene	0.000	73.5 and 75.0
Methane	n-Hexane	0.030/0.040	37.8/104.4
Carbon dioxide	n-Octane	0.080/0.120	40.0/110.0
Carbon dioxide	n-Butanol	0.070/0.095	40.0/110.0
n-Octane	n-Butanol	0.080	109.2 to 115.6
+Carbon dioxide	n-Hexadecane	0.080	60.0
Carbon dioxide	Oleic acid	0.019	35.0 to 76.9
Carbon dioxide	mono-Olein	-0.030	35.0 to 60.0
Carbon dioxide	Naphthalene	0.0976	35.0
Carbon dioxide	n-Dotriacontane	0.10	35.0

* Estimate - see text

+ The pressure composition diagram at 60°C for this system is given in reference 1 and the work on which it is based is to be described shortly.

which cannot be fitted to this equation of state are not necessarily inconsistent, they must be viewed with caution if similar systems are known to obey the equation well. Results obtained for the system Ethane/Toluene were considered to be suspect for this reason and are not reported here.

The system Ethylene/Benzene was studied both using the static cell and using the vapour/liquid recirculation equipment, in order to check their operation. The results (Fig. 9b) are in good mutual agreement and are closely represented by the Redlich-Kwong equation with a deviation parameter (ℓ_{12}) of zero, which is the same as the value required for the ethylene/toluene system. (Except in the close vicinity of the critical point the predicted phase envelope for 75.0°C is virtually coincident with that for 73.5°C). The vapour and liquid compositions obtained are also in good agreement (5) with earlier work by Harris (3) and Ellis and Valteris (4). The vapour phase compositions agree well with those reported by Zhuze (14) though the liquid phase at low pressures was found to be up to 0.03 mole fraction richer in ethylene. Some test results obtained for the Methane/Hexane system at 104.4°C and pressures up to 100 bar agree within a mole fraction of 0.01 with those of other workers (Fig.11).

Vapour pressure data for oleic acid and mono-olein were not available at the temperatures studied, so the Redlich-Kwong parameters a and b for these components could not be obtained by Zudkevitch's method (9). The values used were estimated as follows:

(1) b was obtained from the liquid molar volume at 20°C using correlations proposed by Brunner, his correlation for carboxylic acids (16) being used for oleic acid and his "general correlation" (17) for mono-olein. (A similar procedure was used for triolein, though the results are not shown in detail here).

(2) the value of a at each temperature was arbitrarily adjusted until a value was found such that it was possible to fit both the vapour and the liquid phase data using the same value for ℓ_{12}. It was found (Table 3) that the values of ℓ_{12} arrived at were quite normal suggesting that the procedure was working reasonably well.

The parameters obtained in this way gave a satisfactory fit to the data for the Carbon dioxide/Oleic acid system (Fig. 14) though the fit obtained for glyceride containing systems, such as that shown in Fig.15, was not so satisfactory.

The vapour phase compositions given in Table 1 and shown

Table 4

Carbon dioxide solubilities at normal pressure in some solvents of low volatility

(x^1 is the mole fraction solubility of carbon dioxide for a partial pressure of one atmosphere)

Solvent	Temp (K)	Solubility $(x^1) \times 10^4$	Henry's constant atm/molefraction
n-Hexadecane	299.15	139	72
	303.15	130	77
	313.15	120	83
	323.15	108	93
	333.15	100	100
	343.15	94	106
	353.15	88	114
Diphenylmethane	303.15	79	126
	313.15	71	141
	323.15	63	160
	333.15	58	173
	343.15	53	187
	353.15	50	201
Diphenylethane	333.15	76	131
	338.15	71	141
	343.15	67	149
	353.15	59	169
	358.15	55	181
Ethylbenzene	296.15	72	139
	298.15	71	140
	303.15	69	145
	313.15	63	164
	333.15	55	181
Triolein	313.15	385	26
	333.15	323	30
	353.15	255	39
Monolein	313.15	147	68
	333.15	123	81
	353.15	97	103
Glycerol	313.15	11.2	890
	333.15	9.2	1100
	353.15	8.1	1200

Except for the solubilities in glycerol, when the uncertainty is about 3%, the above values are considered trustworthy to about 1% of the stated figure. The equipment used for the solubility measurements was similar to one described by Ali (31). The solubility work is to be described in more detail later.

in Figs 14 and 15 for the systems Carbon dioxide/Oleic acid and Carbon dioxide/mono-Olein were obtained using the vapour recirculation equipment with vapour/liquid contactor. Substantial precautions are taken to prevent entrainment with this equipment, the contactor being surmounted by a 6" depth of knitmesh packing. For this reason vapour compositions obtained with this equipment for systems (such as Carbon dioxide/mono-Olein) where very little heavy component is dissolved in the vapour are considered to be more reliable than ones obtained using the vapour/liquid recirculation equipment, which was designed with more "normal" systems in mind. The latter tended to be richer in extract at the lower end of the pressure range.

Liquid compositions at elevated pressure for the system Carbon dioxide/mono-Olein at $60°C$ and $100°C$ were obtained using the vapour/liquid recirculation equipment while those shown in Fig.14 for the system Carbon dioxide/Oleic acid at $60°C$ were obtained by interpolation from data reported by Brunner (17). The terminal slopes of the liquid composition curves for this system at $35.0°C$ and $60.0°C$ were obtained from the gas solubility observations of Ouellet and Dubois (18) while those for the system Carbon dioxide/mono-Olein were obtained from gas solubility measurements at normal pressure carried out as part of the present work (Table 4).

The vapour phase compositions for the system Carbon dioxide/Oleic acid are consistent with work reported by Brunner (16). They are less rich in oleic acid than was suggested by some preliminary work by Fallaha (2). It is thought that this is due to decomposition of his sample, of which there was evidence (2).

Ternary Systems

Ethylene/iso-Octane/Toluene and Ethane/iso-Octane/Toluene

The pressure/composition diagram for the binary system Toluene/iso-Octane at $100°C$ (the only temperature at which complete isothermal data appear to be available) is shown in Fig.16. The data points shown are from the work of Ramalho and Delmas (12). It is seen that the Redlich-Kwong equation provides a good fit to these data and that the relative volatility of iso-octane to toluene is poor, particularly in the iso-octane rich part of the composition range. This type of system is difficult to separate by normal distillation and extractive distillation (10) or solvent extraction (4) are often used for this reason. One reason for the tests on the ternary systems of toluene and iso-octane with ethane and with ethylene was to see whether critical fluid extraction

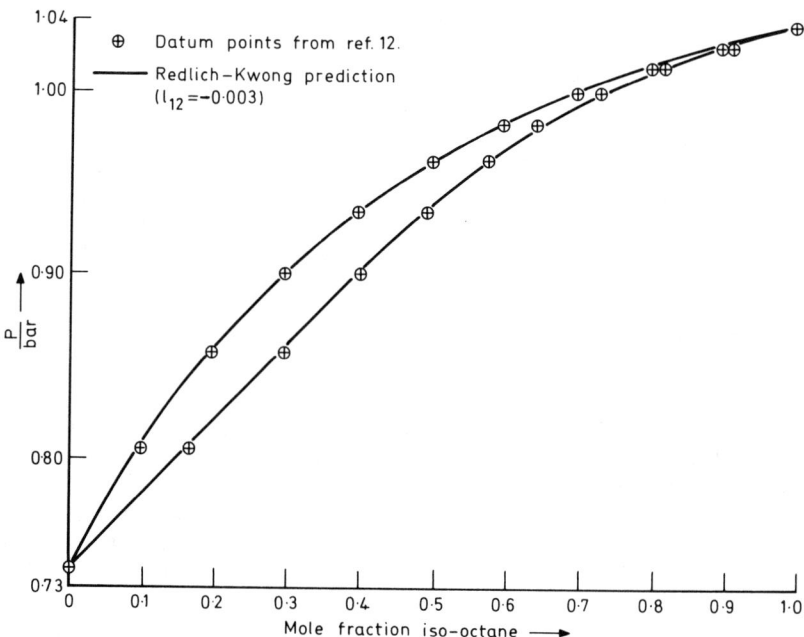

Figure 16. Pressure/composition diagram for system iso-Octane/Toluene at 100°C.

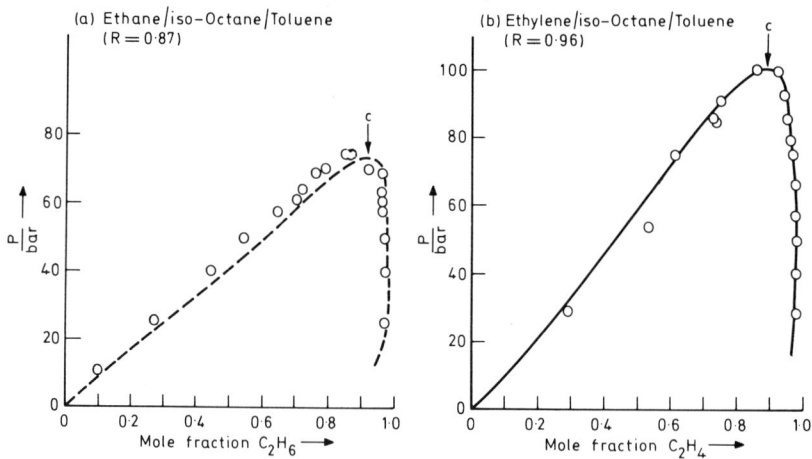

Figure 17. Pressure/composition diagrams at 73.5°C for the ternary systems Ethane/iso-Octane/Toluene (for R = 0.87) and Ethylene/iso-Octane/Toluene (for R = 0.96).

with a suitable solvent might provide an alternative separation method in such cases. Approximately equimolar mixtures of iso-octane and toluene were contacted with compressed ethane and with compressed ethylene. These tests were carried out at 73.5°C using the static stirred cell, the mole ratio of iso-octane to toluene in the vapour and liquid phases being determined by G.L.C. (2). The results are given in Table 5 and figures 17 and 18. Partly because the extraction of the heavy components into the vapour phase was small and partly because the relative volality of iso-octane to toluene did not substantially exceed unity, the mole ratio of iso-octane to toluene in the liquid phase (R) remained virtually constant during each set of experiments (Table 5). Since (R) is constant, the number of degrees of freedom is reduced by one and the compositions of the equilibrium vapour and liquid phases are, according to the phase rule, unique functions of pressure at given temperature. The mole fractions of ethane in the vapour and liquid phases of the system Ethane/Toluene/iso-Octane at 73.5°C, R = 0.87 are shown as functions of pressure in Fig.17a; Fig.17b is the corresponding diagram for the system Ethylene/Toluene/iso-Octane for an iso-octane to toluene ratio of 0.96. These diagrams are virtually identical in form to the familiar pressure/composition diagrams for binary systems. The continuous curves are those predicted by the Redlich-Kwong equation. In the case of the ethylene system (Fig.17b), the values of ℓ_{12} used for the ternary prediction are those used to fit the data for the constituent binary systems (Table 3) and it is apparent that a good prediction is obtained. Unfortunately, consistent data are not available for the Ethane/Toluene binary system: the predicted curve for the Ethane/iso-Octane/Toluene system was obtained by arbitrarily setting ℓ_{ij} = 0 for the ethane/toluene interactions, while using the values of ℓ_{ij} listed in Table 3 for the ethane/iso-octane and iso-octane/toluene interactions.

Figure 18 shows predicted and experimentally determined relative volatilities $\alpha_{o/T}$ of iso-octane to toluene as functions of pressure in the systems Ethane/iso-Octane/Toluene and Ethylene/iso-Octane/Toluene. Only data obtained at pressures above about 60 bar are shown. At lower pressures the amounts of vapour condensate (which had been subjected to freezing during the sampling process) were so small that homogenisation and the subsequent removal of representative samples for analysis was difficult. The relative volatility is sensitive to experimental errors in the determination of the vapour and liquid compositions, 2% errors in these can lead to a worst error of 8% in α. The experimentally determined relative volatilities for the ethylene ternary fall around the predicted curve within a scatter of this order of magnitude In view also of the good fit obtained for the ethylene mole fractions

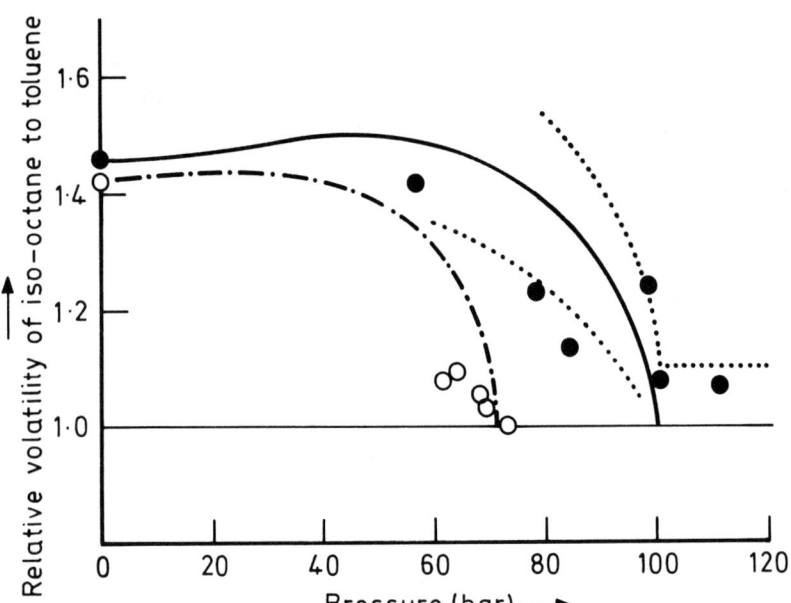

Figure 18. Relative volatilities of iso-Octane to Toluene in ternary systems Ethane/iso-Octane/Toluene (for R = 0.87) and Ethylene/iso-Octane/Toluene (for R = 0.96) at 73.5°C. (R is the mole ratio of iso-Octane to toluene in the liquid phase).

Table 5

Vapour/Liquid Equilibrium Measurements in Ternary Systems

(A) Data obtained with static cell

(x_1 and y_1 are the mole fractions of the lightest component (the extractant) in the liquid and vapour phases; X_2 and Y_2 are the mole fractions of the "heavy" component 2 in the liquid and vapour phases expressed on an extractant-free basis. Interpolated values are shown in brackets).

P (bar)	x_1	y_1	y_1 (calc)	X_2	Y_2
Ethylene (1)/Toluene (2)/iso-Octane (3) at 73.5°C					
112.7	0.933*	0.939*	s	0.528	0.511
102.3	0.940*	0.949*	s	0.528	0.509
99.9	0.862	0.927	0.913	0.526	0.473
93.1	-	0.945	0.947	(0.529)	-
91.1	0.751	-	0.952	0.530	-
85.5	0.732	0.957	0.961	0.529	0.498
84.8	0.734	-	0.962	0.529	-
79.2	-	0.965	0.967	(0.531)	0.479
75.1	0.617	0.976	0.970	-	-
66.3	-	0.980	0.975	(0.534)	0.475
57.6	-	0.980	0.978	(0.536)	0.450
53.8	0.534	-	0.978	0.537	-
50.4	-	0.985	0.978	-	-
41.5	-	0.981	0.978	-	-
29.2	0.295	0.989	0.976	-	-
Ethane (1)/Toluene (2)/iso-Octane (3) at 73.5°C					
74.5	0.856	0.860	s	0.510	0.511
70.0	0.799	0.929	0.962	0.511	0.500
69.1	0.768	0.961	0.963	0.511	0.499
63.6	0.726	0.967	0.972	0.511	0.490
61.6	0.709	0.961	0.974	0.512	0.493
57.9	0.648	0.961	0.975	-	-
49.1	0.549	0.970	0.978	-	-
40.1	0.446	0.971	0.979	-	-
25.3	0.275	0.970	0.975	-	-
10.3	0.097	-	-	-	-

* probably single phase

s denotes single phase

Table 5

(B) Data obtained with vapour/liquid recirculation equipment

As in Table 1B, the pressures for the vapour and liquid samples are $(\bar{P} + \Delta P)$ and $(\bar{P} - \Delta P)$ respectively and V^G and V^L are the molar volumes of the vapour and liquid phases.

$\bar{P}$ (bar)	ΔP (bar)	x_1	y_1	y_1 (calc)	X_2	Y_2	V^G (cm³/mol)	V^L (cm³/mol)
\multicolumn{9}{c}{Carbon dioxide (1)/n-Butanol (2)/n-Octane (3) at 110.0 °C}
140.60	−0.00	0.840	0.877	s	0.31	0.34	103.1	96.8
130.30	−0.30	0.796	0.930	0.924	0.30	0.34	130.9	96.2
98.95	0.05	0.608	0.937	0.954	0.29	0.35	209.1	104.5
71.65	−0.15	0.477	0.935	0.965	0.29	0.36	305.2	124.3
40.15	0.05	0.310	0.931	0.960				
129.35	−0.85	0.709	0.936	0.930	0.55	0.52	139.0	88.4
102.10	−0.10	0.578	0.936	0.956	0.56	0.57	197.7	96.0
70.65	−0.15	0.423	0.934	0.964	0.55	0.56	323.0	111.4
41.75	0.05	0.265	0.925	0.958				
138.40	0.60	0.686	0.936	0.926	0.79	0.76	121.5	84.2
106.65	−0.05	0.526	0.934	0.957	0.80	0.75	189.4	87.6
72.55	−0.15	0.362	0.932	0.965	0.80	0.74	323.9	98.3
47.85	0.05	0.219	0.920	0.962				

x_1, y_1, X_2 and Y_2 are defined in Table 5A.

(Fig.17b) it would appear that the Redlich-Kwong prediction for this ternary is reliable. A slight imprevement in the iso-octane/toluene relative volatility is predicted as the pressure rises to about 50 bar, but this quantity falls off at higher pressures as the critical point is approached. In the ethane ternary, $\alpha_{o/T}$ is predicted to remain constant upto about 40 bar and then to fall off. Of the two extractants, ethylene would appear to be the more suitable since;

(a) the critical pressure of the ternary system is higher, giving a wider range over which gas extraction effects are appreciable.

(b) it does provide a small improvement in the iso-octane/ toluene relative volatility.

Broadly speaking however, the relative volatility in the gas extraction situation appears to be very similar to that for distillation until the critical point is approached.

Carbon dioxide/n-Butanol/n-Octane

Three butanol/octane mixtures, containing butanol mole fractions of 0.30, 0.55 and 0.80 were contacted with compressed carbon dioxide at 110°C in the vapour/liquid recirculation equipment. The results are given in Table 5 and figures 20 through 22. As before the liquid composition remained virtually constant during each set of experiments.

The "y/x" diagram for the binary system n-Butanol/ n-Octane is given in Fig.19. It shows an azeotrope at a butanol mole fraction of about 0.55 (15). A good fit could not be obtained to these data using the Redlich-Kwong equation with the simple mixing rules (2) and (3), possibly because of the polarity of the butanol. The best fit that could be obtained was with a deviation paramter ℓ_{ij} of 0.08: this gave a butanol mole fraction of 0.62 at the azeotrope. Using this and the other values for ℓ_{ij} given in Table 2, the carbon dioxide mole fractions in the liquid and vapour phases have been calculated for the three mixtures studied, taking the composition of the liquid phase on a CO_2-free basis to be constant. The predicted curves are compared with the experimental data in Figures 20 and 21. Although qualitatively acceptable, the prediction is not as accurate as for the Ethylene/iso-Octane/Toluene system; at pressures below 100 bar the vapour is predicted to contain less of the heavy components than was observed and the predicted critical pressures appear to be about 5 bar too low.

The "y/x" diagrams (on a CO_2-free basis) at 70, 100 and

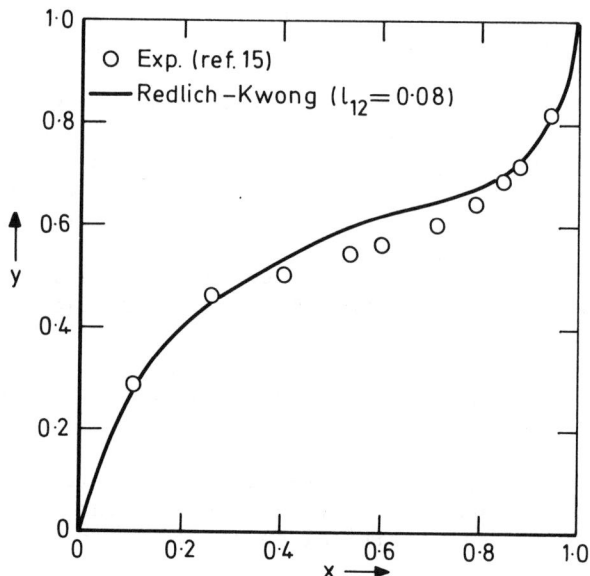

Figure 19. y/x diagram for system n-Butanol/n-Octane at 1 atm. (y and x are the mole fractions of butanol in vapour and liquid phases respectively.)

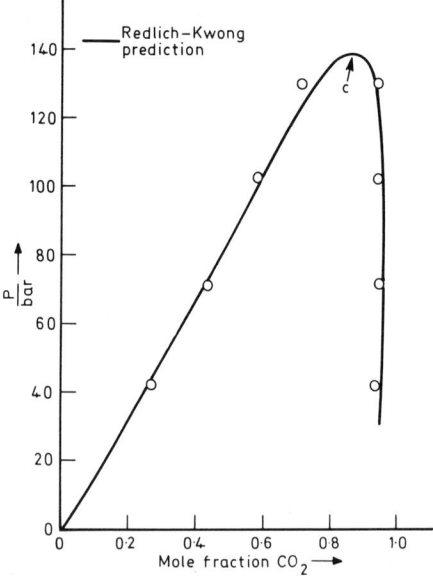

Figure 20. Pressure/composition diagram at 110°C for ternary system Carbon dioxide/n-Butane/n-Octane for a butanol to octane mole ratio in liquid of 1.22.

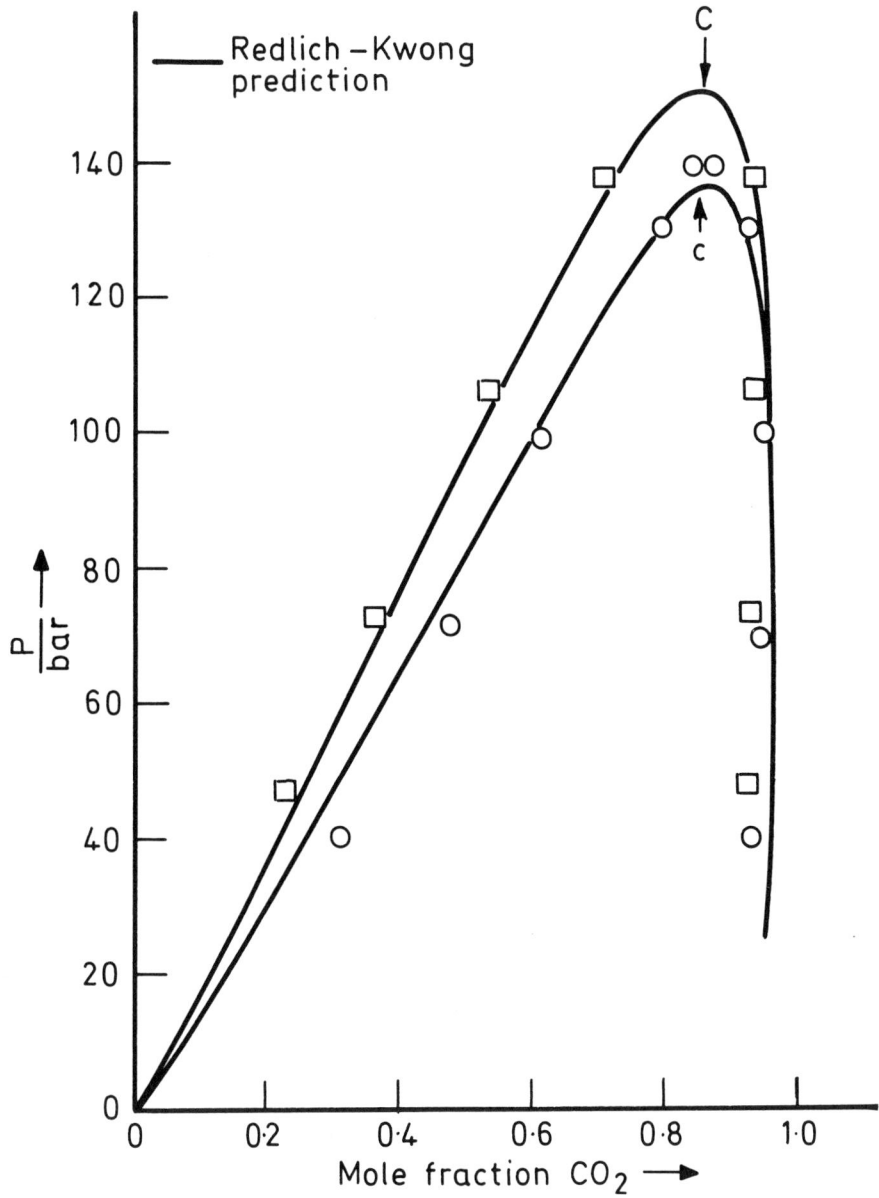

Figure 21. Pressure/composition diagrams at 110°C for system Carbon dioxide/n-Butanol/n-Octane for butanol to octane mole ratios in liquid of 0.43 (O) and 4.00 (□).

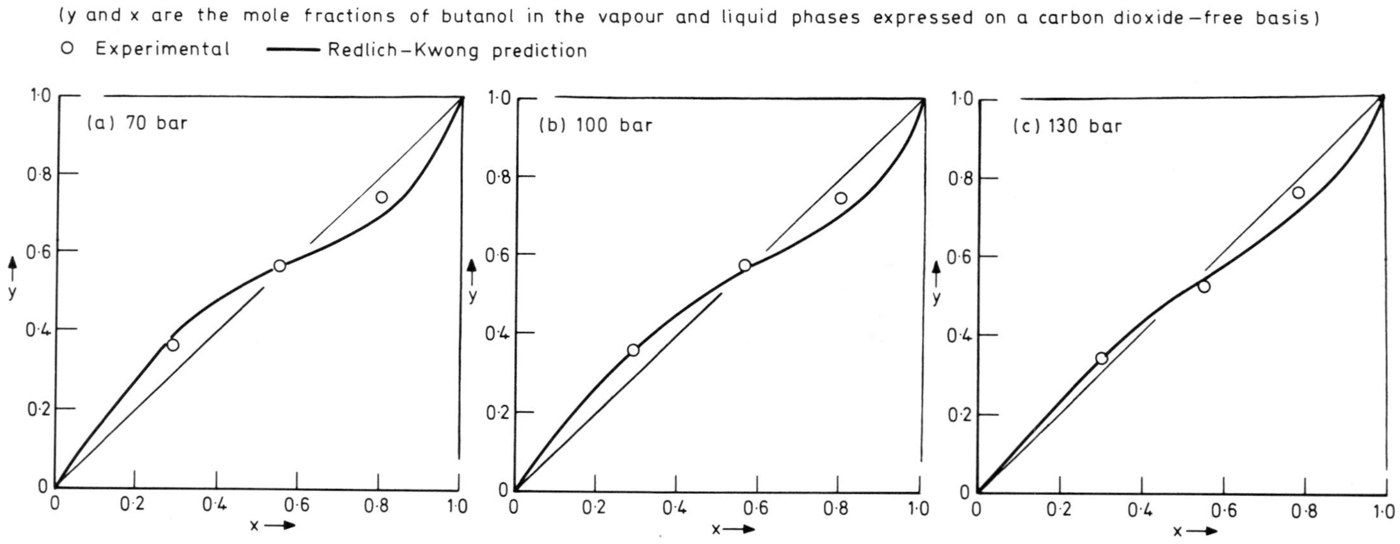

Figure 22. "y/x" diagrams for n-Butanol/n-Octane mixtures at 110°C contacted with carbon dioxide. (a) at 70 bar, (b) at 100 bar and (c) at 130 bar.

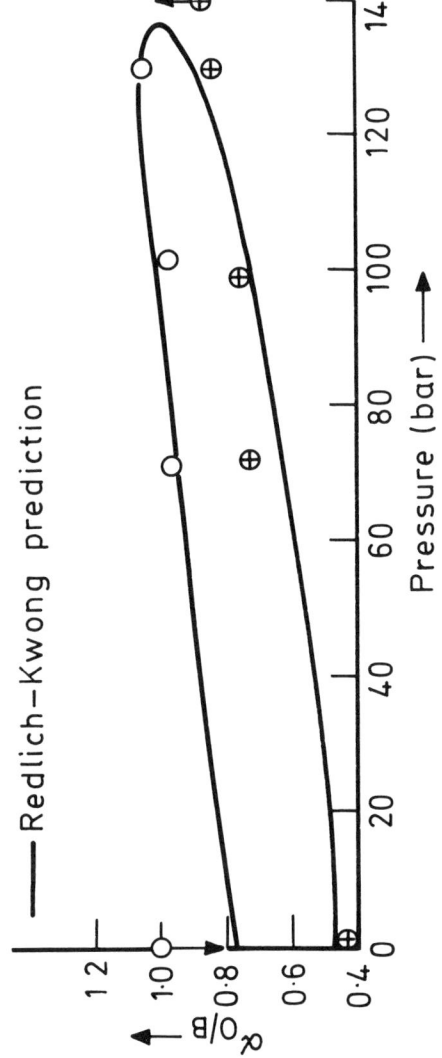

Figure 23. Relative volatilities of Octane to Butanol ($\alpha_{O/B}$) as functions of pressure in system Carbon dioxide/n-Butanol/n-Octane at 110°C, for butanol to octane mole ratios in liquid of 0.43 (O) and 1.22 (⊕).

130 bar are compared with the predicted ones in Fig.22 (data at pressures below about 50 bar are omitted for the reasons given above). Agreement is as close as can be expected in view of the poor fit obtained to the distillation data at normal pressure. The predicted situation differs from that in the systems previously considered in that the octane to butanol relative volatility initially rises with pressure for all the compositions shown, though returning to unity at the critical point (Fig.23).

Vapour/Solid Equilibria

Systems studied (Table 6) included Carbon dioxide/Naphthalene and Carbon dioxide/n-Dotriacontane (at 35.0°C). The system Carbon dioxide/Citric acid was also studied. In this case no detectable extract was obtained at 35°C and pressures upto 240 bar or at 50°C at pressures upto 160 bar. It was concluded (21) that the equilibrium mole fraction of citric acid in the vapour was less than 10^{-5} over the above range. The vapour recirculation equipment was used for the three systems.

The data for the solubilities of naphthalene in carbon dioxide were correlated by applying the Redlich-Kwong equation (with mixing rules 2 and 3) to the vapour phase and using this to calculate the fugacity f_2^G of naphthalein in this phase as a function of the naphthalene mole fraction y_2 at the temperature T and pressure P of interest. y_2 at given temperature and pressure was obtained to satisfy the equation

$$f_2^G (y_2) = p_2^o \exp (V_2^s (P-p_2^o)/RT) \tag{6}$$

p_2^o is the vapour pressure of solid naphthalene and V_2^s is its molar volume. $f_2^G (y_2)$ was calculated from the Redlich-Kwong equation using the equation

$$\ln (f_2^G/y_2 P) = \ln \frac{RT}{(V_m - b_m)P} + \frac{b_{22}}{V_m - b_m}$$

$$- \frac{a_m b_{22}}{R(T)^{3/2}} \left(\frac{1}{b_m(V_m + b_m)} + \frac{1}{b_m^2} \ln \frac{V_m}{V_m + b_m} \right)$$

$$+ \frac{2(y_2 a_{22} + y_1 a_{12})}{R(T)^{3/2} b_m} \ln \frac{V_m}{V_m + b_m}$$

Table 6

Vapour/Solid Equilibrium Measurements

y_2 is the mole fraction of the extracted component in the vapour phase. V is the specific volume of the extracting phase and V_{CO_2} is that of CO_2 under the same conditions.

P (bar)	$y_2 \times 10^4$	$V (cm^3/g)$	V/V_{CO_2}
\multicolumn{4}{c}{Carbon dioxide/Naphthalene at 35.0°C (For diagram, see ref.1)}			
125.1	115	1.23	0.93
119.6	103	-	-
112.7	86	1.26	0.92
102.4	80	1.34	0.95
97.5	58	1.34	0.93
94.8	39	1.40	0.95
90.6	58	1.43	0.92
81.0	34	2.90	1.3
\multicolumn{4}{c}{Carbon dioxide/n-Dotriacontane ($C_{32} H_{66}$) at 35.0°C}			
161.0	0.7	1.20	0.99
146.5	0.5	1.23	0.99
125.1	0.7	1.27	0.97
108.7	0.8	1.35	0.98
108.6	0.8	1.33	0.96
103.8	0.2	1.38	0.99
91.9	0.3	1.49	0.96

V_m is the molar volume of the gas phase (found by trial and error solution of the Redlich-Kwong equation for each trial value of y_2) and $y_2 = (1 - y_1)$. a_m and b_m are given by equations (2) and (3). The parameters a_{22} and b_{22} for naphthalein could not be calculated directly using Zudkevitch's method since this requires the component to be liquid at the temperature of interest. Two sets of values were tried, the first (given in Table 2) from Soave's correlation (20) and the second obtained using extrapolated liquid phase properties in conjunction with Zudkevitch's method. The former values were found to result in a better fit to the combination of the data obtained in the present work (21) and that of Tsekhanskaya et al (19). (Fig.6 of the accompanying publication (1)).

The solubilities of n-Dotriacontane in compressed carbon dioxide were much smaller than those of naphthalene and could barely be measured. The data show considerable scatter. They can be correlated using the Redlich-Kwong equation (Fig.24) though estimated values of the parameters a and b are required (as in the case of naphthalene) and the vapour pressure of the solid (p_s^o) also has to be estimated. A value of 2.07×10^{-14} bar was used. This was arrived at (21) by using extrapolated "corresponding states" relationships to obtain the triple point pressure (1.8×10^{-10} bar) and then using the relationships

$$\left(d\ln p_s^o / d(1/T) \right) = -L^{S/V} / R$$

and

$$L^{S/V} = (L^{S/V})_{tp} + \int_{tp}^{t} (C_p^o - C_p^s) \, dT$$

to obtain p_s^o at 35°C.

$L^{S/V}$ is the latent heat of sublimation of the solid and C_p^s is its specific heat (22). C_p^o, the ideal vapour heat capacity, was calculated using the method recommended in API project 44 (23). The subscript tp denotes triple point temperature. $(L^{S/V})_{tp}$ was obtained by adding the latent heat of vaporisation at the triple point (136 kJ/mol, estimated from extrapolated corresponding states relationships (24)) to the latent heat of fusion (76.8 kJ/mol, obtained by interpolation of data listed in Timmermans (25) for n-alkanes with carbon numbers 26, 28 and 36).

When discussing the solubilities of solids in compressed gases it is sometimes convenient to work in terms of the enhancement factor E. This is the enhancement of the actual mole fraction solubility of the solid (y_2) over the "ideal"

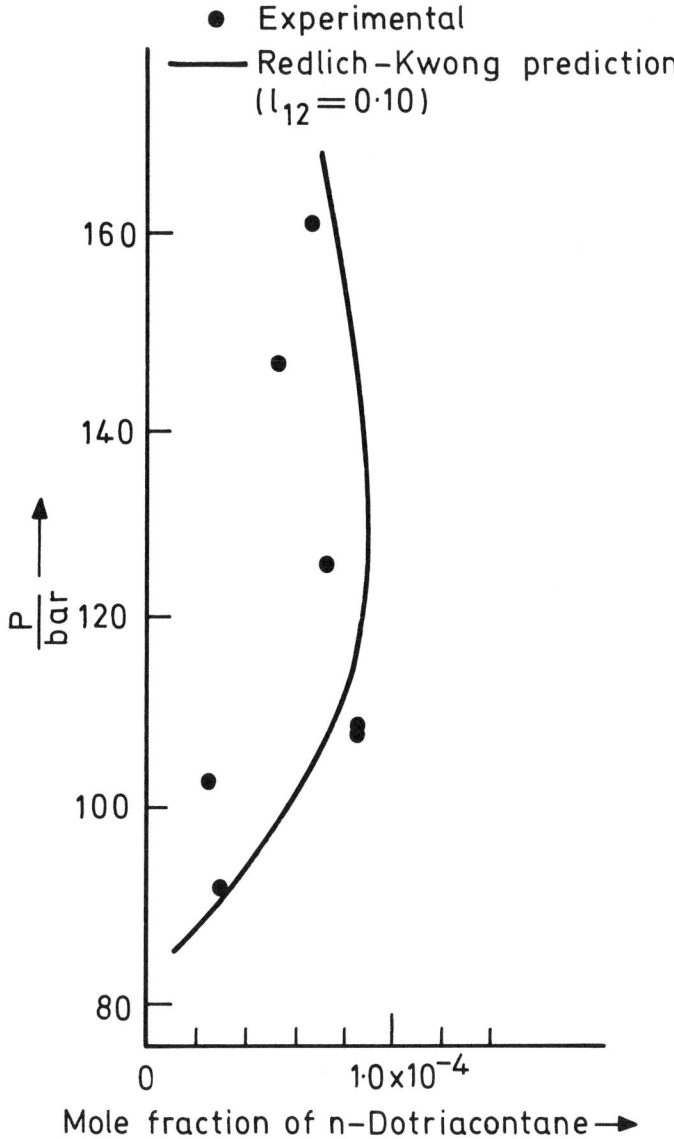

Figure 24. Mole fraction solubility of n-Dotriacontane in compressed carbon dioxide at 35.0°C.

value (p_2^o/P). In terms of this factor

$$y_2 = (Ep_2^o/P)$$

where p_2^o is the vapour pressure of the solid and P the total pressure. From Eq. 6

$$E = \left[\exp\left(V_2^s(P-p_2^o)/RT\right)\right] \times \left(\frac{1}{\psi_2^G}\right)$$

$$= E_{Poynting} \times E_{gas} \quad (7)$$

ψ_2^G is the fugacity coefficient for the solid component in the gas phase

$$\psi_2^G = f_2^G/(y_2 P)$$

Eq. 7 gives the enhancement factor as the product of two terms. The first ($E_{Poynting}$) arises from the influence of pressure on the fugacity of the solid phase. This effect was predicted by Poynting many years ago (7). The second (E_{gas}) arises from the interactions in the vapour phase and is given by the reciprocal of the fugacity coefficient for the solid component (2) in this phase. E_{gas} is normally substantially the larger of the two factors. For example, for naphthalene contacted with carbon dioxide at 60°C and 200 bar,

$$E_{Poynting} = 2.2$$

and

$$E_{gas} = 4000/2.2 = 1800$$

The term E_{gas} is strongly dependent on the density of the extractant gas. Fig. 25 shows the enhancement factor for naphthalein contacted with carbon dioxide as a function of the density of pure carbon dioxide under the given conditions. The relationship between log E and density is approximately linear over the range shown.

DISCUSSION

(1) Of the pieces of equipment used for equilibrium studies, those in which the vapour phase recirculates were found to be much the most rapid in operation. A close approach to equilibrium was often achieved after the circulation of a gas volume equal to about twice the system volume, ie. after

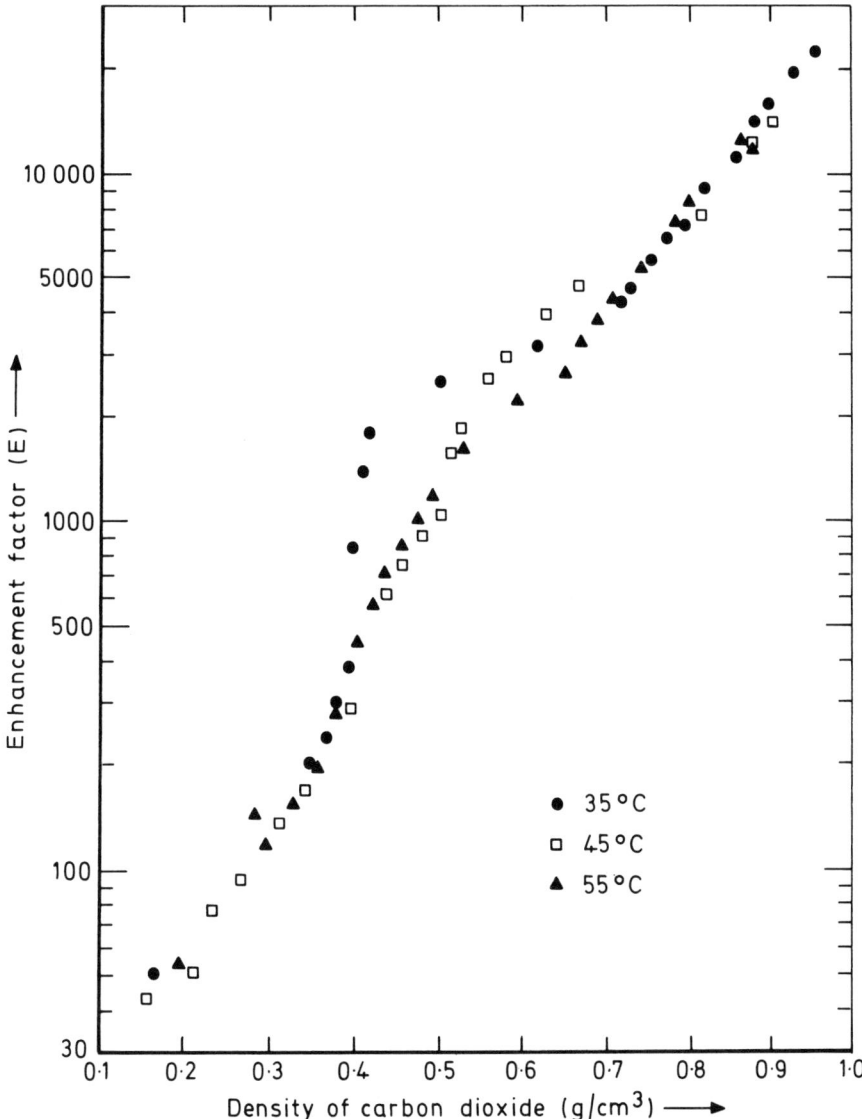

Figure 25. The enhancement factor (E) for naphthalene in compressed carbon dioxide as function of density of extractant.

about 15 minutes. The static stirred cell on the other hand sometimes required days in which to equilibrate.

(With slight adaption the vapour/liquid recirculation equipment can be used also for straightforward VLE studies under subcritical conditions. When used for this purpose the recirculating vapour phase is condensed by cooling before passing to the vapour sampling bomb and pump. It is then re-vapourised before entering the equilibrium bomb).

(2) The Redlich-Kwong equation with the simple mixing rules (2) and (3) is surprisingly accurate in fitting data for many systems of interest in gas extraction work. It promises to be a useful tool for preliminary feasibility studies.

(3) The extraction of solids and of liquids of low volatility depends strongly on the density of the extracting fluid (Fig. 25) and only becomes appreciable when the critical density is exceeded. For carbon dioxide at 35.0°C this corresponds to a lower pressure limit of about 70 bar while at 60.0°C the limit is about 100 bar (Figs. 14 and 15).

(4) Selective extraction effects on liquid mixtures of appreciable volatility forming "closed loop" pressure composition diagrams with the extractant gas (Figs. 9 to 13, 17, 20 and 21) are appreciable at lower pressures and are substantially dependent on the effects of the dissolved gas on the properties of the liquid phase. In the cases investigated, the initial effect of raising pressure appeared to be to raise the relative volatility of the component with the higher deviation parameter (ℓ_{12}) with the extracting gas. As the critical pressure is approached, however, the relative volatility tends to unity in all cases. In the systems studied these changes in the relative volatilities were not markedly advantageous to the separation of the heavy components. However, it may well be possible to develop separation processes for other systems based on relative volatility changes brought about in this way. Such separation processes can only take place at pressures below the system critical pressure (c in figures 17, 20 and 21). If all that is required is the total extraction of a group of solutes from a virtually insoluble matrix, it is quite possible to operate under conditions such that the extractable components, viewed as a group, are completely miscible with the solvent. It is then possible to carry out the extraction at pressures above the critical pressures of the systems formed by the extracted components with the extracting gas.

(5) In the food and related industries there is currently an upsurge of interest in the possibility of using supercritical,

or marginally subcritical, carbon dioxide for extraction purposes. In the course of simple "once through" extraction tests using carbon dioxide at 20 and 35°C we have extracted from natural products components such as nicotine, methyl palmitate, neophytadiene, dehydroabietic and isopimaric acids and the glycerides of fatty acids ranging from C_{12} to C_{22}. It appeared that equilibrium was quite closely approached in these tests in which the residence time was typically between 3 and 6 minutes. Although carbon dioxide has obvious advantages as a solvent in the food industry (1) it has inconveniently high triple point and critical pressures. As a result extraction pressures with this solvent tend to be high and also it cannot be fed into an equipment as a liquid at normal pressure. This makes it impossible to form a slurry between the material to be extracted and the solvent at normal pressure, a procedure which is sometimes advantageous (1).

(6) In conjunction with the above extractions we have carried out phase equilibrium tests on the systems formed by carbon dioxide with glycerol mono-oleate and glycerol trioleate (mono-olein and triolein). These systems form "open loop" pressure/composition diagrams at the temperatures studied (Fig.15). The vapour loadings are not high for these systems. For example, about 1 kg of carbon dioxide is required to dissolve 1 g of glycerol mono-oleate at a temperature of 35°C and a pressure of 150 bar. This loading can be increased by the use of a suitable entrainer (30). Fig.26 shows the enhancement which was obtained in the vapour loading when 10 cm^3 of CCl_4 was added to a batch of 150 cm^3 of an industrial mixture of glycerides which was contacted with carbon dioxide (this mixture consisted of 69% mono-oleates, 29% dioleates, 2% trioleates, together with some glycerol).

Stringent precautions were necessary in order to obtain reproducible results for the binary systems of carbon dioxide with the glycerol oleates. Over the long contact times used in the equilibrium tests there was sometimes evidence of breakdown and resynthesis reactions taking place even at temperatures as low as 35°C. If such reactions are allowed to occur they lead to erroneous results for the "binary" system under study. It is believed that the vapour loadings for this type of system obtained in preliminary tests by Fallaha (2) and Alderson (5) were mostly too high, at least in part for this reason. In order to guard against such effects, analysis of all extracts and residues is required. In the work on the few systems of this type reported in this paper, such analyses have been carried out. Work on this type of system continues and we hope to submit a more comprehensive account of it in due course.

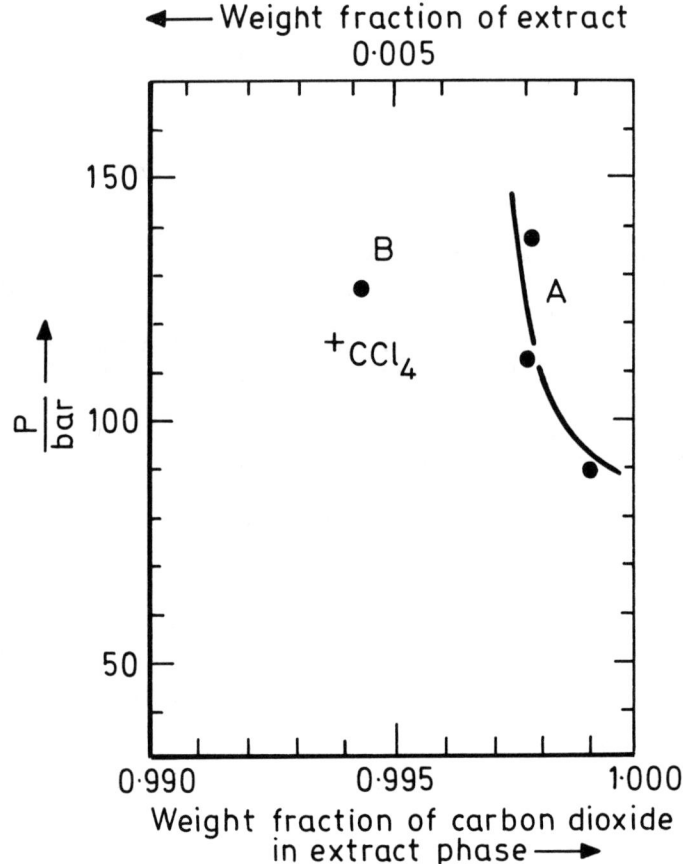

Figure 26. Enhancement of weight fraction of extract from commercial mixture of glycerides (see text) by addition of CCl_4.

(7) There seems to be little doubt that the extraction of natural products (often containing water) with both supercritical and subcritical carbon dioxide will come into increasing use, at least in batch and "semi-continuous" processes. Before the full potential of such processes can be realised, however, a better understanding of the extraction mechanisms, including the role sometimes played by water (1), is required.

REFERENCES

(1) M.B. King and T.R. Bott, Separation Science and Technology, 17, 119 (1982).

(2) F.H. Fallaha, PhD Thesis, University of Birmingham (1974).

(3) G.J. Harris, PhD Thesis, University of Birmingham (1964).

(4) S.R.M. Ellis and R.L. Valteris, Chemistry and Industry (London), 442 (1966).

(5) D.A. Alderson, PhD Thesis, University of Birmingham (1975).

(6) B.H. Sage, Ind.Eng.Chem., 43, 2595 (1951).

(7) J.J. Poynting, Phil.Mag., 5, 12, 32 (1881).

(8) D. Ambrose, "Vapour-Liquid Critical Properties", NPL Report Chem.107, (1980).

(9) J. Joffe, M. Schroeder and D. Zudkevitch, AIChE J., 16, 496 (1970).

(10) R.E. Treybal, "Mass Transfer Operations", Second Edition pp.393-395, McGraw-Hill 1968.

(11) E.P. Schoch, A.E. Hoffman and F.D. Mayfield, Ind.Eng. Chem., 33, 688 (1941).

(12) R.S. Ramalho and J. Delmas, J.Chem.Eng. Data, 13, 161 (1968).

(13) D.B. Todd and J.C. Elgin, AIChE J. 1, 20 (1955).

(14) J.P. Zhuze, Izvest Akad. Nauk S.S.S.R., O.K.N., 4, 361 (1960).

(15) V.B. Kogan, V.M. Fridman and T.G. Romanova, Zh.Fiz.Khim., 33, 1521 (1959).

(16) G. Brunner, PhD Dissentation, University of Erlangen (1978).

(17) G. Brunner and H. Hederer, "High Pressure Science and Technology", (Ed. Timmerhaus, K.D.), Vol.1, p.527, Plenum Press (1978).

(18) C. Ouellet and J.T. Dubois, Can.J.Res., $\underline{26B}$, 54 (1948).

(19) Y.V. Tsekhanskaya, M.B. Iomtev and E.V. Mushkina, Russ. J.Phys.Chem. $\underline{38}$, 1173 (1964).

(20) G.Soave, Chem.Eng.Sci., $\underline{27}$, 1197 (1972).

(21) R.S. Mahmud, PhD Thesis, University of Birmingham (1980).

(22) G.S. Parks, G.E. Moore, M.L. Renquist, B.F. Naylor, L.A. McClaine, P.S. Fujii and J.A. Hatton, J.Am.Chem. Soc., $\underline{71}$, 3386 (1949).

(23) Selected Values of Properties of Hydrocarbons and Related Compounds, API research project 44, Thermodynamics Research Centre, Dept.of Chemistry, Texas A. and M. University, 1971.

(24) G.F. Carruth and R. Kobayashi, Ind.Eng.Chem. Fundementals $\underline{11}$, 509 (1972).

(25) J. Timmermans, Physico-chemical Constants of Pure Organic Compounds, Vol.2, Elsevier (1965).

(26) R.D. Gunn, J.J. McKetta and N. Ata, AIChE J., $\underline{20}$, 347 (1947).

(27) J. Shim and J.P. Kohn, J.Chem.Eng. Data, $\underline{7}$, 3 (1962).

(28) R.S. Poston and J.J. McKetta, J.Chem.Eng. Data, $\underline{11}$, 362 (1966).

(29) B.H. Sage, D.C, Webster and W.N. Lacey, Ind.Eng.Chem., $\underline{28}$, 1045 (1936).

(30) S. Peter and G. Brunner, Angew.Chem.Int. Ed. Engl. $\underline{17}$, 746 (1978).

(31) J.K. Ali, PhD Thesis, University of Birmingham (1975).

CHAPTER 3

THE RAPID DEPRESSURIZATION OF
HOT, HIGH PRESSURE LIQUIDS OR
SUPERCRITICAL FLUIDS

M. E. Kim-E
   Department of Chemical
     Engineering, M.I.T

R. C. Reid
   Department of Chemical
     Engineering, M.I.T.

ABSTRACT

When a hot, high pressure liquid or supercritical fluid is very rapidly depressurized as the result of an accident, it is suggested that normal flashing will be delayed and a superheated liquid formed. If this superheated liquid attains a spinodal state, homogeneous nucleation would then result with the possibility of concomitant shock waves. A thermodynamic model has been developed to indicate what ranges of initial pressure and temperature would allow a spinodal state to be attained upon rapid depressurization. Simple experiments are described using both saturated liquid and supercritical carbon dioxide where depressurization did not lead to an explosion. The reason for these failures is discussed. Some reported industrial accidents are described and analyzed as to whether they did result from the system reaching a spinodal state.

INTRODUCTION

The chemical industry routinely handles liquids at high pressures. Carbon dioxide, propane (and LPG), sulfur dioxide, ammonia, and many refrigerants are examples of liquids that are normally transported and stored at ambient temperature and at the prevailing saturation pressure. In many chemical process operations, reactors and separators containing significant amounts of liquid are operated at elevated pressures and temperatures. In the growing area of supercritical extraction, one works with a fluid phase above its critical temperature and pressure.

While such operations at high pressure do involve a real element of hazard, design and operational considerations minimize this risk as attested to by the excellent safety record of high-pressure operations.

Our interest lies in the rare accident involving a pressurized liquid which could lead to very rapid venting of the vapor phase. Even in this case, serious consequences are uncommon unless the rapid acceleration of the liquid-vapor mixture results in a damaging "water-hammer" (Ogiso et al., 1972). Or, of course, if the uncontrolled venting leads to toxic, inflammable, or explosive clouds, serious results are then possible. We would like to limit our discussion to a small subset of such accidents wherein the pressure decay is so rapid that normal flashing plays no role. A weld failure that "unzips" a tank or the failure of a head on a pressurized vessel can drop the internal pressure in a very short (< 1 s) time. In such instances, one can conceive of having a liquid phase at essentially ambient pressure but at a temperature significantly above that corresponding to equilibrium. The liquid could then exist briefly in a metastable state as a superheated liquid. It is well known (Reid, 1976, 1978) that if the thermodynamic state of such a superheated liquid is such that the spinodal curve is reached, homogeneous nucleation then occurs within the bulk liquid. Such nucleation events occur in a microsecond time domain and often lead to strong and damaging shock waves.

In this paper, we explore first the thermodynamics behind the formation of such superheated liquids so as to show under what initial conditions the attainment of a spinodal state is possible by rapid depressurization. We then present the results of a few experiments which, while yielding negative results in terms of depressurization explosions, do indicate some important characteristics of such systems. We close by citing several recorded accidents which appear to have resulted from the hypothesized depressurization phenomenon noted above.

THERMODYNAMIC MODEL

We choose as our system a homogenous liquid at some initial pressure P_i and temperature T_i. (The term liquid could also include a fluid phase if the initial state were supercritical.) P_i may be the vapor pressure of the liquid at T_i or it may exceed this value. We assume that the initial system undergoes a very rapid pressure drop to some lower value P_f. No heat transfer occurs during this brief period, and the process is essentially isentropic, i.e., the specific entropy remains constant. No vapor forms unless bulk homogeneous nucleation occurs. In the liquid phase we have assumed no solid surfaces initiate nucleation.

The liquid phase is not incompressible and would expand in an isentropic manner during depressurization; the temperature also falls. As described below, we are interested in following the temperature-pressure trajectory and, in particular, in determining if the system ever attains the limit-of-stability where homogeneous nucleation must occur. This limit of stability is a spinodal state and results when, for a pure material, $(\partial P/\partial V)_T = 0$. The criterion for the spinodal state is different for mixtures as shown by Beegle et al. (1974).

We can illustrate the general thermodynamic model for a pure substance in several different ways. Consider Figure 1. Here the saturation vapor pressure of a pure substance is plotted as a function of temperature. Also shown is the spinodal curve which represents the locus of liquid states where $(\partial P/\partial V)_T = 0$. Between these two curves we have a superheated liquid. Also shown is an isentropic curve along which the specific entropy is constant. If the initial state of T_i, P_i were on this isentrope, then, during rapid depressurization, with no flashing, the initially subcooled liquid would soon become superheated and eventually would intersect the spinodal curve where homogeneous nucleation must occur. The same process is shown in Figure 2 on a P-V plane. The isentropic path from T_i, P_i intersects the spinodal curve at $T_f (T_f < T_i)$. Note that isotherms T_i and T_f have a zero slope at the spinodal to match the criterion $(\partial P/\partial V)_T = 0$. Finally, in Figure 3, we show the process on the T-S plane. Again the saturated liquid and spinodal curves are shown. In this case the initial state is assumed to be a saturated liquid. With rapid depressurization, homogeneous nucleation occurs when the spinodal curve is reached at T_f. One other point to note in Figure 3 is that the isobars have zero slopes when intersecting the spinodal curve. This results because, as shown by Beegle et al. (1974), for a pure material, an alternate criterion for the spinodal state is $(\partial T/\partial S)_p = 0$.

Saturated liquid states are limited to those with positive pressure. Superheated liquids (including states on the spinodal curve) have no such limitation and can exist at both positive and negative pressures (Reid, 1978). However, in any real accident scenario, the lowest pressure which could be attained in a depressurized vessel would be near ambient (1 bar) -- or perhaps slightly below if the venting vapor develops appreciable inertia. Therefore, for realistic situations, $P_f > 0$. If, in the thermodynamic analysis, the spinodal curve is attained at negative pressures, then homogeneous nucleation cannot occur and no depressurization explosion is possible.

Also, no such explosions would be expected if the initial state were such that the entropy exceeded the critical-point value. In such cases, a rapid, isentropic depressurization would lead to the formation of a subcooled vapor. If the final pressure were sufficiently small, the system could attain an unstable state on the vapor branch of the spinodal curve. Prompt condensation would then result. While this event might

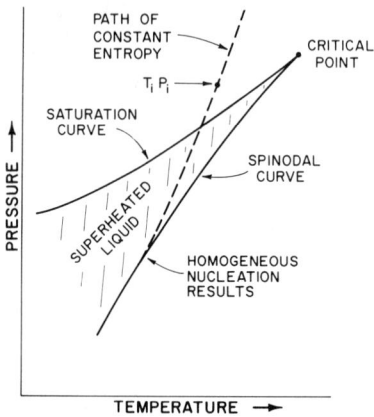

Figure 1. Schematic of an Isentropic Depressurization in T-P Space.

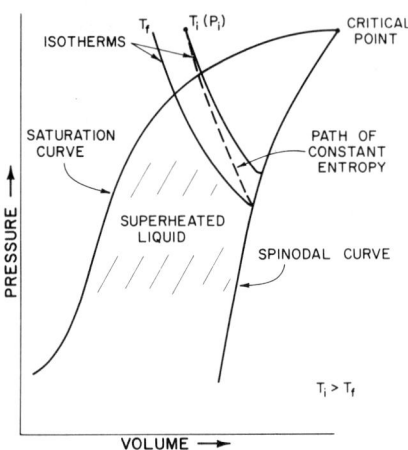

Figure 2. Schematic of an Isentropic Depressurization in P-V Space.

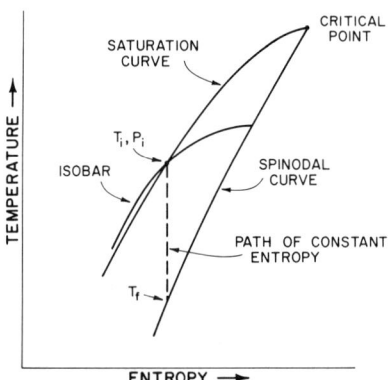

Figure 3. Schematic of an Isentropic Depressurization in T-S Space.

lead to implosion effects, there would not be expected to be any strong vapor-generated shock waves.

To carry out the necessary calculations to produce curves similar to those shown in Figures 1-3 for a real material, an equation of state applicable to the liquid (and superheated-liquid) region as well as to the supercritical region was necessary. No completely satisfactory equation is available. We selected the Peng-Robinson (1976) as it provides a reasonable approximation for the P-V-T properties of a liquid. This equation of state is given as:

$$P = \frac{RT}{V-b} - \frac{a}{V(V+b) + b(V-b)} \quad (1)$$

where
$$a = a(T_c)\alpha(T_r,\omega) \quad (2)$$
$$a(T_c) = 0.45724(R^2 T_c^2/P_c) \quad (3)$$
$$b = 0.07780(RT_c/P_c) \quad (4)$$
$$\alpha(T_r,\omega) = [1 + \kappa(1 - T_r^{1/2})]^2 \quad (5)$$
$$\kappa = 0.37464 + 1.54426\,\omega - 0.2699\,\omega^2 \quad (6)$$

For mixtures, the parameters a and b are also functions of composition.

With the criterion for a spinodal state,

$$(\partial P/\partial V)_T = 0 \quad (7)$$

and Eq. (1),

$$(\partial P/\partial V)_T = 0 = -\frac{RT}{(V-b)^2} + \frac{2a(V+b)}{[(V(V+b) + b(V-b)]^2} \quad (8)$$

Thus knowing the parameters a and b, the P-V-T properties of spinodal states may be determined.

To introduce entropy, two different techniques were used. In both, the single-phase isothermal-entropy departure function was utilized.

$$\Delta S_T = \int_{V_1}^{V_2} (\partial S/\partial V)_T \, dV \quad (9)$$

which, for the Peng-Robinson equations leads to

$$\Delta S_T = R \ln \left\{\frac{V_2 - b}{V_1 - b}\right\} + \frac{a(T_c)\alpha^{1/2}\kappa}{(TT_c)^{1/2} 2\sqrt{2}b} \times$$

$$\ln\left\{\frac{[V_2 + b(1 - \sqrt{2})]}{[V_2 + b(1 + \sqrt{2})]} \frac{[V_1 + b(1 + \sqrt{2})]}{[V_1 + b(1 - \sqrt{2})]}\right\} \quad (10)$$

In method I, we used literature values for saturated liquid volumes and entropies. From any state on the saturated liquid curve, Eq. (10) was employed to calculate entropies along isotherms. Thus the entropy of spinodal states could be found as could entropies in the subcooled-and superheated liquid and supercritical regions.

In method II a more ambitious procedure was followed. The entropy reference state was chosen as an ideal gas, and temperature effects on entropy were determined using literature values of ideal-gas heat capacities. An arbitrary entropy base state was selected at some convenient temperature (and pressure) in the ideal-gas state. Isothermal entropy changes from the ideal-gas state to the real state, saturated vapor were then found by using a combination of the ideal-gas entropy departure equation and Eq. (10) to proceed from the ideal gas state to a hypothetical state, $P \to 0$, and then to the saturated vapor state. The Peng-Robinson equation was also used to identify the saturated vapor and liquid P-V-T properties using a fugacity equality approach. From the saturated-vapor state, the entropy of the saturated liquid was found from

$$S_{sat\ liq} = S_{sat\ vap} - \Delta S_{vap} \quad (11)$$

$$\Delta S_{vap} = \Delta V_{vap} (dP/dT)_{sat} \quad (12)$$

with $(dP/dT)_{sat}$ determined by numerical differentiation of saturated P, T values and

$$\Delta V_{vap} = V_{sat\ vap} - V_{sat\ liq} \quad (13)$$

Volumes were found with Eq. (1).

Once the entropy of the saturated liquid state was obtained, the same procedure was used as in Method I to find entropies of supercritical states and in the liquid region. More details on the calculational procedure are available elsewhere (Kim-E, 1981).

Methods I and II gave entropy values in the liquid and

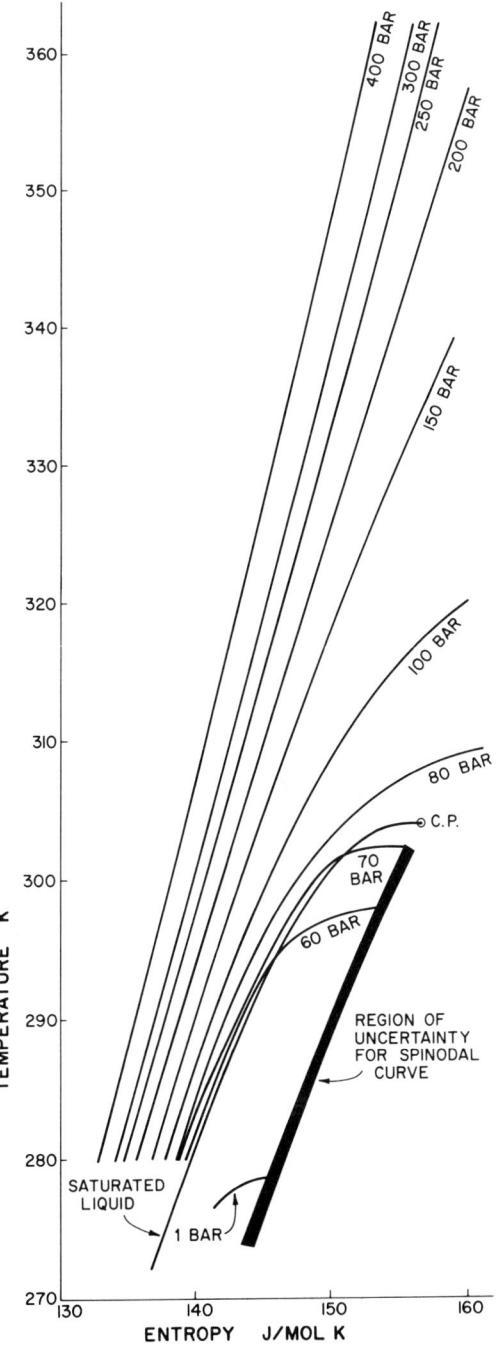

Figure 4. T-S Diagram for CO_2 Showing Superheated Liquid Region.

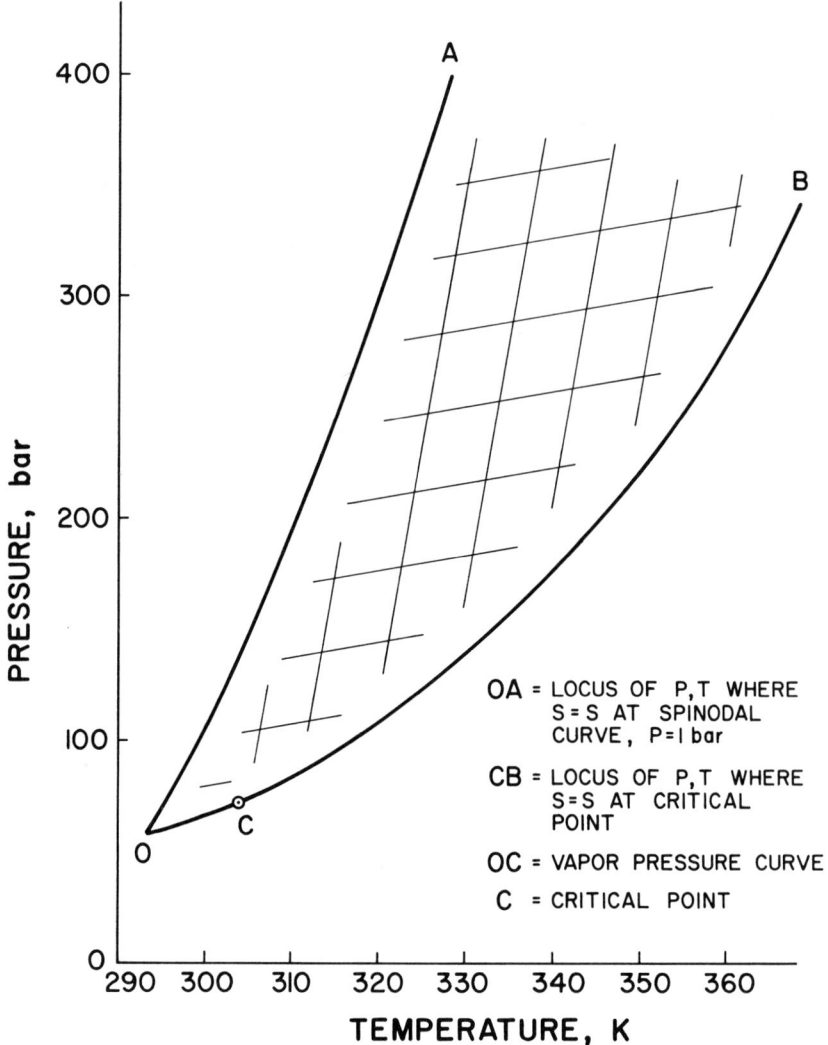

Figure 5. CO_2 P-T Domain Which Could Reach a Spinodal State Upon Rapid, Isentropic Depressurization.

supercritical regions which were in good agreement with each other in most cases. The small errors in predicted saturated liquid volumes from Eq. (1) did, however, lead to some uncertainty in locating the entropies along the spinodal curve.

We show in Figure 4 some calculated results for carbon dioxide. Temperature is plotted as a function of entropy from 280 to 360 K for pressures up to 400 bar. The range was extended to high temperatures to encompass regions which have been considered for supercritical CO_2 extraction. The entropies shown at various values of P,T match those from IUPAC (Angus et al., 1976) except in the superheated liquid region where they were, necessarily, calculated. The spinodal curve is shown with a range of uncertainty as calculated by Methods I and II. The one-bar isobar intersects the spinodal curve with an entropy of 145-146 J/mol K. If this state had been attained by the rapid depressurization of saturated liquid carbon dioxide, a vertical line from the 1-bar spinodal state intersects the saturated liquid curve between about 292 to 294 K where the vapor pressure is 56 to 58 bar. Thus one would conclude that rapid depressurization of saturated liquid carbon dioxide between temperatures of 292 K and the critical (304.2 K) could lead to superheated liquid states that could reach the spinodal and homogeneously nucleate. In addition, it might be concluded that if the saturated liquid carbon dioxide were at a temperature below 292 K, depressurization to one-bar (or higher) would never allow attainment of a spinodal state.

Generalizing the comments made above and taking the supercritical initial states into account, we show in Figure 5 the P-T ranges for CO_2 where one might expect homogeneous nucleation to result during a very rapid depressurization. The curve marked OA represents those P-T states with a value of entropy equal to the estimated entropy on the spinodal curve at one bar. Curve CB contains the locus of P, T states with an entropy equal to that at the critical point. Branch OC represents the saturated liquid state discussed above. Any initial P, T state within the cross-hatched region could attain a spinodal state by isentropic depressurization.

EXPERIMENTAL

A few experiments were conducted wherein a vessel containing liquid (or supercritical) carbon dioxide was rapidly vented. A schematic of the apparatus is shown in Figure 6 and a more detailed view of the pressure vessel is given in Figure 7. The vessel was made from 6-inch nominal diameter stainless steel pipe with welded end caps. The capacity was 7 liters. A jacket was provided and, in some runs, coolant was circulated to reduce the wall temperatures. Heating tapes were also used to attain temperatures higher than ambient.

Venting was achieved by driving a pointed spike verti-

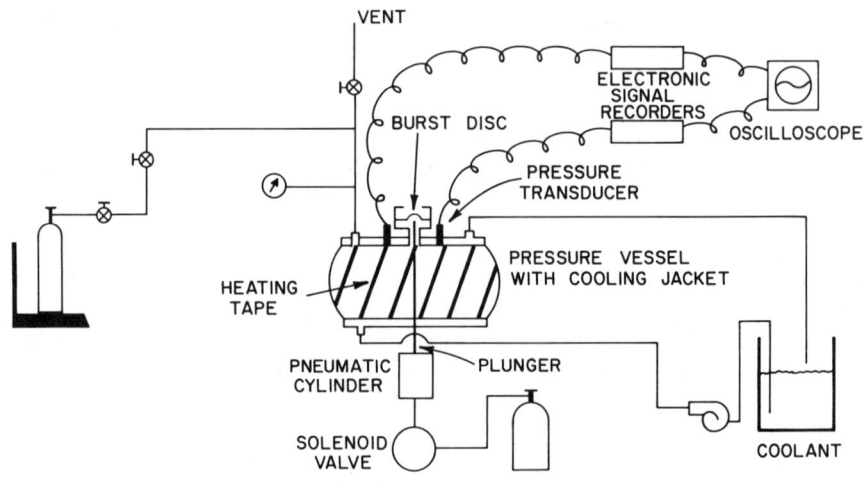

Figure 6. Experimental Apparatus.

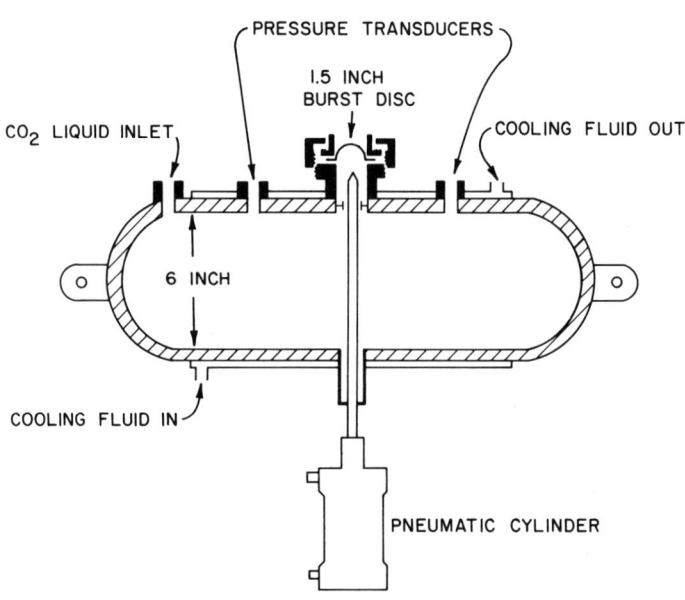

Figure 7. Details of the Pressure Vessel.

cally against a 1.5 inch rupture disc.

Pressure changes were monitored by two pressure transducers (PCB Piezotronics, models 102A01 and 102A04) mounted in the top of the vessel. These transducers were reported to have a rise time of one microsecond, but they could only be used to measure pressure changes, not absolute pressure. In fact, if the pressure were steady (or varied slowly) for a period of time longer than the discharge period for the transducer's capacitor (circa 1 s), then a zero output voltage resulted. We were, however, only interested in pressure variations over a much shorter time period.

The outputs of the transducers were fed to electronic signal recorders (Ballentine, model 7050 A) with a 20 ms tape recorder operating continuously in a write/erase mode. Permanent recording was triggered by a signal from the transducer that exceeded a preset critical value of $\pm dP/dt$. The 20 ms tape could be viewed on an oscilloscope after a test and permanent photographs made.

The burst (or rupture) discs were obtained from BS & B Safety Systems, type B. In all cases, when broken they opened completely.

The pressure vessel was located in a walk-in hood with one inch armor plate steel walls. Lights were blast resistant and a high-capacity ventilation system installed. (This cell was formerly used to study the performance of small hypergolic rockets.)

Nitrogen gas was used in preliminary experiments. In this particular case, since the properties of nitrogen are well approximated by an ideal gas, one can readily derive an equation to predict the rate of pressure decay as a function of time. The relation is (Kim-E, 1981),

$$\frac{P}{P_i} = \left[1 - \frac{CA}{V}\left(\frac{RT_i \gamma^3 K}{M}\right)^{1/2}\left(\frac{1-\gamma}{2\gamma}\right) t\right]^{2\gamma/(1-\gamma)} \quad (14)$$

where P is the pressure at time t and $K = [2/(\gamma + 1)]^{(\gamma+1)/(\gamma-1)}$ The other terms are defined in the Notation.

The agreement between measured and calculated values of P/P_i is shown for one test in Figure 8. The value for the orifice discharge coefficient C used to draw the solid curve was 0.26. This value is less than normally used for sharp edged orifices, but the presence of the plunger rod and short inlet stub may have affected the value of C. In other tests using different initial pressures, equally good agreement was obtained, but optimum values of C varied slightly between about 0.26 and 0.31.

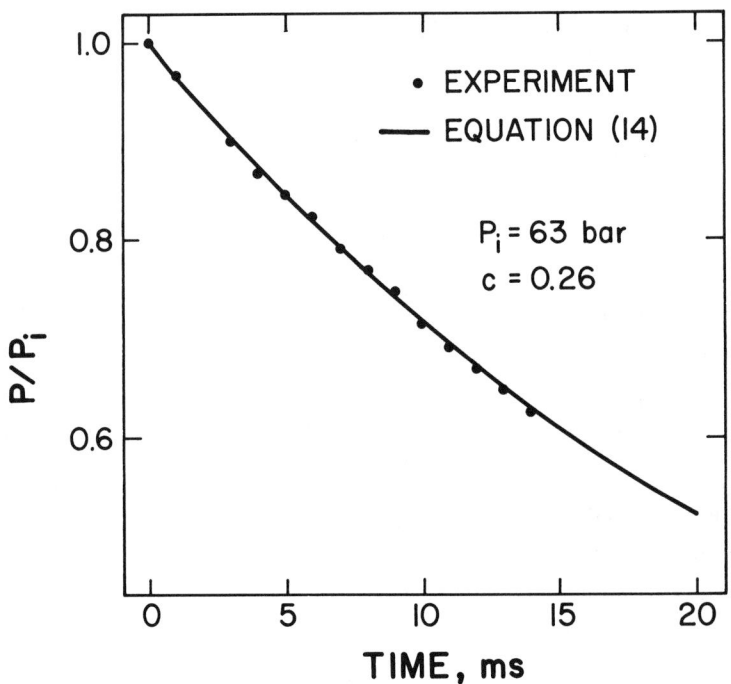

Figure 8. Blow-Down Experiment With N_2 Gas.

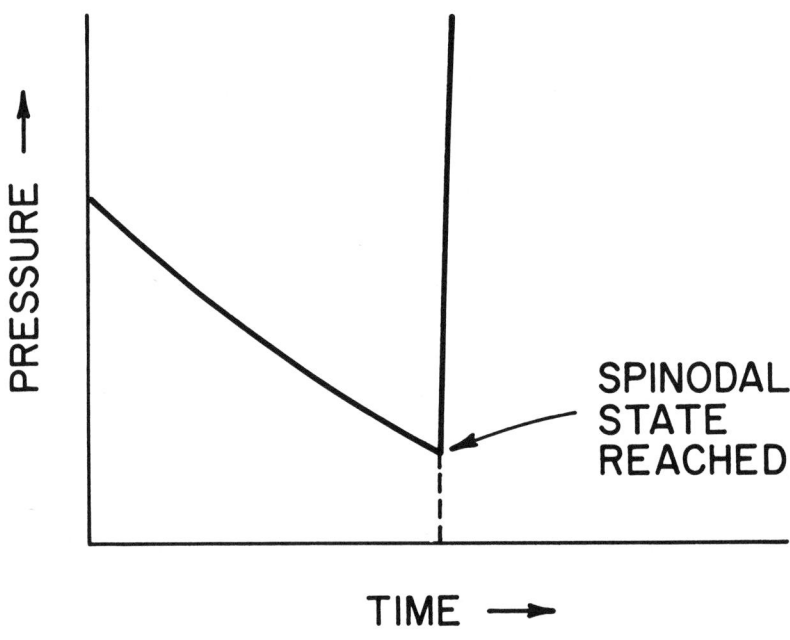

Figure 9. Expected Pressure for a Depressurization Explosion.

RESULTS

Thirteen experiments were conducted with carbon dioxide. The conditions are shown in Table 1. The initial pressure was a measured value. The liquid volume fraction (or specific volume if supercritical) was obtained by back-calculation knowing the vessel volume and the carbon dioxide mass added. CO_2 properties were obtained from the IUPAC tables (Angus et al., 1976).

The experiments were divided into two groups, i.e., those which began with saturated liquid carbon dioxide (runs 1-9) and those with the carbon dioxide initially in a supercritical state (runs 10-13).

For the saturated liquid tests, runs 1-3 were designed so that, if no boiling occurred during depressurization, the spinodal state would be attained at a pressure of about 24 bar. In runs 4-6, the initial state was such that the spinodal state would not be achieved until the pressure reached about ambient, while in runs 7-9, it would have been impossible to attain a spinodal state ($P_{spinodal} < 0$). The predicted pressures in the spinodal state are determined from the thermodynamic calculations described earlier and should only be considered as estimates.

For the supercritical tests (runs 10-13), all were carried out just above the critical temperature (304.2 K). The best estimate of the spinodal state pressures are shown in Table 1. For tests 11 and 12, these spinodal pressures are near the critical pressure (73.8 bar).

In runs 3 and 6, the saturated liquid carbon dioxide was first heated to a given temperature (with heating tapes). Then, coolant at 263 K was pumped through the vessel jacket. The burst disc was broken while the system temperature was decreasing. By this technique we hoped to minimize the boiling on the walls during depressurization.

If a spinodal state had been achieved during depressurization, we would have expected the measured pressure trace to have been similar to that shown in Figure 9. That is, after an initial pressure drop, the prompt homogeneous nucleation at the spinodal state would be expected to cause a sharp pressure spike. In no experiment was such a trace observed. Instead, typical pressure traces were as shown in Figure 10 for runs 1 and 3. Pressures never attained the spinodal value ($P/P_i \sim 0.4$) over the 20 ms recording time. A more rapid pressure decay was noted for the case where the wall was cooled (run 3), and this suggests that vapor formation, even in the early phase of depressurization, prevented a rapid decay in vessel pressure. Even for the supercritical tests, no sharp pressure rise was noted. In Figure 11 we show the data for runs 12 and 13. For run 12, the projected spinodal state should have been attained at $P/P_i \sim 0.9$ but there is no indication that this

Table I. Experimental Conditions

Run	Initial Pressure, bar	Initial Volume Fraction Liquid	Initial Specific Volume, cm^3/mol	Initial Temperature, K	Wall Cooling	Initial State	Calculated Pressure When Spinodal State attained, bar (P/P$_i$)
1	63.1	0.99		297	No	Saturated	~24 (0.4)
2		0.75			No		
3		0.90			Yes		
4	56.2	0.99		292.5	No		~1 (0.02)
5		0.73			No		
6		0.69			Yes		
7	45.8	0.98		284	No		< 0
8		0.84			No		
9		0.72			No		
10	90.7		59.0	~306	No	Supercritical	~30 (0.3)
11	75.2		133.4	~306	No		~66 (0.9)
12	75.2		98.6	~305	No		~66 (0.9)
13	76.9		81.9	~305	Yes		~54 (0.7)

†See text for method to cool wall before burst disc was broken

occurred.

Unless the theory (or calculated spinodal pressures) is in error, we simply have to conclude that in the small pressure vessel used, vapor formation was immediate and reduced the rate of pressure drop significantly. The vapor could have been formed on the wall or plunger rod or, more likely, the reflected shock wave resulting from the breaking of the burst disc led to vapor formation in the partially superheated liquids.

DISCUSSION

In view of the fact that theory and experiment did not agree in this project, one might well ask if there is other experimental evidence which might confirm the existence of rapid phase transitions from a rapid pressure decay of a hot liquid (or supercritical phase)?

One instance where this type of explosion appears to have been proved was in a vinyl chloride monomer autoclave that overpressurized and exploded in 1977. The violence of the event blew a two-ton section of the shell 1600 feet! At the time of the event, the bulk-liquid vinyl chloride temperature exceeded 110°C which is above the spinodal state temperature at one bar. Metallurgical examination of the metal fragments suggested that the initial vessel failure occurred in the top head and resulted from simple overpressure - a finding reinforced by an eyewitness account. (Burst discs and safety valves operated properly, but the rate of heat release from the polymerization reaction was such that the pressure continued to rise.) Subsequently the vessel failed catastrophically in a brittle fracture mode with failure originating in at least twelve sites in the lower shell of the autoclave. This second failure occurred in milliseconds causing high-speed cracking followed by non-ductile shear failure in the upper one-third of the vessel which was constructed of stainless steel. After a detailed analysis of the accident, company engineers concluded that the explosion was due to rapid, homogeneous nucleation of the bulk vinyl chloride as a result of the rapid pressure release.

BLEVEs*, which often result when pressurized liquefied petroleum gas (LPG) tank cars are caught in a fire, have been suggested to result from metal failure with very rapid depressurization of the LPG (Reid, 1979).

Liquid carbon dioxide presents an interesting situation. Fire extinguishers containing this liquid are very common. Yet from the thermodynamic analysis discussed earlier (and from the experimental plan), it would appear that depressurization explosions are possible if a carbon dioxide extinguish-

---
*Boiling Liquid Expanding Vapor Explosion.

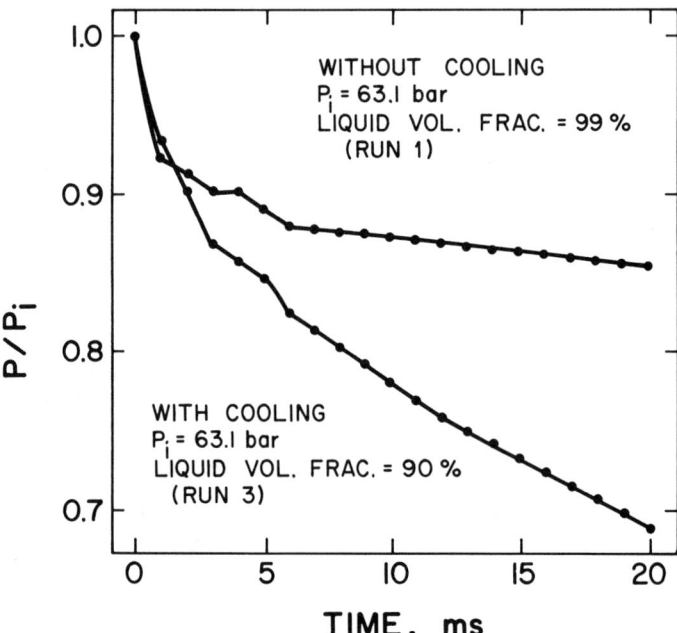

Figure 10. The Effect of Wall Cooling on a Depressurization Experiment.

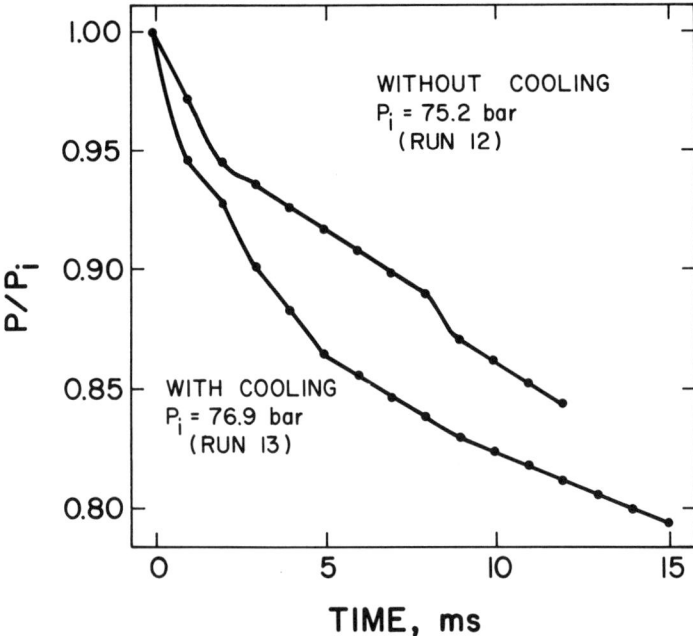

Figure 11. The Effect of Wall Cooling on a Supercritical Depressurization Experiment.

Figure 12. Damage in Laboratory Resulting from Explosion of CO_2 Fire Extinguisher. Extinguisher was Mounted on End of Workbench in Center of Photograph. (From Burghard and Wangler, 1979.)

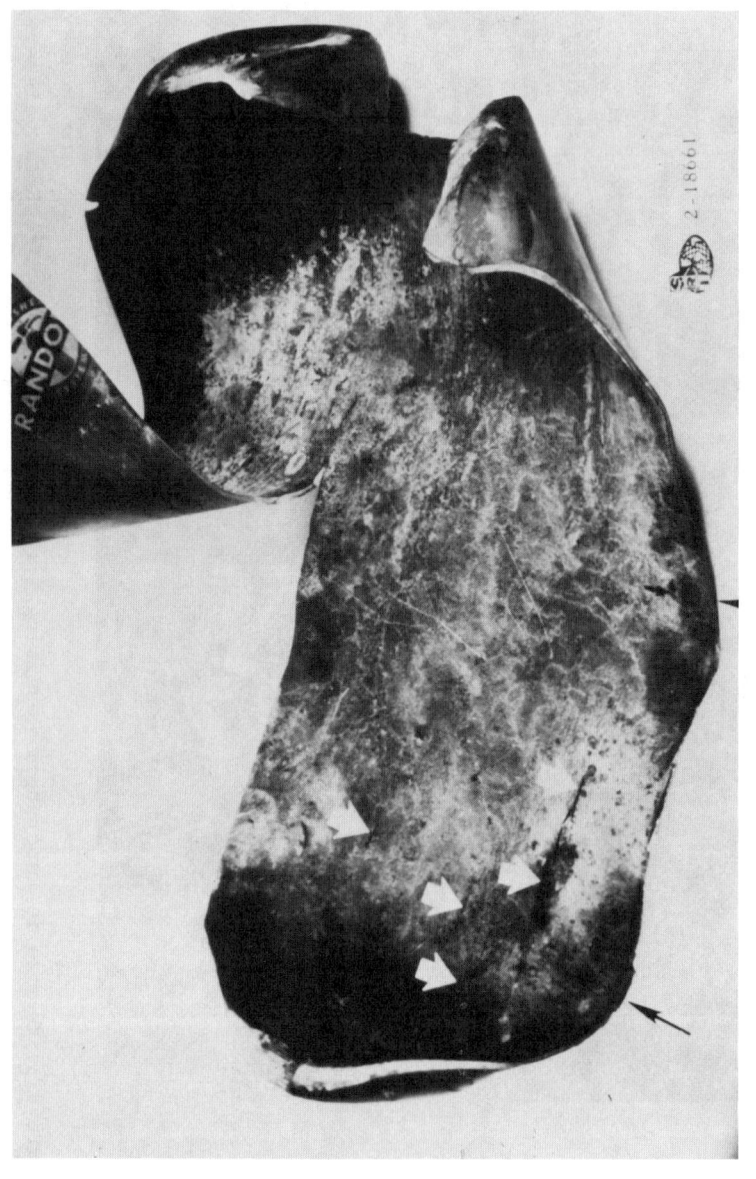

Figure 13. Ruptured Fire Extinguisher Cylinder. (From Burghard and Wangler, 1979.)

er were to fail and very rapidly depressurize the contents. One such incident has been reported -- although it is not known whether this accident was the result of the liquid CO_2 attaining a spinodal state. The extinguisher in question was located on a laboratory wall at the Southwest Research Institute. As later shown the vessel failed very rapidly as a result of stress-corrosion cracking that may have been caused by moisture remaining in the extinguisher after hydrostatic testing. The bursting caused extensive damage to the building and furnishings. Figures 12 and 13 taken from a paper by Burghard and Wangler (1979) give some idea of the force of the explosion. These same authors state that "The National Association of Fire Equipment Distributors has documented numerous other catastrophic failures of fire extinguishers of all types. It is our impression from literature and technical reports that this problem exists throughout the industry. An ultrasonic screening program involving a large sampling of CO_2 cylinders supplemented by metallographic examination appears to be warranted."

Other instances of violent explosions resulting from failure of CO_2 tanks are discussed by Copeland (1976) and Lieber (1980).

It would appear that, in most instances, a depressurization explosion occurred <u>after</u> vessel failure from some other means. If this is true, one cannot but question the methodology of <u>chemical blasting</u> as a viable means of disposing of chemicals involved in a hazardous situation (Anon., 1980).

In any case, until the hypothesis advanced in this paper is proved - or disproved, it would be prudent to consider this unusual hazard when operating facilities which contain appreciable quantities of liquids at high pressures and temperatures. The same warning should be directed to those who are planning on conducting supercritical extraction experiments.

NOTATION

    a Peng-Robinson parameter, Eq. (2)
    A area
    b Peng-Robinson parameter, Eq. (4)
    C orifice coefficient
    M molecular weight
    P pressure
    R gas constant
    S entropy
    T temperature
    V specific volume; <u>V</u> total volume

<u>GREEK</u>

    α Peng-Robinson parameter, Eq. (5)

γ C_p/C_v
κ Peng-Robinson parameter, Eq. (6)

SUBSCRIPTS

    c critical
    f final
    i initial
    sat saturated state
    sat vap saturated vapor
    sat liq saturated liquid
    vap vaporization

LITERATURE CITED

Angus, S., B. Armstrong, and K.M. de Reuch, "International Thermodynamic Tables of the Fluid State-Carbon Dioxide," Permagon Press, New York, 1976.

Anon., *Chem. Week*, Oct. 15, 1980, p. 52.

Beegle, B.L., M. Modell, and R.C. Reid, *AIChE J. 20*, 1200 (1974).

Burghard, H. and R.B. Wangler, Private Communication, Southwest Res. Inst., San Antonio, TX, 1979.

Copeland, E.H., "Pressure Safety Aspects of Liquid CO_2 Storage - A Case History", Report SAND-79-0125, Sandia Nat. Laboratory, 1976.

Kim-E, M.E., M.S. Thesis in Chemical Engineering, Mass. Inst. of Tech., Cambridge, MA 1981.

Leiber, C.O., *J. Occ. Accid. 3*, 21 (1980).

Ogiso, C., N. Takagi, and T. Kitagawa, paper presented at the Loss Prevention and Safety Symposium, PACHEC, Japan, 1972.

Peng, D.-Y. and D.B. Robinson, *Ind. Eng. Chem. Fundam. 15* 59(1976).

Reid, R.C., *Am. Sci. 64*, 146(1976).

Reid, R.C., *J. Chem. Eng. Ed. 12*(2), 60; (3), 108; (4), 194 (1978).

Reid, R.C., *Science 203*(23), 1263(1979).

CHAPTER 4

SOLUBILITY OF OXYGENATED HYDRO-
CARBONS IN SUPERCRITICAL
CARBON DIOXIDE

M. S. Kuk and J. C. Montagna
Gulf Research & Development Company
P.O. Drawer 2038
Pittsburgh, PA 15230

ABSTRACT

Solubility isotherms and selectivities of carbon dioxide for oxygenated hydrocarbons over water were measured as a function of the concentration of oxygenated compound in feed, the temperature and pressure in the supercritical region of carbon dioxide. Results on two carbon dioxide-water-alcohol systems are presented.

The dehydration characteristics of i-propanol and ethanol via supercritical carbon dioxide extraction are compared. The results suggest that the dehydration to the near atmospheric azeotropic composition is feasible.

INTRODUCTION

In the past few years, interest in separation techniques that utilize the solvent characteristics of supercritical fluids has been increasing rapidly. This is resulting in a number of useful applications in chemical engineering -- deasphalting heavy residuum oil with supercritical propane or pentane [1,2], decaffeination [3], regeneration of activated carbon [4], fluid-liquid chromatography [5] and soy oil extraction [6] with supercritical carbon dioxide. The potential application of this separation technique, which is often called supercritical gas (fluid) extraction, also exists for tertiary oil recovery [7,8] and conversion of oil from fossil fuels [9,10,11]. Dehydration of methyl ethyl ketone via supercritical extraction using ethylene, which was proposed by Elgin and Weinstock [12] is a classical example of the application mentioned above. The driving force in using these applications are potentially significant energy savings and the ability to separate heat labile compounds at temperatures well below their thermal decomposition points. Supercritical gas extraction is viewed as an alternative to energy intensive separation techniques such as distillation.

In 1954, Francis [13] reported a number of ternary systems of liquid-liquid equilibrium with liquid CO_2. The reported equilibrium was, however, qualitative and limited to a temperature of 25°C. Supercritical carbon dioxide has been extensively considered for the dehydration of alcohols. However, the ternary equilibrium data for alcohols-water-critical carbon dioxide have been scarcely reported. In our lab we have measured solubilities of oxygenated hydrocarbons in supercritical carbon dioxide. In this paper the solubility isotherms of supercritical CO_2 in aqueous ethanol and loadings of iso-propanol in supercritical CO_2 are reported.

EXPERIMENTAL SECTION

The experimental approach used in this work for measuring the solubility data was a dynamic method in which coexisting phases were continuously and countercurrently cycled. The phases are sampled after equilibrium is established. Recently Tsang and Streett [14] reported V-L-E data at elevated pressure which were obtained with a dynamic equilibrium apparatus similar to that used in this work. As presented in Figure 1, the experimental apparatus essentially consisted of five parts:

- the equilibrium cell with view-glasses (a stainless steel Jerguson gauge which holds a volume of approximately 60 ml of liquids).

- recycling system for the coexisting phases: a HPLC pump (Altex Scientific Company, Model 110A) for circulating the supercritical fluid phase, and a Milton-Roy high pressure liquid pump for the liquid phase.

- in-situ sampling devices using Valco high-pressure switching valves (sample volume 1 μl) which are directly connected to a GC column.

- pressure control system comprising of a validyne pressure transducer, a Moore pneumatic valve actuator and a Badge flow control valve.

- temperature control system consists of a Haake thermostat and an oil bath. The entire system with exception of the pumps is immersed in the temperature controlled oil bath.

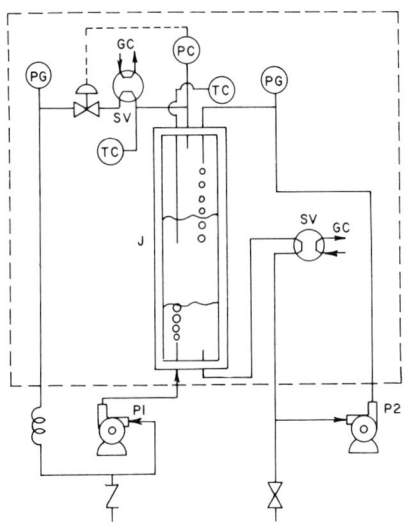

Figure 1. Schematic Diagram for Experimental Apparatus

The letters used in Figures 1 represent the following: J for Equilibrium Cell, P for Solvent Recycle Pump, P_2 Liquid Phase Recycle Pump, PC Pressure Controller, PG Pressure Gauge and TC Thermocouple.

A typical experimental run is started with the evacuation of the apparatus. A known composition and premeasured amount of water-alcohol mixture is then charged, liquid carbon dioxide is pumped into the equilibrium cell to obtain the desired equilibrium pressure. The temperature is controlled via oil bath. After the desired equilibrium condition is attained, the fluid and liquid samples are obtained from the recycling lines by use of the Valco switching valves and are directly injected into the GC. The charged amount of CO_2 is measured by a calibrated gas meter whenever required.

The pressure in the equilibrium cell is measured within an accuracy of ± 10 psia with a Validyne transducer demodulator and a Bordon-type Heise gauge (range 0-5000 psi) which was calibrated with a dead-weight tester. The temperature inside the equilibrium cell was measured within an accuracy of $\pm 0.1°C$ with an iron-constantan thermocouple which was calibrated with a NBS-calibrated platinum resistance thermometer. During the experimental runs the equilibrium temperature and pressure were maintained at a constant value within the accuracy limit of the equipment as stated above. GC analyses of the ternary system were confirmed by use of synthesized samples.

Coleman as well as C.P. grade of liquid carbon dioxide were used. Pharmacopoeia grade anhydrous ethanol purchased from Publicker Chemical and reagent grade iso-propanol from Baker Chemical were also used without further purification.

EXPERIMENTAL RESULTS AND DISCUSSIONS

The mutual solubility of CO_2 in water (distilled) near and above its critical point were measured and compared with the reported values in the literature; Wiebe [15], Wiebe and Gaddy [16], and Dodds et al. [17]. An excellent agreement between the measurements and the literature values was observed (see Figure 2).

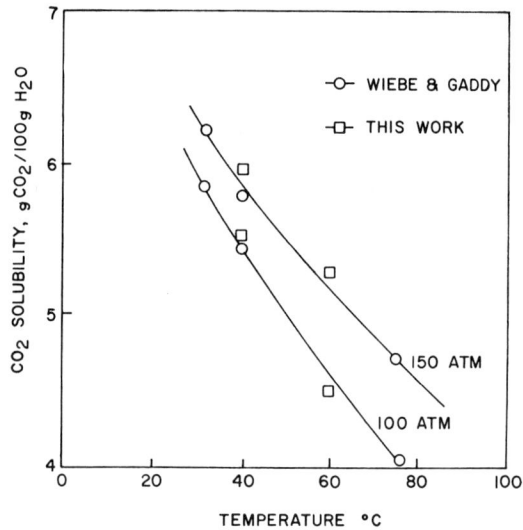

Figure 2. Carbon Dioxide Solubility in Water

Experimental solubility isotherms of carbon dioxide in aqueous ethanol, measured as a function of ethanol concentration at 40°C and 60°C, and pressures between 75 and 200 atmospheres, are presented in Figures 3 and 4. Ethanol loadings in the supercritical CO_2 and selectivities of CO_2 for ethanol over water at the measured conditions are listed in Tables 1 and 2. Examination of the reported results in Tables 1 and 2 for ethanol loadings reveals that a solubility inversion occurs between 1500 and 3000 psi along the 40°C and 60°C isotherms (i.e., the solubility isotherm of ethanol at 40°C consistently crosses that of 60°C between 1500 and

3000 psia). The solubility inversion for phenol in supercritical CO_2 was observed by Van Leer and Paulaitis [20]. In the above mentioned region, the solubility curves as a function of pressure are much steeper (larger slope) at the higher temperature. Therefore, at the higher temperatures significantly larger changes in loadings can occur with similar changes in pressure. This type of equilibrium data is imperative in the selection of the optimum conditions for the extraction and solvent regeneration steps of a process.

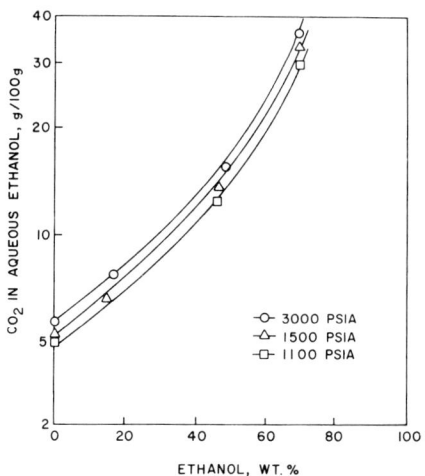

Figure 3. Supercritical Carbon Dioxide Solubility in Aqueous Ethanol at $40°C$

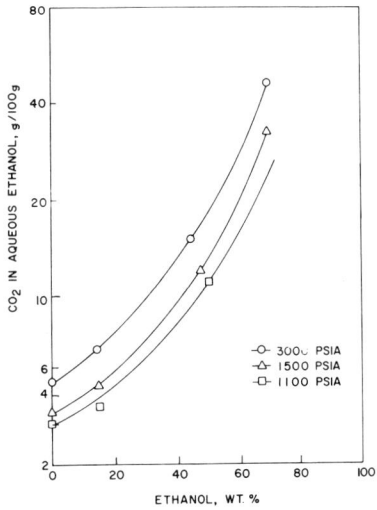

Figure 4. Supercritical Carbon Dioxide Solubility in Aqueous Ethanol at $60°C$

Table 1. Ethanol Loadings in Supercritical CO_2 at 40°C

Ethanol (wt. %) in Feed	Pressure (PSIA)	Ethanol Loading (wt. %) in Supercritical CO_2	Selectivity (β*)
15	1100	< 0.5	–
	1500	1.3	25
	3000	1.8	26
30	1100	< 1.0	–
	1500	2.8	18
	3000	3.6	22
50	1100	1.2	2.5
	1500	5.0	6
	3000	7.7	8
70	1100	2.5	1.5
	1500	8.7	4.0
	3000	12.5	4.2

*β = [Ethanol/water] at fluid phase/ [Ethanol/water] at liquid phase

Table 2. Ethanol Loadings in Supercritical CO_2 at 60°C

Ethanol (wt. %) in Feed	Pressure (PSIA)	Ethanol Loading (wt. %) in Supercritical CO_2	Selectivity (β*)
15	1500	1.0	15
	3000	2.2	20
30	1500	2.3	13
	3000	4.0	16
50	1500	3.0	2
	3000	8.6	8
70	1500	4.8	1.5
	3000	15.1	3

In addition, Tables 1 and 2 illustrate that ethanol loadings are not significant near the critical region. However, in the region where the density of supercritical CO_2 is comparable to that of liquids (e.g., supercritical region 3000 psia and temperatures between 40°C and 60°C), the ethanol loadings are significantly greater. The selectivities for ethanol at constant pressure are more favorable at lower temperatures as illustrated by the results presented in Tables 1 and 2.

General ternary phase behavior of the ethanol-water-CO_2 system at temperatures between 25°C and 60°C, and pressures between 850 psia and 3000 psia were observed as presented in Figure 5. The appearance of L-L-V (water-rich and CO_2-rich liquid phase with gas phase) was observed in the subcritical and the near critical regions of carbon dioxide. The CO_2-rich liquid phase merges with the gas phase to form the fluid phase as the corresponding critical point is attained. The phase behavior in this region has also recently been described by Paulaitis, et al. [19].

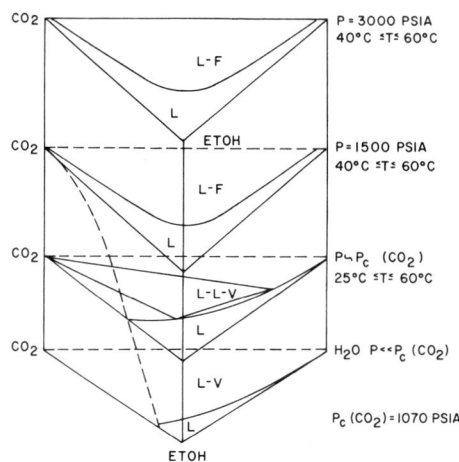

Figure 5. General Phase Behavior of Ethanol-Water-Carbon Dioxide System

In the iso-propanol-water-carbon dioxide system, the three phases (L-L-V) were observed at 40°C and 1500 psia. The loadings of iso-propyl alcohol are compared to those of ethanol in carbon dioxide at 1500 psi and 40°C in Table 3. The loadings of C_3 alcohol were found to be significantly greater than those of ethanol in carbon dioxide. The maximum purity of iso-propanol in the supercritical CO_2 was 86 wt. % (CO_2-free basis), which is slightly lower than the atmospheric azeotropic composition for water-iso-propanol.

Table 3. A Comparison of Loadings of iso-Propanol and Ethanol in Supercritical CO_2 at 1500 psi and 40°C

Alcohol wt. % in Feed	iso-Propanol Loading (wt. %)	Ethanol Loading (wt. %)
10	1.8	-
30	8.5	2.8
50	15.6	5.0
70	17.4	8.7

In the ethanol-water-CO_2 system, the three phases (L-L-V) also appear at 40°C and pressures between 1150 and 1200 psia with 75 mole % ethanol (CO_2-free basis) in the feed mixture. Figure 6 represents the ternary phase diagram of the ethanol-water-CO_2 system at 1500 psia and 40°C. A point on the coexistence curve at 1750 psia and 40°C was interpolated from Baker and Anderson's [18] results. It was found to agree with this work as shown in Figure 6. The tie lines illustrated in Figure 5 and ethanol loadings in the solvent as given in Table 1 indicate that dehydration of ethanol by the supercritical-CO_2 extraction is technically feasible. Because dehydration via supercritical extraction would displace a very energy intensive distillation step, a substantial amount of amount savings in the requirement for process energy can be expected. Based on the results presented in Figure 6, at 1500 psia and 40°C, ca. 90 wt. % ethanol can be theoretically produced from feed mixtures of ca. 10 wt. % ethanol by supercritical-CO_2 extraction. In the process region where the ethanol loadings are significant, the near atmospheric azeotrope composition was not attained in this work. (Under none of the explored conditions, was the atmospheric azeotrope for ethanol circumvented.)

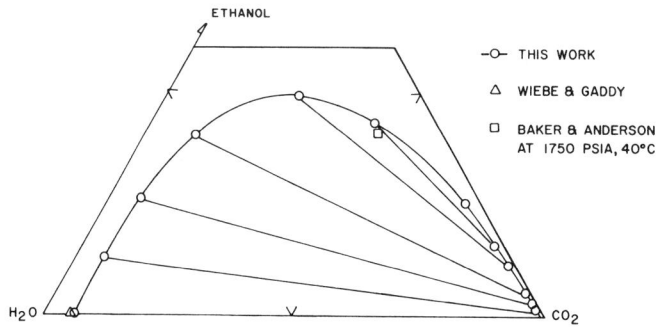

Figure 6. Ternary Phase Equilibrium Diagram for Ethanol-Water-Carbon Dioxide at 1500 PSIA and 40°C

In recovering the solvent and alcohol, the density of the solvent is the controllable variable. The information given in Tables 1 and 2 indicates that ethanol may be recovered by reducing the density of the supercritical extract either by lowering the pressure or raising the temperature. As mentioned earlier, it was found that at the higher temperature (60°C) a smaller drop in pressure is necessary to affect a similar change in alcohol loading. Although, it is not within the scope of this paper to develop optimum process-type conditions, it is clear that higher ethanol loadings can be achieved in the solvent at higher supercritical pressure.

SUMMARY AND CONCLUSIONS

In summary this paper represents:

a) A dynamic equilibrium system has been used to measure ternary equilibria of two alcohol-water-carbon dioxide systems.

b) Equilibrium data on the ethanol-water-carbon dioxide system at 40°C and 60°C and at pressures of 1100, 1500 and 3000 psia are reported: (i) The presence of the solubility inversion phenomena for the ethanol system was observed; (ii) Results suggest that ethanol can be dehydrated from ca. 10 wt. % to ca. 90 wt. % via supercritical extraction with CO_2.

c) The loadings of ethanol and iso-propanol in supercritical CO_2 were measured at a number of conditions and compared at 1500 psia and 40°C. The loadings of ethanol were significantly lower than those of iso-propanol.

REFERENCES

(1) Solomon. H. J., "Propane Deasphalting in the Neighborhood of the Critical Point of Propane", presented at the ACS Meeting Washington, DC, Sept. (1971).

(2) Gearhart, J. C., and Garwin, L., "A New Economical Approach to Residuum Processing", presented at the National Petroleum Refiners Assoc. Meeting, San Antonio, Texas, March (1976).

(3) Vitzhum, O. and Hubert, P., "Process for Decaffeination of Coffee", German Patent 2,357,590 (1975).

(4) Model, M., deFillipi, R. P., and Krukonis, V., "Regeneration of Activated Carbon with Supercritical Carbon Dioxide", presented at the ACS Meeting, Miami, Florida, Sept. (1978).

(5) Giddings, J. C., Myers, M. N., McLaren, L. and Keller, R. A., "High Pressure Gas Chromatography of Nonvolatile Species", Science 162, 67 (1968).

(6) Freidrich, J. P., and List, G. R., "Characterization of Soybean Oil Extracted by Supercritical CO_2 and Hexane", J. Agricultural Food Chem. to be published.

(7) Holm, L. W., "A Comparison of Propane and Carbon Dioxide Solvent Flooding Processes", J. A. I. Ch. E. 7,179 (1961).

(8) Holm, L. W., and Jasendal, J. A., J. Petl. Tech. 26, 1427 (1974).

(9) Bott, T. R., "Supercritical Gas Extraction", Chem. Ind. (London) 11,228 (1980).

(10) Whitehead, J. C., and Williams, D. F., "Solvent Extraction of Coal by Supercritical Gases", J. Institute of Fuel 48,182 (1975).

(11) McCollum, J. D. and Quick. L. M., U. S. Patent 3,948,754 (1976).

(12) Elgin, J. C., and Weinstock, J. J., "Phase Equilibria-Molecular Transport Thermodynamics", J. Chem Eng. Data 4,3 (1959).

(13) Francis, A. W., "Ternary Systems of Liquid Carbon Dioxide", J. Phys. Chem., 58,1099 (1954).

(14) Tsang, C. Y. and Streett, W. G., "Vapor-Liquid

Equilibrium in the System Carbon Dioxide/Dimethyl Ether", J. Chem. Eng. Data 26,155 (1981).

(15) Wiebe, R., "The Binary System Carbon Dioxide-Water Under Pressure", Chem. Revs. 29,475 (1941).

(16) Wiebe, R., and Gaddy, V. L., "Vapor Phase Composition of Carbon Dioxide-Water Mixtures at Various Temperatures and at Pressures to 700 Atmosphere", J. Amer. Chem. Soc. 63,475 (1941).

(17) Dodds, W. S., Stutzman, L. F., and Sollami, B. J., "Carbon Dioxide Solubility in Water", J. Chem. Eng. Data 1,92(1956).

(18) Baker, L. C. W., Anderson, T. F., "Some Phase Relationships in the Three-Component Liquid System CO_2-H_2O-C_2H_5OH at High Pressures", J. Amer. Chem. Soc. 79,2071 (1957).

(19) Paulaitis, M. E., Gilbert, M. L., and C. A. Nash, "Separation of Ethanol-Water Mixtures with Supercritical Fluids", presented at World Congress of Chemical Engineering, Montreal, Canada, October (1981).

(20) Van Leer, R. A., and Paulaitis, M. E., "Solubilities of Phenol and Chlorinated Phenols in Supercritical Carbon Dioxide", J. Chem. Eng. Data 25,257 (1980).

CHAPTER 5

HIGH PRESSURE FLUID PHASE
EQUILIBRIA OF ALCOHOL-WATER-
SUPERCRITICAL SOLVENT MIXTURES

Mark A. McHugh
Mark W. Mallett
James P. Kohn
  Department of Chemical
    Engineering
  University of Notre Dame

ABSTRACT

Experimental data are presented on the high pressure fluid phase behavior of the ternary ethanol-water-ethane system as well as the binary ethanol-ethane system. A thermodynamic model is developed based on the Peng-Robinson equation of state. The experimentally observed phase behavior together with computer generated phase diagrams are used to determine the possibility of separating ethanol/water mixtures with supercritical ethane.

INTRODUCTION

Recently ethanol (2) as well as other alcohols (3) have been suggested as potential fuel sources. At present however, alcohols have not been fully exploited as an energy source due to the prohibitive energy demands involved in separating water from alcohol/water mixtures (2). To circumvent these excessive energy requirements various novel separation schemes, ranging from selective separations with membranes (4) to solvent extraction using organic liquids (5) including alkanes which are representative of gasoline (6), have been proposed. At present, these separation techniques are either still at an early stage of development, as with the membrane work, or they have not produced the type of significant results necessary for large scale application.

One separations technique which as yet has not been extensively tested is supercritical solvent extraction - that is, extraction with a solvent which is at a temperature above its critical temperature and a pressure above its critical pressure (7,9). Recent applications of supercritical solvent extraction range from the decaffeination of green coffee beans (8) to the deashing of coal liquids (10). A unique advantage of a supercritical solvent is that solvent power, which is a strong function of density, can be easily varied with small changes in pressure as well as temperature when operating near the solvent critical point. In addition, as shown in Table I, certain physico-chemical properties of a supercritical fluid make it an ideal solvent. Notice that while a supercritical fluid has a liquid-like density (15), hence solvent loading comparable to a liquid, the diffusivity (16) and viscosity are intermediate to that of a gas and a liquid. Since the mass transfer characteristics of a supercritical fluid are enhanced, separations should be more efficient.

Table I. Properties of a Gas, a Liquid, and a Supercritical Fluid (SCF).

PROPERTY	PHASE		
	SCF	GAS	LIQUID
DENSITY (G/CC)	0.7	10^{-3}	1.0
VISCOSITY (CP)	10^{-2}	$10^{-3}-10^{-2}$	10^{-1}
DIFFUSION COEFFICIENT (cm^2/s)	10^{-3}	10^{-1}	10^{-5}

Experimental work in the late 1950's by Elgin and Weinstock (11) on ternary mixtures consisting of an organic liquid, water, and supercritical ethylene suggests the possibility of using a supercritical fluid to dehydrate organic/water mixtures in an energy efficient manner. The authors suggest that to fully exploit the high pressure fluid phase behavior exhibited by the ternary system, it is necessary to understand the phase behavior of the binary systems which compose the ternary mixture. This binary phase behavior by itself can be quite complex as shown in Figure 1 (12).

Depicted in Figure la is the pressure-temperature projection for a binary mixture where the two pure component vapor-liquid equilibrium lines terminate at the pure

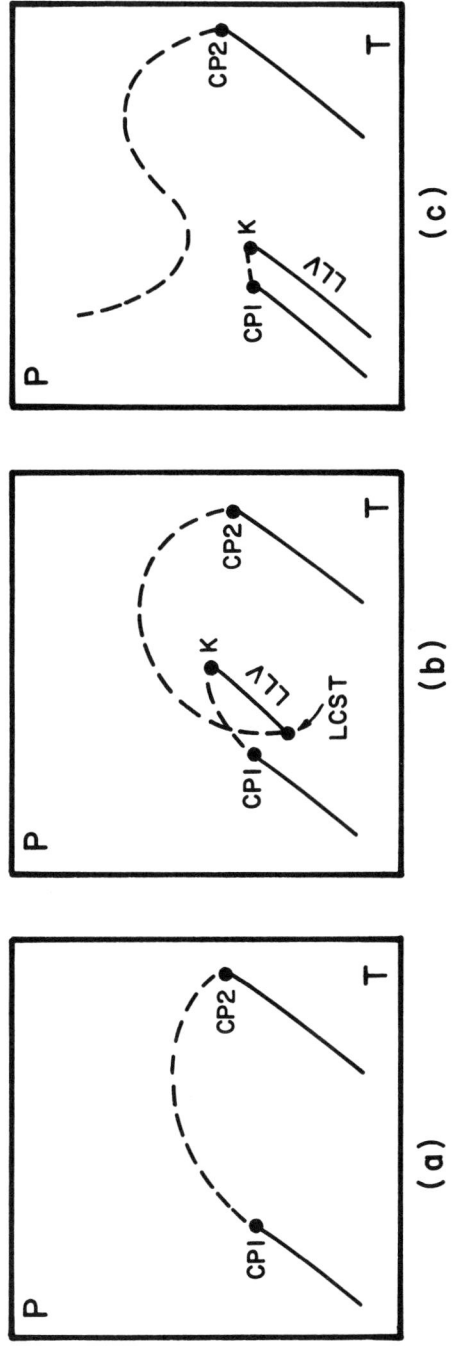

FIGURE 1. Examples of the various types of phase behavior exhibited by binary mixtures.

component critical points CP1 and CP2. The dashed line represents the binary critical mixture curve which in this instance runs continuously from the critical point of the heavier component to the critical point of the lighter component. This type of phase behavior is representative of that exhibited by the ethanol-water system (13).

Depicted in Figure 1b is a slightly more complex situation. In this instance the critical mixture curve intersects a region of liquid-liquid immiscibility (i.e., the L-L-V line) at two locations: the lower critical solution temperature (LCST) and a type K critical end point. At the lower critical solution temperature, the two immiscible liquid phases become critically identical in the presence of vapor while at the type K point one of the liquid phases becomes critically identical with the vapor phase while in the presence of the other liquid phase. This is the type of phase behavior exhibited by the ethane-ethanol binary system (1).

In Figure 1c, the case where the three phase L-L-V line is intersected only once by the critical mixture curve at a type K point is shown. The other branch of the critical mixture curve which starts at the critical point of the heavier component never meets either the L-L-V line or the critical point of the lighter component. Knowing the type of phase behavior exhibited by the binary mixture enables one to make reasonable guesses as to location of immiscibility regions of the ternary system.

The objectives of this work are to study the high pressure fluid phase behavior of the ternary system ethanol-water-ethane to determine the feasibility of separating ethanol/water mixtures using supercritical ethane. Extensive literature information is available for two of the binary mixtures which compose the ternary ethanol-water-ethane system -- that is, ethanol-water (13) and ethane-water (14). However, only a limited amount of information exists for the ethane-ethanol binary (1). Therefore, we experimentally determined the phase behavior for the ethane-ethanol system before studying the ternary ethanol-water-ethane system. The resultant phase behavior is modeled using the Peng-Robinson equation of state (23). The feasibility of efficiently separating water from ethanol/water mixturs using supercritical ethane is determined using the experimentally determined phase behavior as well as model generated phase diagrams.

EXPERIMENTAL APPARATUS AND PROCEDURE

Shown in Figure 2 is a schematic diagram of the experimental apparatus used in this study to obtain dew and bubble point data as well as various phase boundary curves. Since the experimental apparatus and techniques used in this study are described in detail elsewhere (17,27), only a brief description will be given.

The experimental system consists of three basic components: a calibrated mercury displacement pump (Ruska Corp.), gas reservoir, and a calibrated equilibrium view cell. Initially a known amount of the condensed phase, either pure ethanol or an ethanol/water mixture, is introduced into the equilibrium view cell. Air is then purged from the system with ethane. Then ethane is metered into the equilibrium view cell from the constant temperature gas reservoir using the calibrated mercury displacement pump. Ethane is continually added to the view cell until either all the liquid in the view cell disappears thus obtaining a dew point, or until the liquid phase expands in the view cell leaving only a negligible amount of the gas phase thus obtaining a bubble point. In either situation the composition of the resultant saturated phase is determined from a mass balance. In addition, since the equilibrium view cell is calibrated, the molar volume of resultant phase is also precisely determined.

To obtain the ethanol-ethane liquid-liquid-vapor line, the system is operated such that one of the liquid phases present is infinitesimally small when compared to the other liquid phase. In this instance, two separate runs are required to determine the properties of both liquid phases. The vapor phase in equilibrium with the two liquid phases is considered to be pure ethane except in the vicinity of the Type K point.

The pressure of the equilibrium view cell is measured with a Bourdon-tube Heise gauge which is accurate to within ± 1 psi and the temperature of the cell is measured with a platinum resistance thermometer accurate to within $\pm 0.02°K$ Absolute ethanol (Commercial Solvents Corp.) is used without further purification while ethane (Linde Co., 99±%) is purified using activated charcoal and 13A molecular sieves to remove any low boiling impurities.

EXPERIMENTAL RESULTS

Ethanol-Ethane

In Figure 3 the pressure-temperature projection of the

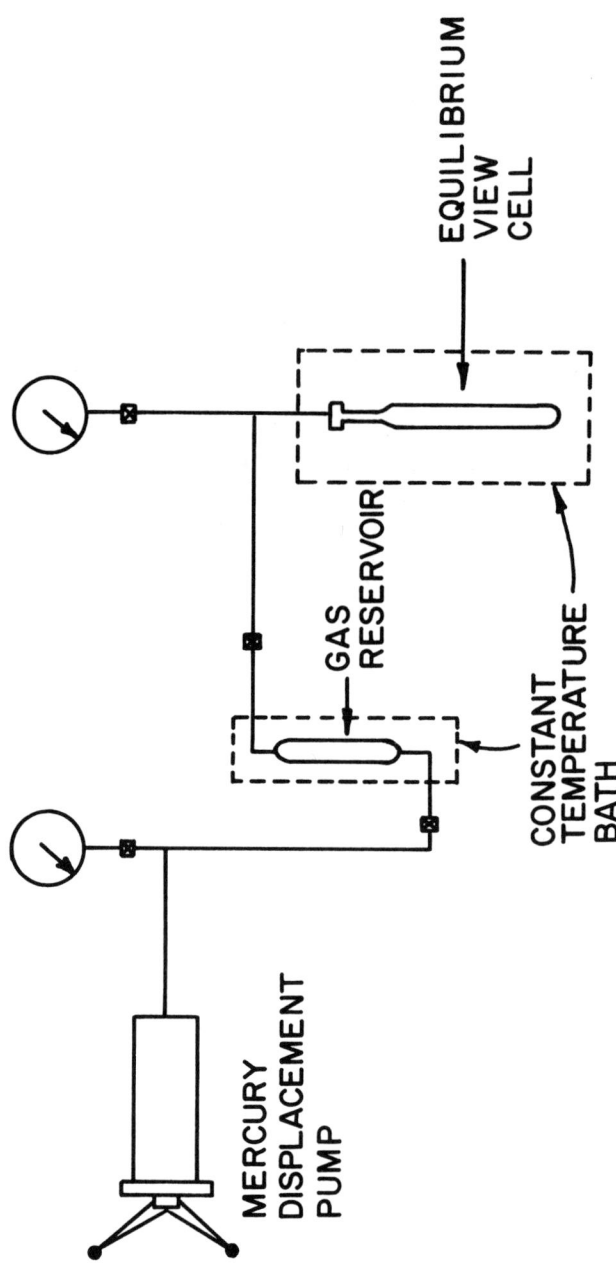

FIGURE 2. Schematic diagram of the experimental apparatus used in this study to obtain dew point and bubble point data.

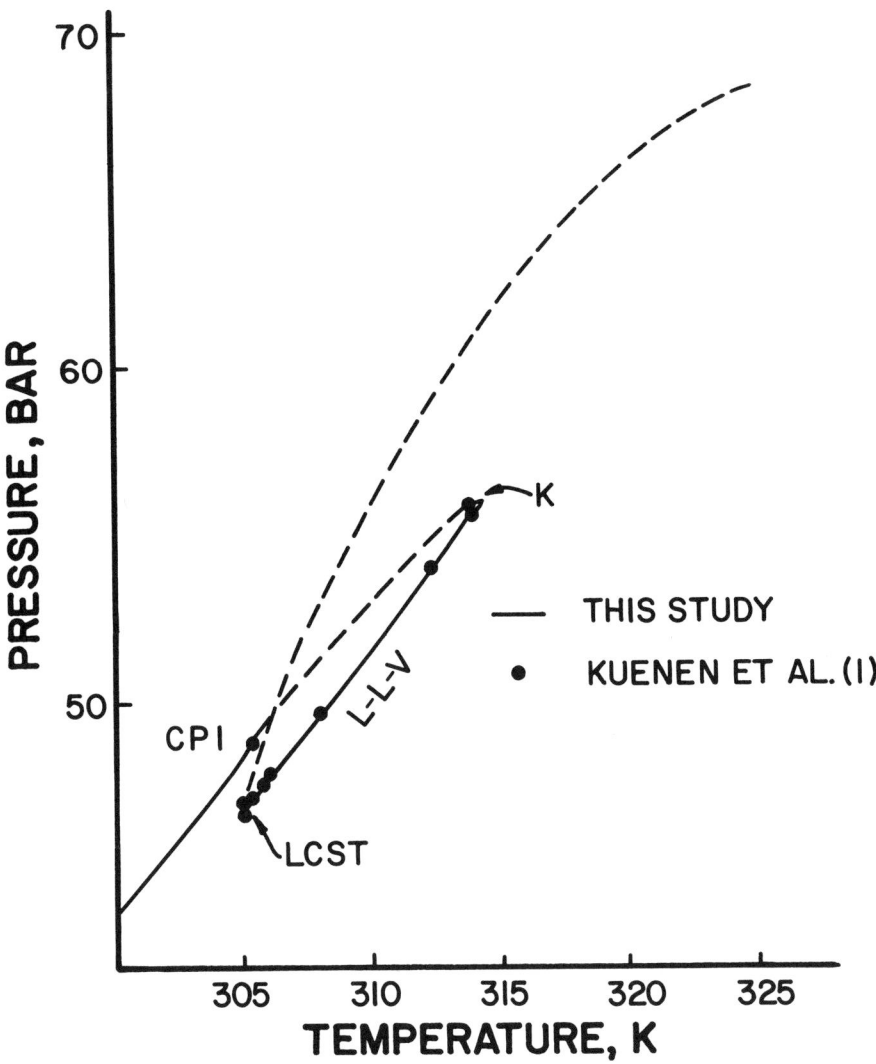

FIGURE 3. Comparison of the P-T projection of the ethanol-ethane L-L-V line obtained in this study to that of Kuenen et al. (1).

ethanol-ethane system obtained in this study is compared to the work of Kuenen and Robson (1). The solid L-L-V line shown in this diagram represents our data which are listed in Table II. The agreement of our results with those of Kuenen and Robson (1) is excellent. The dashed line shown in Figure 3 represents the suggested shape of the binary critical mixture curve as determined by the LCST and the type K point.

Table II. Smoothed Values of Compositions Along the L-L-V Line from Data Obtained in this Study.

Temp. (K)	Press. (Bar)	Mole Fraction Ethane L_1	L_2	L_1 L_2 (ml/gmol)	
314.20[a]	55.80	.580	.975	69.0	130.0
313.75	55.20	.587	.936	69.5	110.5
312.50	54.00	.610	.910	71.0	99.0
311.25	52.70	.623	.885	72.0	93.0
310.00	51.50	.638	.865	73.0	88.0
308.75	50.30	.655	.845	73.5	85.0
307.50	49.00	.670	.825	74.0	82.0
305.00[b]	46.80	(.740)		(76.5)	

(a) Type K point, (b) LCST measured by Kuenen et al. (1), values in parenthesis represent best estimates from data from this study. Full tables of data available on request to M. A. McHugh.

Figure 4 shows the composition versus temperature diagram of the three equilibrium phases along the L-L-V line. The composition of the gas phase is essentially pure ethane except near the K point.

Depicted in Figure 5 is the variation of the molar volume of the two liquid phases along the L-L-V line. Notice that the molar volumes of the L_1 and L_2 phases are very close to each other as the lower critical solution temperature is approached. This leads to experimental difficulties in accurately determining the lower critical solution temperature.

In Figure 6 the dew and bubble point lines at 310.25°K and 323.15°K are shown. In both cases mixture critical points are not precisely determined.

Ethanol-Water-Ethane

For the ternary ethanol-water-ethane system, experimental information are obtained at 313.15°K and 323.15°K and

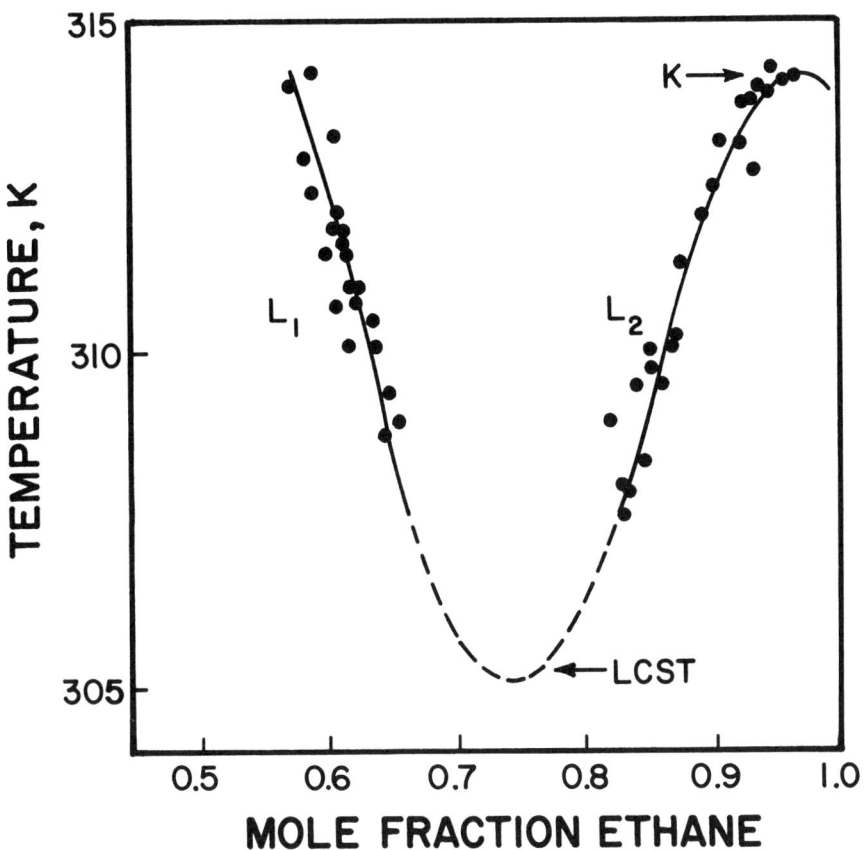

FIGURE 4. Compositions of the three equilibrium phases along the ethanol-ethane L-L-V line obtained in this study.

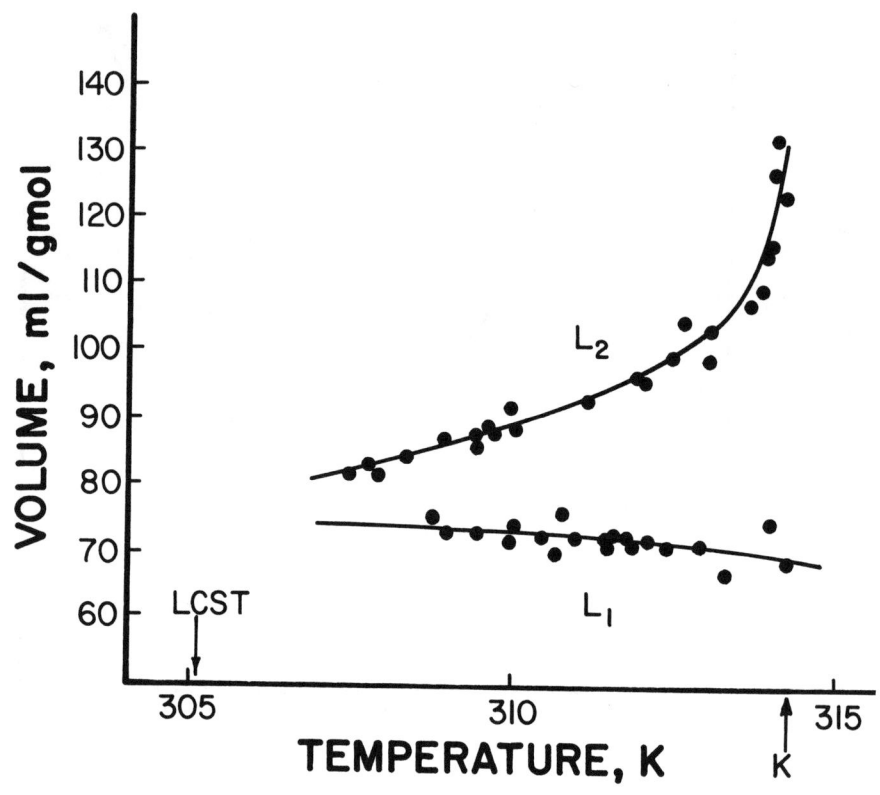

FIGURE 5. Molar volumes of the two liquid phases along the ethanol-ethane L-L-V line obtained in this study.

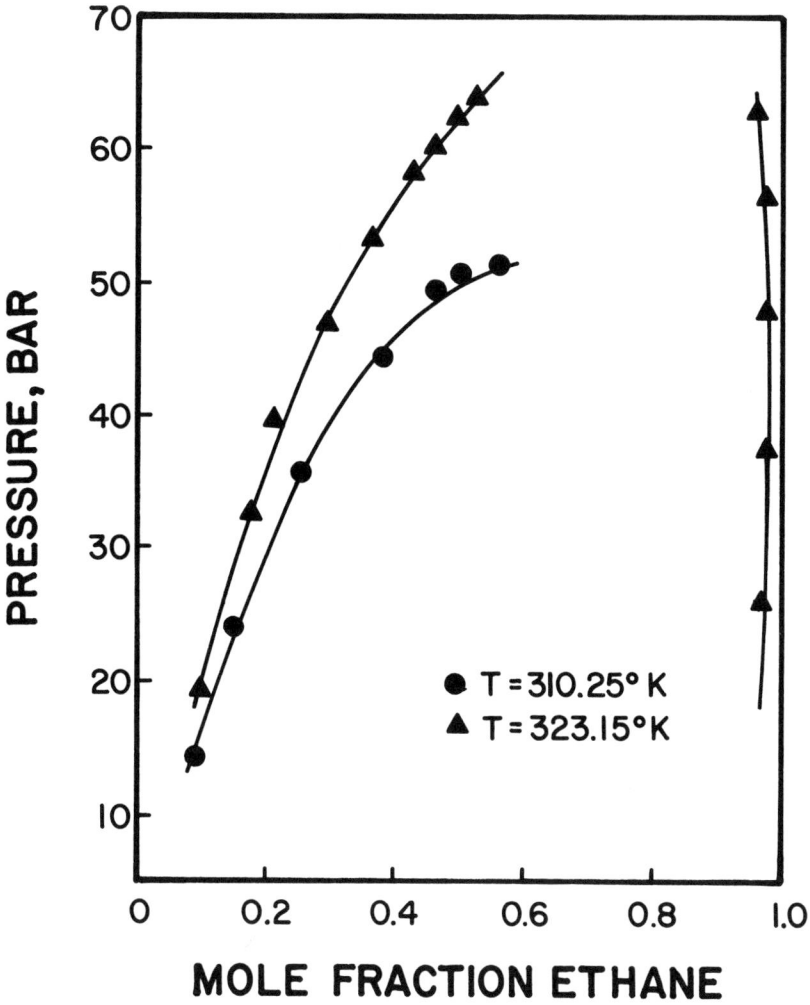

FIGURE 6. Dew-point and bubble-point data for the ethanol-ethane system obtained in this study at constant temperatures of 310.25°K and 323.15°K.

at pressures ranging from 0 to 300 bar.

Shown in Figure 7 are the dew and bubble point lines for ternary mixtures of varying overall weight ratios of ethanol to water obtained at 313.15°K i.e., these weight ratios represent the compositions of the liquid mixtures which are initially loaded into the equilibrium view cell. The gas phase compositions for overall weight ratios of 77 and 50% are essentially pure ethane.

Depicted in Figure 8 are the dew and bubble point lines for ternary mixtures of varying overall weight ratios of ethanol to water obtained at 323.15°K. Again gas phase compositions for weight ratios of 77 and 50% are essentially pure ethane.

By constructing constant pressure tie-lines in Figures 7 and 8, the ternary mixture data can be represented on triangular diagrams. Data obtained in this manner are shown in Figures 9 and 10 and are listed in Table III.

Table III. Smoothed Dew and Bubble Point Data for the Ethanol-Water-Ethane System Obtained in this Study.

Temperature = 313.15°K
Pressure = 50 Bar

Bubble Point			Dew Point		
x_{H_2O}	x_{ETOH}	$x_{C_2H_6}$	y_{H_2O}	y_{ETOH}	$y_{C_2H_6}$
.7108	.2792	.0100	.0122	.0048	.9830
.4036	.5074	.0890	.0074	.0096	.9830
.1764	.6216	.2020	.0038	.0132	.9830
.0804	.5956	.3240	.0020	.0150	.9830
.0403	.6097	.3500	.0011	.0159	.9830

Temperature = 313.5°K
Pressure = 80 Bar

.7072	.2778	.0150	.0223	.0087	.9690
.3797	.4973	.1230	.0134	.0176	.9690
.1551	.5469	.2980	.0069	.0241	.9690
.0632	.4678	.4690	.0071	.0529	.9400
.0198	.3002	.6800	.0112	.1638	.8200

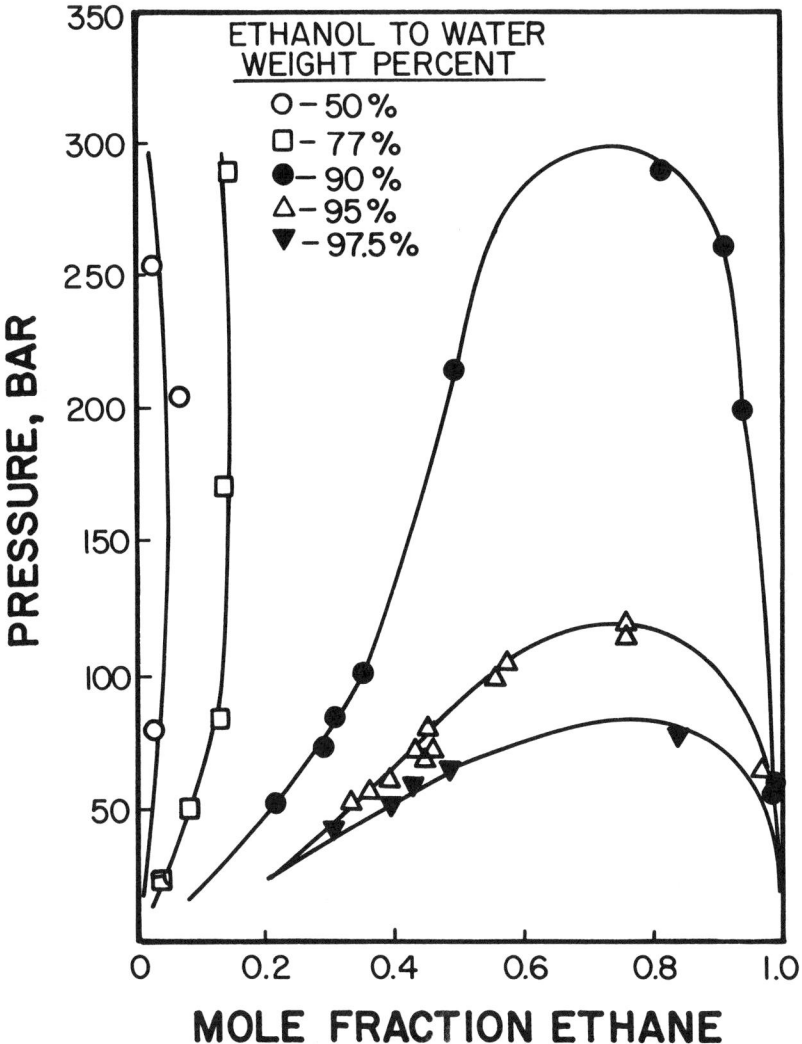

FIGURE 7. Dew-point and bubble-point data for the ethanol-water-ethane system obtained in this study at a constant temperature of 313.15°K. Data are shown for various mixtures of overall fixed ethanol to water weight ratios.

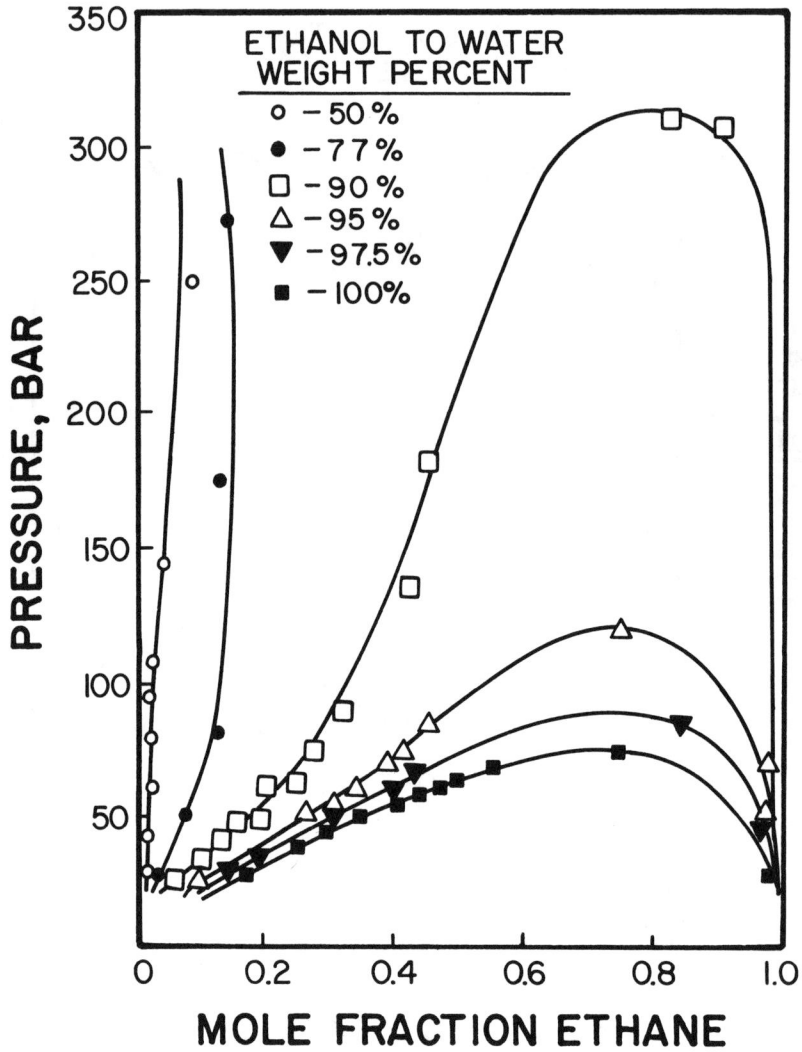

FIGURE 8. Dew-point and bubble-point data for the ethanol-water-ethane system obtained in this study at a constant temperature of 323.15°K. Data are shown for various mixtures of overall fixed ethanol to water weight ratios.

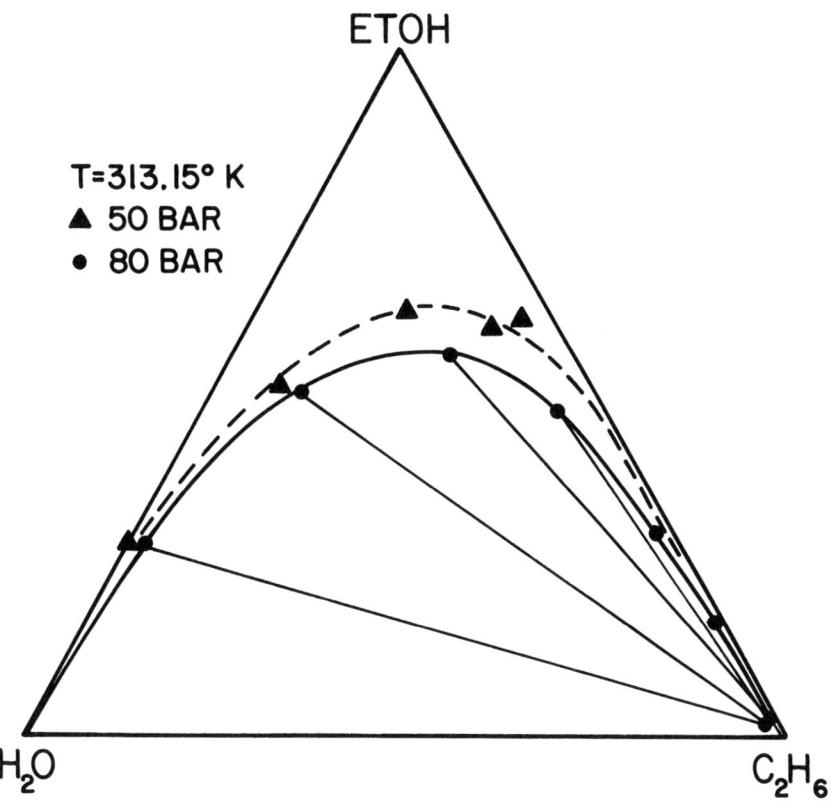

FIGURE 9. Tie lines for the ethanol-water-ethane system obtained in this study at a constant temperature of 313.15°K and at pressures of 50 and 80 bar.

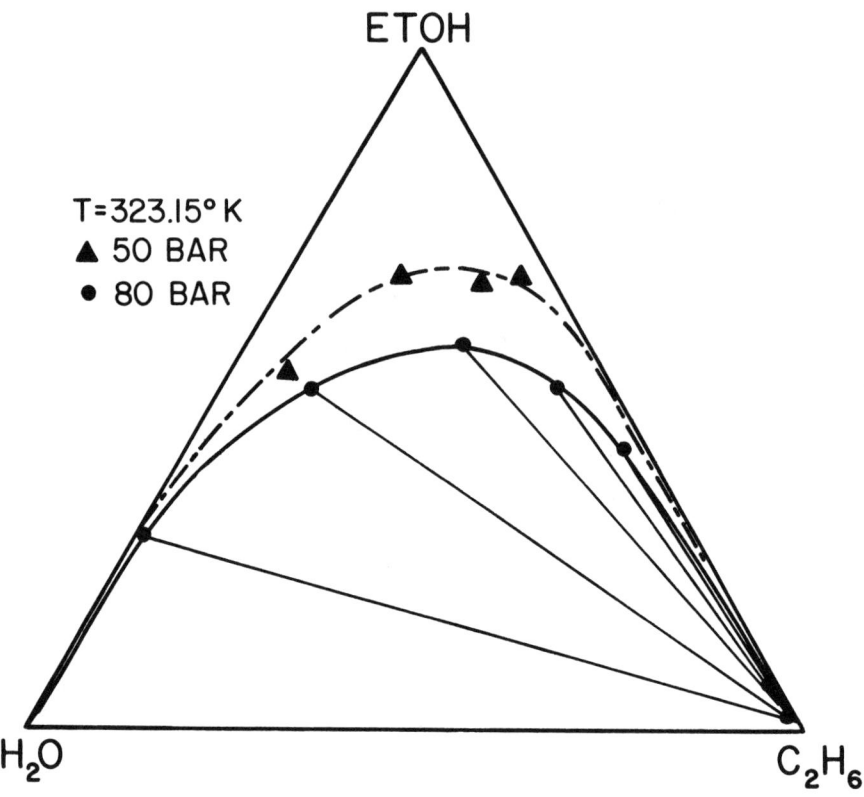

FIGURE 10. Tie lines for the ethanol-water ethane system obtained in this study at a constant temperature of 323.15°K and at pressures of 50 and 80 bar.

Bubble Point			Dew Point		
x_{H_2O}	x_{ETOH}	$x_{C_2H_6}$	y_{H_2O}	y_{ETOH}	$y_{C_2H_6}$
		Temperature = 323.15°K			
		Pressure = 50 Bar			
.7094	.2786	.0120	.0108	.0042	.9850
.3962	.5188	.0850	.0065	.0085	.9850
.1867	.6583	.1550	.0033	.0117	.9850
.0893	.6607	.2500	.0018	.0132	.9850
.0438	.6632	.2930	.0009	.0141	.9850
		Temperature = 323.15°K			
		Pressure = 80 Bar			
.7060	.2770	.0170	.0158	.0062	.9780
.3802	.4978	.1220	.0095	.0125	.9780
.1602	.5648	.2750	.0049	.0171	.9780
.0678	.5022	.4300	.0036	.0264	.9700
.0270	.4080	.5650	.0047	.0703	.9250

Shown in Figure 9 are the ternary mixture data obtained from Figure 7 at 313.15°K and at two separate pressures of 50 and 80 bar. The tie-lines at 80 bar are shown in Figure 9 while those at 50 bar are left off the diagram since they are essentially parallel to the 80 bar tie-lines.

Shown in Figure 10 are the ternary mixture data obtained from Figure 8 at 323.15°K and at two separate pressures of 50 and 80 bar. Again only tie-lines at 80 bar are shown.

Three phase liquid-liquid-vapor equilibria is observed for the ternary system in the range of 298 to 313°K and approximately 715 psia for varying overall water concentrations. However, no attempt is made to accurately determine the three phase region or to determine the composition of the equilibrium phases.

DISCUSSION OF RESULTS

The experimental results which are obtained for the ethanol-ethane binary exhibited certain trends which are also exhibited by the ternary ethanol-water-ethane system. Consider the ethanol-ethane vapor-liquid equilibrium curves shown in Figure 6. At 323.15°K, the concentration of ethane in the liquid phase increases at a much faster rate than the concentration of ethanol in the vapor phase as the pressure increases. In fact, the concentration of ethanol in the vapor phase is virtually insensitive to increasing pressure.

Our immediate goal in this study is to determine whether supercritical ethane would separate ethanol/water mixtures while operating in the two phase liquid-fluid region for the ternary ethanol-water-ethane system. Therefore, to avoid any regions of liquid-liquid immiscibility for the ethanol-water-ethane ternary, the experimental investigation is limited to temperatures of 313.15°K and above.

The general trends of the ternary mixture data are exhibited in Figure 9. The data in this figure are at 313.15°K, a temperature which is only slightly higher than the critical temperature of pure ethane (305.45°K). Notice that at 50 bar, very close to the critical pressure of pure ethane (48.8 bar), regardless of the overall composition of the mixture, the loading of ethanol in the ethane-rich supercritical fluid phase remained at approximately 1.4 mole %. At a higher pressure of 80 bar, again regardless of the overall composition of the mixture, the loading of ethanol increases only slightly to approximately 2.0 mole %. Also the selectivity of ethanol to water in the ethane-rich supercritical fluid phase is insensitive to pressure in the range of 50 to 80 bar.

The experimental tie-lines at 80 bar depicted in Figure 9 increase in slope as the amount of ethanol in the overall mixture increases. This results in the experimental tie-lines becoming essentially parallel with the azeotropic ethanol/water composition thus making it impossible to break the ethanol/water azeotrope. These experimental trends are also apparent at a constant temperature of 323.15°K as shown in Figure 10.

For an extraction process using supercritical ethane, a large amount of ethane is needed in the process due to the low loading of ethanol in the supercritical solvent phase. Although selectivities are very good with supercritical ethane as compared to say supercritical CO_2 (26), the ethanol/water azeotrope cannot be broken.

Paulaitis et al. (26) determine the energy requirements for a simple countercurrent extraction process to upgrade the product from a fermentor using either supercritical carbon dioxide or ethylene. The results of their analysis show that even with supercritical ethylene which has low ethanol loadings, a substantial energy savings can be realized with supercritical extraction when compared to distillation. Since the experimental results obtained in this study are similar to those obtained by Paulaitis et al. (26), it is reasonable to assume that supercritical ethane can also be considered a candidate solvent for separating ethanol/water mixtures.

THERMODYNAMIC MODEL

The objective in modeling the ternary ethanol-water-ethane phase behavior is to extend the experimental effort by generating phase diagrams in pressure-temperature regions which are not experimentally investigated, especially the three phase L-L-V region which occurs near the critical point of pure ethane.

Various modeling procedures are proposed in the literature to predict the phase behavior of mixtures which contain water as well as a supercritical component (18-22). The most computationally straightforward and thermodynamically consistent method is to model the equilibrium liquid and vapor phases using the following equilibrium relationship

$$y_i \phi_i P = x_i \phi_i P \tag{1}$$

where fugacity coefficients are defined by the thermodynamic relationship

$$\ln \phi_i = \frac{1}{RT} \int_V^\infty \left[\left(\frac{\partial P}{\partial n_i} \right)_{T,V,n_j} - \frac{RT}{V} \right] dV - \ln Z$$

The Peng-Robinson equation of state (23) is chosen to evaluate fugacity coefficients, ϕ_i. This choice is based on the success Peng and Robinson have using their equation to calculate the phase behavior of ternary systems containing water (18,19).

The pure component properties of ethanol along its vapor-liquid equilibrium curve (24) are in good agreement with calculated values when using the relationships given by Peng and Robinson for the pure component parameters a and b (22). However, as might be expected, the molar volume of water along its vapor-liquid equilibrium curve (25) is overestimated by approximately 18% using the suggested value of 18.99 cm^3/mol for the pure component b parameter. Using a variation of a method proposed by Wenzel et al. (20) to calculate the liquid volume of water, b is adjusted to a constant value of 16.31 cm^3/mol to obtain a good fit of the molar volume of water without a loss in accuracy in other pure component properties.

Information on the binaries which comprise the ethanol-water-ethane ternary is used to obtain suitable binary interaction coefficients, i.e., K_{ij}'s. The vapor-liquid equilibrium data obtained in this study for the ethane-ethanol system are used to obtain a constant value of K_{ij} of 0.0525. The data of Culberson et al. (14) are used to determine the ethane-water K_{ij} which is found to be highly temperature dependent in agreement with the results of Peng and Robinson (19) for hydrocarbon-water systems. The ethane-water K_{ij} is calculated from the following equation

$$K_{ij} = A + B/T \tag{3}$$

where T is absolute temperature in degrees Kelvin, A equals 0.6115 and B equals $-2.3984 \cdot 10^2$.

The ethanol-water system is particularly troublesome to model in that literature vapor-liquid equilibrium data (13) can only be modeled in a qualitative manner. It is found that a constant value of K_{ij} equal to -0.140 reproduced the characteristics of the vapor-liquid equilibrium data, however, this value tends to overestimate the concentration of ethanol in the vapor phase by about a factor of two. This value of K_{ij} is in close agreement with that found by Paulaitis et al. (26). Using only binary interaction coefficients, ternary diagrams for the ethanol-water-ethane system at various temperatures and pressures are generated. Representative results of these calculations are shown in Figures 11 and 12.

The results for a constant temperature of 313.15°K and a constant pressure of 80 bar are shown in Figure 11. Also shown in this figure are experimental data represented as a dashed line. Only calculated tie-lines are shown. Comparing this figure to Figure 9, we see that the computed tie-lines are parallel with those experimentally observed although the actual compositions of the equilibrium phases are not quantitatively predicted.

The results shown in Figure 12, which depict phase behavior at 323.15°K, are very similar to those found at 313.15°K. The calculated tie-lines are parallel to experimental tie-lines, while again the experimental and calculated concentrations of the equilibrium phases agree only semi-quantitatively.

Since it was experimentally observed that the ethanol-water azeotrope cannot be split in the two phase region, and that ethanol loadings in supercritical ethane in this region are quite low, the model is extended to estimate the separations and loadings possible in the three phase region which

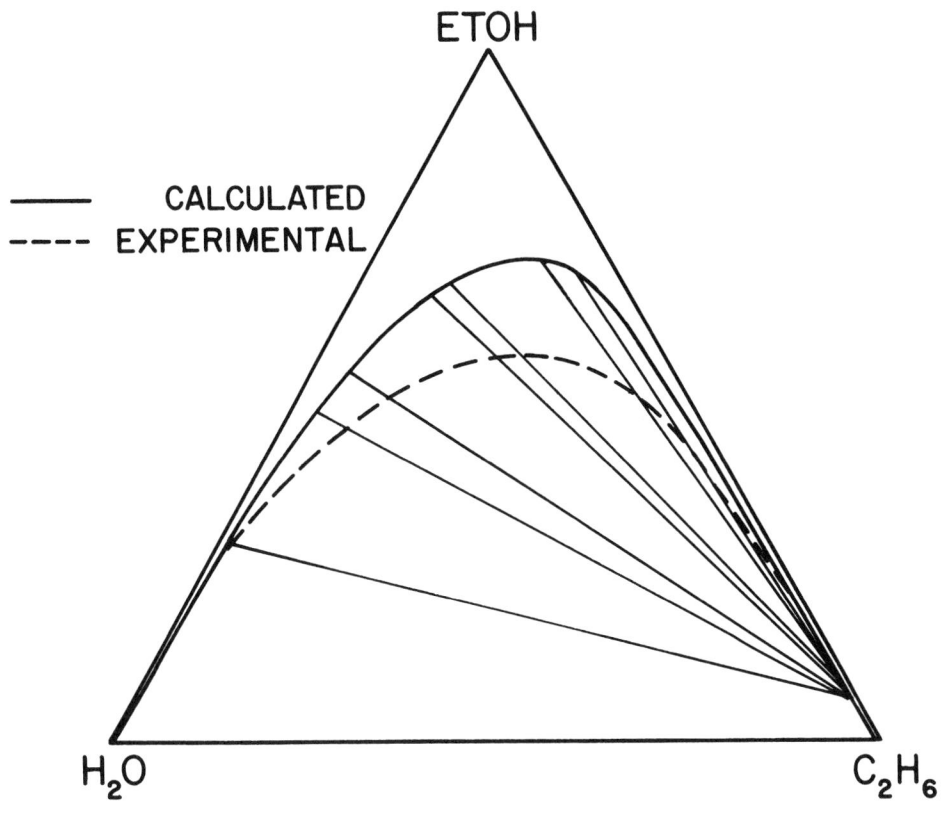

FIGURE 11. Comparison of calculated tie lines with experimental data obtained at 313.15°K and 80 bar.

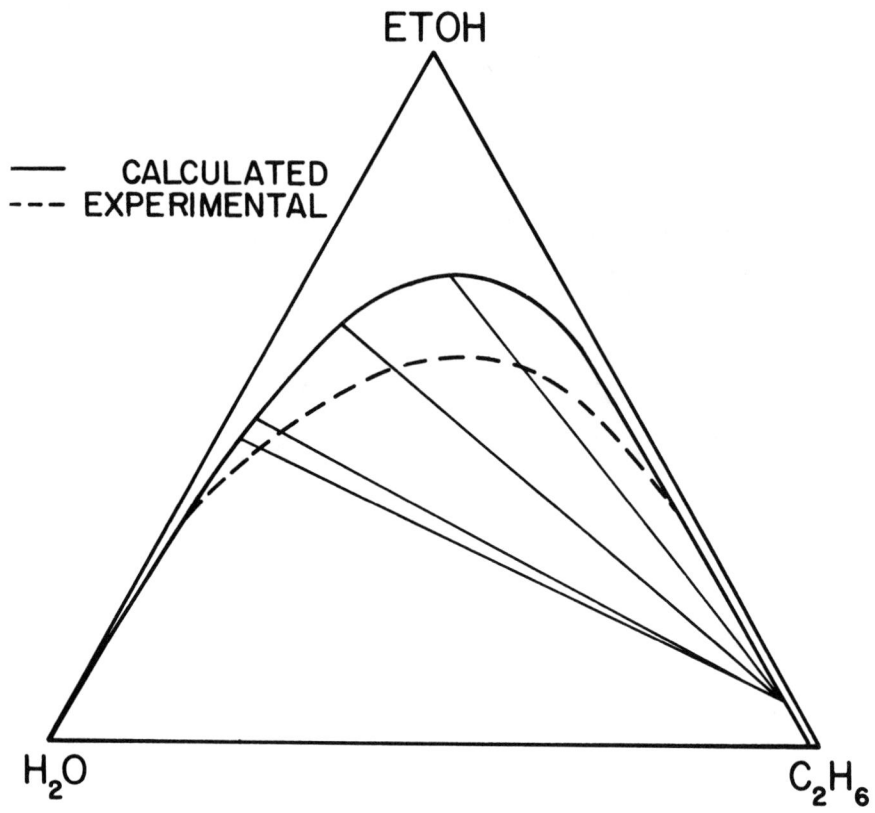

FIGURE 12. Comparison of calculated tie lines with experimental data obtained at 323.15°K and 80 bar.

exist in the vicinity of 25°C to 40°C and at pressures near 715 psia. The algorithms used for these calculations are those developed by Peng and Robinson (19).

Unfortunately, the model fails to predict three phase equilibria in the region near the critical temperature of pure ethane. This is probably due to the very large negative value of the ethanol - H_2O K_{ij} since the other binary pairs, ethanol-ethane and ethane-water, are reasonably represented by the model.

CONCLUSION

The experimental phase behavior of the ethanol-water-ethane system as well as the ethane-ethanol system is determined. The results of this study indicate that extraction of ethanol from ethanol/water mixtures using supercritical ethane is possible, although large amounts of ethane are needed due to the low loading of ethanol in the supercritical fluid phase. However, it is not possible to break the ethanol/water azeotrope operating in the two phase liquid-fluid region of the ternary mixture.

LITERATURE CITED

1. Kuenen, J. P. and W. G. Robson, Phil. Mag. [VI], 48, 180 (1899).

2. Chem. Engng., p. 101, October 23, 1978.

3. Ibid, p. 78, February 26, 1979.

4. Hartline, F., Science, 206, 41 (1979).

5. Roddy, J. W. and C. F. Coleman, I & EC Fund., 20, 250 (1981).

6. Roddy, J. W., Ibid, 20, 104 (1981).

7. Peter, S. and G. Brunner, in 'Extraction with Supercritical Gases,' p. 141, G. M. Schneider, E. Stahl, and G. Wilke, eds., Verlag Chemie, Deerfield Beach, Florida, 1980.

8. Zosel, K., Ibid, p. 1.

9. Paulaitis, M. E. and M. A. McHugh, paper presented at the 'High Pressure as a Reagent and an Environment,' 17th State-of-the-Art Symposium, ACS, Washington, D. C., June 1981.

10. Knebel, A. H. and D. E. Rhodes, paper presented at the 71st Annual Meeting AIChE, Miami, Florida, November 1978.

12. Rowlinson, J. S., 'Liquids and Liquid Mixtures,' 2nd ed., Butterworth, London 1969.

13. Barr-David, F. and B. F. Dodge, J. Chem. Engng. Data, 4, 107 (1959).

14. Culberson, O. L. and J. J. McKetta, Petrol. Trans. AIME, 192, 223 (1951).

15. Schneider, G. M., Agnew. Chem. Int. Ed. (Eng.), 17, 716 (1978).

16. Swaid, I. and G. M. Schneider, Ber. Bunsenges, Phys. Chem., 83, 969 (1979).

17. Kohn, J. P., AIChE J., 7, 514 (1961).

18. Robinson, D. B. and D.-Y. Peng, paper presented at the Proceedings of the 2nd International Conference of the European Federation of Chemical Engineering, Berlin, March 1980.

19. Peng, D.-Y. and D. B. Robinson, Can. J. Chem. Engng., 54, 595 (1976).

20. Wenzel, H. and W. Rupp, Chem. Engng. Sci., 33, 683 (1978).

21. Gmehling, J., D. D. Liu, and J. M. Prausnitz, Chem. Engng. Sci., 34, 951 (1979).

22. deSantis, R., G.J.F. Breedveld, and J. M. Prausnitz, I & EC Process Des. Develop., 13, 374 (1974).

23. Peng. D.-Y., and D. B. Robinson, I & EC Fund., 15, 59 (1976).

24. International Critical Tables, Vol. III, p. 238, McGraw-Hill (1933).

25. Sandler, S. I., 'Chemical and Engineering Thermodynamics,' John Wiley, 1977.

26. Paulaitis, M. E., M. L. Gilbert, and C. A. Nash, paper presented at the 2nd World Congress of Chemical Engineering, Montreal, Canada, October 1981.

27. Huie, N. C., Luks, K. D., and Kohn, J. P., J. Chem. Engng. DATA, 18, 311 (1973).

CHAPTER 6

SOLID SOLUBILITIES IN SUPER-
CRITICAL FLUIDS AT ELEVATED
PRESSURES

M. E. Paulaitis
M. A. McHugh
C. P. Chai
   Department of Chemical
      Engineering
   University of Delaware

ABSTRACT

Experimental solubilities for biphenyl in supercritical CO_2 at elevated pressures are presented in addition to the pressure-temperature projection of the three-phase, solid-liquid-gas equilibrium line for this mixture. Interpretation of the experimental phase equilibrium behavior is accomplished using a thermodynamic model based upon the Peng-Robinson equation of state. The model is then used to predict phase equilibrium behavior to describe a promising method for separating solids dissolved in supercritical-fluid solvents.

INTRODUCTION

Solvent extraction with supercritical or near-critical fluids has been proposed as an alternative to conventional separation processes, such as distillation and liquid-liquid extraction [1,2]. Several potential advantages are often cited for considering supercritical-fluid (SCF) solvents. One advantage is the ability to continuously vary solvent density and thus control solvent power, which can enhance the selectivity of the solvent. The solvent/solute separation can also be easily achieved, and fractionation of multiple solutes is possible by stepwise changes in solvent density [3]. Another advantage is the desirable mass transfer properties of SCF solvents. While the density of a supercritical fluid is comparable to liquid-like densities, SCF viscosities and diffusivities are intermediate to those properties for liquids and gases [4]. Thus supercritical fluids have the solvent power of liquids with better mass transfer characteristics than typical liquid solvents, and separation efficiencies for SCF solvent extractions can be appreciably higher than those for liquid solvent extractions [5,6]. Supercritical carbon dioxide is one solvent that has many unique advantages over other SCF and liquid solvents. In particular, CO_2 is nontoxic, nonflammable, environmentally acceptable, and relatively inexpensive. Furthermore, the critical temperature of CO_2 is approximately 31°C, and extractions can be accomplished at moderate temperatures where thermal degradation of heat-labile extracts is minimal.

Although SCF solvent extraction processes of practical interest would involve multicomponent mixtures, the fundamental concept of a SCF solvent and the separations that can be achieved with such solvents have been developed by considering phase equilibria at elevated pressures for model two- and three-component systems. Solid solubilities in supercritical fluids for simple two-component systems are frequently used to exemplify both the direct correlation between solvent density and solvent power, and the influence of the critical region on solubility behavior. The solubility of solid naphthalene in supercritical ethylene (T_c = 9.21°C; P_c = 49.66 atm), as shown in Figure 1, represents one specific example that has been used extensively [7-9]. From Figure 1, the effect of isothermally increasing pressure above 50 atm is to enhance naphthalene solubilities in supercritical ethylene by several orders of magnitude. At constant pressures just above 50 atm, reducing temperature from 45° to 25°C will also enhance naphthalene solubilities by approximately one order of magnitude. Both phenomena are directly related to the increased density of ethylene at higher pressures and lower temperatures. In the latter case, the increase in ethylene density more than compensates for the reduced volatility of naphthalene when decreasing temperature. This solubility behavior for highly compressible ethylene in the vicinity of its critical point can be compared to behavior for ethylene at liquid-like densities. For example, over the pressure range from 200 to 250 atm, the effect of pressure on naphthalene solubilities is relatively insignificant, and raising temperature causes solid solubilities to increase--the expected behavior for solids dissolved in typical liquid solvents.

SOLID SOLUBILITIES IN SUPERCRITICAL FLUIDS

The solubility behavior for naphthalene in supercritical ethylene is not the general case for solids dissolved in supercritical fluids, but is characteric behavior for binary

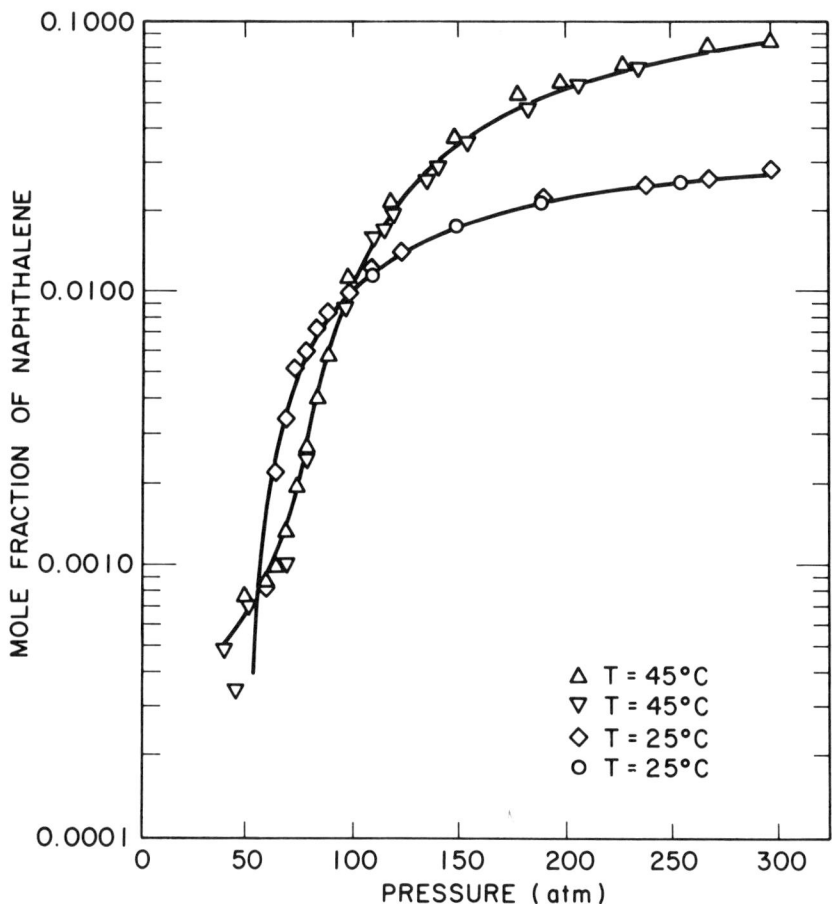

Figure 1. Solid Solubilities for Naphthalene in Ethylene at 25° and 45°C [7,8].

mixtures with constituents that are relatively different in molecular nature--i.e., asymmetric mixtures. The various types of phase equilibrium diagrams for binary mixtures are discussed in detail elsewhere [10,11].

For asymmetric binary mixtures, the phase equilibrium behavior can be characterized by the pressure-temperature (PT) projection of the three-dimensional pressure-temperature-composition diagram, as shown in Figure 2. The following features are depicted in this particular projection: (1) vapor pressure curves terminating at critical points for both pure components, (2) the melting curve (S=L) and a portion of the sublimation curve for component 2, (3) the gas-liquid critical curve (L=G) for mixtures of the two

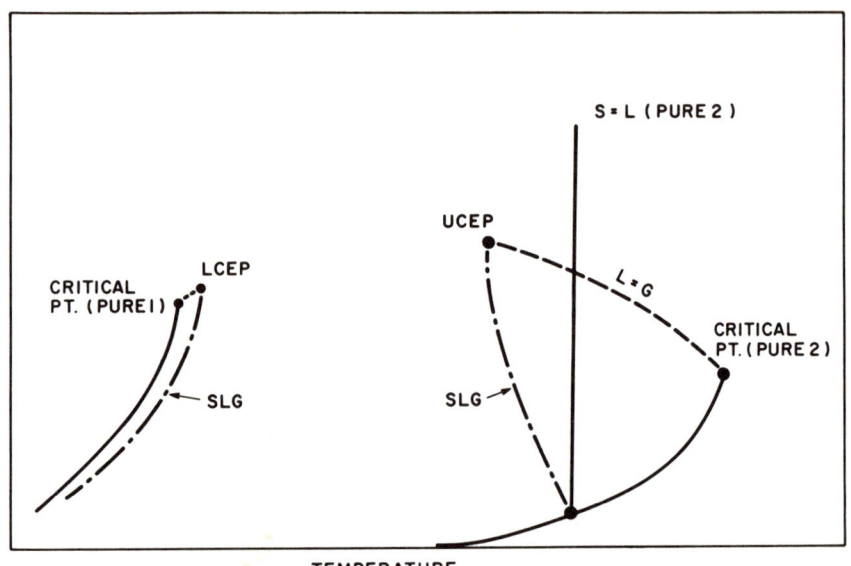

Figure 2. Schematic PT Projection of the PTx Diagram for a Binary Mixture.

components, and (4) the three-phase, solid-liquid-gas equilibrium line (SLG). One obvious characteristic of this two-component system is that the triple point temperature of component 2 is significantly greater than the critical temperature of component 1. Another characteristic is the limited solubility of component 1 in liquid component 2, and the relatively small freezing point depression for component 2 in the presence of the supercritical component, even at elevated pressures. As a consequence, the SLG equilibrium line originating at the triple point of component 2 extends to relatively high pressures and intersects the gas-liquid critical curve at two points, the upper critical end point (UCEP) and the lower critical end point (LCEP). At temperatures between the LCEP and UCEP temperatures, solid-fluid equilibrium will exist at all pressures up to very high pressures and high fluid densities. Both the LCEP and UCEP represent mixture critical points in the presence of excess solid, although in many cases, the LCEP is reasonably close to the critical point for component 1, as shown in Table I. The dramatic solubility behavior presented in Figure 1 actually represents behavior in the vicinity of the LCEP, and similar behavior should occur near the UCEP for this system. Figure 3 represents a portion of the experimental PT projection for naphthalene-ethylene including the SLG line, part of the gas-liquid critical curve, and the UCEP at T = 52.1°C and P = 174 atm [12]. The binary mixture composition at the UCEP is 17 mol% naphthalene [12]. The solubility behavior in the vicinity of this UCEP is illustrated in

Table I. Experimental LCEPs for Binary Mixtures Compared with the Critical Points for the Light Components

System		Critical Point of Cmpt 1		LCEP for the Binary Mixture		LCEP Reference
Cmpt 1	Cmpt 2	T(°C)	P(atm)	T(°C)	P(atm)	
ethylene+naphthalene		9.21	49.66	10.7	51.15	12
ethylene+biphenyl		9.21	49.66	11.2	50.95	13
ethylene+anthracene		9.21	49.66	9.4	50.5	12
ethylene+naphthalene		32.28	48.16	36.8	51.5	14
				36.8	51.6	15
methane+cyclohexane		-82.60	45.44	-80.4	47.5	16
methane+n-octane		-82.60	45.44	-82.0	45.9	16

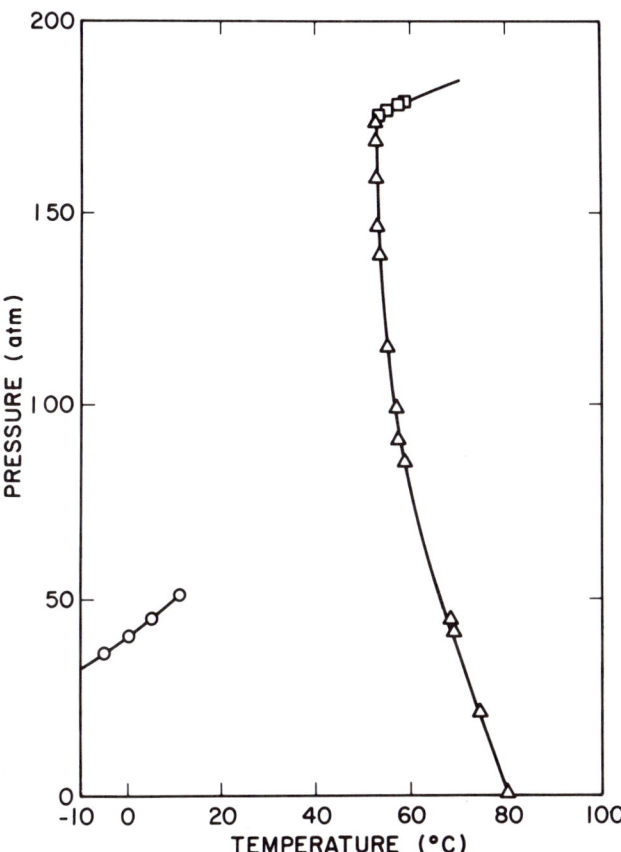

Figure 3. Experimental PT Projection of the PTx Diagram for Naphthalene-Ethylene [12,18].

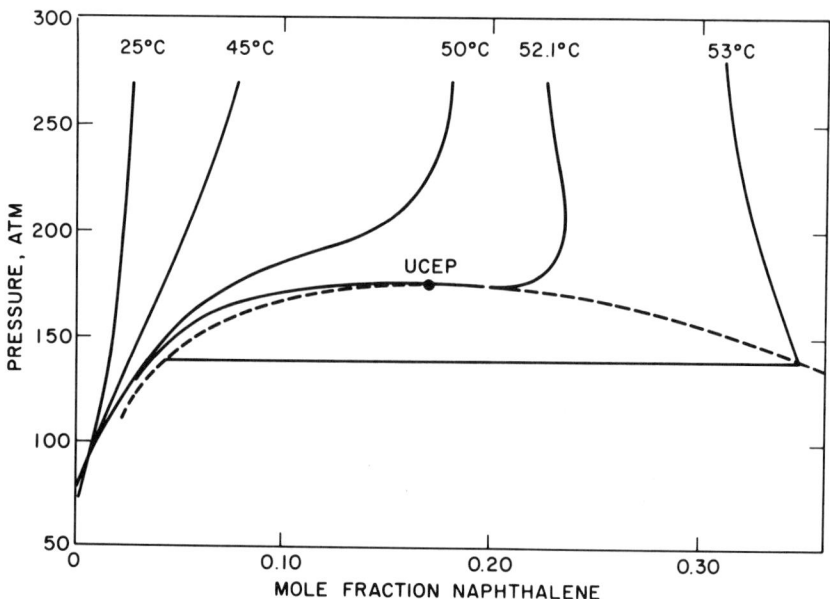

Figure 4. Smoothed Fit of the Experimental Solubilities of Diepen and Scheffer for Naphthalene in Ethylene [7].

Figure 4. From standard thermodynamic relations, the 52.1°C solubility isotherm must have an inflection point at the UCEP pressure. The physical significance of this curvature is a dramatic enhancement of solid solubilities in the supercritical fluid with increasing pressure. Solubility isotherms at temperatures just below the UCEP temperature will also exhibit appreciable curvature as a result of the influence of the UCEP, and therefore significant solubility enhancements with increasing pressure will be obtained when approaching the UCEP pressure. Experimental naphthalene solubilities at 25°, 45°, and 50°C are also shown in Figure 5. At 50°C and pressures near 174 atm, naphthalene solubilities in supercritical ethylene are enhanced appreciably with increasing pressure. Moderate temperature changes on the order of 5°C also produce significant solubility changes at pressures above 174 atm.

Comparing the solubility behavior in Figures 1 and 5 shows that solid solubilities are sensitive to temperature and pressure variations at conditions near both the LCEP and UCEP, however, higher SCF solvent loadings are obtained in the vicinity of the UCEP. This combination of high solvent loadings and variable solubilities should be of practical interest for SCF solvent extractions, and therefore the

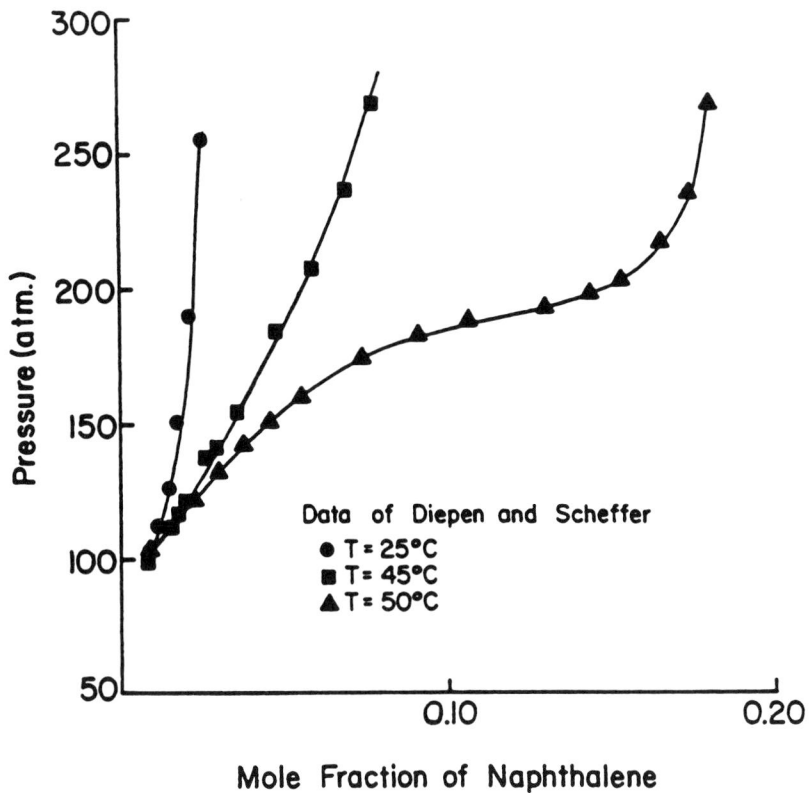

Figure 5. Experimental Solubilities of Diepen and Scheffer for Naphthalene in Ethylene [7].

location of the UCEP and solubility behavior in the region around this critical point should be of interest for other systems. In this paper, an investigation of solubility behavior near the UCEP is described for biphenyl-CO_2. The phase equilibrium behavior has several important differences compared to the naphthalene-ethylene system. One interesting application of this behavior is a new method for separating solids dissolved in SCF solvents.

EXPERIMENTAL RESULTS

A schematic diagram of the experimental apparatus for measuring solubilities in supercritical fluids is shown in Figure 6. A flow technique is used to contact supercritical CO_2 with biphenyl packed in the two equilibrium cells, and sampling of the saturated fluid-phase mixture is accomplished

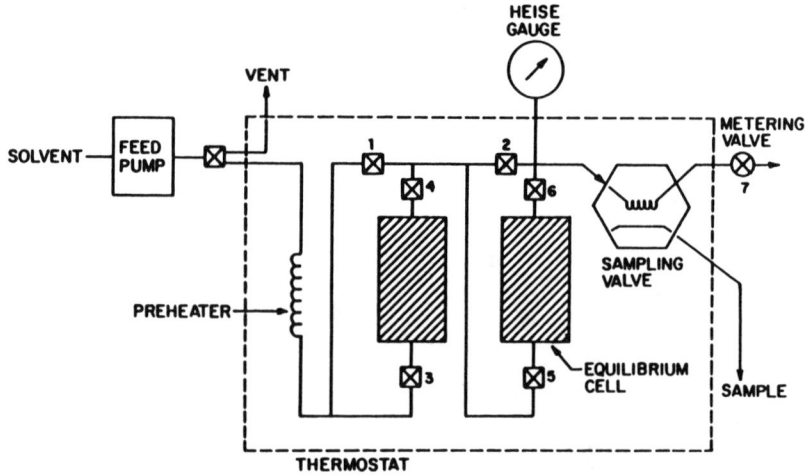

Figure 6. Schematic Diagram of the Experimental Apparatus for Measuring Solubilities.

with the switching valve between the second equilibrium cell and the metering valve. Details of the experimental method and experimental solubilities for biphenyl in CO_2 at 35.8°, 45.4°, 49.5°, 55.2°, and 57.5°C and at pressures up to 500 atm have been reported previously [19]. These experimental results are presented in Figure 7. Additional biphenyl

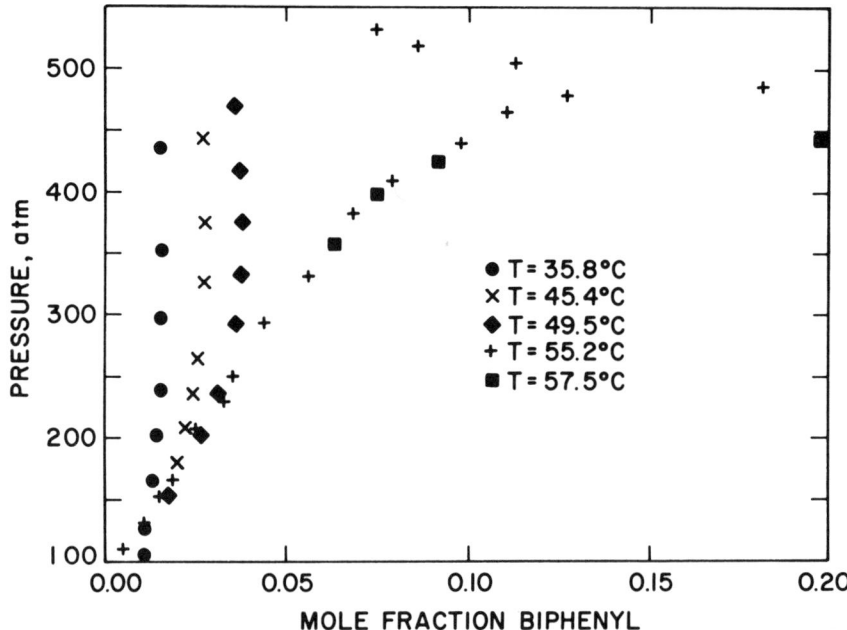

Figure 7. Experimental Solubilities for Biphenyl in CO_2 at various temperatures.

solubilities in CO_2 at 55.2°C and at pressures above 500 atm have been obtained and are given in Table II and also shown

Table II. Experimental Solubilities for Biphenyl in Supercritical CO_2 at 55.2°C

Pressure (Atm)	Biphenyl Mole Fraction
503.4	.11277
517.0	.08611
530.6	.07503

in Figure 7. Comparing the 45.4°C solubility isotherm in Figure 7 with naphthalene solubilities in ethylene at 45°C from Figure 5, indicates that, at a given pressure, biphenyl solubilities in supercritical CO_2 are generally less than those for naphthalene in supercritical ethylene. However, solid solubility behavior for both systems is essentially the same at the lower temperatures--e.g., 25° and 45°C for naphthalene in ethylene and 35.8°, 45.4°, 49.5°C for biphenyl in CO_2. At 55.2°C and pressures below 500 atm, biphenyl solubilities in CO_2 exhibit an enhancement comparable to that for the 50°C isotherm for naphthalene in ethylene at pressures approaching the UCEP pressure. From the previous discussion on the naphthalene-ethylene system, it would be reasonable to assume that this enhancement is characteristic of solubility behavior near the UCEP for biphenyl-CO_2. A reasonable estimate of the UCEP temperature and pressure for this system would be 56°C and 485 atm, respectively. At 55.2°C and pressures above 500 atm, however, biphenyl solubilities in CO_2 decrease rapidly with pressure, while naphthalene solubilities in ethylene at 50°C and pressures above the UCEP pressure approach a limiting mole fraction as pressure increases.

To obtain additional experimental information on the phase equilibrium behavior for biphenyl-CO_2, a simple static view cell was used to measure the PT projection of the SLG equilibrium line by observing the solid-liquid transition for biphenyl in the presence of compressed CO_2. The view cell is depicted in Figure 8, and consists of a 1 mm I.D. x 7 mm O.D. x 154 mm long borosilicate glass tube coupled through an adaptor to a high pressure delivery system. The high pressure seal is made with a Buna-N o-ring that fits around the cell after insertion into the adaptor recess. A steel collar is also epoxied to the glass tube to prevent it from slipping out of the coupling under pressure. Details of the experimental apparatus and procedure are given elsewhere [20].

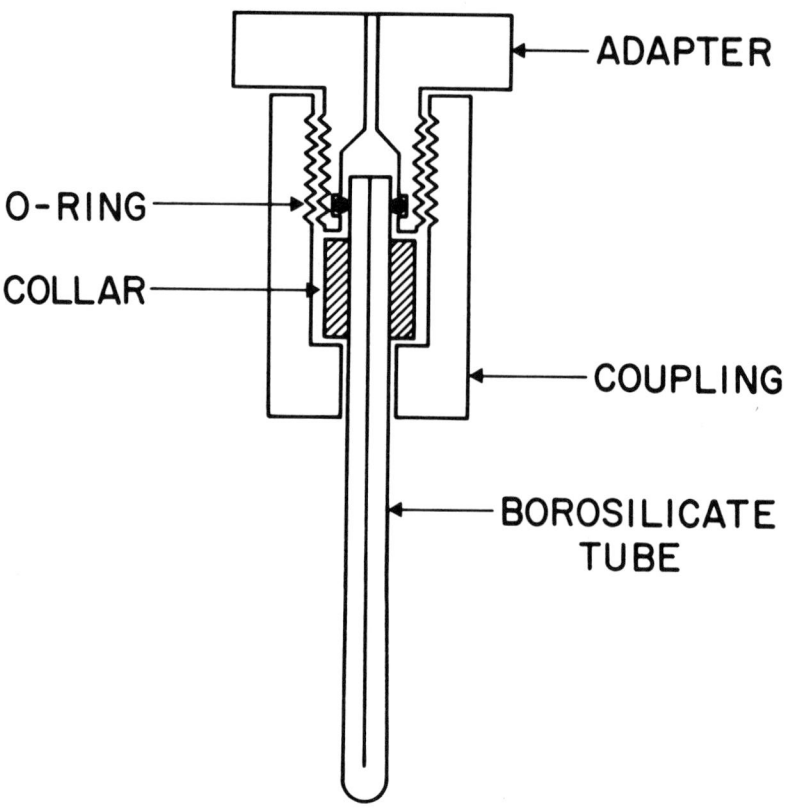

Figure 8. Schematic Diagram of High-Pressure Glass View Cell.

The PT projections of the SLG equilibrium lines have been measured for three binary mixtures: naphthalene-ethylene, naphthalene-CO_2, and biphenyl-CO_2. These results are presented in Figures 9-11. Tabulated data for the two CO_2-hydrocarbon mixtures are also given in Table III.

The results for naphthalene-ethylene obtained in this study are compared with data from previous experimental investigations in Figure 9. The agreement indicates that the experimental method described above is a reliable and accurate one. In contrast to this SLG equilibrium line, the three-phase equilibrium lines for naphthalene-CO_2 (Figure 10) and biphenyl-CO_2 (Figure 11) exhibit temperature minimums. In both cases, these results are in reasonable agreement with previous experimental work [21], although this work is quite limited. The UCEP temperature and pressure can also be determined with this experimental apparatus. For naphthalene-CO_2, an UCEP was obtained at 60°C and 240 atm,

Table III. Experimental Pressure-Temperature Data on the Solid-Liquid-Gas Equilibrium Lines for Naphthalene and for Biphenyl with Carbon Dioxide

Naphthalene-CO_2		Biphenyl-CO_2	
Pressure (Atm)	Temperature (°C)	Pressure (Atm)	Temperature (°C)
1.0	80.0	1.0	69.5
19.7	75.2	26.9	63.5
24.5	74.5	32.3	61.5
26.2	74.4	50.0	58.5
27.5	73.5	63.9	54.9
39.8	72.1	75.2	52.5
49.0	70.3	88.1	49.8
58.0	68.6	98.3	49.1
77.9	63.4	121.1	48.5
93.9	60.0	130.3	48.6
104.7	58.5	150.8	48.5
119.2	57.7	172.1	48.9
137.7	58.1	109.5	49.2
157.5	58.3	210.9	49.7
174.5	58.7	227.7	49.8
196.6	59.2	251.5	50.2
202.7	59.5	275.5	51.0
224.5	59.8		
228.6	60.3		
237.7	60.0		
239.8	60.0		
242.5	60.0		
244.5	60.0		
245.9	60.0		
247.3	60.0		
250.0	60.0		

respectively, which is in agreement with the UCEP estimated from experimental solubilities obtained previously [19]. The UCEP for biphenyl-CO_2 could not be determined since the maximum operating pressure of the equipment was reached before an UCEP was observed. However, a linear extrapolation of the PT data to higher pressures does extend to approximately 56°C and 458 atm--the UCEP estimated from the solubility data.

The characteristic shape of the solid-liquid-gas equilibrium lines for these binary mixtures depends strongly on the mutual solubility of the two component in the liquid mixture. For naphthalene-ethylene, the freezing point of naphthalene decreases rapidly with increasing pressure as

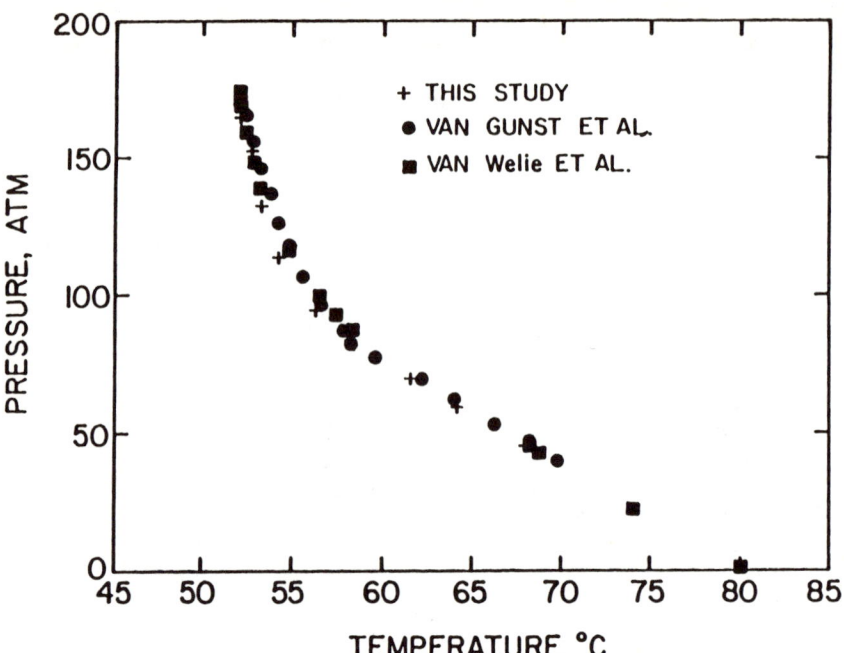

Figure 9. Comparison of the PT Projection of the Naphthalene-Ethylene SLG Line Obtained in This Study With Previous Results [12,18].

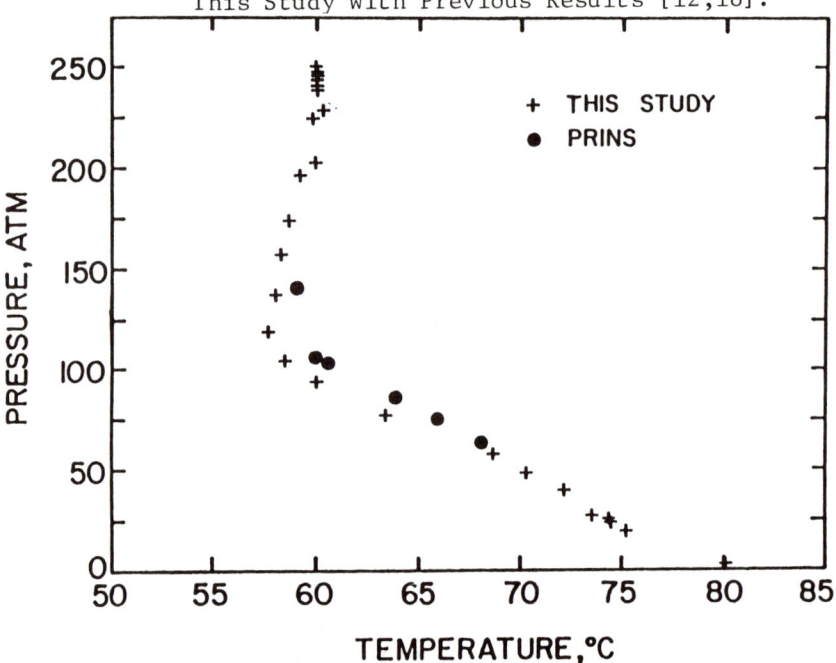

Figure 10. The PT Projection of the Naphthalene-CO_2 SLG Line Obtained in This Study Compared With the Results of Prins [21].

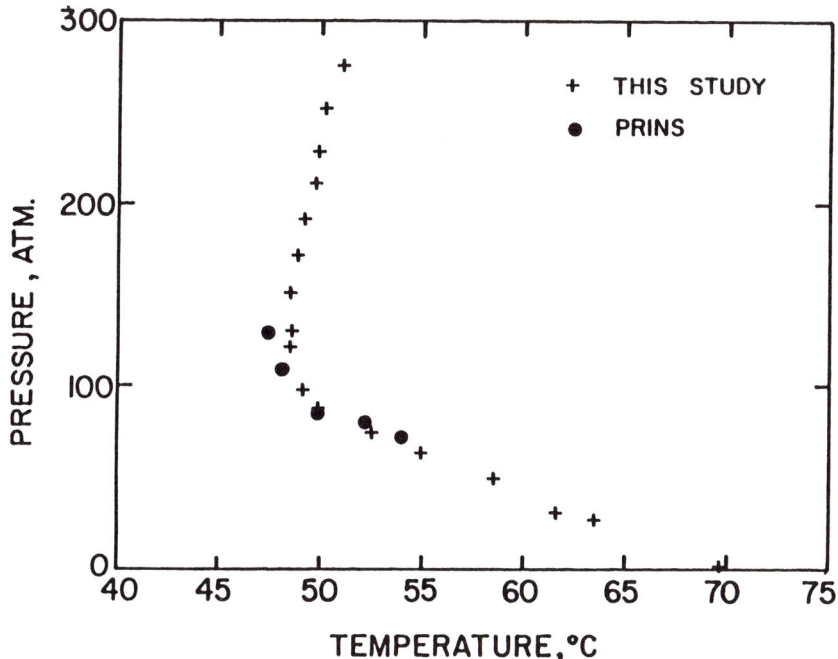

Figure 11. The PT Projection of the Biphenyl-CO_2 SLG Line Obtained in This Study Compared with the Results of Prins [21].

more ethylene dissolves in the liquid phase. This freezing point depression at elevated pressures more than compensates for the increase in freezing temperature with pressure for pure naphthalene. For naphthalene-CO_2 and biphenyl-CO_2, the initial freezing point depression is also significant due to the solubility of CO_2 in the liquid phase. However, CO_2 solubilities in the liquid mixture will approach a limiting solubility with increasing pressure. At this point, SLG equilibrium temperatures will increase with increasing pressure because the dominant effect becomes the increase in freezing temperature with pressure for the pure hydrocarbons. Thus the solid-liquid-gas equilibrium lines for these systems exhibit a temperature minimum with increasing pressure. Such temperatures minimums have also been observed for other systems--for example, naphthalene-methane [22].

THERMODYNAMIC MODELING

The experimental solid-liquid-gas equilibrium line for biphenyl-CO_2 has important implications in interpretting the solubility behavior for this system in Figure 7. In particular, some of these solubility isotherms represent liquid-gas equilibrium rather than solid-fluid equilibrium. To test the consistency of the experimental results in Figures 7 and 11, and to gain further insights into this solubility behavior, a computer simulation has been developed to model the phase equilibrium behavior for biphenyl-CO_2 mixtures.

The computer simulation is based upon a thermodynamic model utilizing the Peng-Robinson equation of state [23] for calculating fluid-phase fugacities. One adjustable parameter δ_{ij} is required for each pair of mixture constituents. For a binary mixture, this parameter is defined by,

$$\delta_{12} = 1 - \frac{a_{12}}{(a_{11}a_{22})^{\frac{1}{2}}}$$

where the a_{ij}'s are standard equation-of-state parameters characteristic of interactions between components i and j. For biphenyl-CO_2, a reasonable estimate of $\delta_{12} = 0.09$ is used in all calculations. Details of the thermodynamic calculations and computational methods are given elsewhere [24].

The calculated PT projection of the solid-liquid-gas equilibrium line for biphenyl-CO_2 is compared with experimental results in Figure 12. Although the agreement in Figure 12 is only semi-quantitative, the calculated three-phase line does exhibit the characteristic temperature minimum. This suggests that the thermodynamic model can describe the essential features of the phase equilibrium behavior for biphenyl and CO_2, and therefore can provide a correct interpretation of the solubility behavior in Figure 7.

Calculated isothermal pressure-composition (Px) diagrams for biphenyl-CO_2 at four temperatures in the vicinity of the calculated UCEP are presented in Figure 13. In all calculations, the solid phase is assumed to be pure biphenyl. At 41.5°C, the Px diagram in Figure 13A shows solid-fluid equilibrium at all pressures. This calculated fluid-phase solubility behavior at a temperature just below the calculated temperature minimum in Figure 12 would correspond to the experimental solubility behavior at 35.8° and 45.4°C in Figure 7. At 42.5°C, the Px diagram in Figure 13B shows solid-fluid equilibrium at all pressures except for an intermediate pressure range where solid-liquid equilibrium or liquid-fluid equilibrium will exist depending on the overall mixture composition. This temperature is between the

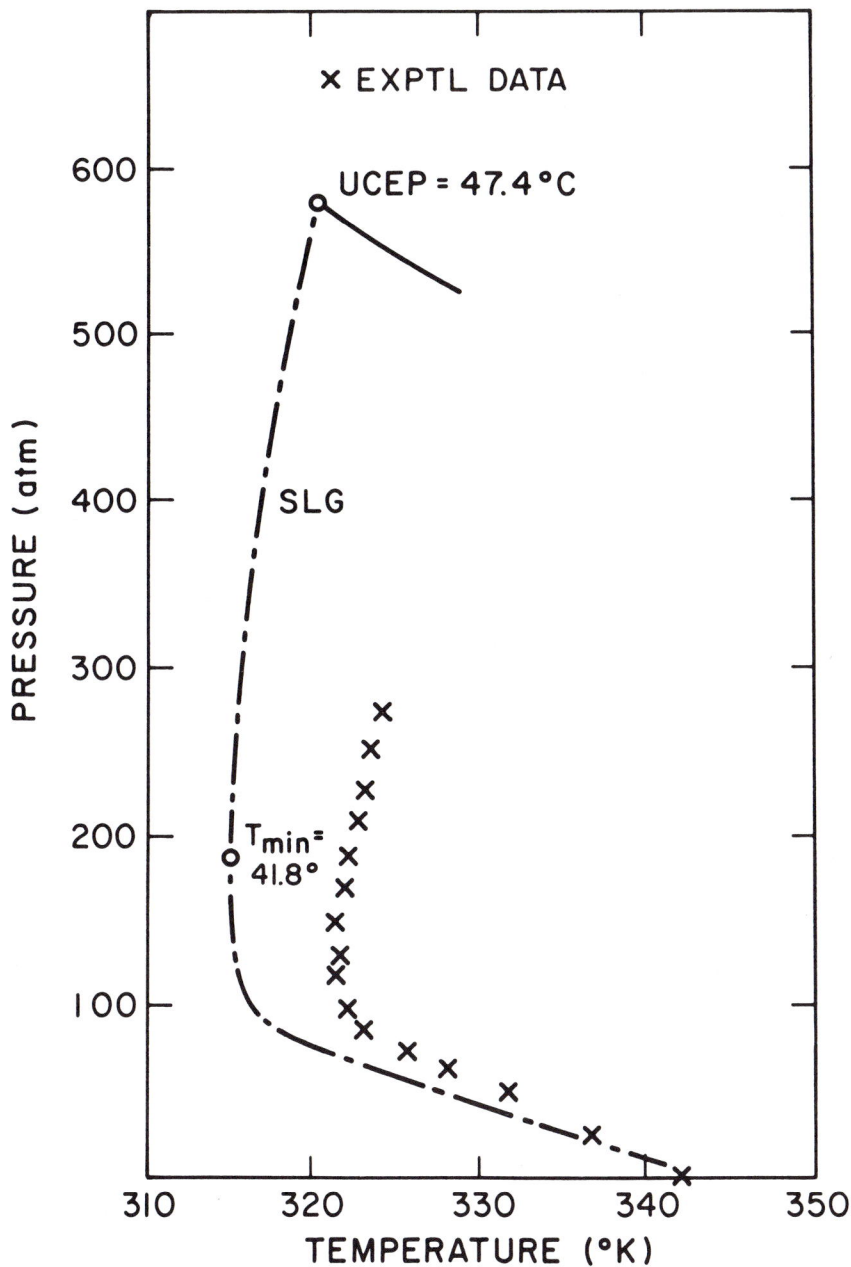

Figure 12. Comparison of the Calculated SLG Line for Biphenyl-CO_2 With the Experimental Results Obtained in This Study.

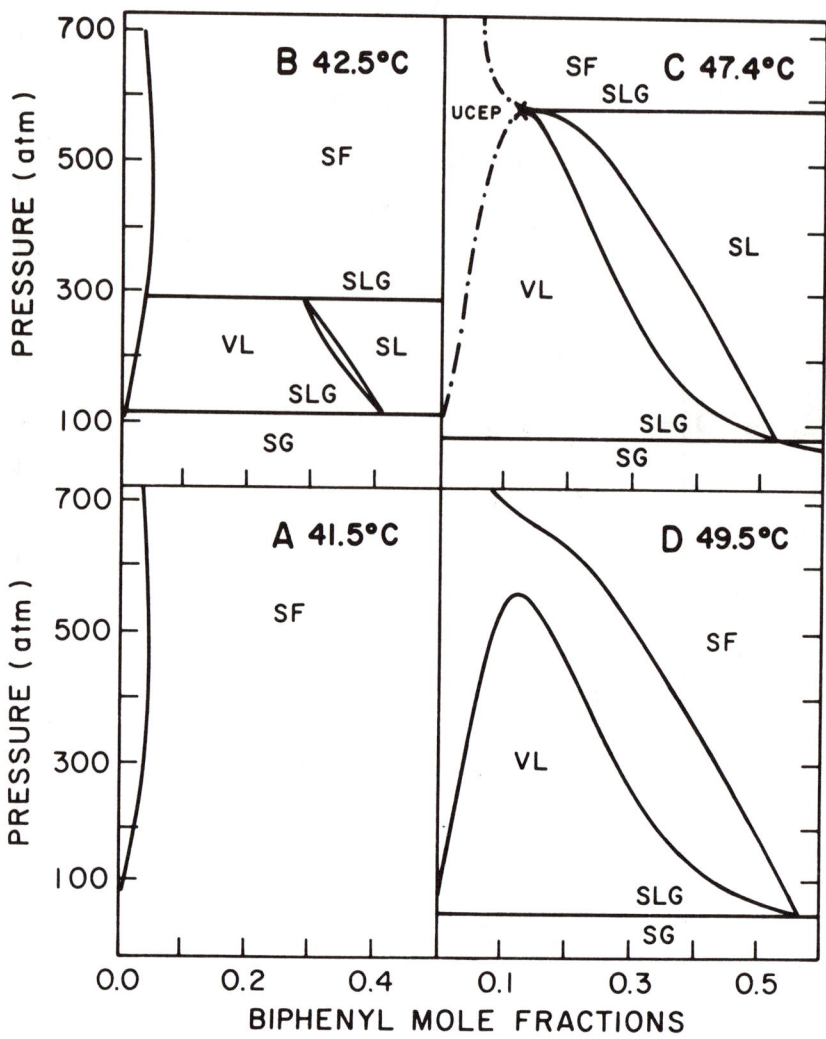

Figure 13. Calculated Px Diagrams for Biphenyl-CO_2 at Four Different Temperatures in the Vicinity of the Calculated UCEP.

calculated UCEP temperature and the temperature minimum in Figure 12, and the calculated fluid-phase solubilities for biphenyl in CO_2 would correspond to the experimentally observed solubility behavior at 49.5°C in Figure 7.

The Px diagram at the calculated UCEP temperature of 47.5°C is given in Figure 13C. Calculated fluid-phase solubilities (dash-dot line) show behavior strikingly similar to the experimental behavior at 55.2°C in Figure 7. A direct comparison between calculated and experimental solubilities is shown more clearly in Figure 14. The dramatic solubility enhancement in the vicinity of the calculated UCEP pressure is a result of the influence of the gas-liquid critical point at the UCEP, as discussed above for the naphthalene-ethylene system. The decreasing biphenyl solubilities at higher pressures, however, reflect both the influence of the UCEP and solubilities for biphenyl in supercritical CO_2 which decrease with increasing pressure due to the size asymmetry of these two components (i.e., free-volume effects). These free-volume effects are also shown in Figure 7 for the 49.5°C isotherm, where experimental biphenyl solubilities decrease with increasing pressure at pressures above approximately 375 atm. At this temperature, the influence of the UCEP is negligible. These calculated results also indicate that the values for the UCEP temperature and pressure estimated from the experimental solubility data--i.e., 56°C and 485 atm--are reasonably accurate.

PRACTICAL APPLICATIONS

The phase equilibrium behavior depicted in Figure 13D suggests an interesting method for separating solids dissolved in SCF solvents. This calculated Px diagram shows that appreciable amounts of biphenyl can be dissolved in CO_2 at 49.5°C and elevated pressures. For example, at 600 atm, the solubility of biphenyl in supercritical CO_2 is greater than 20 mol%. One method for recovery of the biphenyl from solution would be to isothermally change the pressure. Figure 13D indicates that reducing pressure may not be efficient because a considerable pressure change would be required (e.g., to values below approximately 77 atm) to obtain a significant separation, and thus recycling the CO_2 by recompressing it could be expensive. Alternatively, pure solid biphenyl can be obtained from solution by relatively moderate increases in pressure. Recycling the CO_2 would require a comparable pressure decrease, which could be more efficient in terms of solvent recovery and recycle.

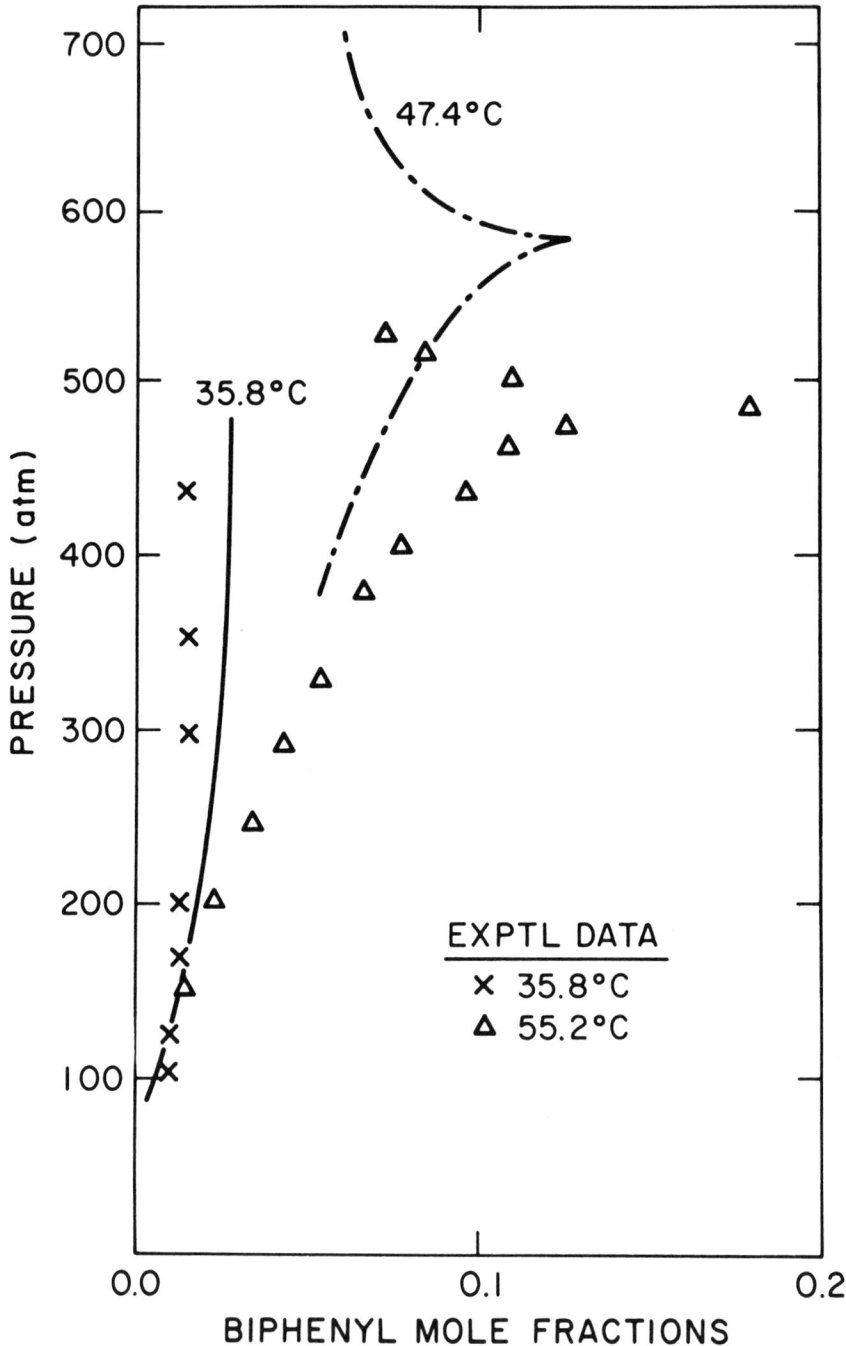

Figure 14. Calculated Fluid-Phase Solubilities for Biphenyl in CO_2 Compared with the Experimental Results at 35.8° and 47.4°C.

Comparing the calculated Px diagrams in Figures 13C and 13D suggests that a similar separation can be accomplished by changing temperature at constant pressure. For example, at a constant pressure of 600 atm, decreasing temperature by approximately 2°C (from 49.5° to 47.4°C) will reduce biphenyl solubilities in supercritical CO_2 from mole fractions greater than 0.20 to mole fractions less than 0.10. Further modest temperature reductions (e.g., to 41.5°C in Figure 13A) would produce even greater decreases in biphenyl solubilities. This separations scheme might be more attractive than the previous one, since isobaric temperature changes could be easily achieved with heat exchangers.

Although operating pressures for this mixture might be high for large-scale commercial separations, operating pressures for other processes of practical interest can be much lower. Separations processes, based on essentially identical phase equilibrium behavior, have been proposed for water desalination [25] and as an energy-efficient method for separating amorphous polymers and their solvents [26,27].

LITERATURE CITED

1. Zozel, K., Angew. Chem. Int. Ed. Engl. 17, 702 (1978).

2. Williams, D. F., Chem. Eng. Sci. 36, 1769 (1981).

3. Jentoft, R. E. and T. H. Gouw, J. Polymer Sci. B7, 811 (1969).

4. Swaid, I. and G. M. Schneider, Ber. Bunsenges. Phys. Chem. 83, 969 (1979).

5. Sie, S. T. and G. W. A. Rijnders, Sep. Sci. 2, 729 and 755 (1967).

6. Gere, D. R., R. Board, and D. McManigill, Anal. Chem. 54, 736 (1982).

7. Diepen, G. A. M. and F. E. C. Scheffer, J. Phys. Chem. 57, 575 (1953).

8. Tsekhanskaya, Y. V., M. B. Iomtev, and E. V. Mushinka, Russ. J. Phys. Chem. 38, 1173 (1964).

9. Masuoka, H. and M. Yorizane, J. Chem. Eng. Japan 15, 5 (1982).

10. Streett, W. B., Icarus 29, 173 (1976).

11. Luks, K. D., Proc. 2nd Int. Conf. on Phase Equilibria and Fluid Properties in the Chemical Industry, Berlin, 1980.

12. van Gunst, C. A., F. E. C. Scheffer, and G. A. M. Diepen, J. Phys. Chem. 57, 578 (1953).

13. Diepen, G. A. M. and F. E. C. Scheffer, J. Am. Chem. Soc. 70, 4081 (1948).

14. van Welie, G. S. A. and G. A. M. Diepen, J. Phys. Chem. 67, 755 (1963).

15. Tiffin, D. L., J. P. Kohn, and K. D. Luks, J. Chem. Eng. Data 24, 98 (1979).

16. Kohn, J. P., K. D. Luks, P. H. Liu, and D. L. Tiffin, J. Chem. Eng. Data 22, 419 (1977).

17. Rowlinson, J. S., "Liquids and Liquid Mixtures," 2nd Ed., Butterworth, London (1969).

18. van Welie, G. S. A. and G. A. M. Diepen, Rec. Trav. Chim. 80, 659, 666, 673, 683, 693 (1961).

19. McHugh, M. and M. E. Paulaitis, J. Chem. Eng. Data 25, 326 (1980).

20. McHugh, M. A., Ph.D. Thesis, University of Delaware, 1981.

21. Prins, A., Proc. Acad. Sci. Amsterdam 17, 1095 (1915).

22. van Hest, J. A. M. and G. A. M. Diepen, Symp. Soc. of Chem. Industry, London, 10 (1963).

23. Peng, D.-Y. and D. B. Robinson, IEC Fund. 15, 59 (1976).

24. Chai, C. P., Ph.D. Thesis, University of Delaware, 1981.

25. Hess, H. V. and F. E. Guptill, U. S. Patent 3,318,805 (1967).

26. Gutowski, T. G. and N. P. Suh, SPE Antek 28, 160 (1982).

27. Anolick, C. and E. P. Goffinet, U. S. Patent 3,553,156 (1971).

CHAPTER 7

AN EXPERIMENTAL METHOD FOR MEASURING
SOLUBILITIES OF HEAVY FOSSIL-FUEL
FRACTIONS IN COMPRESSED GASES
TO 100 BAR AND 300 °C

A. Monge
J. M. Prausnitz
  Chemical Engineering Dept.
  University of California
  Berkeley, CA 94720

A new experimental method has been developed to measure solubilities of narrow-boiling, heavy fossil-fuel fractions in compressed gases. Solubilities are determined from the volume of gas required to vaporize completely a small, measured mass of fossil-fuel sample. This method has been used successfully for several heavy solutes dissolved in compressed methane and in a compressed methane/water mixture.

INTRODUCTION

The high cost of energy has led to a search for new process technologies which allow more efficient utilization of energy resources. As a result, there has been growing interest in processing or upgrading of coal, heavy crudes, heavy petroleum fractions, tar sands, shale, etc. Development of such processes requires quantitative information for equilibrium properties of these materials under processing conditions at elevated temperatures and pressures. Of particular interest here is the solubility of a heavy fossil-fuel mixture in a compressed gas. This solubility is needed for the design of extraction processes with supercritical or near-critical fluids, (e.g. deasphalting, de-ashing and general upgrading of petroleum fractions); for petroleum-reservoir pressurization with light gases, toward removal of high-molecular-weight hydrocarbons left behind after primary recovery; and for coal-gasification process steps (condensation and quenching) where product gas streams often contain high-boiling coal tars.

Fossil-fuel mixtures typically contain very many components. The wide range of properties of the components and the analytical problem to identify these components makes phase equilibrium predictions and experimental measurement extremely difficult. A common procedure is to "divide" the mixture into fractions with fewer components and smaller ranges of properties. Phase equilibrium predictions and process design are then based on average or effective properties of the fractions.

This work reports an experimental technique for measuring vapor-liquid equilibria in systems containing narrow-boiling fossil-fuel fractions and compressed gases. The solubility of a fossil-fuel fraction in a compressed gas (defined as the sum of the vapor-phase concentrations of the heavy fraction's individual components) is determined by measuring the quantity of gas required to vaporize completely, under equilibrium conditions, a measured sample of that fraction. Because stripping effects cause the composition and solubility of the fraction to change during the vaporization process, a computer simulation of the vaporization process is used to relate the total-vaporization data to the solubility of the heavy fraction before its composition and solubility have changed. Approximate characterization of the heavy mixture, required for computer simulation, is obtained from the average molecular weight, hydrogen-to-carbon ratio, and approximate chromatographic analysis of the heavy mixture. A much simpler but less "rigorous" and slightly less accurate procedure is also presented for accounting for stripping effects. The simpler method does not require computer simulation, chromatographic analysis, or average molecular weight, but gives results (solubilities in terms of weight fractions instead of mole fractions) which are within 5% of those obtained by the more "rigorous" data-reduction procedure.

The experimental technique differs from those used previously to measure compressed-gas solubilities of pure heavy hydrocarbons [e.g. Johnston 1981, Kaul 1978, Kurnik 1981, Paulaitis 1980]. First, the total-vaporization technique requires only semi-quantitative analysis of the gas phase. For measuring solubilities of complex mixtures, such as coal-tar fractions which are difficult to analyze quantitatively, this represents an important advantage over conventional methods which require precise quantitative analysis. Second, because only semi-quantitative analysis of the gas phase is required, water vapor can be introduced into the gas phase without significantly affecting the experimental and data-reduction procedures. This capability is important because water is so often present during the processing of heavy fossil fuels. Third, the total-vaporization method allows experiments to be performed with small liquid samples, such as those obtained from fractionation of coal tars or petroleum crudes into narrow-boiling fractions. Whereas sparged gas-liquid contactors, used in conventional methods, often require on the order of 100 cm^3 of liquid, the packed-bed equilibrium cell used here requires less than 0.5 cm^3, per measurement.

Our technique has been tested by measuring solubilities

of hexadecane, and a synthetic mixture containing known amounts of naphthalene, 1-methylnaphthalene, 2-ethylnaphthalene, and diphenylmethane, in compressed methane. It has also been used to measure solubilities of two coal-tar fractions produced from a coal-tar sample obtained from SASOL's Lurgi coal gasifier in South Africa. Solubilities were measured in methane and in a methane/water mixture.

EXPERIMENTAL APPARATUS

Presented in Figure 1 is the experimental apparatus to determine the solubility of a narrow-boiling, heavy-hydrocarbon liquid mixture in a compressed gas. Gas from a compressed gas cylinder is preheated in a constant-temperature air bath and passed through a packed-bed cell where it equilibrates with the hydrocarbon liquid at measured temperature and pressure. Saturated gas leaving the cell is expanded and directed through a heated gas-sampling valve used to take samples intermittently. Gas samples are analyzed with a flame-ionization detector in a gas chromatograph; the results are recorded by an electronic integrator. The saturated gas is then cooled to room temperature and its cumulative volume is measured using a wet test meter. Completion of the vaporization of the hydrocarbon liquid is signaled by a sharp decrease in the chromatogram peak area corresponding to the heavy liquid.

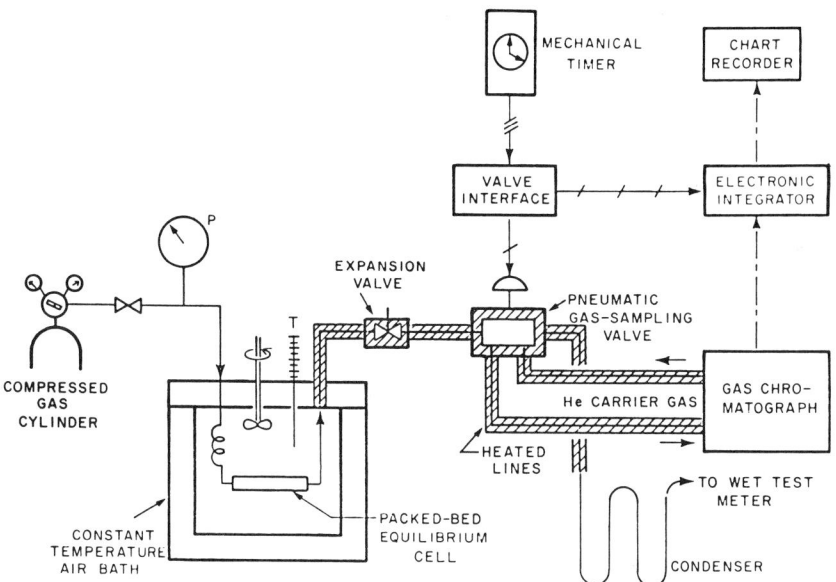

Figure 1. Total-vaporization apparatus for measuring the solubility of a heavy fossil-fuel fraction in a compressed gas.

We use two packed-bed cells in parallel. These packed-bed cells are stainless-steel tubes, 16 cm in length and 0.5-cm ID, packed with 30/60-mesh Chromosorb-P column support, coated with the hydrocarbon liquid of interest. The hydrocarbon liquid is dispersed throughout the cell by syringing into the cell at five evenly spaced points. The mass of hydrocarbon liquid introduced into the cell is determined by weighing the syringe before and after loading.

A specially designed high-temperature, high-pressure, fine-metering valve is used to expand the compressed-gas mixture from the packed-bed cell so that the gas can be sampled and metered. The gas-sampling valve is a VALCO 10-port, high-temperature sampling valve with 0.75-cm^3 sample loops. A 10-port valve with two sample loops is used so that gas streams from two different packed-bed equilibrium cells can be sampled alternately. Sampling is automated using a VALCO digital valve interface, a VALCO 2-position helical-drive air actuator and a simple mechanical timer. A typical run lasts two to three days and samples are taken every 30 minutes.

Gas samples are analyzed using a Perkin-Elmer 990 gas chromatograph with a hydrogen-flame ionization detector under constant-temperature, single-column operation. The chromatographic column used is a 2-m long, 3.2-mm OD stainless-steel tube packed with 3% OV-101 on 80/100-mesh Chromosorb W-HP column support (supplied by Varian). The output from the gas chromatograph is measured and recorded using a Spectra-Physics single-channel, computing integrator, operating in the "simulated distillation" mode (cumulative integration at regular intervals).

Temperature measurements in the constant-temperature air bath are made with an accuracy of $0.2^\circ C$ using Brooklyn precision thermometers calibrated against primary N.B.S.-certified thermometers. The uniformity of the temperature within the baffled, air-stirred bath is better than $0.2\ ^\circ C$. Thermocouples are used for temperature measurements outside the air bath to insure that condensation of heavy hydrocarbon does not take place. Pressure measurements are made using two Heise bourdon-tube gauges with accuracies of 0.2 psi in the 0-200 psi range and 2.0 psi in the 200-2000 psi range. The pressure drop through the system is negligible at gas flow rates (less than 5 cm^3/min) used in this work.

The experimental apparatus has been modified to allow the introduction of water vapor into the gas phase. Shown in Figure 2 are two water-containing sparged cells upstream of the original packed-bed cells. Otherwise, the apparatus in Figure 2 is identical to that in Figure 1.

The sparged cells are stainless-steel tubes, 18 cm in length and 2.2-cm ID. The spargers are sintered-stainless-steel disks. Two-cm thick layers of pyrex-glass wool, supported by perforated stainless-steel disks, are placed at the outlets of each sparged cell to eliminate possible

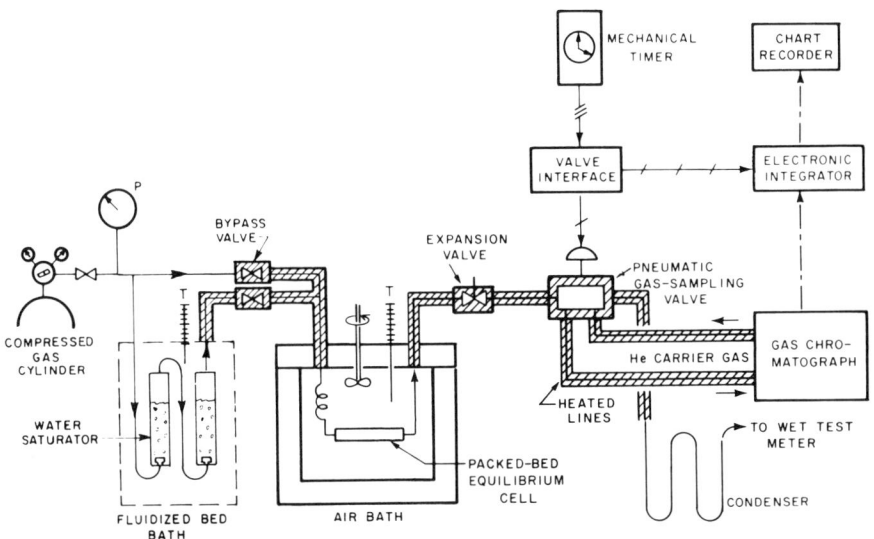

Figure 2. Total-vaporization apparatus for measuring the solubility of a heavy fossil-fuel fraction a in a compressed gas containing water.

entrainment. Two sparged cells are connected in series to assure saturation. These cells are placed in a fluidized-bed bath. Temperatures are measured as described previously.

DATA REDUCTION

Figure 3 presents experimental results of two total-vaporization experiments with normal-hexadecane in methane at two different gas flow rates. The results of intermittent gas-phase analysis are plotted against the corresponding cumulative gas volume (at STP) having passed through the equilibrium cell. For this simple case, where a pure heavy liquid is vaporized, the vapor-phase solubility in methane can be obtained directly from the weight of hexadecane vaporized, $Wc16$, and from that of methane, $Wc1$, required to vaporize completely the hexadecane sample. The solubility, expressed as the equilibrium weight fraction, w, is given by

$$w = Wc16 / (Wc16 + Wc1).$$

Results obtained for this system are in close agreement with those obtained by conventional techniques and establish that equilibrium is achieved in the equilibrium cells at flow rates less that 5 cm^3 per minute.

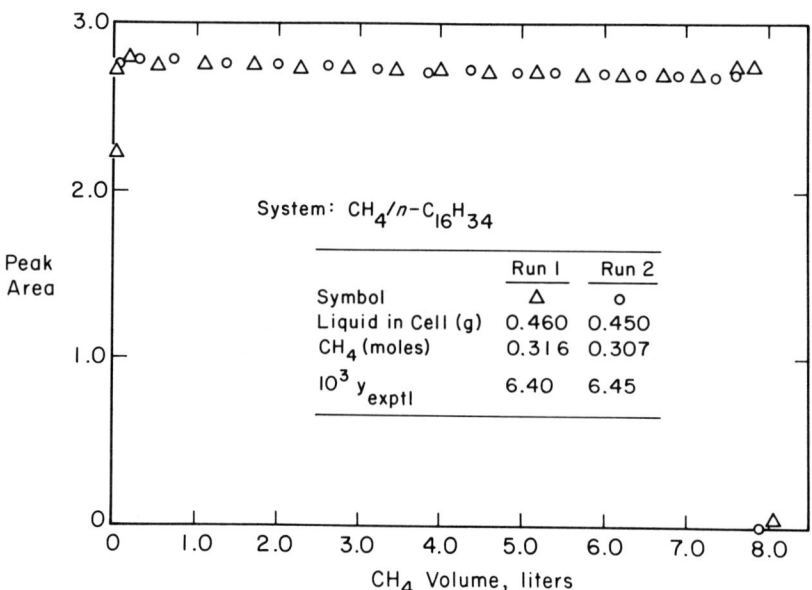

Figure 3. Total-vaporization measurement at 150°C and 2.36 bar.

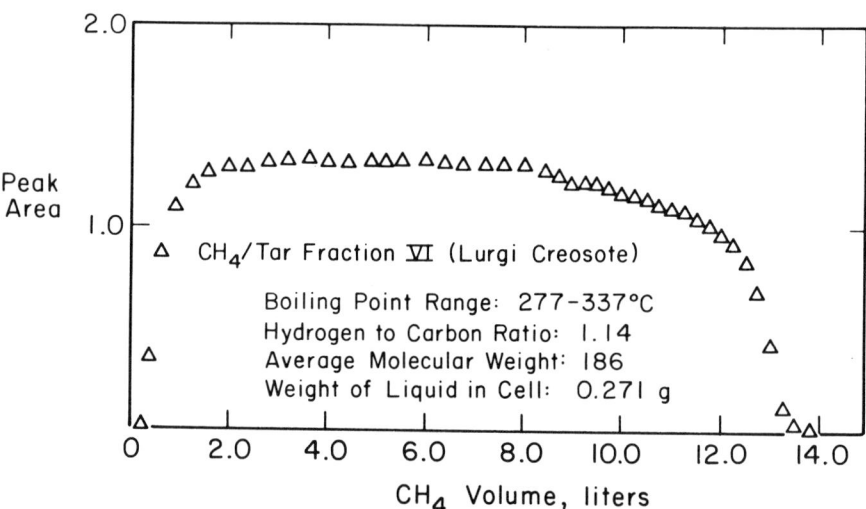

Figure 4. Total-vaporization measurement at 200°C and 40.8 bar.

Experimental results from the vaporization of a multicomponent liquid mixture from a packed bed cannot be interpreted as easily. Stripping effects cause the composition of the heavy liquid to change during the experiment. Figure 4 shows a typical set of data from a total-vaporization experiment with a Lurgi coal-tar fraction and methane. Figure 4 indicates three features characteristic of experiments with multicomponent mixtures;

1. Non-equilibrium start-up: the heavy-hydrocarbon content of the gas increases steeply from zero to the steady equilibrium value. This feature is more pronounced at elevated pressures where the effect of the dead volume in the apparatus is significant;
2. Steady equilibrium plateau: gas leaving the cell is saturated with heavy liquid whose composition has not changed from the original composition;
3. Transient equilibrium: during this period of gradually decreasing heavy hydrocarbon content, the gas is equilibrating with liquid whose composition is changing due to stripping effects where light components are preferentially removed from the equilibrium cell. This period ends with a sharper decrease in heavy-hydrocarbon content, signaling total vaporization.

Experimental conditions (e.g. initial load of heavy liquid) can be chosen to reduce to less than 2%, uncertainties arising from the start-up effects of Stage 1.

During Stage 3, equilibration takes place with a heavy liquid whose composition has been altered by preferential stripping of lighter components. Since the composition and overall properties of the mixture have been altered, the vaporization process is now more complex. The experimentally observed gas volume required for total vaporization, V, does not provide a direct measure of the solubility of the _original_ heavy liquid. To relate the experimentally observed V to y, the solubility of the original, _unaltered_ heavy liquid, we use the results of a computer simulation of the packed-bed multicomponent vaporization process.

In the total-vaporization computer simulation, the packed bed is modeled as 20 mixed cells (well-stirred vessels) in series. Vaporization occurs as about 100 gas increments successively equilibrate with the liquid (distributed equally between the cells initially) in each of the mixed cells. In the required, successive, flash calculations, it is assumed that for the components of a narrow-boiling mixture, the phase-equilibrium nonidealities can be lumped into a single, constant parameter, PI. The vapor-phase mole fraction of each of the mixture components, (y_i) is then

$$y_i = x_i * PSAT_i * PI / P$$

and the solubility of the entire mixture, y is;

$$y = \sum_i (x_i * PSAT_i * PI / P)$$

where x_i and $PSAT_i$ are the liquid-phase mole fraction and vapor pressure of component i respectively, and P is the pressure. Details of the computer simulation program and assumptions are presented elsewhere [Monge 1982].

The inputs to the simulation are: (1) a guess for PI or the solubility of the heavy fraction in the gas (an iterative procedure is used to determine the solubility which results in the best fit of the experimentally observed data showing peak area versus gas volume), (2) the area under the data showing observed peak area versus gas volume for the total-vaporization experiment, (3) the weight of the heavy liquid placed in the equilibrium cell, (4) the number-average molecular weight of the fraction, determined from freezing point depression measurements using a Model 5008 Petroleum Cryoscope obtained from Precision Systems Inc. [Alexander 1982], (5) the number and relative composition of the components in the fraction, approximated from gas chromatography of the fraction, as the number and relative areas of the "key" peaks or groups of peaks, and (6) the vapor pressure of the "keys", calculated [Smith 1976] from the aromaticity (assumed to be the same as that for the entire fraction and approximated from the fraction's hydrogen-to-carbon ratio obtained from elemental analysis) and approximate normal boiling points obtained by comparing the "key-peak" retention times to retention times for a calibration mixture containing compounds with known boiling points.

Figure 5 shows an example of the chromatographic analysis described in (5) and (6) for a Lurgi coal-tar fraction. The gas-chromatography column and detector used are those described in the Experimental Apparatus section. For narrow-boiling fractions, the column retention and detector response are assumed not to vary significantly with component normal-boiling point and aromaticity. Tests of the computer simulation indicate that the results are not very sensitive to the assumed number of mixed cells, number of "key" components or the absolute values of component normal-boiling points.

Figure 6 summarizes the procedure for data reduction. It has been tested on the experimental results of a total-vaporization experiment with a synthetic mixture containing known amounts of naphthalene, 2-methylnaphthalene, 1-ethylnaphthalene, and diphenylmethane in compressed methane. The experimental and simulation results are shown in Figure 7. Results uncorrected for stripping effects differ by 20% from those calculated from literature values for the properties of the individual components. Application of the proposed data-reduction procedure reduces this to less than 2%.

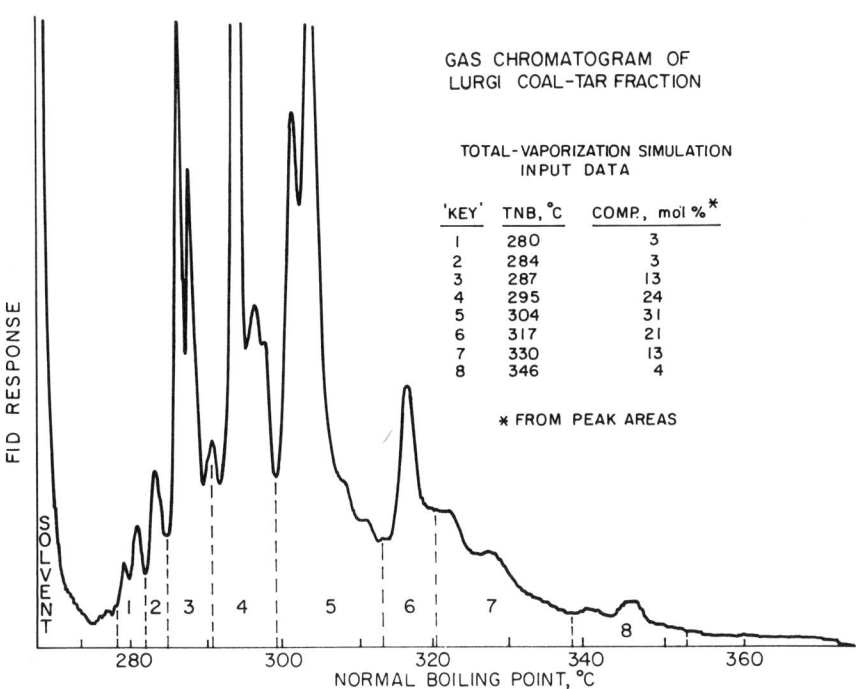

Figure 5. Gas-chromatography analysis of a Lurgi coal-tar fraction: input data for total-vaporization simulation and "rigorous" data reduction procedure.

DATA REDUCTION

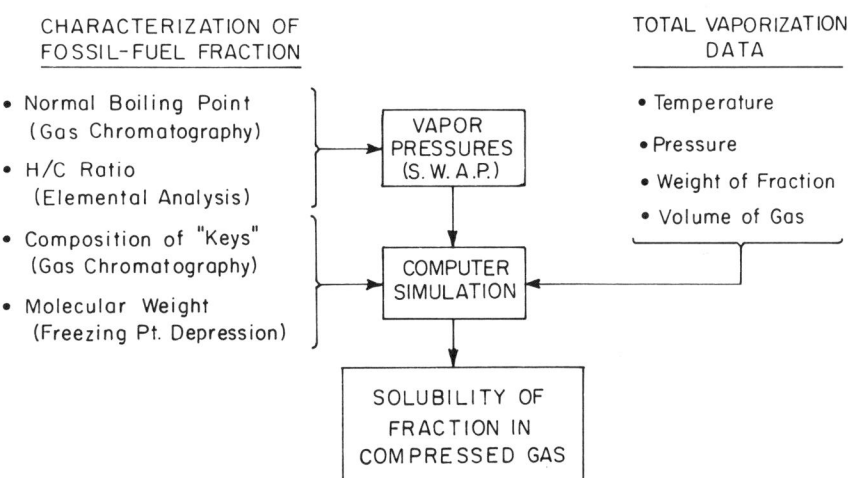

Figure 6. "Rigorous" data-reduction procedure for obtaining solubilities from total-vaporization data.

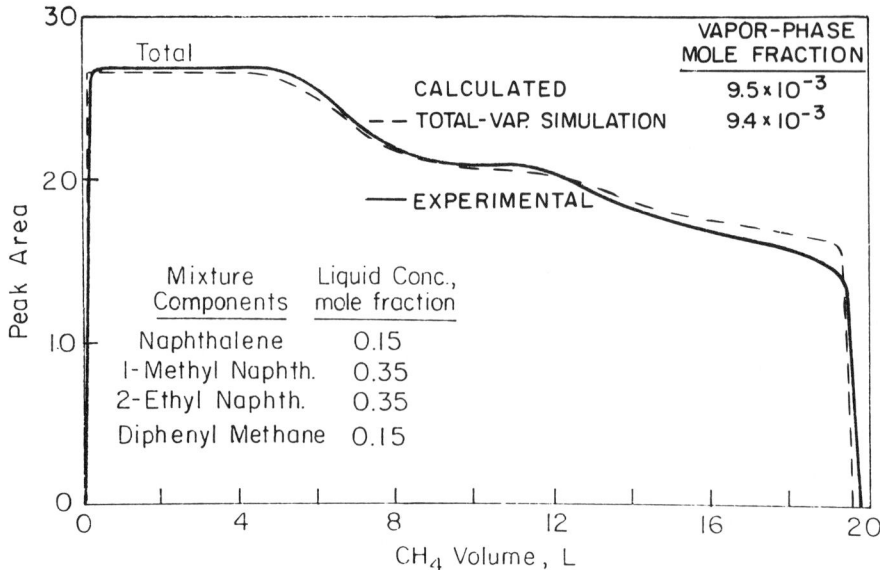

Figure 7. Comparison of total-vaporization experimental data and simulation results for a synthetic mixture at 125°C and 3.07 bar.

If a slightly less accurate solubility (in terms of weight per volume instead of mole fraction) is acceptable, it can be obtained directly from data showing the observed peak area versus gas volume without the computer simulation, molecular-weight measurement, or chromatographic analysis described above. In the approximate procedure, a simple equal-area analysis is used to determine the volume of gas that would be required for total vaporization of the heavy liquid sample if the composition and solubility of the liquid (and the observed peak area) had not changed from their initial, steady value. In Figure 8 the equal-area procedure has been applied to experimental results for the total-vaporization experiment for the synthetic mixture described previously. The "equivalent" volume of gas, Ve, and the weight of the heavy liquid sample, Wl, can be used to approximate directly the solubility of the heavy liquid (expressed as the weight of heavy liquid per volume of gas, v);

$$v = Wl \, / \, Ve.$$

Solubility v can be converted to mole fraction using the density of the gas and the molecular weight of the heavy liquid if it has been determined.

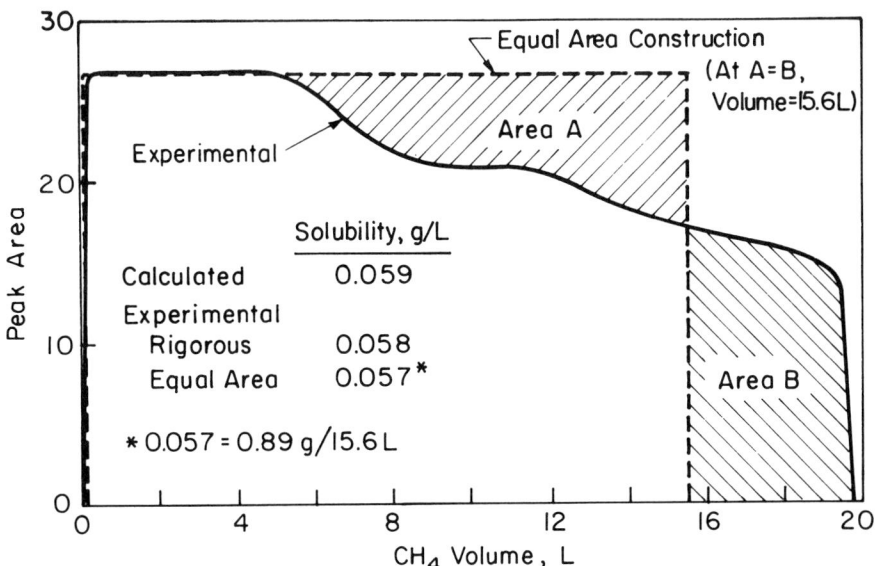

Figure 8. Equal-area construction for approximate reduction of total-vaporization data for a synthetic mixture at 125 °C and 3.07 bar: comparison with results of "rigorous" data reduction and calculated (predicted) results.

Although this approximate analysis has little theoretical basis, it has been found to give results which are within 5% of those obtained (for the synthetic mixture and actual coal-tar fractions) using the more rigorous data reduction procedure described earlier.

CONCLUSIONS

An experimental method has been developed for measuring solubilities of narrow-boiling heavy fossil-fuel fractions in compressed gases. The total-vaporization method, apparatus and data-reduction procedure have been tested with a pure heavy compound and a heavy synthetic mixture in methane. Experimental results for a pure liquid and for a synthetic mixture agree, to within 3%, with those calculated. Figure 9 shows the results of experimental measurements for two Lurgi coal-tar fractions in compressed methane. It also shows predicted results [Monge 1982]. The measurement for Fraction 1 at 42 bar and 200 °C has been repeated for a gas mixture containing 25 mol % water in methane. The effect of the water on the solubility was found to be negligible compared to experimental uncertainties. Further measurements are needed to determine the effects of water on solubilities of this and other mixtures at higher pressures and higher water concentrations.

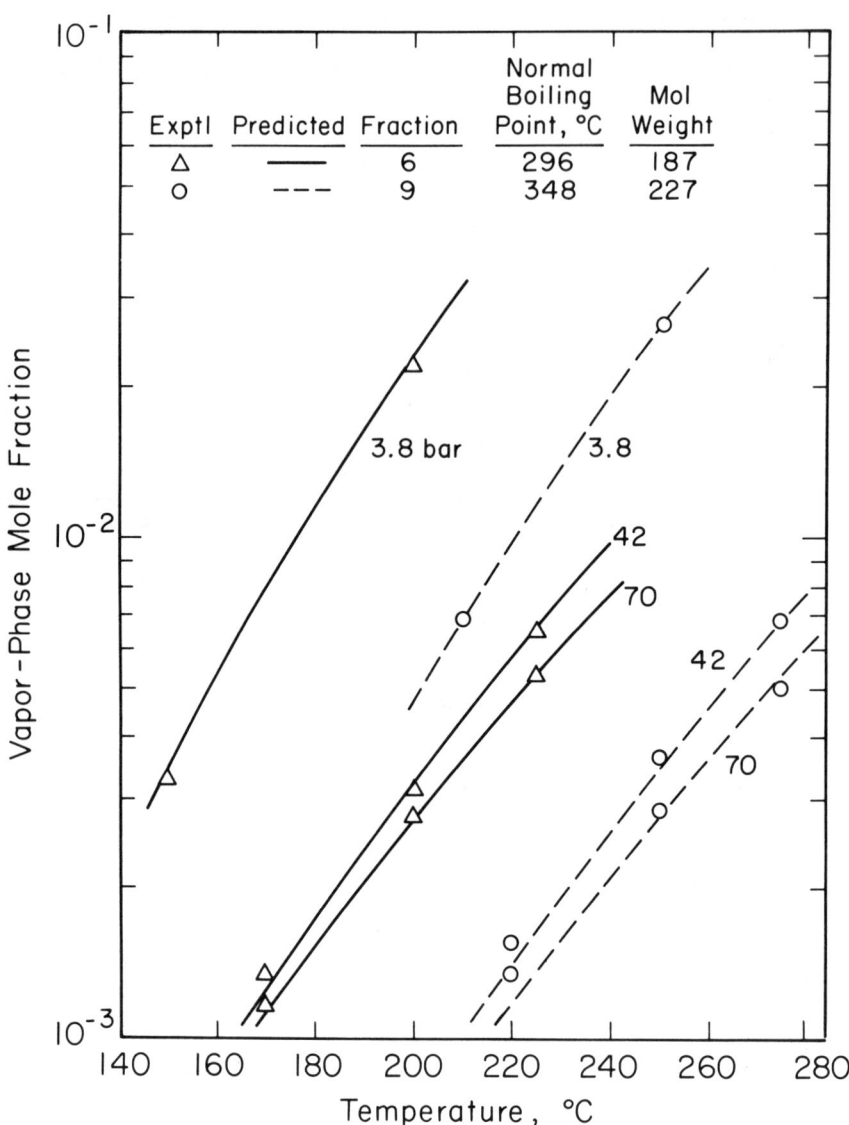

Figure 9. Solubilities of two Lurgi coal-tar fractions in methane.

ACKNOWLEDGMENTS

For financial support, the authors are grateful to the Fossil Energy Program, Assistant Secretary of Energy Technology, United States Department of Energy and the Materials and Molecular Research Division, Lawrence Berkeley Laboratory.

LITERATURE CITED

Alexander, G., and Prausnitz, J. M., AIChE Annual Meeting, November 1981, New Orleans, Louisiana

Johnston, K. P. and Eckert, C. A., "An Analytical-Carnahan-Starling-van der Waals Model for Solubility of Hydrocarbon Solids in Supercritical Fluids," AIChEJ., 27,773 (1981)

Kaul, B. K. and Prausnitz, J. M.,"Solubilities of Heavy Hydrocarbons in Compressed Methane, Ethane, and Ethylene: Dew-Point Temperatures for Gas Mixtures Containing Small and Large Molecules," AIChEJ., 24,223 (1978)

Kurnik,R. T., Holla, S. J., and Reid, R. C., "Solubilities of Solids in Supercritical Carbon Dioxide and Ethylene," J. Chem. Eng. Data, 26,47 (1981)

Monge, A., PhD Dissertation, University of California, Berkeley, 1982

Smith, G., Winnick, J., Abrahms, D. S., and Prausnitz, J. M., "Vapor Pressures of High-Boiling Complex Hydrocarbons," Can. J. Chem. Eng., 54,337 (1976)

Van Leer, R.A. and Paulaitis, M. E., "Solubilities of Phenol and Chlorinated Phenols in Supercritical Carbon Dioxide," J. Chem. Eng. Data ,25,257 (1980)

CHAPTER 8

MEASURING THE PROPERTIES OF
PETROLEUM RESERVOIR FLUIDS UP
TO 20,000 PSIA (138 MPa) AND
400°F (200°C)

R. Simon
Chevron Oil Field Research Company
La Habra, California

INTRODUCTION

The petroleum industry is extending its search for oil and gas to deeper reservoirs having higher pressures and temperatures. Recent discoveries have reservoir pressures up to 18,000 psia (124 MPa) and temperatures up to 370°F (188°C) (SI units will be used hereafter in this paper). In order to optimize production from such reservoirs, it is necessary to know the fluid properties at reservoir conditions.

Existing industrial equipment for measuring fluid properties is generally limited to 70-80 MPa and 150°C. Thus, there is a need for equipment that can make measurements at much higher pressures and temperatures to provide engineering data for reservoir studies and also extend research capabilities.

To satisfy this need, Chevron Oil Field Research Company has built a system for operation at 138 MPa and 200°C. The following describes the highlights of this system, and the methods used for measurements. All of this is related to the subject of this symposium: "Chemical Engineering at Supercritical Fluid Conditions."

DESCRIPTION

Chevron's system (Figure 1) is designed to measure the quantities, compositions, densities, viscosities, and interfacial tensions of the vapor and liquid phases being studied. It can also be used to prepare gases and oils for displacement tests.

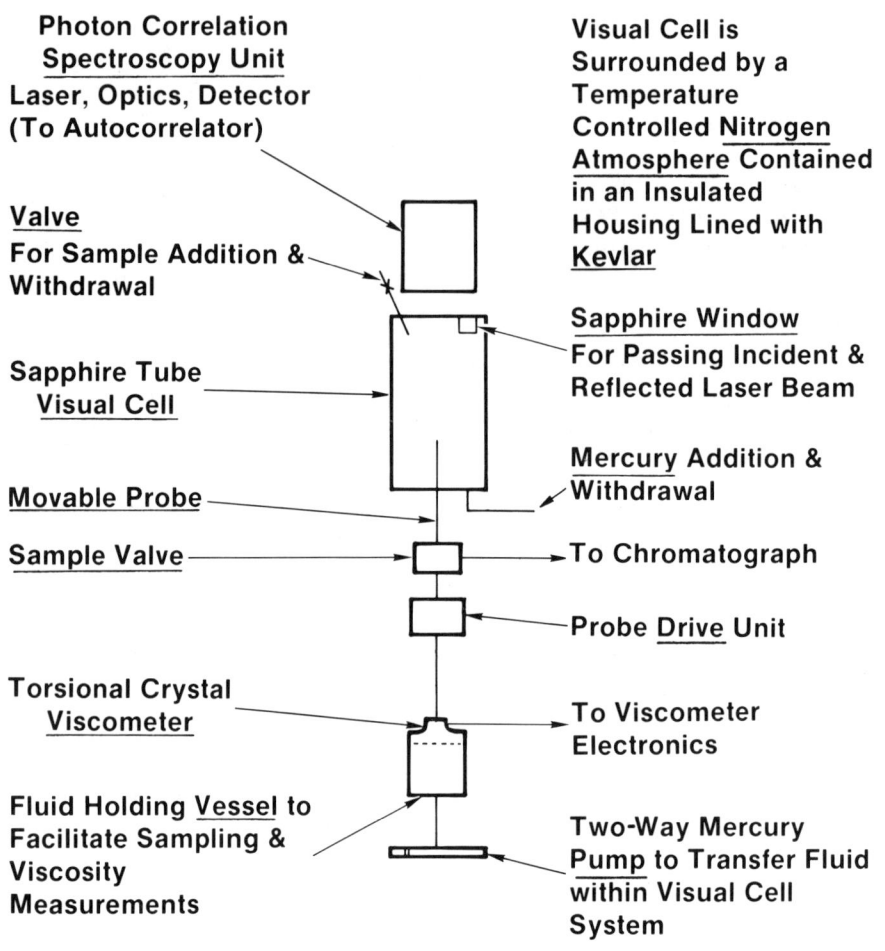

Figure 1. Schematic Drawing of Chevron Oil Field Research Co. Visual Cell System.

The heart of the system is a cell whose contents are visible. This is important for condensate systems which tend to occur in high pressure reservoirs. Such systems cannot be accurately studied in a "blind" cell. In addition, the visual cell is valuable for measuring relative phase volumes, finding bubble and dew points, examining multiple liquid phases, detecting solids that form (such as asphaltenes and hydrates), looking for emulsions that complicate measurement, and - perhaps most important - observing the unexpected.

Visual Cell Construction

A visual cell meeting our requirements was not commercially available. To increase our chances of developing one we devised three different types (Figure 2a, b, c):

The first (2a), which received the most attention, is an unsupported sapphire tube with an internal volume of 50 cubic centimeters. It is essential to have a sapphire tube manufactured from a single crystal, of 0° orientation. The tube is specially cored from the boule, drilled along the central axis, and polished by mechanical and chemical methods. It is glazed to eliminate surface imperfections, add strength, and minimize the adverse effects of moisture and acid gases. Surface defects would decrease the tensile strength of the sapphire cylinder. We plan a separate paper describing the details of a sapphire tube suitable for 138-MPa operation.

The cell is sealed with conventional "O" and "backup" rings. Extreme care is required in fitting, polishing, and assembling the parts of these seals. We have learned that the best arrangement for us is to seal the ends of the tube. Sealing the inside diameter restricted the path of the laser beam discussed later. Sealing the outside results in excessive force on the ends of the cell assembly, stretching it and making sealing difficult.

The effective volume of the cell is controlled with a mercury piston. The vapor pressure of the mercury and its mutual solubilities with reservoir oils and gases are neglible when obtaining petroleum engineering data.

The second type of cell (2b) has sapphire windows and is sealed with gaskets. The windows are considered optically flat. The long axis of the window is the 0° axis of the sapphire crystal. The gasket is copper. Thermal expansion problems are minimized by using Incoloy 903 with an expansion coefficient near that of sapphire.

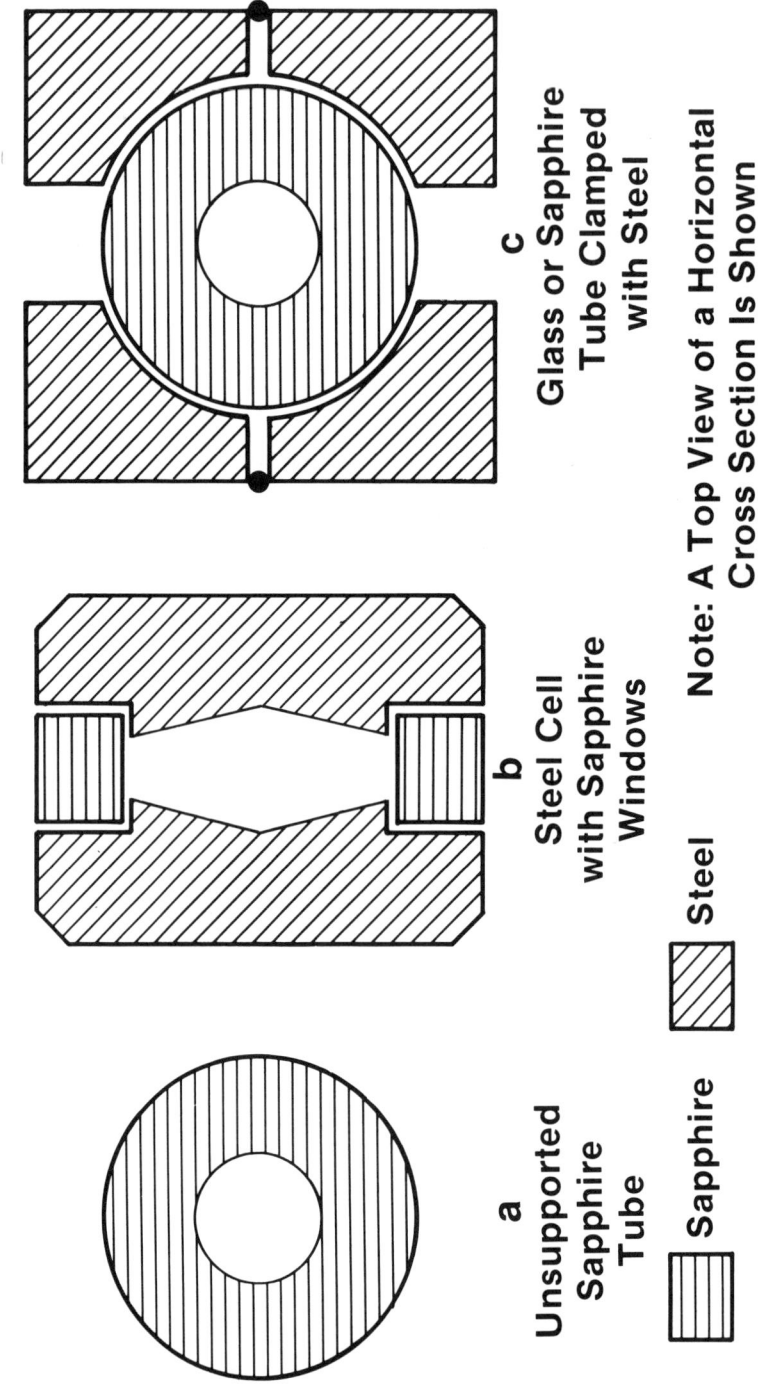

Figure 2. Principal Features of 3 Types of Visual Cells.

The third cell (2c) configuration was originally suggested by Prof. E. U. Franck of the University of Karlsruhe, and modified by A. B. Mason of Chevron. The clamping system is designed to reduce the tensile stress on the cell. It is still being evaluated.

Methods of Measurement

Compositions of the phases are measured by using mercury to displace the fluid samples (usually about 3 µl) through a specially designed probe and valve, then conveying the sample with helium or special solvents to a gas chromatograph adjacent to and integrated with the visual cell system. The sampling probe is designed to move to any height within the cell and withdraw a sample from any phase. The sampling valve has two main features: fluid from the visual cell can be flushed through to minimize contamination from previous samples; and the sample volume can be adjusted from 1 to 300 µl. The sample probe will bend under axial load and tend to cut its packing. A satisfactory arrangement has been devised by superpolishing the probe, chrome plating it, and using Polypak packing. The details of the sampling valve are covered in a patent application.

The chromatograph gives the compositions of vapors directly. For oils containing a $C_{33}+$ fraction, chromatographic data are combined with measurements on a C_7+ liquid to calculate the $C_{33}+$ mole weight and density. Using microliter samples minimizes changes in overall composition and reduces the need to recharge the cell for subsequent measurements. Useful sampling ideas come from Legret et al. (1981).

Densities of the phases in the visual cell were originally intended to be measured by gamma ray transmission (Levert, et al. (1973), Drotning (1979)). The intensity of gamma rays transmitted through the cell is a function of the electron density (therefore the actual density) of the material in the cell. This technique would not require withdrawal of any fluid; however, its use was postponed to expedite completion of construction. The interim procedure for measuring density will require withdrawing a known volume of fluid at cell conditions and weighing the displaced material.

To measure viscosities, we rejected the rolling ball, capillary tube, and torsional disk methods because they require relatively large samples and separate pieces of equipment. Instead we focused on a torsional crystal located within the visual cell system. This technique has been reported by the U.S. National Bureau of Standards and California Institute of Technology (Haynes (1973), Strumpf et al. (1974)). It employs a piezoelectric quartz crystal in a tuned, resonant electronic circuit. The resonance of the circuit is related to the viscosity of the fluid surrounding the crystal. The viscosity provides an additional damping force on the crystal oscillation which changes the parameters of the resonant circuit.

In operation, fluids are transferred from the visual cell to the viscometer chamber (see Figure 1), where the crystal and its associated electronics provide the information needed to calculate viscosity. After the measurement, the fluid is returned to the visual cell, to avoid disturbing the overall fluid composition.

Devising a crystal operable in a mercury environment was an additional problem that we plan to describe in a separate paper. Data obtained thus far have agreed satisfactorily with those from other techniques, and are suitable for engineering purposes.

Interfacial tension between phases is measured by the pendent drop technique (described in detail by Schoettle and Jennings (1968)). Measurements between gas and oil require inverting the visual cell, displacing a drop of oil from the tip of the probe, photographing the drop, and measuring the volume. This plus the gas and oil densities are used to calculate the interfacial tension. Gas-water and oil-water interfacial tensions can be determined in a similar manner without inverting the visual cell. After the measurement, the fluids are returned to the cell, so as not to alter the overall composition. We are testing another method: counting bubbles of the less-dense fluid rising through the more-dense fluid at a constant flow rate (Ohsawa and Ozaki (1981)). This can be done in the existing cell.

Independent measurements of viscosity and interfacial tension will also be made with photon correlation spectroscopy (Langevin et al. (1976), Herpin et al. (1974)). In this method a helium-neon laser beam is focused and split before being transmitted into the visual cell through a special sapphire window (see Figure 1). The beams that are scattered and reflected from the vapor-liquid interface leave the cell through the same window and enter the PCS detection system.

The PCS electronic system measures the intensity fluctuations of the light scattered by the movement of the vapor-liquid interface, and provides the power vs. frequency spectrum used to calculate viscosity and interfacial tension. It is necessary to know the densities of both phases. Chevron's PCS unit was fabricated by Langley-Ford Instruments, Inc. Photon correlation spectroscopy does not alter the composition of the material in the cell.

To our knowledge this method has not been previously used for petroleum reservoir fluids at reservoir conditions. Our results to date are entirely suitable for reservoir engineering purposes. We also plan to describe this application of photon correlation spectroscopy in a separate paper.

Fluid <u>pressure</u> in the visual cell is controlled by changing the fluid volume with a mercury piston. In this widely used procedure mercury is either pumped in or withdrawn from the cell as required.

Fluid <u>temperature</u> is controlled with a gaseous nitrogen bath surrounding the cell. The nitrogen is heated or cooled to the desired temperature and circulated by a blower within an insulated housing (oven).

Safety Features

Safety features of the visual cell system warrant special mention. The main objective is to guard against explosive failure of the pressurized equipment and the subsequent release of projectiles and combustible hydrocarbons.

At the start of each experiment the system will be tested above the maximum working pressure. This reduces the likelihood of sudden failure of the cell, heads, seals, tubes, valves, etc.

The temperature bath has a nitrogen atmosphere rather than air, to prevent a combustible or explosive mixture from forming in the oven surrounding the cell. Also a blowout panel in the oven wall minimizes the chance that the doors will blow off.

The entire oven wall, except for the windows, is lined with duPont's Kevlar. This "bullet-proof" fabric reduces the likelihood of parts being shot out by an explosion and endangering the operator.

The nitrogen atmosphere surrounding the visual cell is kept under a slight vacuum to prevent releasing mercury vapors into the laboratory.

STATUS - November 1981

The separate parts of the system have operated successfully. We are currently bringing them together to handle Chevron's oil field fluid property technical service and research efficiently over the design range of conditions.

FUTURE

The continuing evolution of fluid property measurements will involve:

> Miniaturizing equipment and samples to reduce the problems inherent in constructing massive equipment and preparing large samples.

> Measuring with nondestructive methods to prevent altering the overall composition of the fluids. This improves accuracy and reduces the time required for a series of measurements that start with the same fluid mixture.

> Increasing efficiency, which is important for all laboratories, particularly commercial organizations. This can be accomplished by integrating various measurements into one system and providing appropriate computer support.

The improvements mentioned above do not come easily; but require years of dedicated, painstaking work. This is a valid field for the attention of government laboratories such as the National Bureau of Standards. The results will be valuable to many American industries.

LITERATURE CITED

Drotning, W. D., "Thermal Expansion and Density Measurements of Molten and Solid Materials at High Temperatures by the Gamma Attenuation Technique," Report 79-0074, Sandia Laboratories (1979).

Haynes, W. M., "Viscosity of Gaseous and Liquid Argon," Physics, 67, 440 (1973).

Herpin, J. C. and J. Meunier, "Spectral Study of Light Scattering from the Thermal Fluctuations of the Liquid-Vapor Interface of CO_2 Near its Critical Point. Surface Tension and Viscosity Measurements," J. de Phys., 35, 857 (1974).

Langevin, D. and J. Meunier, "Light Scattering by Liquid Surfaces," in Photon Correlation Spectroscopy and Velocimetry, ed by H. Z. Cummins and E. R. Pike, Plenum Press, NY (1976).

Legret, D., D. Richon, and H. Renon, "Vapor Liquid Equilibria up to 100 MPa: A New Apparatus," AIChE Jour., 27 (2), 203-207 (1981).

Levert, F. E., I. G. Dillon, and H. J. Tarng, "High Temperature Density Apparatus Using the Photon Attenuation Technique," Review Sci. Instruments, 44, 313 (1973).

Ohsawa, T., and T. Ozaki, "New Method for Determination of Surface Tension of Liquids," Review Sci. Instruments, 52(4), 590 (1981).

Schoettle, V. and H. Y. Jennings, "High-Pressure High-Temperature Visual Cell for Interfacial Tension Measurements," Review of Scientific Instruments, 39, 386 (1968).

Strumpf, H. J., A. F. Collings, and C. J. Pings, "Viscosity of Xenon and Ethane in the Critical Region," J. Chem. Phy., 60, 3109 (1974).

OVERVIEW

PART II. THERMODYNAMIC THEORIES
AND EQUATIONS OF STATE

Ralph D. Gray, Jr.
Exxon Research and Engineering Co.
Florham Park, NJ 07932

In Part II, the emphasis shifts from experimental work to theoretical developments. Chapters 9 through 16 deal with efforts to improve understanding of the complex fluid phase equilibria of interest in supercritical extraction processes. These efforts range from fundamental studies of the underlying physical phenomena to applications-directed work on representation of fluid-phase equilibria. These eight chapters provide a representative cross-section of the direction of current work in this area.

Chapter 9 reviews recent progress in understanding the nature of tricritical points, which occur in mixtures involving three or more components. Chapter 10 describes somewhat related work in adapting the concept of field variables (originally developed to characterize mixture critical phenomena) to represent phase equilibria in a much more extended region near the mixture critical locus. These two chapters contribute significantly to improved understanding of the details of phase behavior in the mixture critical region. This behavior is of considerable practical interest because it sets limits on operating conditions for supercritical separation processes.

Chapter 11 is devoted to the application of perturbation theory to study the dilute solution behavior of a supercritical gas dissolved in a liquid, as well as that of a liquid dissolved in a supercritical gas. This technique is applied to examine the role of several types of intermolecular forces (central, dipolar, and quadrupolar), as well as the effect of molecular shape. Chapter 12 describes the application of another statistical thermodynamic model, that of a mean field lattice gas, with particular emphasis on adjustments necessary to represent gas-gas equilibria. These papers demonstrate progress toward developing a molecular thermodynamic basis for representation of phase behavior of systems involving supercritical fluids.

Chapter 13 is an exhaustive study of the usefulness of the popular Peng-Robinson cubic equation of state in representing phase equilibria to evaluate supercritical extraction processes. In this work, the equation is used to generate semiquantitative P-T-x-y diagrams over the entire range of fluid phase behavior. Chapter 14 shows how modifications in mixing rules for a cubic equation of state can improve the accuracy of phase equilibrium calculations, particularly for mixtures with a wide disparity in molecular size that are typical of many supercritical extraction applications. These two chapters illustrate the current emphasis on cubic equations of state in supercritical extraction applications and give a good indication of the advantages and disadvantages of this approach.

In Chapter 15, a four-parameter corresponding states method is used to describe mixture critical phenomena and critical region phase behavior. The primary advantage of this method in comparison with a cubic equation of state is its improved accuracy in predicting density. Chapter 16 describes a new approach to corresponding states theory, in which polymer scaling theories are adapted to correlate equation-of-state data for asymmetric mixtures involving chain molecules at supercritical conditions. These two chapters suggest that corresponding states methods, which have generally been less popular than equation-of-state methods in representing phase behavior in supercritical fluid systems, deserve more consideration.

Taken together, all eight chapters demonstrate impressive continuing progress in understanding the complex fluid phase behavior in supercritical systems and in translating this understanding into methods for process application. They also demonstrate the importance of understanding the intricacies of phase behavior in this systems in order to determine process feasibility and to set process conditions for supercritical extraction. In view of this, applied thermodynamics will undoubtedly continue to play a key role in the development of this emerging process technology.

CHAPTER 9
THREE-PHASE EQUILIBRIUM AND THE TRICRITICAL POINT

B. Widom
Department of Chemistry
Cornell University
Ithaca, New York 14853

I shall tell you of some of the work we have been doing on three-phase equilibrium and the tricritical point, and I shall at the same time tell you something of the history of the subject as I saw it or learned of it.

The modern theoretical ideas about the tricritical point started with Griffiths's[1] explanation of the experimental results on the phase transitions in He^3-He^4 mixtures. In Fig. 1 we see the temperature-composition phase diagram, with x the fraction of He^3. The line of λ-points descends to the top of the coexistence curve, at which the transition becomes first order and below which there is separation into two coexisting phases. Note that the coexistence curve is not rounded at its top; its two branches meet at an angle at the critical solution point. Griffiths was able to account for this and other associated behavior by recognizing that the thermodynamics really require a third dimension in Fig. 1, shown as the axis marked ζ. That ζ is the thermodynamic conjugate of the superfluid order parameter; it may be thought of as a field that couples to that order parameter and distinguishes two different values of it, just as a magnetic field would distinguish two opposite senses of magnetization. Griffiths also showed the equivalent of Fig. 1 in an alternative representation, Fig. 2. Here the composition x has been replaced by its thermodynamic conjugate field, the difference μ_3-μ_4 of the chemical potentials of the two species, now plotted vertically, while the temperature is plotted horizontally. The "wings" are surfaces of first-order transitions, as is that part of the μ_3-μ_4, T plane beneath the phase-transition curve. The curves shown hatched are critical lines; and the critical solution point of Fig. 1 is seen then to be a point of confluence of three critical lines. For this reason it was named the tricritical point by Griffiths, who then showed that the classical, phenomenological theory of such a point, due to Landau [2], could account for the experimental results, including the striking appearance in Fig. 1 of a coexistence curve that is not rounded at its apex (the critical solution point), but is such that its two branches intersect there at an angle less than $180°$.

There the matter stood, for me, in 1972, when my colleague ME Fisher asked me if I knew of any examples in classical, multicomponent fluids of the confluence of three critical lines; i.e., of a tricritical point. I replied that I did not; but I was unsure of how such a phenomenon would manifest itself, and therefore of whether I would have recognized it even if I had seen it. Then, early in 1973, RB Griffiths came to spend some time with us in our Department, and he asked me exactly the same question Fisher had asked me some months before. Again I replied that I knew of no examples. Soon thereafter I saw a paper in the Russian Journal of Physical Chemistry (the English translation of the Zhurnal Fizicheskoi Khimii), reporting a "higher order critical point", a point at which three coexisting phases simultaneously become identical — thus, the extension of the idea of an ordinary critical point, where two previously distinct phases become identical. I asked Griffiths whether this higher order

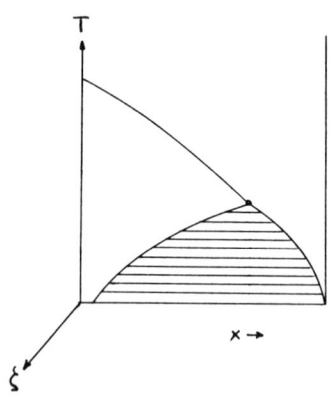

Figure 1.

critical point was anything like what he had in mind when he asked me his question, and he replied that it was <u>exactly</u> what he had in mind! That, then, in one stroke, opened for us the substantial literature and history of what we now call tricritical points, and which we define as the limit of three-phase coexistence, where three previously distinct phases all become identical simultaneously. (Thus, the "tri-" in tricritical now refers to the three phases rather than to the three critical lines. The former is the simpler conception. Fortunately, only its definition changed, but the word itself has remained the same!).

In Fig. 3 we see, schematically, three such phases, α, β, and γ. The tricritical point is that at which the $\alpha\beta$ and $\beta\gamma$ interfaces simultaneously disappear. Returning to Fig. 2, the region between the two wings is now marked β, the region beneath the far wing, at negative ζ, is called γ, and the region beneath the near wing, at positive ζ, is called α. Those are one-phase regions, and the previously identified first-order-phase-transition surfaces are the two-phase coexistence surfaces that separate those regions. The curve in which they intersect, previously thought of as the locus of first-order transitions, is now recognized to be the line of triple points of $\alpha\beta\gamma$ phase equilibrium. The critical lines, shown hatched, are the loci of critical points of $\alpha\beta$, $\beta\gamma$, and $\alpha\gamma$ phase equilibria.

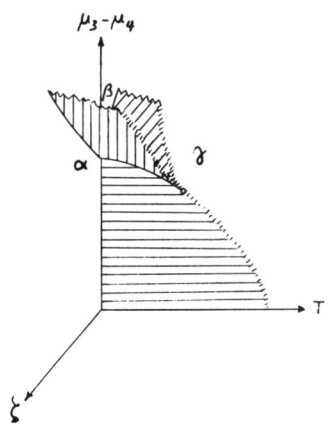

Figure 2.

It is as though the field distinguished two mirror-image superfluid phases, in which the values of the order parameter were equal in magnitude but opposite in sign; and, therefore, as though the two-phase region in the T-x plane of Fig. 1 were really a three-phase region (corresponding to what we have now identified as the locus of triple points in Fig. 2).

From the Russian literature we found references to a theoretical discussion of the possibility of such tricritical points as early as 1926, in the <u>Handbuch</u> article of Kohnstamm [3], but I have sinced learned from Professor EGD Cohen that the idea goes back to van der Waals, and may be found in van der Waals-Kohnstamm [4] in 1912. Wherever I have worked in the subject of fluids — dense liquids, critical points, or interfaces — I have found that the basic ideas all go back to van der Waals!

Van der Waals and Kohnstamm recognized that for a fluid mixture to have a tricritical point it must consist of at least three independent chemical components. A general argument along these lines for a critical point of pth order, where p phases become identical, was given by J Zernike [5] in 1949. (He is not the F Zernike of the Ornstein-Zernike theory. J Zernike, also Dutch, was the Professor of Chemistry at the University of Dacca, East Pakistan, and was later the author of the important work, "Chemical phase theory"[6]). According to the Gibbs phase rule, the number of thermodynamic degrees of freedom, f, the number of independent chemical components, c, and the number of coexisting phases, p, are related by $f = c - p + 2$; but, if the p phases are all to be identical, we have p - 1 additional constraints: there are p - 1 interfaces that have to be specified as disappearing, i.e., p-1 conditions of criticality. Therefore,

$$f = c - p + 2 - (p - 1) = c - 2p + 3 .$$

But $f \geq 0$, so $c \geq 2p - 3$.

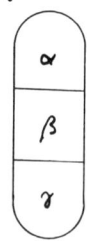

Figure 3.

Thus, for an ordinary, two-phase critical point we must have at least one component, which is true, but trivial; for a tricritical point we must have at least three chemical components, as van der Waals and Kohnstamm knew; for a fourth-order critical point we must have $c \geq 5$; etc.

Through the 1960's and early 1970's the experimental study of tricritical points in classical fluid mixtures was very much a Russian subject — but, even there, had two independent lines of development. The experimental discovery of tricritical points has been ascribed to GD Efremova (who died in 1972, at the age of 56), in an obituary that was written by her colleagues, among whom was Professor IR Krichevskii [7]. In 1961, Efremova and Pryanikova [8] had seen both liquid-gas and liquid-liquid critical end points in mixtures of acetic acid, water, and n-butane, at temperatures above the critical temperature of pure butane. Referring to Fig. 3, we may think of α as vapor and β and γ as two liquids (though, of course, at or near the tricritical point there is no fundamental distinction between liquid and vapor). When one pair of these phases is critical in the presence of the third phase, which remains distinct, that thermodynamic state is said to be a critical end point. What Efremova and Pryanikova had observed, then, were $\beta\gamma$ (liquid-liquid) and $\alpha\beta$ (liquid-vapor) critical end points, at the former of which the $\beta\gamma$ interface disappears as the β and γ phases become identical, and at the latter of which the $\alpha\beta$ interface disappears as the α and β phases become identical. Then, in their historic 1963 paper [9], Krichevskii et al. reported a liquid-liquid critical end point at $189.2°$, and a liquid-gas critical end point at $189.5°$, at both of which all three phases were simultaneously opalescent. This showed the close proximity of the tricritical point, where the two critical end points would coincide. They estimated the tricritical-point temperature to be $190°$. It was clear that these researchers understood the basic phase relations, and it was from them also that we had the reference to Kohnstamm and thence to van der Waals.

Probably even earlier, and in what appears, remarkably, to have been an independent development, tricritical points were also found by another group in the Soviet Union. We have seen that we must have at least three chemical components to have a tricritical point; but if we have only three, then the system is invariant at its tricritical point, so that

the temperature, pressure, and chemical composition are all
uniquely determined. If there were an additional chemical
component, c = 4, then we could have a tricritical end point,
at which three phases were at their tricritical point while in
equilibrium with a fourth, distinct phase. In this way we could
have a liquid-liquid-liquid tricritical point, in which the tricritical fluid was an ordinary, dense liquid in equilibrium
with its vapor. It was just such a tricritical point that was
reported in another historic paper, in 1962, by Radyshevskaya,
Nikurashina, and Mertslin [10]. They noted that Mochalov
had earlier observed three liquid phases in benzene-ethanol-
water-ammonium sulfate mixtures (as reported by Mertslin and
Mochalov [11] in 1959); and they then reported locating the tricritical point in that system. (I was informed by KK Il'in, who
was working under the supervision of NI Nikurashina, having
formerly been a student of the late Professor Mertslin (who
died in 1971), that the Mertslin group had already studied the
three-phase region in that system, isothermally, as early as
1950.) This system, which we learned of from the 1962 paper
of the Russian group, is the one in which we did our own first
studies of the tricritical point.

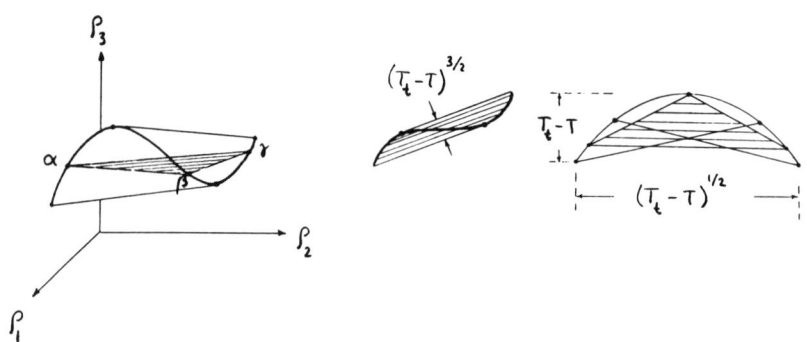

Figure 4.

Griffiths [12] made the appropriate extension of the
classical Landau theory to these multicomponent fluids, in
which, unlike in the He^3-He^4 mixtures, there are no special
symmetries so that the thermodynamic space is four-dimensional. When $c = 3$ the independent thermodynamic variables
may be taken to be the three chemical potentials and the
temperature, for example, or three densities, ρ_1, ρ_2, ρ_3, and
the temperature. For a fixed temperature the region of
three-phase coexistence may be pictured in the ρ_1, ρ_2, ρ_3 space
as in the first part of Fig. 4. The points labeled α, β, and γ
give the compositions of one set of coexisting phases as the
vertices of a triangle, which is shown shaded in the figure.
The whole three-phase region at this temperature consists of
a stack of such triangles. The curve in the figure is the locus
of their vertices. As one moves upward in the stack of triangles
(in the direction of increasing ρ_3, as the figure is drawn), the
length of the $\alpha\beta$ side of the triangle shrinks to a point — the
$\alpha\beta$ critical end point — and the triangle there degenerates to a
line, as shown, which is the critical-end-point tieline between
the $\alpha\beta$ critical end point and the distinct γ phase with which
that critical phase is in equilibrium. If, instead, we move
downward in the stack of triangles, it is the $\beta\gamma$ side of the triangle that shrinks to a point — the $\beta\gamma$ critical end point — and
there the triangle has degenerated to another critical-end-point tieline, connecting a distinct α phase with the $\beta\gamma$ critical
phase. The whole three-phase region in the isothermal composition space is then bounded by the $\alpha\beta\gamma$ locus and the two critical-end-point tielines, as shown.

In the second part of Fig. 4 we view the stack of triangles edge-on, so that they appear as a set of parallel lines,
and the locus of the vertices of the triangles appears as the
inflected (cubic) curve. The two dots on the curve are the
two critical end points. From this perspective we see the
thickness of the stack of triangles. According to the Griffiths
theory this thickness vanishes proportionally to $(T_t - T)^{3/2}$, where
T_t is the tricritical-point temperature. Viewed from a direction
perpendicular to that from which we just viewed it, the three-phase region appears as shown in the third part of Fig. 4. The
locus of the vertices of the composition triangles now appears
parabolic. We show in this view one of the triangles — the
isosceles triangle, in the middle of the stack — and the two

critical-end-point tielines. The width of this figure, according to the Griffiths theory, vanishes proportionally to $(T_t-T)^{1/2}$, and its height vanishes proportionally to T_t-T. The three different dimensions of the isothermal three-phase region shown in the first part of Fig. 4 are thus predicted to vanish as $(T_t-T)^{3/2}$, $(T_t-T)^1$, and $(T_t-T)^{1/2}$. This defines three tricritical-point exponents associated with the shrinking of the three-phase region on approach to the tricritical point, with 1/2, 1, and 3/2 the predicted values of these exponents. These predictions have been fairly well verified in the measurements that we and others have made.

I shall now tell you of some of the particular systems we and others have worked with in our qualitative and quantitative studies of tricritical points.

In our laboratory we have studied three systems so far, all of them four-component liquids in equilibrium with vapor. The first of these was the system benzene-ethanol-water-ammonium sulfate, which we had learned of from the Russian work I mentioned before. Our experiments were designed and carried out by Dr. John C Lang [13]. The principle here is that the first, second, and third components on that list are related to each other in mutual solubility in the same way as are the second, third, and fourth. Thus, benzene and water are nearly insoluble in each other while ethanol is a good common solvent for both; while ethanol and the salt are highly insoluble in each other, with water a good common solvent for both. Note that the ethanol and the water interchange their roles in the two ternary subsystems. The second of our systems was that studied by Dr. Peter Bocko [14]: n-hexane-benzene-acetonitrile-water. We have here again the same solubility relations as before: of the first three components, benzene is a good solvent for both hexane and acetonitrile, which are relatively insoluble in each other, while acetonitrile, in turn, is soluble in both benzene and water, which, alone, would be virtually insoluble in each other. The third system, studied by Mr Michael Lovellette [15], was isooctane-n-butanol-ethanol-water. Here, fulfilling the analogy, the butanol and water are relatively insoluble in each other while ethanol is a good solvent for both; but, in apparent violation of the analogy, isooctane and ethanol, while indeed both soluble in butanol, are also soluble in each other — that is what "gasahol" is all about! We were therefore relieved to discover that there is a reported [16] upper critical solution temperature for isooctane-ethanol mixtures at $-70°$; so that, even though the isooctane and ethanol are soluble in each other in all proportions at the higher temperatures at which we studied the system, there is indeed a latent insolubility which makes this system again understandable and in the same class with the others.

Further studies of analogous systems are being made in our laboratory by Mr G Sundar, and we hope as the result of his work to have greatly extended the catalog of known tricritical points in four-component liquids.

I would like also to call your attention to the beautiful experiments of Creek, Knobler, and Scott [17] on the three-component mixture methane+ 2,2-dimethylbutane (or n-pentane)+ 2,3-dimethylbutane. Here methane and 2,3-dimethylbutane, by themselves, have a small three-phase region near the critical point of pure methane, with two critical endpoints that lie near each other; while methane with 2,2-dimethylbutane (or with n-pentane) does not have a three-phase region; so with the right proportions of the two higher hydrocarbons one can approach the condition in which the extent of the three-phase region just becomes vanishingly small, the two critical end points then coinciding at a tricritical point.

When three phases are in equilibrium in the neighborhood of a tricritical point all three are simultaneously opalescent. That is the same as the critical opalescence due to fluctuations in composition or density that one sees near an ordinary critical point of two-phase equilibrium. In photographs of the opalescent phases in the systems studied by Lang, by Bocko, and by Creek, Knobler, and Scott, we see the characteristic blue-white opalescence due to reflected light or to light that is observed transversely to the direction from which it is incident, and also the characteristic yellow-brown opalescence as viewed by transmitted light. In every case the middle phase is the one that is most strongly opalescent, an observation of some theoretical significance, as we shall see.

There are three important questions connected with tricritical points that are at present the subject of active research and that will probably be the foci of concern and active interest for some years to come. The first is the nature of the light scattering — its intensity and angular dissymmetry — in the neighborhood of the tricritical point.

Gollub, Koenig, and Huang [18], in 1976, and Wu [19], in 1978, published early reports of light-scattering studies; and then, more recently, Kim, Goldburg, Esfandiari, Levelt Sengers, and Wu [20] published in 1980 the results of a very extensive study. Griffiths et al. [21] had derived from the earlier Griffiths theory [12] two sum rules for the "susceptibilities" (osmotic compressibilities) χ_α, χ_β, and χ_γ of the three coexisting phases:

$$\sqrt{\chi_\alpha} + \sqrt{\chi_\gamma} = \sqrt{\chi_\beta}$$

$$\frac{1}{\sqrt{\chi_\alpha}} + \frac{1}{\sqrt{\chi_\gamma}} - \frac{1}{\sqrt{\chi_\beta}} \sim T_t - T$$

These susceptibilities, in turn, determine the intensity, I, and the coherence length, ξ, of composition fluctuations, so Griffiths's sum rules predict relations among the I and ξ of the three phases. Note from the first of these relations that χ_β is always the greatest of the three susceptibilities, so the β phase, which is always that in the middle, is the most opalescent, in agreement with observation. But these relations are not otherwise quantitatively verified in the 1980 experiments. The sum rules are so fundamental an expression of the present theory that such a discrepancy, if it proves to be real, may require a fundamental change in our ideas about the nature of the tricritical point. Clearly further experimentation along these lines would be most welcome.

A second issue of great current interest is that of the behavior of the surface tensions on approach to a tricritical point. It had already been predicted by Papoular [22] that the tension of the interface between the two liquid phases in the He^3-He^4 mixtures should vanish proportionally to $(T_t - T)^2$, and this has been verified in experiments by Leiderer et al. [23]. The same prediction, $\sigma \sim (T_t - T)^2$, has been made also for the vanishing of the tensions $\sigma_{\alpha\beta}, \sigma_{\beta\gamma}$, and $\sigma_{\alpha\gamma}$ of the three possible interfaces in three-phase equilibrium near the tricritical point of classical multicomponent fluids [24]. This has not yet been tested by experiment, but we may surely expect that it will be, sometime in this decade. F Ramos Gómez [25] has in the meantime derived an important theoretical relation between these surface tensions and the susceptibilities:

$$\sigma_{\alpha\beta}/\sigma_{\beta\gamma} = (\chi_\alpha^{-1} - \chi_\beta^{-1})/(\chi_\gamma^{-1} - \chi_\beta^{-1})$$

This makes possible a most interesting check of the theory by combining two entirely different kinds of experiments, surface-tension measurements and light scattering. Ramos Gómez's formula comes from combining the Griffiths thermodynamics of the three-phase equilibrium with the van der Waals surface-tension theory.

The last problem I shall mention that will obviously also be of great interest in the coming years is that of providing a convincing account of the critical-tricritical crossover. Near every tricritical point there are two one-dimensional loci of critical end points. A critical end point is just an ordinary, two-phase critical point, at which the behavior is known to be nonclassical, in the sense of being characterized by critical-point exponents that have values different from those given by the classical, mean-field theories. The tricritical point, by contrast, is believed to be characterized by classical, Landau-Griffiths exponents (with additional, nonclassical logarithmic factors [26], but these do not affect the formal values of the exponents). Any formula for any property of the fluid must then be such as to give one set of exponents when the tricritical point is approached, and another, different, set when any of the infinity of critical end points in the immediate neighborhood of that tricritical point is approached. The problem of deriving or devising such expressions is a very challenging one, which is as yet far from being solved, though some important progress has been made by Fox [27] and by Fisher, Griffiths, and their coworkers [28]. We may hope for and expect significant progress on this question, too, in the 1980s.

REFERENCES

1. RB Griffiths, Phys Rev Lett 24, 715 (1970); Phys Rev B 7, 545 (1973).
2. LD Landau, Phys Zeit d Sowjetunion 11, 26 (1937); LD Landau and EM Lifshitz, Statistical Physics, translated by E Peierls and RF Peierls (Pergamon, 1958), § 138, pp. 452-456.

3. Ph Kohnstamm, Handbuch der Physik (Springer, 1926) vol. 10, chap. 4, § 45.
4. JD van der Waals and Ph Kohnstamm, Lehrbuch der Thermodynamik (Barth, Leipzig, 1912) 2er Teil, pp. 39-40.
5. J Zernike, Rec Trav Chim Pays-Bas 68, 585 (1949).
6. J Zernike, Chemical Phase Theory (EE Kluwer, Deventer, 1955).
7. IR Krichevskii, NE Khazanova, YuV Tsekhanskaya, LR Linshits, RO Pryanikova, ES Sokolova, SM Khodeeva, AV Shvarts, GA Sorina, and AI Semenova, Russ J Phys Chem 46, 1411 (1972).
8. GD Efremova and RO Pryanikova, Khim Prom 8, 47 (1961).
9. IR Krichevskii, GD Efremova, RO Pryanikova, and AV Serebryakova, Russ J Phys Chem 37, 1046 (1963) (Zh Fiz Khim 37, 1924 (1963)).
10. GS Radyshevskaya, NI Nikurashina, and RV Mertslin, J Gen Chem USSR (Zh Obshch Khim) 32, 673 (1962).
11. RV Mertslin and KI Mochalov, Zh Obshch Khim 29, 3172 (1959).
12. RB Griffiths, J Chem Phys 60, 195 (1974).
13. JC Lang, Jr and B Widom, Physica 81A, 190 (1975).
14. P Bocko, Physica 103A, 140 (1980).
15. M Lovellette, J Phys Chem 85, 1266 (1981).
16. AW Francis, Critical Solution Temperatures, Advances in Chemistry Series, no. 31 (ACS, 1961), p. 82. (Isooctane is 2,2,4-trimethylpentane).
17. JL Creek, CM Knobler, and RL Scott, J Chem Phys 67, 366 (1977).
18. JP Gollub, AA Koenig, and JS Huang, J Chem Phys 65, 639 (1976).
19. ES Wu, Phys Rev A 18, 1641 (1978).
20. MW Kim, WI Goldburg, P Esfandiari, JMH Levelt Sengers, and ES Wu, Phys Rev Lett 44, 80 (1980).
21. M Kaufman, KK Bardhan, and RB Griffiths, Phys Rev Lett 44, 77 (1980).
22. M Papoular, Phys Fluids 17, 1038 (1974).
23. P Leiderer, H Poisel, and M Wanner, J Low Temp Phys 28, 167 (1977).
24. B Widom, J Chem Phys 62, 1332 (1975).
25. F Ramos Gómez, J Chem Phys 74, 4737 (1981).
26. EK Riedel and FJ Wegner, Phys Rev Lett 29, 349 (1972); MJ Stephen, E Abrahams, and JP Straley, Phys Rev B 12, 256 (1975); MJ Stephen, Phys Rev B 12, 1015 (1975).
27. JR Fox, J Stat Phys 21, 243 (1979).

28. ME Fisher and S Sarbach, Phys Rev Lett $\underline{41}$, 1127 (1978); S Sarbach and ME Fisher, Phys Rev B $\underline{20}$, 2797 (1979); M Kaufman, RB Griffiths, JM Yeomans, and ME Fisher, Phys. Rev B $\underline{23}$, 3448 (1981); JM Yeomans and ME Fisher, Phys. Rev. B $\underline{24}$, 2825 (1981).

Acknowledgments. This work was supported by the National Science Foundation and the Cornell University Materials Science Center. This paper appears also in Kinam (A) $\underline{3}$ (1981) 143-157, and is reprinted with permission.

CHAPTER 10

THERMODYNAMIC MODELS FOR
FLUID MIXTURES NEAR
CRITICAL CONDITIONS

J. C. Rainwater
  National Bureau of
    Standards
  Boulder, CO 80303

M. R. Moldover
  National Bureau of
    Standards
  Washington, D.C. 20234

ABSTRACT
   We use model thermodynamic potentials for correlating VLE data for mixtures in near-cricital conditions. The model potentials are obtained by "interpolation" between the thermodynamic potentials of the pure components. The interpolation is carried out parallel to the critical locus using "field" variables which are ratios of fugacities. This contrasts with more conventional schemes in which "density" variables such as mole fractions play a key role. Our model potentials accurately describe the phase equilibria data in the critical region of CH_4-N_2 and C_4H_{10}-C_8H_{18}. These mixtures have coexisting phases whose compositions differ substantially from each other. Previous applications of the field variable concept to correlating critical region data have been limited to mixtures whose coexisting phases have quite similar compositions.

INTRODUCTION
   While highly successful methods to describe the thermodynamic properties and vapor-liquid equilibrium (VLE) of fluid mixtures over a wide range of thermodynamic variables are presently available (e.g., the Peng-Robinson equation [1]), there remains a need for accurate predictive and correlative techniques in the critical region, within approximately ten percent of the critical temperature. In addition to its importance for traditional distillation processes, knowledge of mixture VLE behavior in the critical region is important for newer technologies such as supercritical extraction.

The difficulties in describing the critical region of mixtures parallel those of pure fluids, where it is known that "classical" equations of state, including the various modifications of the van der Waals equation, do not predict thermodynamic behavior properly near the critical point. Recently alternate thermodynamic models of binary fluid mixtures in the critical region have been developed which incorporate ideas from modern theories of the critical behavior of pure fluids. The first of these, due to Leung and Griffiths [2], used the concept of field variables and was based on a thermodynamic potential which has a scaling form appropriate to the critical region. The Leung-Griffiths model was designed for a very "ideal" mixture: ^{3}He-^{4}He. Subsequently, Moldover and Gallagher [3] modified the Leung-Griffiths method to include concepts of corresponding states; they were successful in correlating VLE data of binary mixtures in which the composition difference between the liquid and the vapor phases is small.

In this paper we further extend the Leung-Griffiths method to mixtures such as CH_4-N_2 and C_4H_{10}-C_8H_{18} in which the liquid-vapor composition difference is large in the equimolar region but not in the parts of the coexistence region very near the pure fluids. This latter qualification excludes, for example, propane-octane since large composition differences are present for dilute solutions of octane in propane. We demonstrate the success of our new method in fitting the methane-nitrogen VLE data of Bloomer and Parent [4] as well as the (normal) butane-octane data of Kay et. al. [5]. To date we have not obtained a satisfactory representation of the propane-octane VLE data of Kay et. al. [5].

The remainder of this paper is organized as follows: In Section 2 we briefly describe the coexistence region of mixtures. In effect, we define the problem we are addressing in more detail. In Section 3, previous efforts are discussed. In Section 4 the present modifications to previous work are discussed. In Section 5 we discuss our successful results for methane-nitrogen and butane-octane mixtures while in Section 6 we present some concluding remarks including a comment on the difficulties that remain in dealing with the propane-octane system.

2. THE COEXISTENCE REGION

In general, the phase equilibrium behavior of binary mixtures can be quite complicated [6]. Van Konynenburg and Scott [7] have extensively classified the various phase equilibrium possibilities for binary mixtures of fluids obeying the van der Waals mixture equation of state. For the simplest type, with which we will be concerned, a critical line proceeds continuously from one pure fluid critical point to the other. These simple mixtures may be azeotropic.

We do not exclude the possibility of liquid-liquid immiscibility, but assume that, if it is present, it occurs in a region sufficiently removed from the critical locus to be ignored here.

Figure 1 shows a pressure (P)-temperature (T) plot for the simple binary mixture methane-nitrogen. In the P-T plane, the coexistence region is bounded from above by the critical line and from the sides by the vapor pressure curves of the pure fluids, as shown. At any (P,T) point within this region and at a particular overall composition, two phases of differing density and composition coexist. The liquid is the more dense phase and is rich in the less volatile component (here methane); the vapor is the less dense phase and is rich in the more volatile component (here nitrogen)

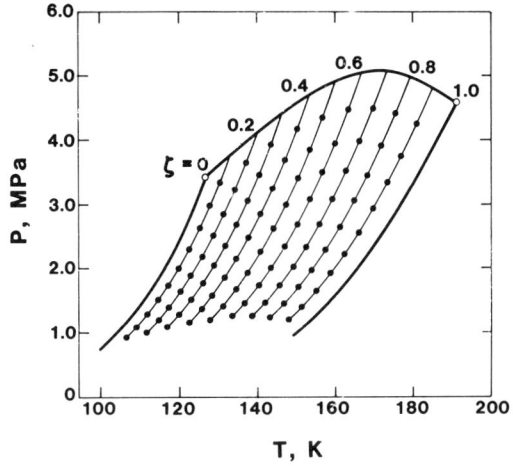

Figure 1. Coexistence region of N_2-CH_4 with vapor pressure curves (pure N_2 at left, pure CH_4 at right), critical locus, and lines of constant ζ according to Eq. (15) in intervals of 0.1. The dots show intervals in t of 0.02.

Our interest is in the critical region, defined roughly to be within ten percent of the critical temperature (the meaning of this for a mixture will be made clear shortly). This is the region where VLE predictions of classical equations of state are in general the least accurate. Note that, since the vapor pressure equation is quite steep at the critical point, i.e., for many fluids

$$\frac{T_c}{P_c} \left(\frac{dP}{dT}\right)_c \approx 6, \qquad (1)$$

a ten percent decrease in temperature from the critical

point can approximately halve the vapor pressure. Therefore, the critical region on a P-T diagram covers approximately the upper half of the coexistence area and, from this viewpoint, is a large region.

In Figs. 2 and 3 we display the VLE data for the CH_4-N_2 system in the region of our interest. A qualitative understanding of the shapes of these curves can be obtained from Ref. 3. An important feature of the present work (as well as its predecessors, Refs. 2 and 3) is that both the P-T data (Fig. 2) and the T-ρ data (Fig. 3) are correlated simultaneously. We are fitting a surface in four-dimensional P-ρ-T-x space to the data which are plotted on two separate two-dimensional subspaces.

3. THERMODYNAMIC DESCRIPTIONS OF THE CRITICAL REGION

Leung and Griffiths [2] introduced a description for the critical region of mixtures which differs from conventional descriptions of mixtures in two respects. First, Leung and Griffiths utilize a thermodynamic field variable to describe mixtures instead of a density variable such as mole fraction. Second, they use scaling-law equations of state for the mixture critical region. We emphasize that the concepts of field variables and scaling-law equations are independent of each other and, in principle, either could be used without the other. We will use both concepts; thus we briefly review them.

The field variable concept has been discussed by Griffiths and Wheeler [8]. They distinguish between field variables, which are continuous across a phase boundary, and density variables, which are discontinuous across a phase boundary. For binary mixtures, there are three independent field variables and three independent density variables. For example, Griffiths and Wheeler [8] chose as independent field variables: T, -P, and $\Delta = \mu_1 - \mu_2$ (the difference in chemical potential between fluids 1 and 2). Within the formalism of Griffiths and Wheeler [8], a fourth field variable is chosen to be the "thermodynamic potential" (e.g., μ_2), and the density variables, discontinuous across the phase boundary, are defined as minus the partial derivatives of the thermodynamic potential with respect to the other field variables. In this example (see Ref. 8, Sec. 2B), the density variables are the entropy per mole s, the molar volume $v = \rho^{-1}$, and the mole fraction x. A set of new field variables, linearly independent functions of the old ones, may also be used as long as certain stability conditions are satisfied [2].

For the purpose of describing phase equilibria, we use the field variables selected by Moldover and Gallagher [3]. Their choice differs slightly from Leung and Griffiths. We take the potential to be $\omega \equiv P/(RT)$ where R is the gas constant.

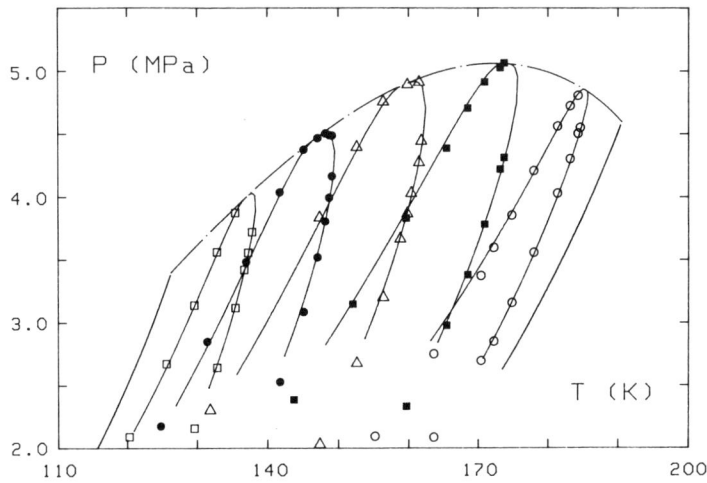

Figure 2. Fit of our model to dew-bubble curves of N_2-CH_4. The solid lines are results of the model with $C_H = -6$, $C_X = 0.3$. Experimental points are from Ref. 4. The mole fractions of N_2 are: 0.1002 ○ ; 0.2897 ■ ; 0.5088 △ ; 0.697 ● ; 0.8422 □ .

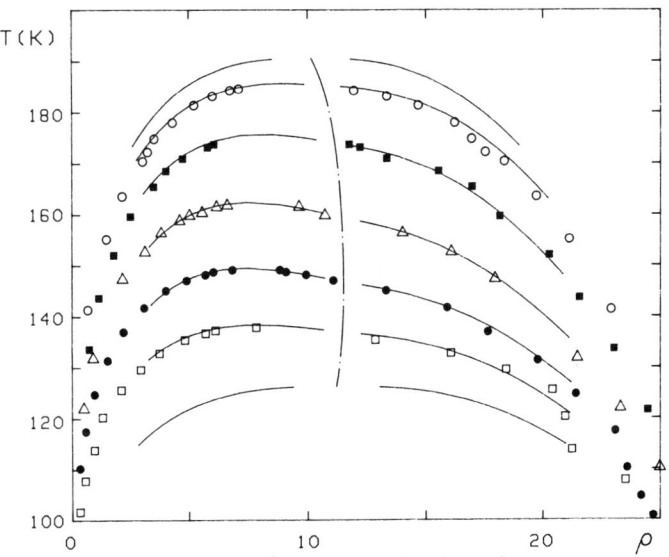

Figure 3. Fit of our model to coexisting density curves of N_2-CH_4. The coexisting density curves of pure N_2 (bottom) and pure CH_4 (top) and the critical locus (broken line) are shown. The solid lines are the results of the model with $C_H = -6$, $C_X = 0.3$. Experimental points are from Ref. 4; the mole fractions are the same as Fig. 2. The units of density are kg-mol/m^3.

The independent field variables are particular combinations of T, μ_1, and μ_2 given by:

$$\zeta = \frac{e^{\mu_1/RT}}{Ke^{\mu_2/RT} + e^{\mu_1/RT}} \tag{5}$$

$$t = \frac{T - T_c(\zeta)}{T_c(\zeta)} \tag{6}$$

$$h = \ln(Ke^{\mu_2/RT} + e^{\mu_1/RT}) - \ln(Ke^{\mu_2^\sigma/RT} + e^{\mu_1^\sigma/RT}) \tag{7}$$

These definitions (Eq. (5)-(7)) require some elaboration. The variable ζ has the convenient property $0 \leq \zeta \leq 1$ for all mixtures. (Because $\mu_i \rightarrow -\infty$ as the mole fraction of fluid i approaches zero it follows that $\zeta = 1$ for pure fluid 1 and by our convention x = 0; similarly, $\zeta = 0$ for pure fluid 2 and x = 1.) Although it is not practical to conduct measurements along paths of constant ζ, it is useful for our purposes to think of a path of constant ζ as defining a particular mixture in a binary system. Indeed we shall place the values of P, T, and the molar density of each species for the coexisting phases of binary mixtures along paths of constant ζ in correspondence with the values P, T, and ρ in the coexisting phases of a pure substance. This correspondence contrasts with the more conventional mixture models in which the properties of a binary system along a path of constant mole fraction are related to those of some (perhaps hypothetical) pure fluid.

The variable t defined by Eq. (6) is a measure of the distance from the critical locus on paths of constant ζ in the two phase region. The significance of ζ and t are illustrated in Fig. (1), where we show the intersection of surfaces of constant ζ with the coexistence surface for the methane-nitrogen system. These intersecting surfaces form a family of nearly parallel curves. The dots spaced along these curves represent points at distances from the critical locus given by: t=0, -0.02, -0.04, -0.06, etc.

The constant K in Eq. (5) is arbitrary. Different values of K are equivalent to different choices for the reference state [2] in which $\mu_2 - \mu_1 = 0$. Hence, K may be adjusted to make $\zeta(x)$ have a convenient functional form along the critical locus. Leung and Griffiths choose K in such a way as to make $1/T_c(x)$ approximately linear in ζ. This choice is satisfactory only when $T_c(x)$ is monotonic. D'Arrigo et al. [9] relaxed this restriction somewhat. Here we follow Moldover and Gallagher in making an implicit choice for K such that x is approximately linear in ζ on the critical locus:

$$x_c = 1 - \zeta_c \tag{8}$$

This choice is satisfactory even when $T_c(x)$ is not monotonic (e.g., CO_2-C_2H_4).

In Eq. (7) μ_1^σ and μ_2^σ are the values μ_1 and μ_2 have at a given value of t and ζ in the region of liquid-vapor coexistence (or its smooth extension). Thus the surface of two-phase coexistence is given by h=0, t<0, and $0 \leq \zeta \leq 1$. It is natural to think of h as a measure of distance away from two-phase coexistence along paths of constant t and ζ. This completes our description of the field variables that we shall use.

The thermodynamic description of the critical region of mixtures parallels that of pure fluids. The potential, ω, is taken to be the sum of two terms. One term is regular, i.e., it has a Taylor series expansion in ζ, t, and h about the critical locus of the mixture. The other term is singular, i.e, it does not have a Taylor series expansion about the critical locus. Following Moldover and Gallagher we write the regular term in the form:

$$\frac{P_{analytic}}{RT} = \frac{P_c}{RT_c}(1 + C_4 t + C_5 t^2 + C_6 t^3) + \rho_c(1 + C_2 t)h \tag{9}$$

The singular term can be written as:

$$\frac{P_{singular}}{RT} = |t|^{2-\alpha} f(h/|t|^{2-\alpha-\beta}) \tag{10}$$

Here and below, T_c, ρ_c, P_c and C_i are considered as known functions of ζ while f is an analytic function of its argument. The non-integer exponent α has a value near 0.1 and, in a pure fluid, is characteristic of the divergence of the constant volume specific heat as the critical point is approached along the critical isochore. We take the non-linear exponent β to have the value 0.355. In pure fluids this exponent appears in approximate expressions describing the vanishing of the difference between the coexisting densities as the critical point is approached: $\rho_{liquid} - \rho_{vapor} = 2\rho_c B(-t)^\beta$. In Eq. (10) the function f is not a function of h and t separately. Instead, it is a function of a suitably scaled combination of these two variables. Thus Eq. (10) leads to a "scaling" equation of state.

The singular part of the potential can be written explicitly in terms of two parametric variables, r and θ. We use a functional form which has been widely used to correlate pure fluid properties near critical points:

$$\frac{P_{sing}}{RT} = \frac{P_c C_3}{RT_c (0.1692)} r^{1.9}(0.4753 - 0.7561\ \theta^2 + 0.45\ \theta^4) \tag{11}$$

$$h = \frac{P_c}{RT_c} \frac{C_3}{\rho_c C_1 (0.1692)} r^{1.545}\ \theta(1-\theta^2) \tag{12}$$

$$t = r(1-1.3909\theta^2)/0.3909 \tag{13}$$

The numerical values for the exponents and coefficients in Eq. (11)-(13) were taken from the correlation of pure fluid data by Levelt Sengers et al. [10].

The nonanalytic term in the potential, Eq. (11)-(13) has been constructed in such a manner as to provide a thermodynamically consistent description of phase equilibria for mixtures with two non-integer exponents. These exponents appear in the following expressions for the coexisting densities and the pressure:

$$\rho/\rho_c(\zeta) = 1 \pm C_1(-t)^{0.355} + C_2 t \tag{14}$$

$$\frac{P}{T} \frac{T_c(\zeta)}{P_c(\zeta)} = 1 + C_3(-t)^{1.9} + C_4 t + C_5 t^2 + C_6 t^3 \tag{15}$$

Equation (14) is an economical representation of the coexisting densities of pure fluids which has been used for many years. Equation (15) is consistent with the divergent C_v which has been observed in 11 pure fluids.

Equations (9)-(13) are written entirely in units reduced by T_c, P_c, and ρ_c. Thus if corresponding states were obeyed exactly, the C_i would take on universal values at $\zeta=0$ and $\zeta=1$. In Ref. 3 the C_i for a variety of fluids have been tabulated. They are indeed approximately constant from fluid to fluid, except for C_2, the slope of the rectilinear diameter, and C_6, the coefficient of the highest-order term in a truncated polynomial series for the vapor pressure.

Because the C_i were roughly constant for the various pure fluids examined, Moldover and Gallagher were led to approximate $C_i(\zeta)$ by linear interpolation between the pure fluid coefficients:

$$C_i(\zeta) = C_i^{(2)} + \zeta C_i' \tag{16}$$

$$C_i' = C_i^{(1)} - C_i^{(2)} \tag{17}$$

(Note: the superscripts (1) and (2) in Eqs. (16) and (17) were erroneously interchanged in Eq. (8) of Ref. 3. This mistake was typographical and did not affect the results of Ref. 3.) In order to describe methane-nitrogen mixtures and butane-octane mixtures, we have found it necessary to relax this linear interpolation assumption for C_1 (see Section 4).

A description of phase equilibrium requires knowledge of ρ_ℓ, ρ_v, x_ℓ and x_v for each (P,T) or (ζ,t) point within the coexistence region. The expression for coexisting compositions has been derived in this model and is equivalent to that of Leung and Griffiths [2], namely

$$x = (1-\zeta) \left\{ 1-\zeta \left(\frac{\overline{Q}(\zeta,t)}{\rho} - \frac{\overline{Q}(\zeta,0)}{\rho_c} - \overline{H}(\zeta,t) \right) \right\} \quad (18)$$

where

$$\overline{Q}(\zeta,t) = \frac{PT_c}{RTP_c} \frac{d}{d\zeta}\left(\frac{P_c}{T_c}\right) + \frac{P_c}{RT_c} [C_3'(-t)^{1.9} + C_4't + C_5't^2 + C_6't^3]$$

$$+ \frac{P_c}{R} \frac{d(1/T_c)}{d\zeta} (1+t)[-1.9C_3(-t)^{0.9} + C_4 + 2C_5 t + 3C_6 t^2] \quad (19)$$

and T_c, P_c and the C_i are functions of ζ. Substitution of $\rho = \rho_\ell$ and $\rho = \rho_v$ from Eq. (14) into Eq. (18) yields, respectively, x_ℓ and x_v.

The function $\overline{H}(\zeta,t)$ is assumed to be analytic but otherwise unknown. In Ref. 3, Moldover and Gallagher elect to treat $\overline{H}$ as a fitting function, although they show in their Appendix B that it could be determined by standard mixture models. For Eq. (18) to reduce to Eq. (8) on the critical line, it is necessary that $\overline{H}(\zeta,0) = 0$. Furthermore, $\overline{H}$ cannot contain any singular terms, so a simple linear dependence on t should suffice in the critical region. It is shown in Ref. 3, App. B, that a reasonable form for $\overline{H}$ is

$$\overline{H}(\zeta,t) = C_H t \frac{d(\ln(T_c(\zeta)))}{d\zeta} \quad (20)$$

where C_H is a numerical constant. Equations (14)-(20) comprise the most general thermodynamic model used by Moldover and Gallagher.

The model thermodynamic potential, Eqs. (9)-(13) has three classes of parameters which must be obtained from data. The first class contains those which characterize the two pure fluids, namely, the critical pressures, temperatures and densities and the C_i coefficients of Eqs (3) and (4). Since, in general, available pure fluid data are more extensive and precise than mixture data, these parameters can usually be determined with confidence. Because the exponents 1.9 and 2 are very close, C_3 and C_5 cannot be accurately found from a fit of the vapor pressure curve. Instead, we determined C_3 from the critical region correlation of Levelt Sengers et al. [10] through:

$$C_3 = 0.7222 a x_o^{-\beta} \tag{21}$$

where a and x_o are tabulated for a variety of fluids in Ref. 10, Table 34.° With C_3 thus fixed, the vapor pressure data in the critical region are fit to Eq. (4). When a and x_o were not available, we have used a "typical" value $C_3=30$.

The second class of parameters in the model thermodynamic potential contains the parameters which characterize the critical line. Moldover and Gallagher have chosen to use polynomials in x arranged in the following forms convenient for making minor adjustments in the critical line (but see the discussion in Sec. 4),

$$\frac{1}{RT_c(x)} = \frac{1-x}{RT_{c1}} + \frac{x}{RT_{c2}} + x(1-x)[T_1+(1-2x)T_2+(1-2x)^2 T_3] \tag{22}$$

$$\frac{P_c(x)}{RT_c(x)} = \frac{(1-x)P_{c1}}{RT_{c1}} + \frac{xP_{c2}}{RT_{c2}} + x(1-x)[P_1+(1-2x)P_2+(1-2x)^2 P_3] \tag{23}$$

$$\rho_c(x) = (1-x)\rho_{c1} + x\rho_{c2} + x(1-x)[\rho_1+(1-2x)\rho_2+(1-2x)^2 \rho_3] \tag{24}$$

In practice, T_i and P_i can be determined accurately by construction of the envelope of the dew-bubble curves in the P-T diagram, but ρ_i can rarely be found in a similar manner. It is extremely difficult to measure $\rho_c(x)$ directly. In our fit of nitrogen-methane VLE, we started with a rough approximation to $\rho_c(x)$ which was then refined using coexisting density data over the entire critical region.

The third type of parameter which appears in the model thermodynamic potential is a property of the entire coexistence surface of the mixture. In the procedure described above there is only one such parameter, C_H, though the extended method presented below adds up to two new parameters, C_X and C_Z. Moldover and Gallagher [3,12] find that C_H is not necessary (i.e., $C_H = \bar{H} = 0$) for fitting VLE of ^{3}He-^{4}He, CO_2-ethane, and SF_6-propane.

Although we have presented the Moldover-Gallagher method as an operational recipe, we emphasize that it is based upon (Ref. 3, App. A) a thermodynamically sensible potential in the same manner as the method of Leung and Griffiths [2]. Since the latter does not introduce reduced variables, it involves more parameters which, operationally, depend on the mixture coexistence region as a whole.

4. INNOVATIONS

In the present work we have made three innovations in

the scheme just outlined. First, we have relaxed the assumption that $C_1(\zeta)$ is a linear function of ζ. Second we have considered a wider class of functional forms for $\overline{H}(\zeta,t)$ than allowed by Eq. (20). Finally, we have found alternative representations of the critical locus which have substantial practical advantages over Eqs. (22)-(24). The first two innovations require the introduction of additional parameters characteristic of the VLE surface of the whole mixture. These features require some motivation which we now present.

First we rewrite the density variables ρ and x for coexisting phases in terms of averages and differences.

$$\Delta\rho = 2\rho_c(\zeta)C_1(\zeta)(-t)^\beta \tag{25}$$

$$\rho_{average} = \rho_c(\zeta)(1+C_2(\zeta)t) \tag{26}$$

$$\Delta x = \zeta(1-\zeta)|\overline{Q}(\zeta,t)|(\rho_v^{-1}-\rho_\ell^{-1}) \tag{27}$$

$$x_{average} = (1-\zeta)\left\{1-\zeta\left[\frac{\overline{Q}(\zeta,t)}{2\rho_\ell} + \frac{\overline{Q}(\zeta,t)}{2\rho_v} - \frac{\overline{Q}(\zeta,0)}{\rho_c} - \overline{H}(\zeta,t)\right]\right\} \tag{28}$$

where the difference quantities are defined to be positive.

We note that Δx is independent of $\overline{H}(\zeta,t)$, so if Δx is incorrectly predicted by the method of Ref. 3, the prediction cannot be rectified by adjusting $\overline{H}$; corrections instead must be made elsewhere in the model. In a preliminary investigation, Rainwater [13] has interpolated the nitrogen-methane VLE data of Bloomer and Parent [4] to estimate $\Delta x(\zeta)$ along the curve $t = -0.12$, and has shown that the method of Ref. 3 significantly and systematically overestimates Δx, especially in the equimolar region.

Upon reexamination of the model, it is seen that, in reduced units, the phase equilibrium behavior within the model for a pure fluid and a mixture are in fact quite different. When $C_1^{(1)} = C_1^{(2)}$ the reduced density change $\Delta\rho/\rho_c$ is the same function of t for the pure fluid and the mixture, but $\Delta x=0$ for the pure fluid and $\Delta x>0$ for the mixture. Thus the overall measure of phase change as a function of measure of distance from critical conditions t appears to be larger for the mixture than for the pure fluid.

At this point we conjecture that there exists, for constant ζ, an "amount of phase change" A(t) which is some functional combination of $\Delta\rho(t)$ and $\Delta x(t)$, which reduces to $\Delta\rho/\rho_c$ for the pure fluid, and which "obeys corresponding states" within the mixture, i.e., A(t), not $\Delta\rho(t)/\rho_c$, is the same function of t for all ζ if the pure fluids obey corresponding states ($C_1^{(1)} = C_1^{(2)}$).

A small-t asymptotic expansion of Eq. (18) yields, to leading order,

$$\Delta x = 2C_1(\zeta)(-t)^\beta \zeta(1-\zeta)|\overline{Q}(\zeta,0)|/\rho_c(\zeta)$$

$$= \Delta\rho \, \zeta(1-\zeta)|\overline{Q}(\zeta,0)|/\rho_c^2(\zeta) \qquad (29)$$

and thus $\Delta\rho(t)$, $\Delta x(t)$ and presumably $A(t)$ are all proportional to $C_1(\zeta)(-t)^\beta$ for small t.

$A(t)$ could in principle also involve the difference of the third independent density variable s, the entropy per mole. To date we have not explored this possibility. A minimal assumption consistent with intuition is that A must be a monotonically increasing function of both $\Delta\rho$ and Δx. The simplest functional form with this property is:

$$A = \Delta\rho/\rho_c + C_X \Delta x \qquad (30)$$

where the sign of C_X is chosen so that $A > \Delta\rho/\rho_c$. (In Rainwater's preliminary study [13], A was chosen to be quadratic in $\Delta\rho$ and Δx.) Equations (25), (29), and (30) yield for a binary mixture whose pure components obey corresponding states ($C_1^{(1)} = C_1^{(2)}$):

$$C_1(\zeta) = \frac{C_1^{(1)}}{1 + C_X \zeta(1-\zeta)|\overline{Q}(\zeta,0)|/\rho_c(\zeta)} \qquad (31)$$

When $C_1^{(1)} \neq C_1^{(2)}$, the natural generalization of Eq. (30) is

$$C_1(\zeta) = \frac{C_1^{(2)} + \zeta(C_1^{(1)} - C_1^{(2)})}{1 + C_X \zeta(1-\zeta)|\overline{Q}(\zeta,0)|/\rho_c(\zeta)} \qquad (32)$$

Equation (32) embodies the modification of the Moldover-Gallagher method needed to correlate the data for the CH_4-N_2 system. Otherwise we retain all rules discussed in Sec. 3, including Eq. (16) and (17) for $C_i(\zeta)$, $i \neq 1$. In the azeotropic mixtures CO_2-C_2H_6 and SF_6-C_3H_8 and the mixture ^{3}He-^{4}He, Δx is quite small and then the introduction of the parameter C_X is not necessary.

We emphasize that this new model remains thermodynamically consistent. The derivation of phase equilibrium properties from an explicit thermodynamic potential (Ref. 3, App. A) does not depend on the precise form of $C_1(\zeta)$ or $C_2(\zeta)$; the only requirement is that these coefficients are functions of ζ alone. However, a change in $C_i(\zeta)$, $i = 3,\ldots 6$, would require a corresponding change in the expression for $\overline{Q}$, Eq.(19).

The addition of the parameter $C_X \neq 0$ in Eq. (32) is sufficient for correlating the methane-nitrogen data. To deal with butane-octane we found $\overline{H}(\zeta,t)$ as given by Eq. (20) had to be generalized. The ad hoc generalization we have used is

$$\overline{H}(\zeta,t) = C_H t \frac{d(\ln(T_c(\zeta)))}{d\zeta} (1+C_z\zeta) \tag{33}$$

where C_z is a constant to be fitted to the data. We have not implemented the suggestion of Moldover and Gallagher that $\overline{H}(\zeta,t)$ be calculated from standard mixture models.

In the butane-octane system, both $d\rho_c/dx$ and dP_c/dx change very rapidly near x=1 (pure butane). Thus it is not possible to represent accurately $\rho_c(x)$ and $P_c(x)/T_c(x)$ with polynomials of low degree in x such as Eqs. (23) and (24). Furthermore, the experimental data for this system [5] are not spaced uniformly in x. Instead, the data are concentrated at values of x near 1 so that the distribution of the data is roughly uniform in $T_c(x)$ or $1/T_c(x)$. This is evident from the roughly uniform spacing of the dew-bubble curves along the ordinate in Fig. 6. To reflect these features of the butane-octane system we define a variable

$$x_T = \frac{1/T_{c1} - 1/T_c(x)}{1/T_{c1} - 1/T_{c2}} \tag{34}$$

We then replace Eq. (23) and (24) with:

$$\frac{P_c}{RT_c(x)} = \frac{(1-x_T)P_{c1}}{RT_{c1}} + \frac{x_T P_{c2}}{RT_{c2}}$$

$$+ x_T(1-x_T)[\overline{P}_1 + (1-2x_T)\overline{P}_2 + (1-2x_T)^2\overline{P}_3] \tag{35}$$

$$\rho_c(x) = (1-x_T)\rho_{c1} + x_T \rho_{c2}$$

$$+ x_T(1-x_T)[\overline{\rho}_1 + (1-2x_T)\overline{\rho}_2 + (1-2x_T)^2\overline{\rho}_3] \tag{36}$$

We have found that replacing Eq. (23) with Eq. (35) has an additional advantage. The dew-bubble curve data determine P_c/T_c more directly as a function of $1/T_c$ (or x_T) than as a function of x. Thus one can fit the envelope of the dew-bubble curves in the P-T representation by adjusting $\overline{P}_1$, $\overline{P}_2$, and $\overline{P}_3$. Once these parameters are adjusted to yield a satisfactory envelope (the dot-dash curve in Fig. 6) the coefficients T_1, T_2, and T_3 can be adjusted in a separate step to locate properly the individual dew-bubble curves with respect to the data.

5. THE METHANE-NITROGEN AND BUTANE-OCTANE SYSTEMS

We now present our representations of the VLE data of Bloomer and Parent [4] in the critical region of the methane-nitrogen system. Then we will turn to the data for (normal) butane-octane mixtures that were obtained by Kay et al. [5].

The pure fluid parameters T_c, ρ_c, C_1 and C_3 (see Eq. (12)) for nitrogen and methane were obtained from Levelt Sengers et al. [10]. C_2 was determined for methane from the coexisting density data of Bloomer and Parent [4] and also Goodwin [14], and for nitrogen from that of Jacobsen et al. [15]. C_4, C_5, C_6 and P_c were determined for methane from a fit to the vapor pressure data of Prydz and Goodwin [16] and, for nitrogen, from a fit to the vapor pressure data of Weber [17]. The pure fluid parameters thus derived are listed in Table I.

Table 1. CONSTANTS FOR PURE COMPONENTS

Substance Property (units)	N_2	CH_4	$n-C_4H_{10}$	C_8H_{18}
T_c (K)	126.24	190.555	425.38	569.20
ρ_c (kg-mol/m^3)	11.21	10.142	3.936	2.005
P_c (MPa)	3.398	4.595	3.809	2.603
C_1	1.90	1.90	1.991	2.159
C_2	-0.67	-0.71	-0.912	-1.493
C_3	25.00	23.30	30.00	30.00
C_4	5.10	4.99	5.99	6.90
C_5	-22.54	-20.67	-24.42	-20.62
C_6	-7.2	-6.4	0	0

Bloomer and Parent provide VLE data (pressure, temperature and density) for the compositions x=0.1002, 0.2879, 0.5088, 0.6970 and 0.8422, where x=0 is pure methane. They have estimated the critical pressure and temperature for each composition, and we have used their estimates except at x=0.5088, where the choices T_c=159.21 K and P_c=4.888 MPa appear to be more consistent with the data [13]. From the five mixture and two pure fluid critical points, a least-squares fit was made to Eqs. (13) and (14) and the parameters T_i, P_i thereby determined.

Although Bloomer and Parent provide ample coexisting density data in the critical region, they do not quote critical densities for the mixtures. However, they do observe that the critical line appears to be approximately straight on a plot of temperature versus <u>mass</u> density. We find that this linear assumption, and the consequent fit to Eq. (15), is an adequate starting point for the present correlation. The reason is that the derived dew-bubble curves in the P-T plane are insensitive to $\rho_c(x)$, whereas the coexisting density curves are highly sensitive to $\rho_c(x)$. Therefore we start with the parameters derived so far, including ρ_i from the linear assumption, and then search for the parameters intrinsic to the mixture (C_H, C_X) which best fit the P-T data; for this system we set $C_Z=0$. Subsequently, we adjust the critical density, i.e., the ρ_i parameters, to obtain a best fit in the T-ρ plane without appreciably altering the P-T fit. This procedure is iterated.

In our fitting procedure we have found it most convenient to use a minicomputer with automatic graphics packages. The best fit of theory to experiment is decided by a visual examination of the generated graphs, as has been done in all previous applications of methods based on the Leung-Griffiths approach [2,3,9].

Derivation of dew-bubble curves and coexisting density curves at a fixed composition requires numerical inversion of Eq. (9). With an array of t values over the appropriate range, we search for roots of Eq. (9) for ζ as a function of t for fixed x_ℓ and x_v, and once the (ζ,t) points are known, derive corresponding (P,T) points algebraically from the mixture version of Eq. (4). Despite the relatively slow computational speed of the minicomputer we have succeeded in developing a routine which calculates and plots several constant-composition curves per minute. The user can thereby rapidly investigate the prediction of the model for a series of trial values of C_H and C_X.

After finding optimal values of C_H and C_X from the P-T diagram, we observed that agreement in the T-ρ plane could be significantly enhanced if $\rho_c(x)$ were increased from the original linear (T_c vs. mass density) assumption, particularly at the methane-rich end. Thus, ρ_1, ρ_2 and ρ_3 were varied to improve the agreement with the coexisting data. The final critical line parameters are listed in Table 2.

Table 2. MIXTURE PARAMETERS FOR METHANE-NITROGEN AND PROPANE-OCTANE

Parameter (units)	CH_4-N_2	$C_4H_{10}-C_8H_{18}$
ρ_1 or $\bar{\rho}_1$ (kg-mol/m^3)	3.0	1.0
ρ_2 or $\bar{\rho}_2$ (kg-mol/m^3)	0.5	0.5
ρ_3 or $\bar{\rho}_3$ (kg-mol/m^3)	1.0	1.0
P_1 or $\bar{P}_1$ (kg-mol/m^3)	2.483	1.019
P_2 or $\bar{P}_2$ (kg-mol/m^3)	-0.048	0.110
P_3 or $\bar{P}_3$ (kg-mol/m^3)	-0.136	-0.171
T_1 (kg-mol/(MPa·m^3))	-0.1623	-0.06254
T_2 (kg-mol/(MPa·m^3))	0.0156	0.03247
T_3 (kg-mol/(MPa·m^3))	-0.0148	-0.02285
C_H	-6	-12
C_X	0.3	0.3
C_Z	0	1.3

The results of the model for $-0.09 < t < 0$, $C_H = -6$ and $C_X = 0.3$, and the data of Bloomer and Parent, are displayed in a P-T plot, Fig. 2, and a T-ρ plot, Fig. 3. Agreement is excellent for both the dew-bubble and coexisting density curves over the entire critical region of the mixture. It is not possible to obtain a fit of this quality to the methane-nitrogen data using $C_X = 0$.

It is instructive to show how the correlation changes when the mixture parameters are altered from their optimal values; this is done in Figs. 4-5. In general, an increase (decrease) of C_H moves the dew-bubble curves to the right (left) and an increase (decrease) of C_X narrows (widens) the dew-bubble curves.

We remark that the determination of critical densities from a fit to the entire critical region, as opposed to direct experimental measurement, is precisely analogous to the method of determining the critical density of a pure fluid.

The results of fitting our model to the butane-octane [5] data are shown in Fig. 6 and Fig. 7. The parameters used to generate the critical locus and the dew-bubble curves for these figures are listed in Table 2. The model provides an excellent representation to the data near the critical locus. As we remarked above, three parameters characteristic of the entire mixture coexistence surface are required: C_H,

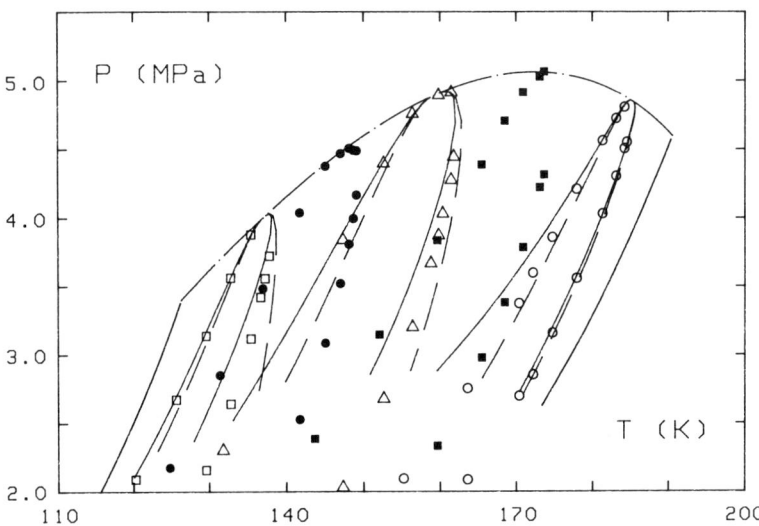

Figure 4. The effect of variations of C_H on Fig. 2. The dashed lines are $C_H = 0$; the solid lines are $C_H = -12$. ($C_X = 0.3$)

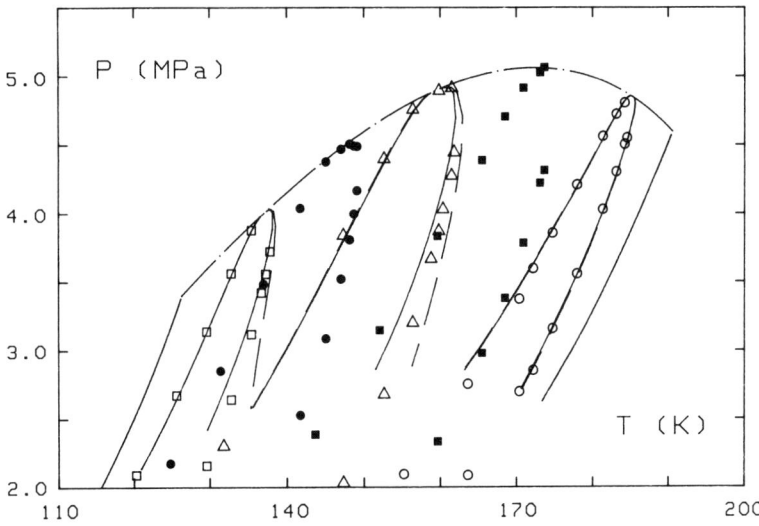

Figure 5. The effect of variations of C_X on Fig. 2. The dashed lines are $C_X = 0$; the solid lines are $C_X = 0.6$. ($C_H = -6$)

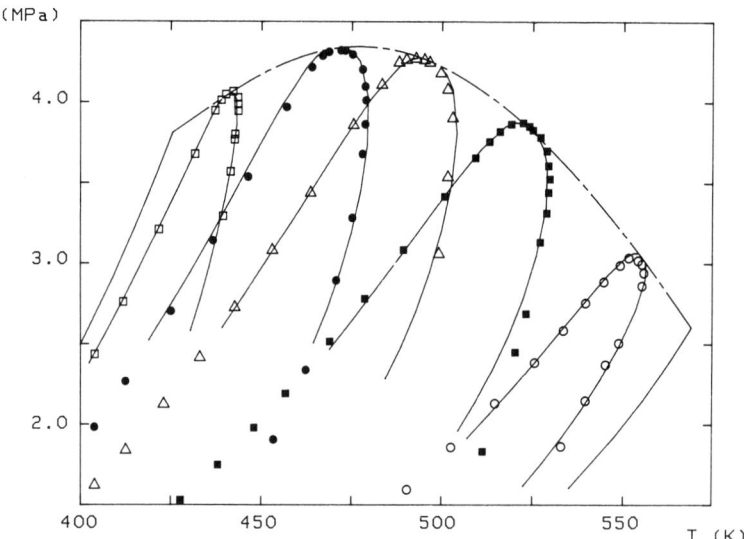

Figure 6. The vapor-pressure curves of n-butane (left) and octane (right) and the critical locus (broken line) are shown. The solid lines are the results of the model with $C_H = -12$, $C_X = 0.3$ and $C_Z = 1.3$. Experimental points are from Ref. 5. The mole fractions of n-butane are: 0.1823 ○ ; 0.4631 ■ ; 0.6707 △ ; 0.8183 ● ; 0.9461 □ .

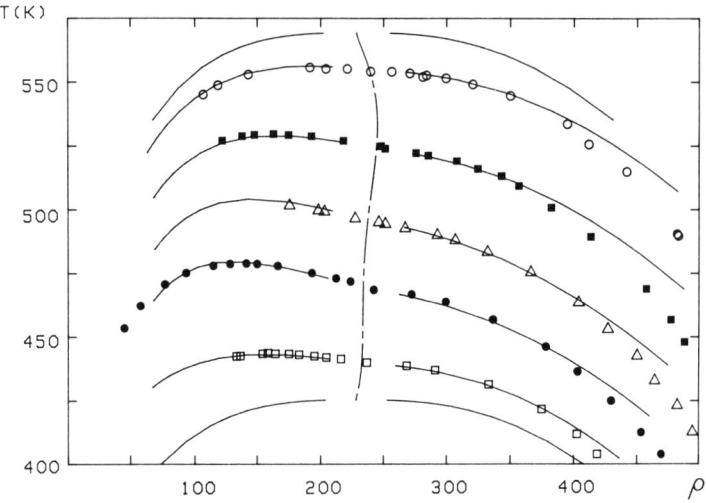

Figure 7. The coexisting density curves of pure n-butane (bottom) and pure octane (top) and the critical locus (broken line) are shown. The solid lines are the results of the model with $C_H = -12$, $C_X = 0.3$ and $C_Z = 1.3$. Experimental points are from Ref. 5; the mole fractions are the same as Fig. 6. The units of density are kg/m^3.

C_X, and C_Z. The roles played by C_H and C_X are similar to the roles these parameters play in the methane-nitrogen system. The role played by C_Z is illustrated in Fig. 8.

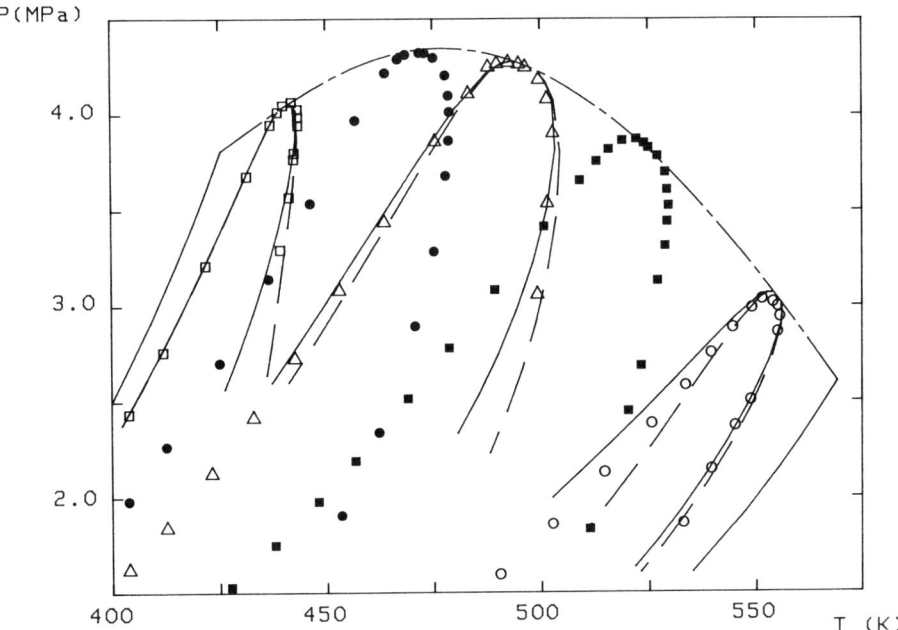

Figure 8. The effect of C_Z on Fig. 6. The solid curves are computed with $C_Z = 0$; the dashed curves are computed with $C_Z = 2.6$.

Frequently, and in contrast to the present examples, coexistence data are taken and correlated along a series of isotherms in the P-x or ρ-x plane. Such correlations are easily done with our method. Successive values of ζ are chosen between 0 and 1 for an isotherm for T less than both pure critical temperatures, or between $(1-x_{max})$ and 1 for T between the two pure critical temperatures. Then for each ζ, $T_c(ζ)$ is found from Eqs. (8) and (22), t for the desired isotherm is found from Eq. (6), the coefficients $C_i(ζ)$ are found from Eq. (16), $i \neq 1$, and (32), the pressure is found from Eq. (15), $ρ_ℓ$ and $ρ_v$ from Eq. (14), and $x_ℓ$ and x_v from Eq. (18). Whereas the calculation of constant-composition curves requires numerical inversion of Eq. (18), no numerical root finding is required for calculation along an isotherm.

6. CONCLUDING REMARKS

We have demonstrated that the concepts of smoothness of the thermodynamic potential as a function of field variables together with the concept of a scaling form for the singularity in the potential at the critical locus are sufficient to correlate PVTx data for the critical region of methane-nitrogen and butane-octane mixtures. These mixtures have coexisting phases whose compositions differ substantially from each other; thus they exhibit more complex behavior than the mixtures that were successfully correlated in the pioneering work of Leung and Griffiths [2] or the subsequent work of D'Arrigio et al. [9], Doiron et al. [18], and Moldover and Gallagher [3]. In order to describe these more complex mixtures we had to use three parameters which are characteristic of the entire PVTx coexistence surface. Two of these paramters, C_H and C_Z, specify properties of the $\bar{H}(\zeta,t)$ surface. The necessity of better characterizing this surface was anticipated in the course of Moldover and Gallagher's unsuccessful attempt to correlate the data for propane-octane mixtures.

We have assumed that $\bar{H}(\zeta,t)$ is an analytic function at the critical point, thus one can hope that standard mixture models can describe this surface sufficiently well near the critical point to obtain C_H and C_Z without recourse to fitting data in the critical region of the mixtures. In future work this hope should be tested. The third parameter that we have used, C_X, was introduced by Rainwater [13]. He observed that a combination of mole fraction and molar density might be a suitable measure of the dissimilarity between coexisting phases (or indeed a suitable "order parameter" in the sense of critical phenomena theories). There is precedent for using measures of dissimilarity that are more complex than density differences. In particular, Balfour et al. [19] found that the PVT data throughout the critical region of steam could be accurately correlated using a scaling thermodynamic potential such that a combination of the density difference and the entropy difference between coexisting phases behaves as $B(-t)^\beta$ to lowest order in t.

We have attempted to use the above-mentioned concepts to correlate the PVTx data for the coexisting phases of propane-octane mixtures obtained by Kay et al. [5]. Even with the innovations of Section 4, we found essentially the same difficulties that Moldover and Gallagher did. In particular we were not able to represent accurately the data for the propane-rich mixtures.

The authors would like to acknowledge stimulating conversations with E. D. Sloan and T. A. Al-Sahhaf and assistance with the minicomputer graphics from J. R. Fox.

7. REFERENCES

1. Peng, D. Y, and Robinson, D. B., *Ind. Eng. Chem. Fund.* 15:59 (1976).
2. Leung, S. S., and Griffiths, R. B., *Phys. Rev. A* 8: 2670 (1973).
3. Moldover, M. R., and Gallagher, J. S., *AICHE J.* 24: 267 (1978).
4. Bloomer, O. T., and Parent, J. D., *Chem. Eng. Prog. Symp. Ser.* 49(6):11(1953).
5. Kay, W. B., Genco, J., and Fichtner, D. A., *J. Chem. Eng. Data* 19:275 (1974).
6. Rowlinson, J. S., *Liquids and Liquid Mixtures* (Plenum Press, New York, 1969).
7. Van Konynenburg, P. H., and Scott, R. L., *Phil. Trans. Roy. Soc. London* 298:495 (1980).
8. Griffiths, R. B., and Wheeler, J. C., *Phys. Rev. A* 2: 1047 (1970).
9. D'Arrigo, G., Mistura, L., and Tartagli, P., *Phys. Rev. A* 12:2587 (1975).
10. Sengers, J. M. H. Levelt, Greer, W. L., and Sengers, J. V., *J. Phys. Chem. Ref. Data* 5:1 (1976).
11. Greer, S. C., and Moldover, M. R., *Ann. Rev. Phys. Chem.* 32:233 (1981).
12. Moldover, M. R., and Gallagher, J. S., in *Phase Equilibria and Fluid Properties in the Chemical Industry*, ACS Symposium Series No. 60, S. I. Sandler and T. J. Storvick, Eds., American Chemical Society, Washington, 1977, p. 498.
13. Rainwater, J. C., "Vapor-Liquid Equilibrium of Binary Mixtures Near the Critical Locus," to be published.
14. Goodwin, R. D., "The Thermophysical Properties of Methane, from 90 to 500 K at Pressures to 700 Bar," Nat. Bur. Stand. Tech. Note 653, (1974).
15. Jacobsen, R. T., Stewart, R. B., McCarty, R. D., and Hanley, H. J. M., "Thermophysical Properties of Nitrogen from the Fusion Line to 3500 R for Pressures to 1500 psia," Nat. Bur. Stand. Tech. Note 648, (1973).
16. Prydz, R., and Goodwin, R. D., *J. Chem. Therm.* 4:127 (1972).
17. Weber, L. A., *J. Chem. Therm.* 2:839 (1970).
18. Doiron, T., Behringer, R. P., and Meyer, H., *J. Low Temp. Physics* 24:345 (1976).
19. Balfour, F. W., Sengers, J. V., Moldover, M. R., and Sengers, J. M. H. Levelt, *Proc. 7th Symposium on Thermophysical Properties*, ed. by A. Cezairliyan, 786 (ASME, New York, 1977).

CHAPTER 11

MOLECULAR THERMODYNAMICS OF
DILUTE SOLUTES IN
SUPERCRITICAL SOLVENTS

D. A. Jonah[†], K. S. Shing, V. Venkatasubramanian, and
K. E. Gubbins

School of Chemical Engineering, Cornell University,
Ithaca, New York 14853

ABSTRACT

The theoretical basis of some commonly used equations of state are briefly reviewed, along with more rigorous perturbation theories for spherical and nonspherical molecules. A particularly successful perturbation theory, which we term the Padé equation of state, is used to investigate the degree of enhancement in supercritical fluid extraction (SFE) systems for various solute and solvent types. Direct comparisons of theory and experiment are made for several systems, using only one adjustable mixture parameter which is independent of temperature. A new form of mean field approximation is also proposed, which leads to a new expression for the solubility of a condensed solute. Comparison with experiment gives good results at high density.

1. INTRODUCTION

Predicting the solubility of condensed nonvolatile solutes in supercritical solvents involves a number of difficulties not normally met with in other phase equilibrium calculations. In particular, the solute and solvent molecules often differ greatly in molecular diameter and in their interaction strengths, leading to highly nonideal mixtures in which simple mixing rules of the van der Waals type

---

[†]Permanent address:
  Department of Mathematics, Fourah Bay College, University of Sierra Leone, Freetown, Sierra Leone, West Africa

are inadequate. Moreover, one must contend with the rapid density changes and anomalous behavior that arise in the critical region — a region that is notoriously difficult for classical equations of state to deal with.

Equations of state that have been used to study supercritical fluid extraction (SFE) systems include (a) the virial expansion [1], (b) the Carnahan-Starling-van der Waals (CSVDW) equation [2], and (c) various empirical equations such as the Redlich-Kwong [3,4], Peng-Robinson [3,5,6], and Lee-Kesler [7] equations. The virial expansion works well at the lower pressures and densities, but cannot describe the dense fluid region. The principal problem with the other equations of state are their reliance on mixing rules of the van der Waals type. Such mixing rules work best for nearly ideal solutions (see Sec. 3 below), but become progressively worse as the molecules become more dissimilar. They are particularly poor when the molecules differ significantly in size, as is generally the case in SFE systems. This defect shows itself in the use of the Redlich-Kwong and Peng-Robinson equations, for example, by the strong temperature dependence of the mixture parameter k_{12}, and in some cases by negative (unphysical) k_{12} values. As a result these equations can correlate existing data, but are of little value for extrapolation or prediction. The CSVDW equation is superior in this respect, because accurate hard sphere mixing rules are built into the hard sphere (CS) part of the equation. Nevertheless the CSVDW is poor at the lower densities, and particularly in the critical region itself. This is expected since the mean field treatment used is only a good approximation at high densities.

In this paper we present some exploratory studies based on thermodynamic perturbation theory. The equation of state is more complex, but offers two advantages over the more empirical approaches: (a) more accurate mixing rules are built into the equations, and (b) electrostatic (and other non-central) intermolecular forces can be accounted for.

2. CLASSIFICATION OF PHASE BEHAVIOR

It is convenient to classify binary phase behavior on the basis of the types of critical and three-phase lines present, and on the way these intersect. For fluid phase equilibria the classification scheme shown in Figure 1 is convenient, and includes all the known binary types. Class I systems are often fairly ideal in a thermodynamic sense, and do not exhibit liquid-liquid immiscibility. The remaining 5 classes display liquid-liquid separation of various kinds. The most common types of behavior are classes I, II, and III, but SFE can involve any of the classes I - V. Detailed discussions of these various types of phase behavior are given elsewhere [8-11]. These classes of phase behavior are

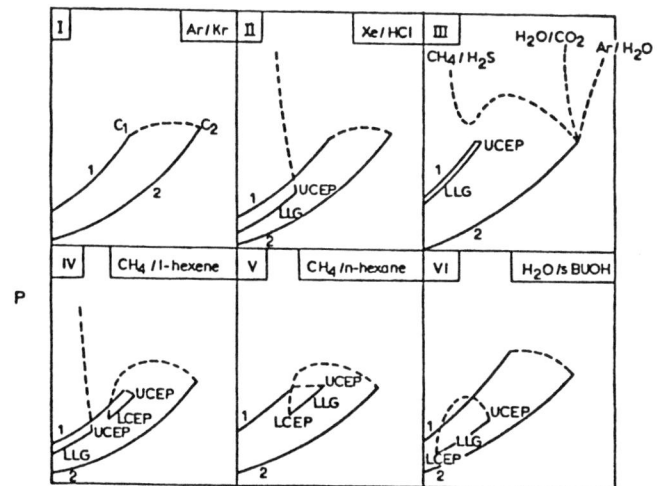

Figure 1. Classification of binary fluid phase diagrams using PT projections. Curves labelled 1 and 2 are pure component vapor pressures, C_1 and C_2 are critical points of pure 1 and 2, dashed lines are critical loci, and LLG is the liquid-liquid-gas line.

further complicated by the presence of solid phases, and many subclasses occur [8-11]. The type of behavior shown in Fig. 2 is that observed for C_2H_4 with naphthalene, anthracene and hexachloroethane, and for C_2H_6 with naphthalene.

3. THEORY

3.1 Empirical Equations of State

In perturbation theory one relates the properties of the fluid of interest (an SFE mixture in this case) to those of a simpler reference fluid whose equation of state and other properties are accurately known. Suitable reference systems include pure fluids (e.g. argon, methane), mixtures of hard spheres, or mixtures of nonspherical hard bodies. The empirical equations of state commonly used by engineers — Redlich-Kwong (RK), Peng-Robinson (PR), etc. — can be regarded as having some molecular foundation and can be derived from a perturbation expansion about a hard sphere mixture [12,13]. It is worthwhile to briefly review this aspect of these expressions, since it throws some light on the reasons for their failure in certain systems, and points the way to possible improvements. For spherical molecules, first-order perturbation theory gives the pressure as

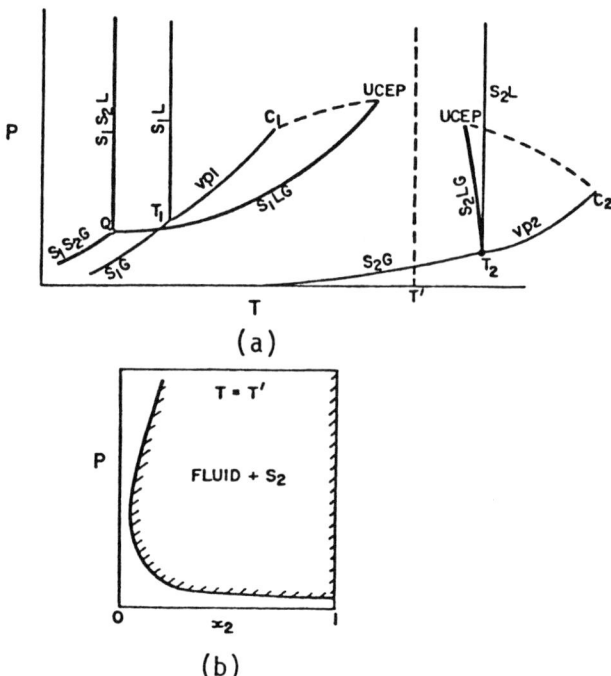

Figure 2. (a) PT diagram for a modified class I system with solid phases present. Here the SLG (solid-liquid-gas) lines intersect the vapor-liquid critical line at critical end points. Three-phase lines are shown bold. Q is the quadruple (S_1S_2LG) point, T are triple points. (b) Px diagram for the system shown in figure (a) at temperature T' between the two ucep's.

$$p = p_o - \frac{\partial}{\partial V} [2\pi \rho^2 V \int_o^\infty u_p(r) g_o(r) r^2 dr] \qquad (1)$$

where p_o is the reference system pressure, $\rho = N/V$ is the number density, $g_o(r)$ is the radial distribution function of the reference system (essentially the probability of finding two molecules a distance r apart), and $u_p \equiv u - u_o$ is the perturbing intermolecular potential for a pair of molecules. To derive the van der Waals (VDW) equation of state from (1) we assume:
(a) the reference system is one of hard spheres,
(b) the integral appearing in (1) is roughly constant (independent of temperature and density) at liquid-like densities (mean field approximation), and
(c) we can use the van der Waals form for p_o, i.e. $p_o = RT/(V-b)$.

This leads to the VDW equation,

$$p = \frac{RT}{V-b} - \frac{a}{V^2} \tag{2}$$

For mixtures this treatment gives the usual mixing rule for a,

$$a = \sum_\alpha \sum_\beta x_\alpha x_\beta a_{\alpha\beta} \tag{3}$$

Eq. (2) is open to several criticisms:

1. As pointed out by Henderson [13] and others, the VDW expression for p_0 is a very poor representation of hard spheres. The Carnahan-Starling (CS) [14] equation is very much better. Moreover the ad hoc mixing rules invoked for b are no longer needed in the CS treatment.
2. The mean field approximation (b) is only reasonable at high density. As the density is lowered towards the critical value it becomes poorer. Also, higher order perturbation terms become important as the density is lowered.
3. The hard sphere fluid is not the best reference system for highly non-spherical or polar molecules.
4. The mixing rule in (3) is likely to be best for mixtures of similar molecules, and poorest for very nonideal mixtures (see Sec. 3.2 below).

The RK and PR equations address problem 2. By introducing a modified temperature and density dependence [12] for the integral in eq. (1) they are better able to describe the first-order perturbation term. Thus, for the RK equation,

$$p = \frac{RT}{V-b} - \frac{a}{T^{1/2} V (V+b)} \tag{4}$$

the last term on the right-hand side is a better representation of the effect of attractive forces than $-a/V^2$ in (2), and probably accounts quite well for the effect of higher order perturbation terms.

The CSVDW equation of Johnston and Eckert [2] addresses problem 1 described above, by replacing the VDW expression for p_0 by the more accurate CS equation. The result is better agreement with experiment at the high densities with $a_{\alpha\beta}$ parameters that are only slightly temperature dependent. However, the mean field treatment and use of the hard sphere reference leaves problems 2 - 4 unresolved.

3.2 Perturbation Theory for Spherical Molecules

For mixtures of spherical or near-spherical molecules the hard sphere perturbation theory referred to above can be used in its rigorous form, or in some simplified way (RK, VDW, CSVDW, etc.). In this section we briefly mention two other theories that have been widely used, and which can form the foundations for the more comprehensive theories described in Sec. 3.3 below.

The first of these theories is the van der Waals 1 fluid (vdW1) theory, a particular form of conformal solution theory [12,15]. It assumes (a) that the molecules are conformal, i.e. they obey the same intermolecular force law, differing only in the values of the potential parameters $\varepsilon_{\alpha\beta}$ and $\sigma_{\alpha\beta}$, and (b) the values of $\varepsilon_{\alpha\beta}$ and $\sigma_{\alpha\beta}$ for the various molecular pairs are not too different from each other. Because of (b) it is possible to expand the Helmholtz free energy A about that of an ideal solution of molecules all of which have the same parameters ε_x and σ_x, and to terminate the series at the first-order term. The expansion parameters in vdW1 theory are the combinations $\varepsilon\sigma^3$ and σ^3, which appear to give more rapid convergence than other choices that have been tried [12]. This leads to

$$A = A_x + R_\varepsilon \sum_\alpha \sum_\beta x_\alpha x_\beta (\varepsilon_{\alpha\beta}\sigma^3_{\alpha\beta} - \varepsilon_x \sigma^3_x)$$

$$+ R_\sigma \sum_\alpha \sum_\beta x_\alpha x_\beta (\sigma^3_{\alpha\beta} - \sigma^3_x) + \cdots \qquad (5)$$

Here R_ε and R_σ are pure fluid integrals for the reference fluid, the precise form of which need not concern us here, A_x is the reference mixture free energy, and σ_x and ε_x are the size and energy parameters for the reference fluid. If ε_x and σ_x are now chosen according to the vdW1 mixing rules,

$$\varepsilon_x \sigma^3_x = \sum_\alpha \sum_\beta x_\alpha x_\beta \varepsilon_{\alpha\beta} \sigma^3_{\alpha\beta} \qquad (6)$$

$$\sigma^3_x = \sum_\alpha \sum_\beta x_\alpha x_\beta \sigma^3_{\alpha\beta} \qquad (7)$$

then the first order term in (5) vanishes, and (5) reduces to the simple result $A = A_x$. This is the basis of corresponding states treatments for mixtures, and for the shape factor methods. It is easy to show that (6) implies the van der Waals mixing rule for a that is given in eq. (3).

Clearly eqns. (5) - (7) can only be expected to apply

for mixtures in which the molecules are not too different, i.e. in which σ_{11}/σ_{22} and $\varepsilon_{11}/\varepsilon_{22}$ are not too different from unity for binary 1/2 mixtures. For binary mixtures near equimolar in composition the vdW1 theory is usually thought to give satisfactory results if these ratios lie within the ranges $\frac{1}{2} \leq (\sigma_{11}/\sigma_{22})^3 \leq 2$ and $\frac{1}{4} \leq \varepsilon_{11}/\varepsilon_{22} \leq 4$. However, for dilute solutions, these ranges of applicability require modification (see below and Figures 3 and 4). If the series in (5) is extended to second order, it is found that these terms involve triple summations over the mole fractions [16]. Thus for mixtures of components of very different critical volumes and/or critical temperatures, as occurs in some SFE mixtures, eq. (6) and hence (3) are likely to be unsatisfactory.

The second theory mentioned in the beginning of this subsection is that due to Mansoori and Leland [17], in which the true mixture radial distribution function $g_{\alpha\beta}(r)$, is replaced by the corresponding function for a pure fluid evaluated at a reduced temperature $kT/\varepsilon_{\alpha\beta}$, and a reduced distance $r/\sigma_{\alpha\beta}$, and a reduced density $\rho\sigma_x^3$ with σ_x^3 given by (7).

A test of these two theories is shown in Figures 3 and 4 for infinitely dilute solutions of 2 in 1, where 1 and 2 are spherical Lennard-Jones molecules. In these figures K_2 is the Henry constant of the solute, defined as the limit of f_2/x_2 as $x_2 \to 0$, where f_2 is the fugacity. The points in these figures are exact computer simulation results for such mixtures [18,19], while the dashed and dotted lines give the results for the two theories. The usual Lorentz-Berthelot rules,

$$\varepsilon_{12} = (\varepsilon_{11}\varepsilon_{22})^{1/2}$$
$$\sigma_{12} = \frac{1}{2}(\sigma_{11} + \sigma_{22})$$
(8)

are used in these calculations. We note that if σ_{12}/σ_{11} or $\varepsilon_{12}/\varepsilon_{11}$ are zero (i.e. σ_{22} and hence σ_{12} are zero, or ε_{22} and hence ε_{12} are zero) the u_{22} intermolecular potential vanishes and the solute 2 is an ideal gas. In that limit the residual chemical potential μ_{2r}, and hence $\ln(K_2/\rho kT)$, vanish. It is seen from these figures that both the Mansoori-Leland approximation (MLA) and vdW1 treatments give quite a good description of the chemical potential for $\frac{1}{4} \leq \varepsilon_{11}/\varepsilon_{22} \leq 4$ (corresponding to critical temperatures that vary by up to a factor of 4) when the molecular sizes are the same; these limits correspond to $\frac{1}{2} \leq \varepsilon_{12}/\varepsilon_{11} \leq 2$. However, neither theory describes the effect of molecular size differences at all well (Figure 4), particularly when the solute is much larger than the solvent molecules. For many SFE systems, therefore, such theories are expected to be

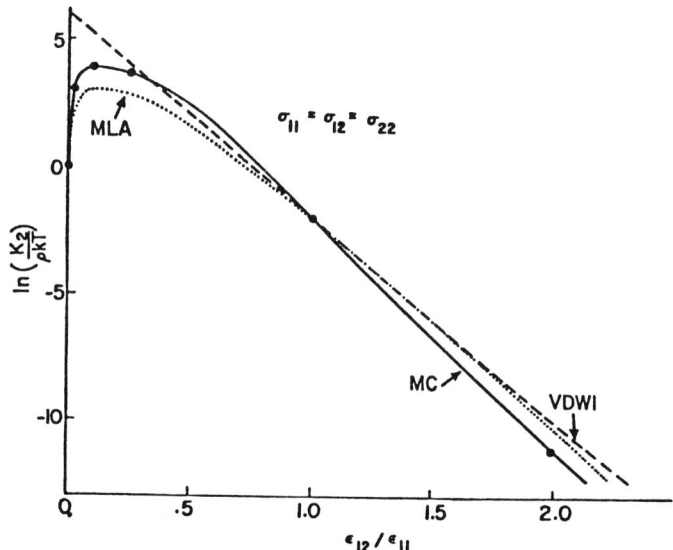

Figure 3. Variation of Henry constant, K_2, with ratio $\varepsilon_{12}/\varepsilon_{11}$ for Lennard-Jones (LJ) mixtures at $kT/\varepsilon_{11} = 1.2$, $\rho\sigma_{11}^3 = 0.7$. Here k is Boltzmann constant, $\rho = N/V$ is number density. The solid line and points are Monte Carlo computer simulation results, and the other lines are theoretical calculations.

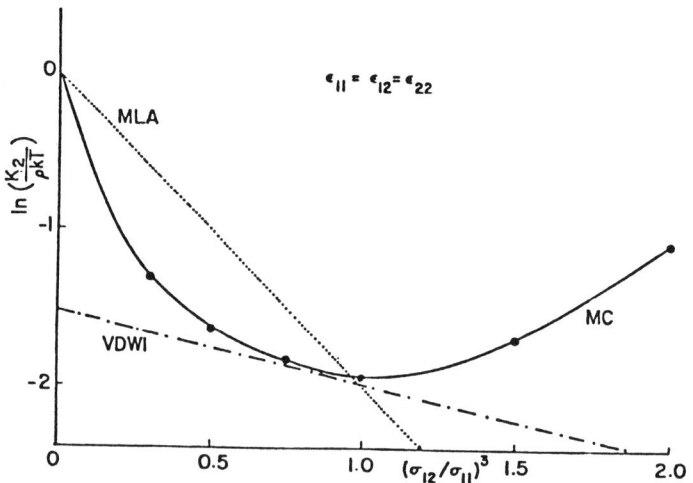

Figure 4. Variation of Henry constant K_2 with size ratio σ_{12}/σ_{11} for LJ mixtures at $kT/\varepsilon_{11} = 1.2$, $\rho\sigma_{11}^3 = 0.7$.

rather poor. Since the RK and related equations of state make use of the vdW1 mixing rules they will have similar defects.

3.3 Perturbation Theory for Nonspherical Molecules

For most fluids the intermolecular forces are strongly orientation-dependent and include both long-range (electrostatic, etc.) and short-range (repulsion, H-bond) contributions. Several forms of perturbation theory have been proposed recently to account for the effects of such forces, but the only one that has been extensively used for calculations is the Padé approximant form due to Stell et al. [20],

$$A = A_0 + A_2 (1 - A_3/A_2)^{-1} \qquad (9)$$

Here A_0 is the reference system free energy, A_2 is the second-order perturbation term, etc. (A_1 vanishes in this series). In this expansion the reference fluid pair potential $u^0_{\alpha\beta}(r)$ is defined by

$$u^0_{\alpha\beta}(r) \equiv \langle u_{\alpha\beta}(\underline{r}\omega_1\omega_2)\rangle_{\omega_1\omega_2} \qquad (10)$$

where α and β are mixture components, ω_i ($=\theta_i\phi_i$ or $\phi_i\theta_i\chi_i$ for linear or non-linear molecules, respectively) denotes the orientation of molecule i, and $\langle\cdots\rangle_{\omega_1\omega_2}$ means an unweighted averaging over molecular orientations. Usually some simple model, such as the Lennard-Jones (12,6) or (n,6), is chosen for u^0. For the (12,6) model both the equation of state and radial distribution function are known from computer simulation studies [21], and it is a simple matter to relate the properties of the (n,6) fluid to those for the (12,6) fluid [22]. Equation (9) has been developed as an equation of state for mixtures by Gubbins, Gray, and co-workers (see refs. 22 and 23 for a review, together with detailed equations). The terms A_2 and A_3 involve two-and three-body integrals for the reference fluid that have been evaluated for a variety of potential forms and fitted to simple functions of temperature and density [21]. Equation (9) agrees very well with computer simulation results for polar and quadrupolar liquids, even when the dipoles or quadrupoles are strong [23].

3.4 A Mean Field Treatment

Finally, we briefly mention a mean field approximation which seems promising, and which is based on a molecular treatment. For a condensed phase in equilibrium with a gaseous phase it is possible to show that y_2 must obey the following equation [24],

$$\frac{(v_2^c - v_1)}{RT} - \left(\frac{\partial \ln y_2}{\partial p}\right)_T - \frac{v_1}{RT} \ln\left[\left(\frac{p_2^{sat} Z_1}{p\, y_2 f_1}\right) + \frac{v_2^c p}{RT}\right] = \frac{K}{T^2} \qquad (11)$$

where v_1 and v_2^c are molar volumes of pure gaseous solvent (1) and condensed phase solute (2), respectively, p_2^{sat} is vapor pressure of 2, Z_1 is the compressibility factor for pure 1, f_1 is its fugacity, and y_2 is mole fraction of solute in the gas phase. For spherical molecules K is given by

$$K = -\frac{N}{k^2} \int_0^1 d\xi \int d\underline{r}\; [g_{12}(r;\xi) u_{12}(r) - g_{11}(r;\xi) u_{11}(r)] \qquad (12)$$

where N is the total number of molecules, k is Boltzmann's constant, ξ is a coupling parameter that couples molecule 1 of component 1 to the system, such that $\xi = 0$ corresponds to uncoupling (molecule 1 does not interact at all) and $\xi = 1$ to full coupling, $g_{ij}(r;\xi)$ is the radial distribution function for the partially coupled system, and $u_{ij}(r)$ is the ij pair potential due to intermolecular forces. The only approximations involved in equation (11) are that the solution is dilute (i.e. in the Henry's law region) and that the potentials are pairwise additive.

The quantity K is difficult to evaluate in general. As a zeroth approximation we propose that at high densities K should be roughly constant. Such a mean field approximation is in the spirit of that used to derive the VDW equation of eq. (2), but in this case it does not involve the neglect of higher order perturbation terms.

4. RESULTS

4.1 Effect of Solute and Solvent Type

In this first series of calculations we use the Padé equation of state, (9), to study the role of various types of intermolecular forces in determining SFE behavior. As solutes we take n-octane as a prototype nonpolar hydrocarbon, and methanol as a prototype polar, hydrogen-bonded substance. These choices are governed by the following factors. First, intermolecular potential models that fit the thermodynamic properties of these two liquids quite well have been given by Gibbs [25]. Second, it is convenient to use the vdW1 conformal solution theory, eqns. (5) - (7), to calculate the reference mixture free energy A_0 in (9), and this is only valid provided that the diameters of the solute and solvent molecules are not very different. We therefore wish to avoid very large solute molecules. The intermolecular potential models used for the C_8H_{18}/C_8H_{18} and MeOH/MeOH interactions are

$$u_{MeOH/MeOH} = u_{(n,6)} + u_{\mu\mu} + u_{\mu Q} + u_{QQ} + u_{ov}^{a} \qquad (13)$$

$$u_{C_8H_{18}/C_8H_{18}} = u_{(12,6)} + u_{QQ} \qquad (14)$$

where $u_{(n,6)}$ and $u_{(12,6)}$ are the (n,6) and Lennard-Jones (12,6) potentials, respectively [26], and are the reference potentials u^o defined by eq. (10); $u_{\mu\mu}$, $u_{\mu Q}$ and u_{QQ} are the dipole-dipole, dipole-quadrupole, and quadrupole-quadrupole potentials, respectively, and u_{ov}^a is an anisotropic overlap (shape) term consisting of $\ell_1\ell_2\ell$ = 202 and 022 harmonics [22,23]. Potential parameters used for methanol and n-octane are given in Table 1. The solubility of the solute 2 in the gas phase was calculated by equating the fugacity of 2 in the condensed and gas phases,

$$f_2^C = f_2^G \qquad (15)$$

Experimental data (vapor pressure, molar volume) was used to calculate f_2^C in the usual way [1], it being assumed that the condensed phase was pure 2. This is believed to be a reasonable approximation in the calculations reported here. Equation (9) was used as the equation of state for the gas phase. In principle it is possible to use eq. (9) for both phases, and this has been done in some comparisons with experimental data (see Sec. 4.2). However, this poses an excessive demand on the equation of state for the reference system, A_0, because of the very large difference in density between the two phases.

Three model solvents were investigated, as prototypes of nonpolar, polar, and quadrupolar fluids, respectively. These models were the Lennard-Jones (LJ), LJ + dipole (LJ + μ, or Stockmayer), and LJ + quadrupole (LJ + Q) potentials,

$$LJ: \quad u = u_{LJ} = 4\varepsilon\left[\left(\frac{\sigma}{r}\right)^{12} - \left(\frac{\sigma}{r}\right)^{6}\right] \qquad (16)$$

$$LJ + \mu: \quad u = u_{LJ} + u_{\mu\mu} \qquad (17)$$

$$LJ + Q: \quad u = u_{LJ} + u_{QQ} \qquad (18)$$

The calculation procedure for f_2^G in eq. (15) was that described by Twu and Gubbins [22,23]. Calculations were carried out at several temperatures to explore the effect of varying the solvent for a given solute, and also to study the effect of varying the solute for a fixed solvent. In most of these calculations the solute was a liquid (because of a lack of experimental data on the solid solute), and the solvent critical temperature was kept fixed by suitably adjusting ε_{11}. In most calculations the solvent critical

Table 1. Potential Parameters for Like-Pair Interactions

Substance	ε/k/K	$\sigma/\text{Å}$	n	δ	$\mu/10^{-18}$ esu	$Q/10^{-26}$ esu	μ^*	Q^*	T_C/K
(a) 'Real' Fluids [a]									
nC_8H_{18} [b]	290.05	6.7818	12	–	0	30.82	0	1.298	–
CH_3OH [b]	276.22	4.207	21.9	–0.17	1.7	9.173	1.002	1.285	–
CH_4 [b] (A)	147.66	3.7456	12	0	0	0	0	0	–
CH_4 [c] (B)	153.95	3.741	13	0	0	0	0	0	–
C_2H_6	208.65	4.3527	12	0	0	5.55	0	0.822	–
(b) Hypothetical Solvents: Octane Solute [d]									
LJ_1	184.15	6.7818	12	0	0	0	0	0	235.5
LJ_2	184.15	9.0424	12	0	0	0	0	0	235.5
$LJ + \mu$	155.50	6.7818	12	0	2.94	0	1.136	0	235.5
$LJ + Q$	100.00	6.7818	12	0	0	19.09	0	1.357	235.5
(c) Hypothetical Solvents: Methanol Solute [d]									
LJ	184.15	4.207	12	0	0	0	0	0	235.5
$LJ + \mu$	100.00	4.207	12	0	0.79	0	0.78	0	235.5
$LJ + \mu$	86.00	4.207	12	0	2.15	0	2.29	0	235.5
$LJ + Q$	100.00	4.207	12	0	0	5.79	0	1.357	235.5

Table 1. Footnotes

a) By 'real fluid' we mean that the potential parameters have been fitted to experimental vapor pressure and saturated liquid density data for the actual liquid.

b) From Gibbs [25].

c) The $CH_4(B)$ model is that of Clancy [27] and is used in the CH_4/H_2 calculations of Fig. 12. The potential model is that of eq. (23) and includes an octopole term. The octopole moment used for methane is $\Omega = 26 \times 10^{-34}$ esu.

d) These are the solvents used in calculations shown in Figures 5 - 10.

temperature was kept fixed at 235.5 K and the temperature of the SFE mixture was 237 K, i.e. 1.5 K above T_{c1}. This temperature is somewhat above the melting-points of octane (216.3 K) and methanol (175.3 K). In the calculations using the LJ solvent, the molecular diameter σ_{11} is the only adjustable parameter (ε_{11} is fixed by requiring T_{c1} = 235.5 K). For the LJ + μ solvent the only free solvent parameters are σ and μ^*, and for the LJ + Q case they are σ and Q^*. Here μ^* and Q^* are dimensionless dipole and quadrupole moment respectively, defined by

$$\mu^* \equiv \mu/(\varepsilon\sigma^3)^{1/2} \tag{19}$$

$$Q^* = Q/(\varepsilon\sigma^5)^{1/2} \tag{20}$$

Values of the potential parameters for these various hypothetical solvents are included in Table 1. In most cases the solvent diameter σ_{11} was set equal to the solute diameter σ_{22}; in such cases the vdW1 theory used to evaluate A_0 should give excellent results.

Results of these calculations are shown in Figures 5-10. In Figures 7, 9 and 10 E is the enhancement, defined as

$$E = \frac{y_2 p}{p_2^{sat}}$$

From Figures 5 and 7 it is seen that the use of a polar or quadrupolar solvent increases the enhancement substantially, even when the critical temperature of the solvent is kept fixed by reducing ε. The densities of the various solvents in these figures are very similar, so that the increased enhancement arises directly from the difference in intermolecular forces rather than from a density effect. For such solvents strong attractions occur between solute and solvent molecules for favorable mutual orientations of the molecules, and this results in increased solubility. Reduced quadrupoles (Q^*) are known to have a greater effect on the orientational ordering of the molecules than for dipoles (μ^*) of the same strength, and this presumably explains the additional enhancement for the quadrupolar solvents. In Figure 6 is displayed the effect of variation in size of the solvent molecules, keeping the size of the solute fixed (and also ε_{11}, ε_{22}, and T_{c1}). We see that an increase in the size of the solvent causes the solubility of the solute to decrease. This is primarily a density effect. For a given pressure at fixed $\varepsilon_{11}, \varepsilon_{22}$, and σ_{22}, an increase in σ_{11} causes the gas number density to decrease, leading to a reduction in solvent power. Calculations for diameter ratios much different from unity cannot be made with confidence

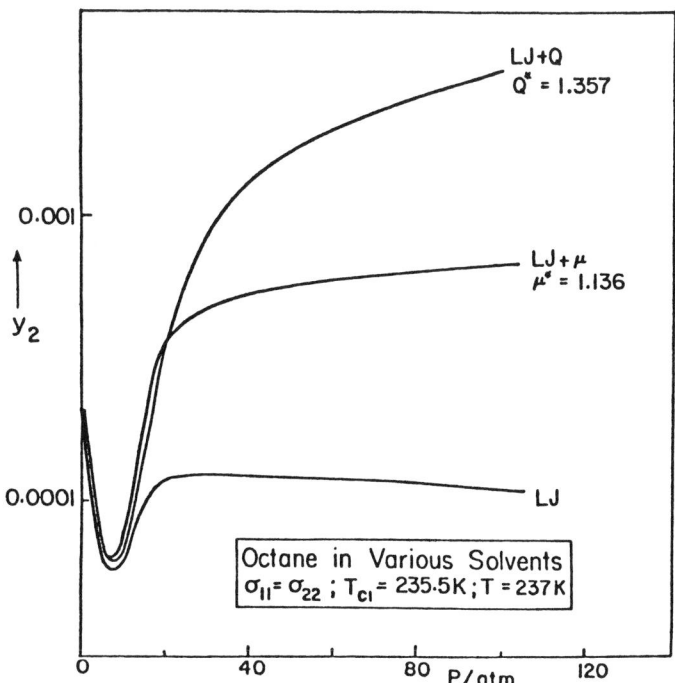

Figure 5. Solubility of octane in various solvents.

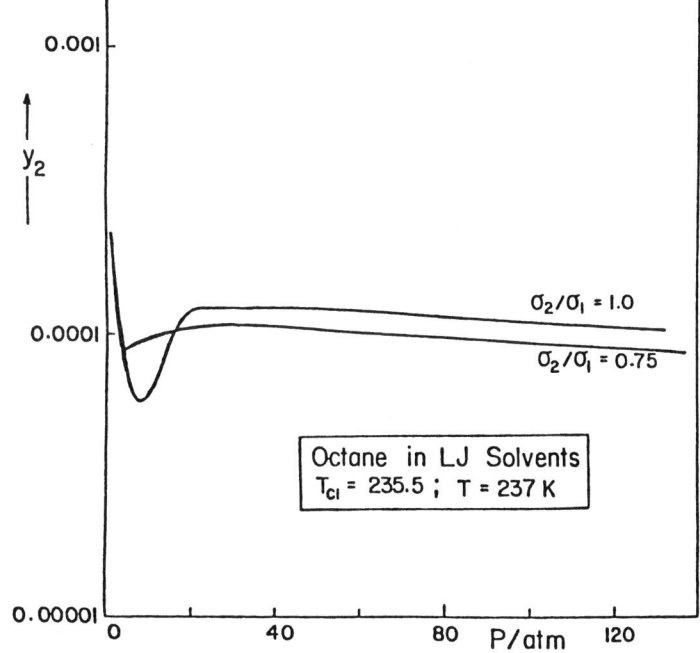

Figure 6. Solubility of octane in LJ solvents.

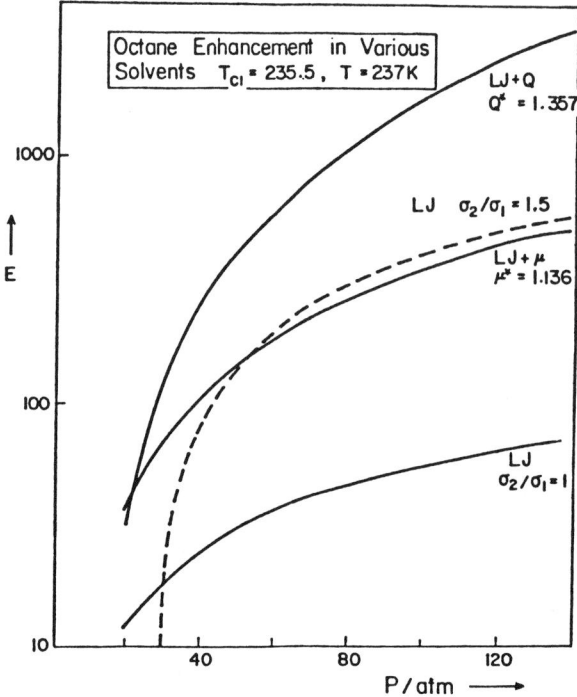

Figure 7. Solubility enhancement of octane.

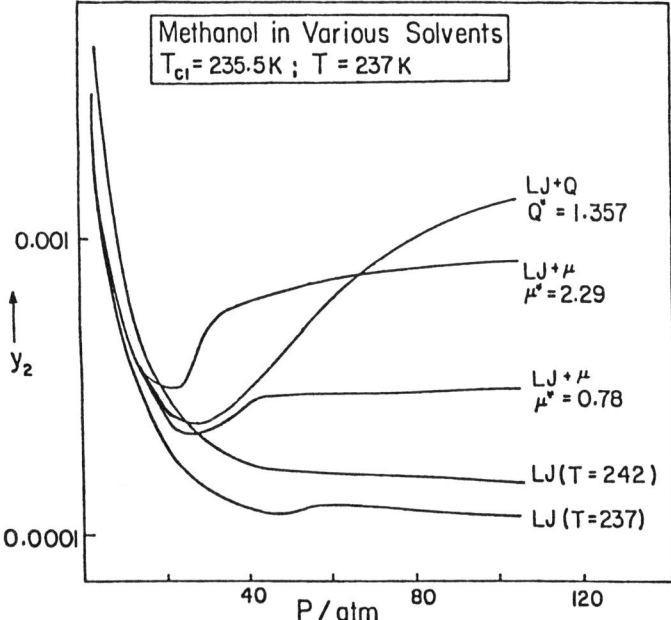

Figure 8. Solubility of methanol in various solvents.

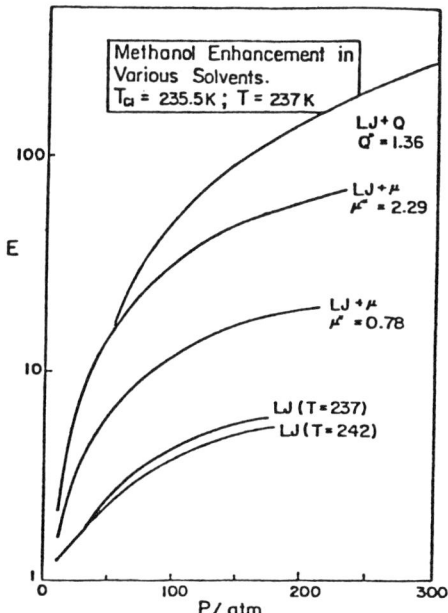

Figure 9. Solubility enhancement of methanol.

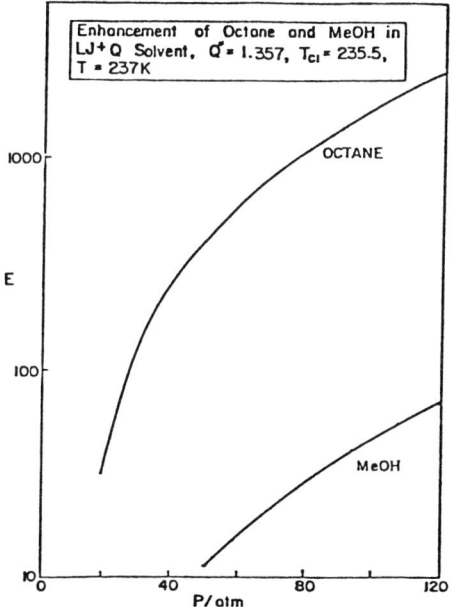

Figure 10. Comparison of enhancements for methanol and octane in a LJ + Q solvent.

because of the failure of the vdW1 model used here (see Fig. 4).

Similar results for methanol as solute are shown in Figures 8 and 9. The trends for various solvents are similar to those for octane, although the enhancements in a given solvent are smaller in the case of methanol (see also Fig. 10). We attribute this to the fact that the electrostatic forces between methanol molecules are stronger than those between octane molecules, so that the dissolving power of a given solvent is relatively less for methanol as solute. Also shown in Figures 8 and 9 is the effect of raising the temperature to 242 K (6.5 K above T_{c1}). Although this increases y_2 the effect is due to the increase in the vapor pressure of methanol, and the enhancement is actually reduced. We note that the reduced densities for the solvents in these figures are quite high, typically $\rho^* = \rho\sigma^3 \sim 0.5 - 0.7$ for 80 atm. or above. This is well above the critical density ($\rho^* \sim 0.35$) of the solvent, and well beyond the range of validity of the virial series. At the higher pressures the density changes only slowly with pressure, and this leads to the flatness of the $y_2 - P$ curves in this region.

4.2 Comparison with Experiment

We first present some results for the direct comparison of the Padé equation of state, (9), with experiment. In Figures 11 and 12 are shown such results for the systems CH_4/CH_3OH and CH_4/H_2. These two systems are of class III. The potential model used for CH_3OH was that of eq. (13). For H_2 the potential model was [27]

$$u_{H_2/H_2} = u_{(n,6)} + u_{QQ} + u_{ov}^a + u_{dis}^a \qquad (21)$$

where u_{dis}^a is an anisotropic dispersion term defined in ref. 27. For CH_4 slightly different potentials were used in the CH_4/CH_3OH calculation reported here to that used by Clancy [27] in the CH_4/H_2 calculation. We refer to these two models as $CH_4(A)$ and $CH_4(B)$. They are

$$u_{CH_4/CH_4}(A) = u_{(12,6)} \qquad (22)$$

$$u_{CH_4/CH_4}(B) = u_{(n,6)} + u_{\Omega\Omega} \qquad (23)$$

where $u_{\Omega\Omega}$ is an octopole-octopole term. The unlike pair potential models used were

$$u_{CH_4/CH_3OH} = u_{(n,6)} \qquad (24)$$

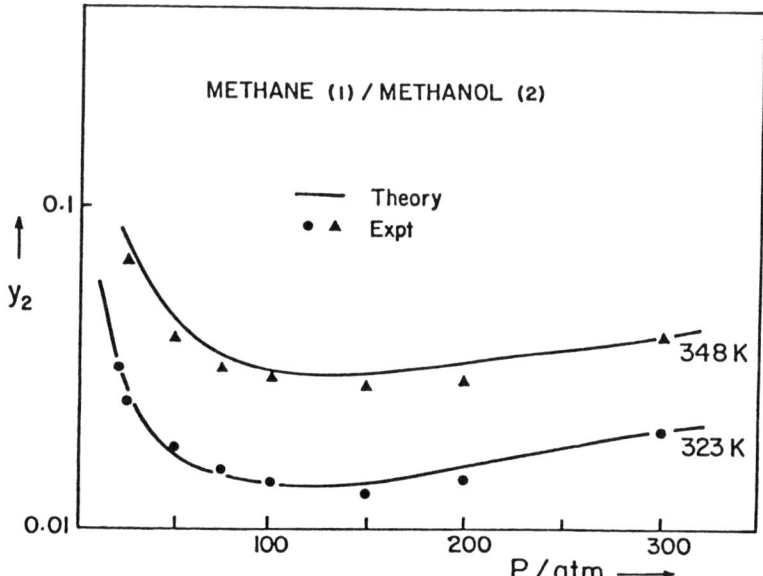

Figure 11. Solubility of methanol in supercritical methane from experiment [28] and from the Padé equation of state, using a single mixture parameter that is temperature-independent.

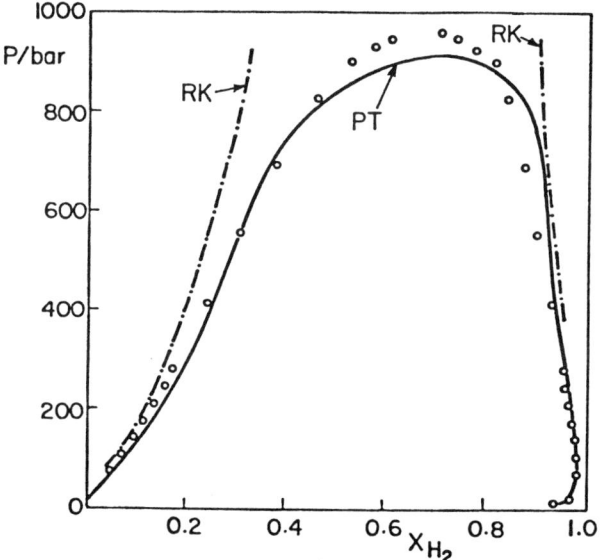

Figure 12. Vapor-liquid equilibria for H_2/CH_4 at 100 K from experiment [29] (points), the Padé equation of state (solid line) including quantum corrections, and the Redlich-Kwong equation of state (dash-dot-line). From Clancy and Gubbins [27].

$$u_{CH_4/H_2} = u_{(n,6)} + u_{Q\Omega} + u_{ov}^a \qquad (25)$$

The unlike pair parameters $n_{\alpha\beta}$, $\sigma_{\alpha\beta}$, $\delta_{\alpha\beta}$ and $\varepsilon_{\alpha\beta}$ were obtained from the approximate combining rules [26]

$$n_{\alpha\beta} = (n_{\alpha\alpha} n_{\beta\beta})^{1/2} \qquad (26)$$

$$\sigma_{\alpha\beta} = \tfrac{1}{2}(\sigma_{\alpha\alpha} + \sigma_{\beta\beta})^{1/2} \qquad (27)$$

$$\delta_{\alpha\beta} = \tfrac{1}{2}(\delta_{\alpha\alpha} + \delta_{\beta\beta}) \qquad (28)$$

$$\varepsilon_{\alpha\beta} = \xi_{\alpha\beta}(\varepsilon_{\alpha\alpha}\varepsilon_{\beta\beta})^{1/2} \qquad (29)$$

The unlike pair parameter $\xi_{\alpha\beta}$ was the only parameter adjusted to the mixture data.

The results for the CH_4/CH_3OH system are shown in Figure 11. The value of ξ_{CH_4/CH_3OH} was 1.5 and was obtained by fitting to the data at 323 K. The same value was used for the calculations for 348 K. In these calculations eq. (15) was used with f_2^C obtained from experimental data as in Sec. 4.1, and f_2^G obtained from the Padé equation of state of eq. (9). As expected from the results of Sec. 4.1 there is only a rather modest solubility enhancement for this system.

In Figure 12 are shown results calculated by Clancy [27] for the H_2/CH_4 mixture at 100 K. In this case the Padé equation of state, (9), was used for both phases in eq. (15), and the resulting value of ξ, $\xi_{H_2/CH_4} = 0.975$, is the best fit to both phases. The Padé calculated values include first order quantum corrections for H_2, which are appreciable at this low temperature, and are substantially better than the Redlich-Kwong results which do not account for quantum effects. Because of this latter defect, no solution of eq. (15) could be found using the RK equation for pressures above about 900 bar.

We have also carried out calculations using the mean field treatment proposed in Sec. 3.4 for solid-fluid SFE systems. The calculations are based on eq. (11) with K taken to be constant. Since the mean field approximation is expected to be best at high temperatures we use two high pressure data points on each isotherm in fitting parameters. Thus, taking values $y_2(p)$ and $y_2(p')$ at two pressures p and p' that are closely spaced enables an estimate of $\partial \ln y_2/\partial p$ to be made at the mean pressure $(p + p')/2$, and also yields the constant K. Values of y_2 at other pressures can then be calculated from eq. (11) by proceeding in small pressure steps to lower pressures, estimating $\partial \ln y_2/\partial p$ by linear interpolation in each step. Such a calculation is shown for ethylene (1)/naphthalene (2) in Figure 13. For each isotherm

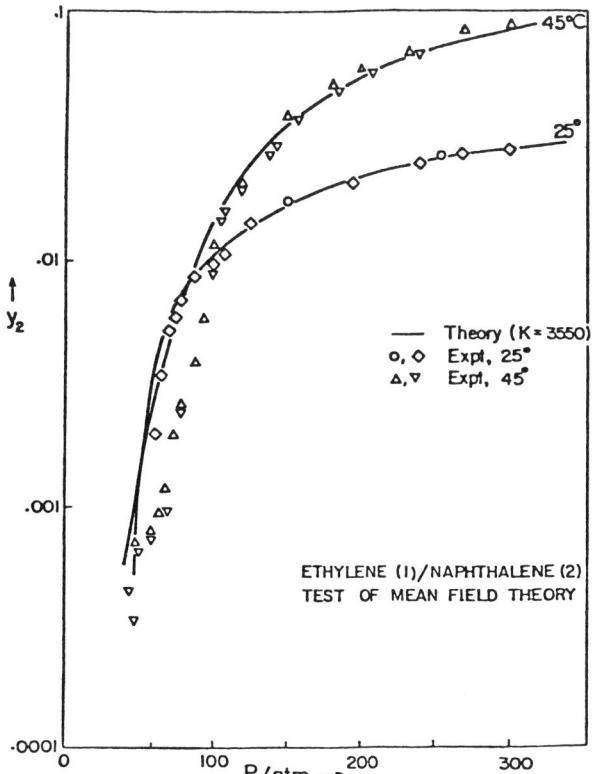

Figure 13. Solubility of naphthalene (2) in ethylene (1) from experiment (points) and from the mean field approximation (lines). Data is from the following sources: $\Diamond$, $\triangle$ Tsekhanskaya et al. [30]; $\mathbf{O}$, ∇ Diepen and Scheffer [13].

the two highest poessure points were used in the fitting. The values of K were:

T/°C	K/atm K^2
25	3508
35	3500
45	3640

The agreement is excellent for pressures above 100 atm, but becomes poorer at the lower pressures, as expected. This may be due to the use of high pressure data alone to fit K, and in part to the mean field approximation. Nevertheless the fit is as good as that from the RK equation, with a mixture parameter that is almost independent of temperature over the range 25 - 45° C. Also, eq. (11) offers a basis for further improvements by introducing better approximations for the K function designed by eq. (12).

5. CONCLUSIONS

Perturbation theory offers a powerful method of developing improved equations of state for SFE systems. In this paper we have presented initial calculations for two such approaches, the Padé equation of state of eq. (9) and the mean field approximation of eq. (11). The first of these can be used to study the role of various types of intermolecular forces in determining the enhancement for prototype solutes and solvents. Such a study has been carried out in Sec. 4.1 of this paper. The Padé equation also offers reasonable agreement with experiment using a single temperature-independent mixture parameter, as does the mean field approximation. These results are in contrast to those obtained using the RK and PR equations of state, where the mixture parameters are usually found to vary rapidly with temperature.

A significant source of error in the present calculations based on the Padé equation is the use of vdW1 theory for A_0. This is known to be poor when the molecular sizes are very different, as is often the case in SFE mixtures. One possible remedy may be to use a hard-sphere perturbation theory to calculate A_0, and we are presently investigating this.

ACKNOWLEDGMENTS

It is a pleasure to thank the National Science Foundation (grant CPE-8016951) and the Gas Research Institute for grants in support of this work.

REFERENCES

[1] For examples see: Prausnitz, J.M. *Molecular Thermodynamics of Fluid-Phase Equilibria* (Englewood Cliffs: Prentice-Hall, 1969), p. 163-173.
[2] Johnston, K.P. and Eckert, C.A. *AIChEJ* 27:773 (1981).
[3] MacKay, M.E. and Paulaitis, M.E. *IEC Fund.* 18:149 (1979).
[4] Dieters, U. and Schneider, G.M. *Ber. Bunsengesell. Phys. Chem.* 80:1316 (1976).
[5] Kurnik, R.T. and Reid, R.C. *AIChEJ* 27:861 (1981).
[6] Kurnik, R.T., Holla, S.J. and Reid, R.C. *J. Chem. Eng. Data* 26:47 (1981).
[7] Francis, D.C. and Paulaitis, M.E. *IEC Proc. Des. Dev.*, submitted 1980.
[8] Rowlinson, J.S. *Liquids and Liquid Mixtures* (London: Butterworths, 1969), Ch. 6.

[9] Schneider, G.M. In: Chemical Thermodynamics, M.L. McGlashan, Ed. (London: Specialist Periodical Report, Chemical Society, 1978); Schneider, G.M. Angew. Chem. Int. Ed. Engl. 17:716 (1978).
[10] Streett, W.B. Can. J. Chem. Eng. 52:92 (1974); Streett, W.B. and Gubbins, K.E. Proceedings of the Eighth Symposium on Thermophysical Properties, Gaithersburg (New York: Amer. Soc. Mech. Eng., 1981), in press.
[11] Streett, W.B., paper presented at the 74th Annual Meeting of the American Institute of Chemical Engineers, New Orleans, November 1981.
[12] Gubbins, K.E. AIChEJ 19:688 (1973).
[13] Henderson, D. In: Equations of State in Engineering and Research, K.C. Chao and R.L. Robinson, Eds. Adv. Chem. Series 182 (Washington, D.C.: American Chemical Society, 1979), p. 1.
[14] Carnahan, N.F. and Starling, K.E. J. Chem. Phys. 51:635 (1969).
[15] Leland, T.W., Rowlinson, J.S. and Sather, G.A. Trans. Faraday Soc. 64:1447 (1968).
[16] Mo, K.C., Gubbins, K.E., Jacucci, G. and McDonald, I.R. Mol. Phys. 27:1173 (1974).
[17] Mansoori, G.A. and Leland, T.W. J. Chem. Soc., Faraday Trans. II 68:320 (1972).
[18] Jonah, D.A., Shing, K.S. and Gubbins, K.E. Proceedings of the Eighth Symposium on Thermophysical Properties, Gaithersburg (New York: Amer. Soc. Mech. Eng., 1981), in press.
[19] Shing, K.S. and Gubbins, K.E. Mol. Phys., in press (1982).
[20] Stell, G., Rasaiah, J.C. and Narang, H. Mol. Phys. 23:393 (1974).
[21] Nicolas, J., Gubbins, K.E., Streett, W.B. and Tildesley, D.J. Mol. Phys. 37:1429 (1979).
[22] Twu, C.H. and Gubbins, K.E. Chem. Eng. Sci. 33:879 (1978).
[23] Gubbins, K.E. and Twu, C.H. Chem. Eng. Sci. 33:863 (1978).
[24] Jonah, D.A., Shing, K.S., Venkatasubramanian, V. and Gubbins, K.E., paper to be published.
[25] Gibbs, G.M. D. Phil. Thesis, University of Oxford (1979).
[26] Reed, T.M. and Gubbins, K.E. Applied Statistical Mechanics (New York: McGraw-Hill, 1973), Ch. 4,5.
[27] Clancy, P. and Gubbins, K.E. Mol. Phys. 44:581 (1981).
[28] Kritschevskii, I.R. and Koroleva, M. Acta Physicochimica, URSS 15:327 (1980).
[29] Tsang, C.Y., Clancy, P., Calado, J.C.G. and Streett, W.B. Chem. Eng. Commun. 6:365 (1980).

[30] Tsekhanskaya, Y.V., Iomtev, M.B. and Mushkina, E.V. Russ. J. Phys. Chem. 38:1173 (1964).
[31] Diepen, G.A.M. and Scheffer, F.E.C. J. Phys. Chem. 57:575 (1953).

CHAPTER 12

MEAN-FIELD LATTICE-GAS
DESCRIPTION OF FLUID-PHASE
EQUILIBRIA

L.A. Kleintjens and
R. Koningsveld
  DSM/Research and Patents,
  Geleen, The Netherlands

SUMMARY

   Fluid-phase behaviour of non-polar and polar, low and high-molecular substances and mixtures can be described nearly quantitatively with a slightly adapted version of the mean-field lattice-gas model, in considerable temperature and pressure ranges. The model is well-applicable to supercritical states and i.a. correctly predicts gas-gas demixing. The simple form of the mathematical expressions in this model allows determination of the parameters on scarce experimental information.

INTRODUCTION

   Lattice models have frequently been used in the description of phase behavior of liquid mixtures. One of the assumptions underlying such treatments is the equality of the partial specific volumes of the components in all liquid phases at constant pressure and temperature. In gas-liquid and supercritical fluid equilibria the densities of the phases may differ a great deal and the conventional rigid-lattice model is less suited for describing such phenomena.
   This problem is circumvented in the lattice-gas model which introduces randomly distributed vacancies in the lattice. The density of a system or phase can be reproduced easily by adjustment of the concentration of vacancies. The

model was developed i.a. by Mermin (1) and by Mulholland and Rehr (2)
for one-component systems on the basis of the hole theory of liquids devised by Altar (5), Cernuschi and Eyring (4), Frenkel (3) and others.

Trappeniers et al (6,7) extended the theory combining Guggenheim's (8) two-component lattice model with the lattice-gas treatment and were able to give a qualitatively correct description of the various forms in which gas-gas demixing occurs. The more-component lattice-gas model has also been used by a number of authors for other purposes (9-12). Recently we have found that quantitative agreement between experiment and predictions on the basis of the model, especially under supercritical conditions, is obtainable if Trappeniers' model is slightly extended. In this paper we review our attempts in the field.

Supercritical extraction is very much coming to the fore nowadays (13). One could speak of a new unit operation which utilises extreme deviations from ideality in the thermodynamic sense. Supercritical phenomena as such have been known for a long time already since Hannay noted the unexpected high solubility of metal halides in supercritical ethanol etc. (14). Here we have one class of supercritical extraction involving equilibria between solid and fluid phases. Another type of supercritical extraction may occur at temperatures above the triple point of the heavy component and involves fluid phases only (15).

Whereas phenomena indicated above relate to states between the two critical points of the two components in a binary system, gas-gas demixing occurring in the latter case is a truly supercritical phenomenon, restricted to conditions close to and above the critical point of the heavier component. The systems dealt with here fall in this category and do not contain solid phases.

EXTENDED LATTICE-GAS MODEL FOR A PURE SUBSTANCE

In the lattice-gas model a pure substance is conceived as a binary mixture of occupied and vacant sites, distributed randomly. Pressure and temperature changes cause variations in the concentration of the vacancies (holes) but v_0, the volume per lattice site (hole size) is constant. The density of a system or phase can be reproduced easily by adjustment of the concentration of vacancies.
Following Staverman (16) we assign surface areas σ_0 and σ_1 to both kinds of sites and further assume only nearest-neighbour interactions to be significant.
Finally, it will prove useful to allow molecules to occupy

more than one site, e.g. m_1 sites. Alternatively, the model also allows m_1 to be determined as a parameter by experimental data.

Standard calculation of ΔF, the Helmholtz free energy of mixing N_0 holes and N_1 molecules 1, each occupying m_1 sites, yields for strictly-regular mixing (17-23):

$$\Delta F/N_\varphi kT = \varphi_0 \ln \varphi_0 + \varphi_1 m_1^{-1} \ln \varphi_1 + \\ [\alpha_1 + g_{11}(1-\gamma_1)(1-\gamma_1 \varphi_1)^{-1}] \varphi_0 \varphi_1 \quad (1)$$

where:
φ_0 and φ_1 are the volume (site) fractions of holes and occupied sites defined by $\varphi_0 = N_0/N_\varphi$; $\varphi_1 = N_1 m_1/N_\varphi$ ($N_\varphi = N_0 + N_1 m_1$).
Further, $g_{11} = -\frac{1}{2} w_{11} \sigma_0/kT$ were w_{11} = the interaction energy per unit contact surface area involved in a 1 - 1 contact,

$\gamma_1 = 1 - \dfrac{\sigma_1}{\sigma_0}$ and α_1 is an empirical entropy correction.

Within this scope the 'concentration' variable φ_1 is directly related to the density d_1 of the system. For a substance with molar mass M_1 and molar volume V_1 we have:

$$\varphi_1 = d_1 v_0 m_1/M_1 = v_0 m_1/V_1 \quad (2)$$

Following Trappeniers etal (6,7) we express the spinodal condition as.

$$(\partial^2 \Delta F/\partial \varphi_1^2)_{V,T} (\partial^2 \Delta F/\partial V^2)_T - (\partial^2 \Delta F/\partial \varphi_1 \partial V)_T = 0 \quad (3)$$

Since $V = N_\varphi v_0$ eq 3 can be rewritten to

$$(\partial^2 \Delta F/\partial \varphi_1^2)_{V,T} = 0 \quad (4a)$$

Application of eq 4a to eq 1 leads to the spinodal expression:

$$\frac{1}{\varphi_0} + \frac{1}{\varphi_1 m_1} = 2\alpha_1 + 2g_{11}(1-\gamma_1)^2 (1-\gamma_1 \varphi_1)^{-3} \quad (4b)$$

which defines the thermodynamic stability limit for fluid systems.

In a similar way we derive for the critical condition

$$(\partial^3 \Delta F/\partial \varphi_1^3)_{V,T} = 0 \quad (5a)$$

which results in

$$\frac{1}{\varphi_0^2} - \frac{1}{m_1 \varphi_1^2} = 6 g_{11} \gamma_1 (1 - \gamma_1)^2 (1 - \gamma_1 \varphi_1)^{-4} \quad (5b)$$

Another standard procedure leads to the equation of state:

$$p = - (\partial \Delta F/\partial V)_{T,n_1} \quad (6a)$$

which expression can be rewritten into

$$p = - \frac{1}{v_0} (\partial \Delta F/\partial N_\varphi)_{T,n_1} \quad (6b)$$

and results in

$$\frac{-pv_0}{kT} = \ln \varphi_0 + (1 - \frac{1}{m_1}) \varphi_1 + \varphi_1^2 \Big\{ \alpha_1 + (1 - \gamma)^2 (1 - \gamma_1 \varphi_1)^{-2} g_{11} \Big\} \quad (6c)$$

Vapour-liquid equilibrium in a pure substance is completely determined by the equality of the pressure and of the chemical potential of the substance in the two phases (' and ").

$$p' = p'' \text{ and} \quad (7a)$$
$$\mu_1' = \mu_1'' \quad (7b)$$

The chemical potential of sites 1 is defined by

$$\mu_1 = (\partial \Delta F/\partial n_1)_{T,N} = m_1 (\partial(\Delta F/N_\varphi)/\partial \varphi_1)_T \quad (7c)$$

Substitution of eq (1) and rearrangement yields for a vapour-liquid equilibrium

$$\frac{1}{m_1} \ln \frac{\varphi_1'}{\varphi_1''} = (1 - \frac{1}{m_1}) (\varphi_1'' - \varphi_1') + \varphi_0''^2 \Big\{ \alpha_1 + g_{11} (1 - \gamma_1) (1 - \gamma_1 \varphi_1'')^{-2} \Big\} - \varphi_0'^2 \Big\{ \alpha_1 + g_{11} (1 - \gamma_1) (1 - \gamma_1 \varphi_1')^{-2} \Big\} \quad (7d)$$

Equations 2 - 7 contain 4 adjustable parameters. Three of them, v_0, γ_1 and g_{11}, have a physical meaning allowing their order of magnitude to be estimated; α_1 however, is a purely empirical parameter. The procedure followed here is due to van der Waals (24) who showed long ago that equation-of-state parameters can be calculated from gas-liquid critical conditions.

For many low molecular-mass substances, literature cites experimental data on gas-liquid equilibria, usually expressed in equilibrium density, temperature and pressure (20). In a number of cases critical temperature, pressure and density have been reported.

The parameters in eqs. 1 - 6 can be calculated from the gas-liquid critical point (p_c, T_c, d_c) of the substance by a suitable combination of eqs. 2, 4, 5 and 6. Since the interaction energy term g_{11} usually showed a temperature-dependence departing slightly from that indicated in eq. 1 additional data are needed, e.g. from vapor-liquid equilibrium at another temperature, to establish this dependence which can be described with

$$g_{11} = g_{11o} + g_{111}/T \qquad (8)$$

BINARY MIXTURES

The Helmholtz free-energy expression for a binary mixture - vacancies and two kinds of sites (1 and 2) - can be derived along analogous lines; it reads

$$\Delta F/N_\varphi kT = \varphi_o \ln \varphi_o + \varphi_1 m_1^{-1} \ln \varphi_1 + \varphi_2 m_2^{-1} \ln \varphi_2 + \qquad (9)$$
$$\varphi_o \varphi_1 [\alpha_1 + g_{11}(1 - \gamma_1)q^{-1}] + \varphi_o \varphi_2 [\alpha_2 + g_{22}(1 - \gamma_2)q^{-1}]$$
$$+ \varphi_1 \varphi_2 [\alpha_m + g_m(1 - \gamma_2)q^{-1}]$$

where $q = 1 - \gamma_1 \varphi_1 - \gamma_2 \varphi_2$; $\gamma_1 = 1 - \dfrac{\sigma_1}{\sigma_0}$; $\gamma_2 = 1 - \dfrac{\sigma_2}{\sigma_0}$

The sets of parameters (α_1, g_{11}, γ_1, m_1) and (α_2, g_{22}, γ_2, m_2) can be deduced from pure-component data as indicated above, and the binary mixture then calls for two more parameters, viz. α_m and g_m. These can be derived from vapor-liquid or fluid-fluid critical data on the mixture. To avoid the need for designing mixing rules for v_o, and the consequent uncertainties introduced when derivatives with respect to concentration are taken, we postulate that all constituents in a mixture shall have the same v_o, independent of the composition of the system. There will be as many values of σ as there are components and these values may differ, in accordance with the shapes of the various molecules.

Expression for chemical potentials, spinodal and critical points can be derived straightforwardly from eq. 9 The resulting equations being complex we only state the binary analogue of eq. 6 here, ref. 17 (available upon request) contains a full report. The equation for the pressure now reads

$$- pv_0/kT = \ln \varphi_0 + (1 - m_1^{-1}) \varphi_1 + (1 - m_2^{-1}) \varphi_2 +$$
$$(\alpha_1 \varphi_1 + \alpha_2 \varphi_2)(\varphi_1 + \varphi_2) - [\alpha_m + g_m(1 - \gamma_2)q^{-2}] \varphi_1 \varphi_2 +$$
$$[g_{11}(1 - \gamma_1) \varphi_1 + g_{22}(1 - \gamma_2) \varphi_2](q - \varphi_0)q^{-2} \qquad (10)$$

A set of binary critical data (T_c, p_c, $x_{2,c}$) can be used for the calculation of α_m and g_m, after which the complete phase diagram of the fluid 1-2 mixture can be computed and compared with experimental data. The mole fraction of the second component x_2 is given by

$$x_2 = [1 + (\varphi_1 m_2 / \varphi_2 m_1)]^{-1} \qquad (11)$$

APPLICATIONS TO PVT-DATA OF PURE SUBSTANCES

In previous work (17,26,27) we have shown that the model is well applicable to gas-liquid equilibrium behaviour in pure substances like CH_4, CO_2, H_2O and n-alkanes. The equations 2 to 7 provide a good agreement with experimental data in a large temperature and pressure range including critical as well as supercritical densities.
Further examples are given here in figs. 1 referring to polar substances like NH_3 and water.

APPLICATION TO BINARY MIXTURES IN THE INTERCRITICAL RANGE

Equations 9 to 11 can be used to calculate the gas-liquid equilibrium in binary mixtures. For each of the components the parameters have already been determined and we only need one binary critical point to establish values for the two binary parameters α_m and g_m using equations for critical point and spinodal.
Figure 2 shows that the gas-liquid phase behaviour of the system CO_2/CH_4 is well covered by equations 9 to 11. For the

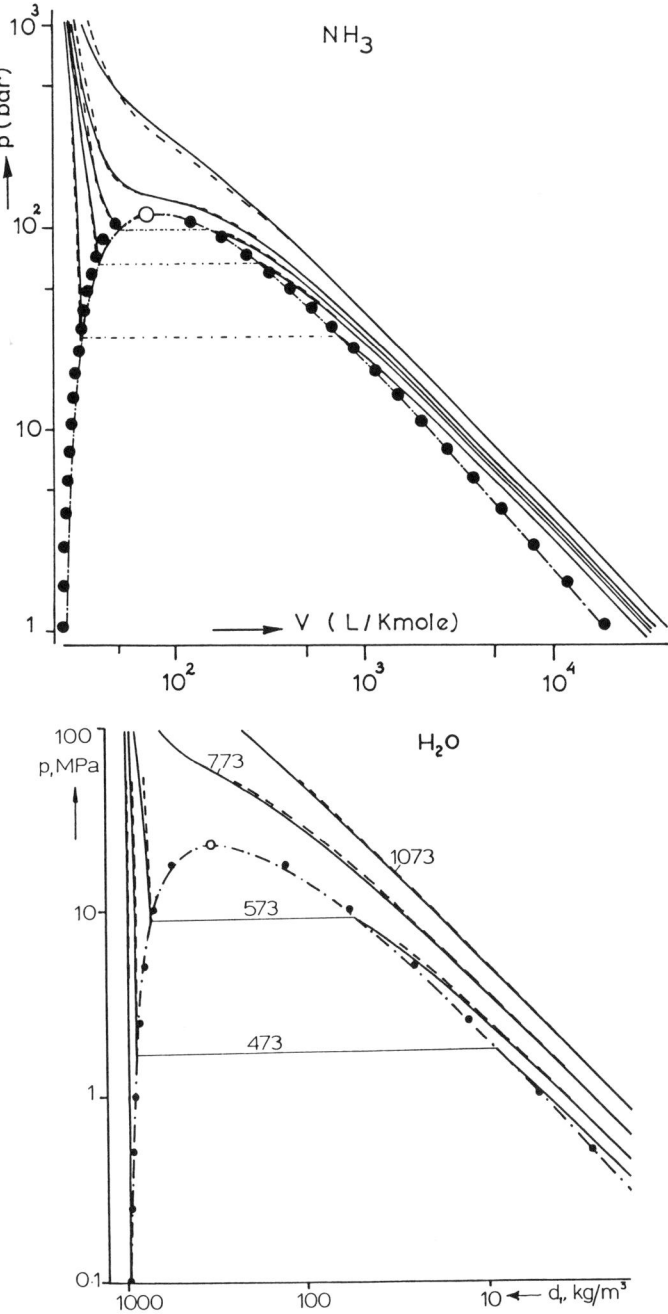

Fig. 1: pVT diagrams for indicated temperatures in K. Calculated; one phase densities (———) and gas-liquid coexistence densities (-.-.-). Experimental (literature) data; smoothed isotherms (----), gas-liquid coexistence (●) and critical point (○).

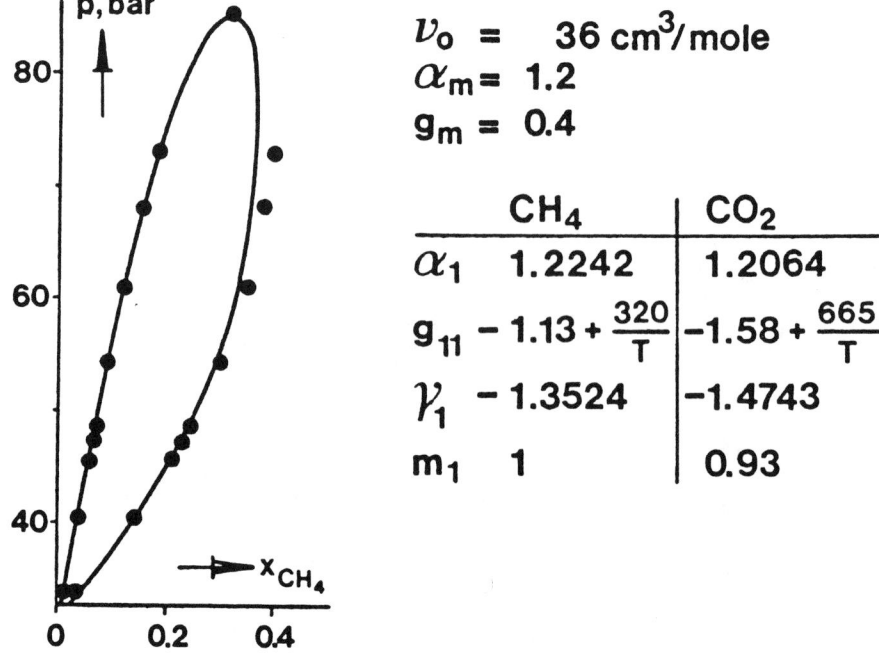

Fig. 2: Gas-liquid coexistence at 271.4 K in the system CO_2-CH_4. Calculated with indicated values of the parameters (———); experimental data (•).

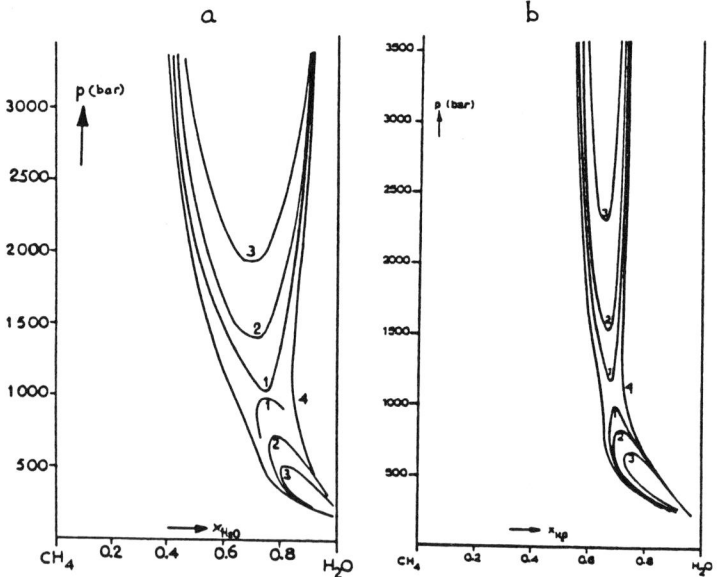

Fig. 3: Gas-gas and gas-liquid demixing in CH_4/H_2O. Left: isotherms from <u>experimental</u> data. Right: spinodals <u>calculated</u> with eq. 1. Temperatures (°C); 1: 353; 2: 355; 3: 360; 4: 350.

temperature considered the mixture can thus be characterized by two binary parameters only. Of course one of these will vary with temperature (g_m).

GAS-GAS DEMIXING

The system CH_4/H_2O exhibits gas-gas demixing of the second kind (28-30). This involves the occurrence of two miscibility gaps in $p(x_2)$ isotherms at temperatures just under the critical temperature of water. One of these is of the upper-critical-miscibility type with a maximum critical pressure, the other – located at higher pressures – is a lower critical miscibility gap with a minimum critical pressure. Determination, as above, of α_m and g_m from some upper critical data proved to suffice for an essentially correct description of the full phase diagram, including the gas-gas lower-critical miscibility gap. Isotherms calculated for this system are illustrated in figure 3b and compared with experimental $p(x_2)$ sections in figure 3a. The calculated curves are spinodals and comprise a narrower $p(x_2)$ region than the corresponding coexistence curves, as they should do. The location of the calculated spinodals is in good agreement with that of the experimental coexistence curves.

Experimental binary critical points and gas-liquid equilibrium densities at given temperature and pressure for the system CO_2/H_2O have been reported in literature (25). Fig. 4 shows some of the existing data. The system shows gas-liquid equilibria but also gas-gas equilibria of the second type (31) at high pressures and temperatures around the critical temperature of water.

We use only some critical points of the gas-liquid equilibrium range of CO_2/H_2O to derive the values of the two binary interaction parameters α_m and $g_m(T)$. From the critical pressure, temperature and composition is it quite straightforward to calculate the critical composition in lattice site volumefractions and the two binary parameters at that temperature, using eqs. 10, 11 and the expressions for spinodal and critical condition of a mixture. The values of α_m and g_m were determined at three different critical temperatures (and of course also different pressure and composition). The value of g_m proved to be temperature dependent, α_m was hardly effected by temperature. The values of g_m and α_m are given in table I.

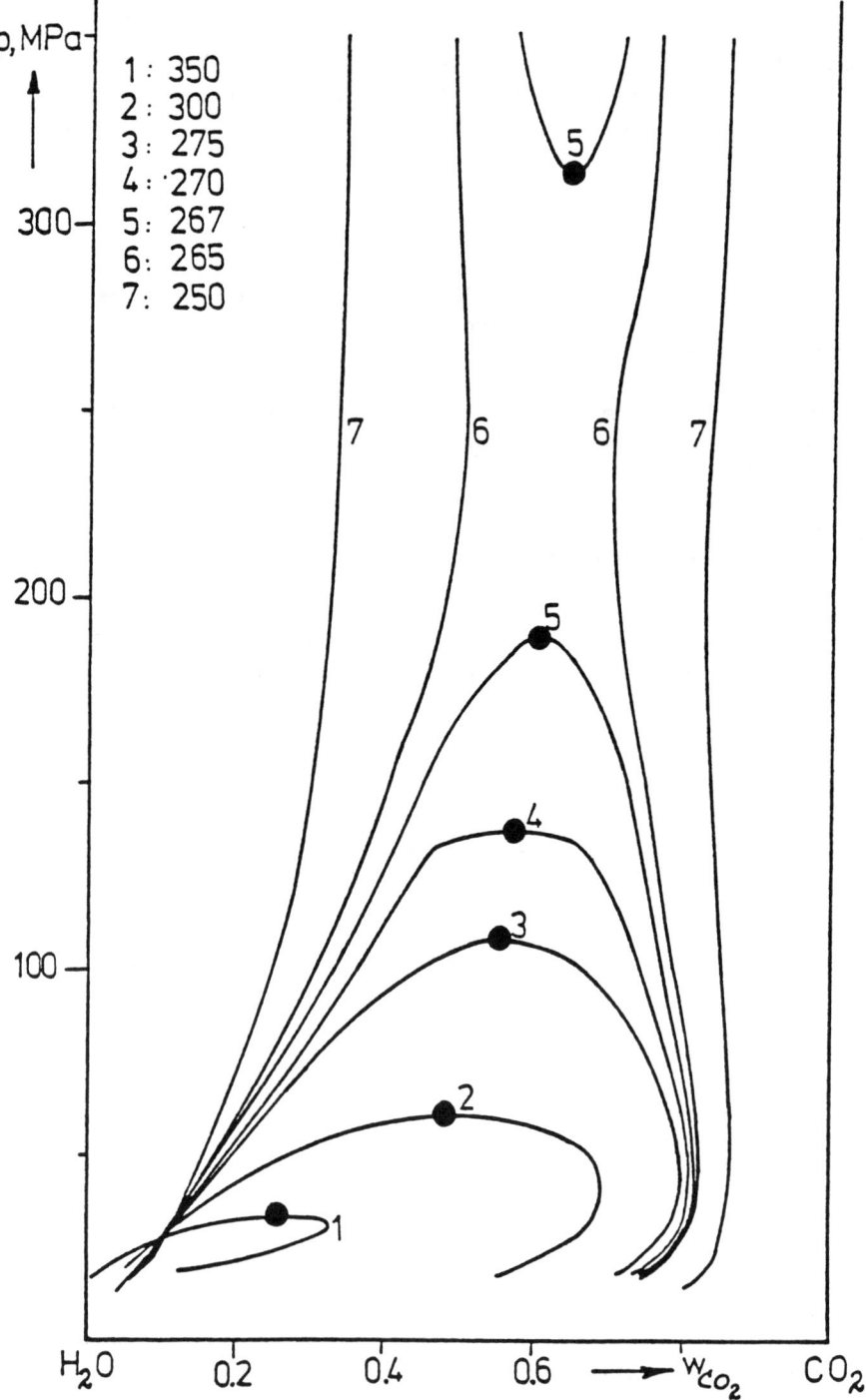

Fig. 4. Experimental gas-liquid and gas-gas equilibrium curves in the system H_2O/CO_2 (25) at indicated temperatures (in °C). ●: critical points.

Table I. Values of the interaction parameters

	CO_2	H_2O	mixture
α	1.1401	1.5467	0.92
γ	-1.6368	-2.469	-
g	$-1.1199+468.25/T$	$-0.5473+297.1/T+3\times10^5/T^2$	$-56.8626+56.8\times10^3/T-129\times10^5/T^2$
m	1.25	0.72	-
v_o	25	25	25

Having derived the values of the binary interaction parameters we can now calculate the complete ternary phase diagram with the relevant expressions for the critical line and the spinodal. The compositions of the coexisting phases are calculated using the expressions of the equality in chemical potentials in both phases for CO_2 and for H_2O together with the relation derived from p' = p" and with eq. 10. Figure 5 shows some 'ternary' lattice-gas representations of CO_2/H_2O mixtures in the pressure and temperature ranges of 0-4000 bar and 250-350 °C. It is seen that the calculation based on critical liquid-gas data predicts gas-gas equilibrium of the second kind to occur. This prediction turns out to be in surprisingly good agreement with the experimental data. A quite similar situation as was encountered with the system CH_4/H_2O.

The ternary lattice-gas representation of the phase behaviour can be converted into the usual binary $P-x_2$ or $T-x_2$ diagrams with eq. 11. Fig. 6 shows the p-w_2 (mass fraction) plot for CO_2/H_2O obtained from fig. 3. To be comparable with fig. 4 we converted the mole fraction into the mass fraction w_2 in the usual way. Although the calculated phase-diagram does not completely fit the experimental findings, one may say that the agreement is quite satisfactory, especially because only three experimental points of the phase diagram were used as information. Better agreement can doubtless be achieved upon fitting all existing PVT-data of the mixture. For the moment the objective was to illustrate that even with a simple model, scarce information and simple computerfacilities an essentially correct description and prediction of PVT-behaviour of binary fluid systems is obtainable.

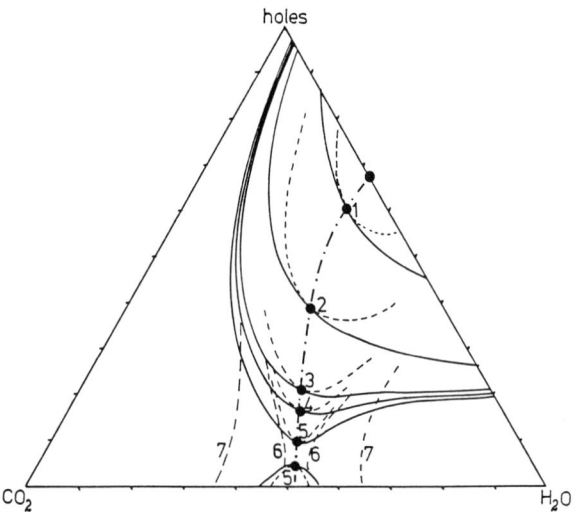

Fig. 5. Calculated 'ternary' lattice gas representation of gas-liquid and gas-gas equilibria in the system H_2O/CO_2; spinodals: -------, critical line: -.-.-, gas-liquid and gas-gas coexistence: ———. Temperatures as indicated in Fig. 4. At temperatures 6 and 7 gas-liquid and gas-gas region coincide to form one continuous two-phase region.

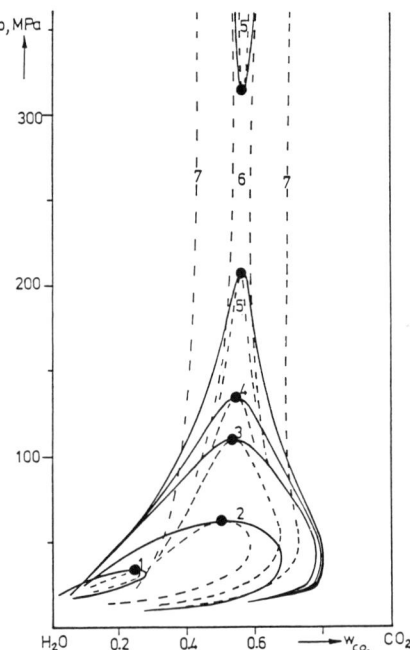

Fig. 6: Calculated gas-liquid and gas-gas coexistence curves: ———, spinodals: ---- and critical points: •. Temperatures as indicated in Fig. 4.

Solution polymerization is an important technological operation which calls for quantitative predictions of possible phase separations that may occur during the polymerization. We have tested the mean-field lattice-gasmodel also here and investigated its applicability e.g. to solutions of linear polyethylene in n-alkanes. For this sytem we introduced a particularly simplifying approximation in setting the two binary parameters α_m and g_m equal to zero. In other words, we identify the C_2H_4-groups in alkane and polyethylene and only need to find values for the parameters of the pure constituents (26).
Within the scope of this model the consequence of assuming $g_m = 0$ is that we should now be able to predict the phase behaviour of solutions of linear polyethylene in n-alkanes since there are no new adaptable parameters for the mixture.

Several authors (32-34) have reported cloud-point curves and critical points for polyethylene in n-alkanes. These systems show a LCST (lower critical solution temperature) which involves phase separation upon heating. The experiments were performed in such a way that the pressure was that of the vapour of the mixture at the temperature of the measurement. This means that the pressure changed slightly along the cloud point curves. For the moment we neglect this pressure variation.

The polymers used in these studies had wide molar-mass distributions which, within the present context, show up in detail in the cloud-point curve. For instance, the second term on the r.h.s. in eq. (9) must then be replaced by $\Sigma \; _im_i^{-1}\ln \; _i$, which sum contains as many terms as there are components in the polymer. As a consequence, the symbol m_1 in eq. 4b will stand for m_w, the mass-average chain length. In eq. 5b it must be replaced by m_w^2/m_z (m_z = z-average chain length).
We avoid complications due to the multicomponent character of polymers using the fact that cloud-points curves and spinodals must have a common tangent at the critical point (26). Thus we can compare measured cloud-point curves and spinodals calculated with the parameters found before for pure alkanes and polyethylene.
The spinodals shown in figs. 7 and 8 were constructed from quasi-ternary isothermal diagrams (solvent/polymer/holes). The locations of the spinodals for constant pressure thus predicted are fairly consistent with those of the experimental cloud-point curves.

To check the usefulness of the present procedure we consider its predictive value with respect to the pressure - and polymer - molar-mass dependence of the cloud-point. Kodama and Swinton (34) reported cloud-point curves in

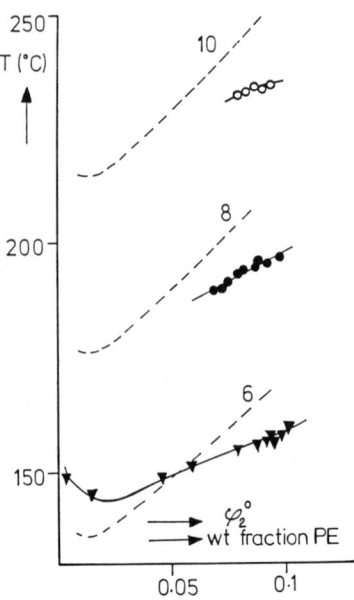

Fig. 7: Comparison of experimental cloud point curves of linear polyethylene (M_w = 177 kg/mole) in n-hexane (▼), n-heptane (●) and n-octane (o), with spinodals (---) predicted at constant indicated pressure (in bar).

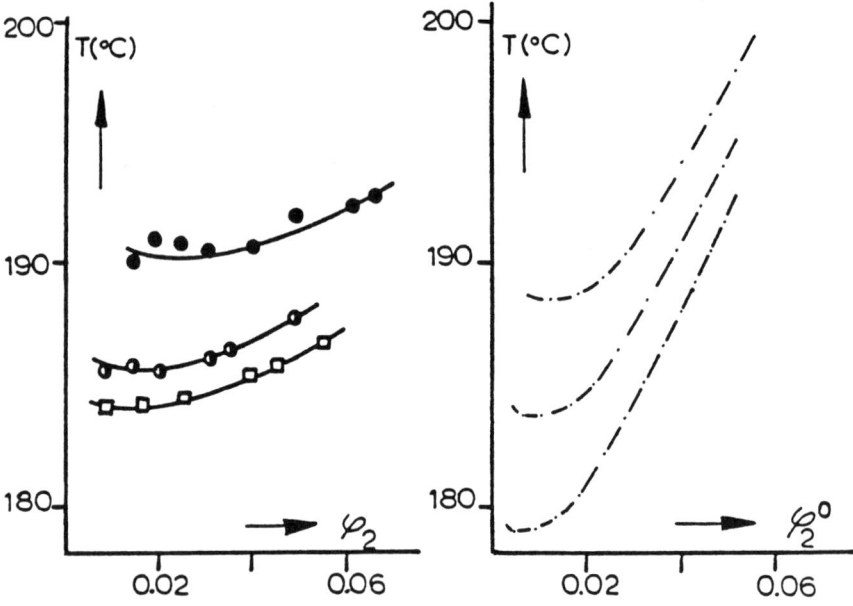

Fig. 8: Comparison of experimental cloud points of polyethylene in n-heptane (39) at several molar masses of the polymer (M_w in kg/mole; ●: 49.3; ◐ : 82.6; □ : 135.9) with spinodals predicted at 8 bar.

n-heptane for polyethylenes varying in mass-average molar mass M_w. As witnessed by fig. 8, the molar-mass-dependence of the spinodal agrees fairly well with that of the miscibility gap found by these authors.

Although the assumption $g_m = 0$ may be considered too simplistic for n-alkane-polyethylene mixtures, we yet obtain a remarkably close agreement with the observed phase behaviour. As was shown elsewhere the model can easily be modified so as to describe solutions of branched polymers as well (17). We may thus conclude the present mean-field-lattice-gas treatment to be useful in dealing with polymer solutions at ambient and elevated pressures. Further conceivable improvements include the introduction of solvent - solute interaction parameters. This is a subject of current study.

We have shown elsewhere that the mean-field lattice gas model also offers a promising framework for dealing with polymer compatibility. Phase behaviour in polymer blends can often be very complex and essentially correct descriptions, albeit qualitative so far, have been obtained (35).

DISCUSSION

The aim of the present study was the development of adequate equations describing fluid-fluid phase behaviour involving a minimum number of adaptable parameters. Using the two-component lattice-gas model as a starting point, we found the introduction of interacting-surface-area ratios to relax the lattice model sufficiently without undue increase in number of adjustable parameters. As yet we did not ascertain whether this favorable situation is either brought about or limited by an important assumption underlying all our calculations, viz. the independence of the lattice cell volume v_0 of concentration, temperature and pressure. This assumption was made to avoid the need to draw up arbitrary mixing rules for v_0, and thus to reduce the number of purely empirical parameters. It should be noted, however, that the choice of the number of lattice sites m, occupied by molecules, fixes the value of v_0 for a substance. Intuitively one would object to values of m deviating considerably from unity when small molecules like CO_2 are concerned. The values we used for v_0 on a number of substances and systems (about 30 cm^3/mole) agree fairly well with the equilibrium value estimated by Frenkel (3).

The surface-area (ratio) σ, number of occupied lattice sites m and cell volume v_0 are parameters open to appraisal on physical grounds. This is also true for the parameter g which is related to the interaction energy between adjacent molecules. It proved necessary, however, to correct the entropy of mixing terms in the Helmholtz free energy expressions (logarithmic terms in eqs. 1 and 9) with merely empirical parameters (α and g_{110}) which have to be accepted at face value. Yet, the fact that the ΔF expressions developed here can be fitted so well to experimental data gives some credence to the mathematical form of the equations. It is probably the interacting surface-area ratio which mainly provides for the versatility of the present treatment.

In this study we derived values for the various parameters from a minimum number of experimental data, in order to be able to test the predictive value of the procedure. Alternatively, one might optimise the parameters using all experimental data available and thus possibly obtain better descriptions. Apart from phase-equilibrium data, the present objective, one could also use compressibilities and thermal expansion coefficients. This is a subject of current study.

ACKNOWLEDGEMENT

The authors are much indebted to mr. A.J.G. Offermans who skilfully carried out most of the calculations used in this paper.

REFERENCES

1. Mermin, N.D. Phys. Rev. Letters, 26, 957 (1971).

2. Mulholland, G.W. and Rehr, J.J., J. Chem. Phys., 60, 1297 (1974).

3. Frenkel, J., Kinetic Theory of Liquids, Dover Publ., New York, 1947.

4. Cernuschi, F. and Eyring, H., J. Chem. Phys., 7 547 (1939).

5. Altar, J., J. Chem Phys., 5, 577 (1937).

6. Trappeniers, N.H., Schouten, J.A. and ten Seldam, C.A., Chem. Phys. Letters, 5, 541 (1970).

7. Schouten, J.A., ten Seldam, C.A. and Trappeniers, N.J., Physica, 73, 556 (1974)

8. Guggenheim, E.A., Mixtures, Clarendon Press, Oxford, 1952.

9. Kanig, G., Kolloid Z. & Z. Polymere, 190, 1 (1963); 223, 829 (1969).

10. Bartis, J.T. and Hall C.K., Physica, 78, 1 (1974).

11. Sanchez, I.C. and Lacombe, R.H., J. Phys. Chem., 80, 2352, 2568 (1976)

12. Arai, Y. and Saito, S., J. Chem. Eng. Japan, 5, 9 (1972).

13. Various authors, Angew. Chem. 17, 701 (1978).

14. Hannay, J.B. and Hogarth, J., Proc. R. Soc. London, 29, 324 (1879).

15. Schneider, G., Chem. Thermodyn., 2, 105 (1978).

16. Staverman, A.J., Rec. Trav. Chim., 56, 885 (1937).

17. Kleintjens, L.A., Ph. D. Thesis, Essex U.K. 1979.

18. Guggenheim, E.A., 'Mixtures', Clarendonpress, Oxford 1952.

19. Orofino, T.A. and Flory, P.J., J. Chem. Phys, 26, 1067 (1957).

20. Koningsveld, R. and Kleintjens, L.A., Macromolecules, 4, 637, (1971).

21. Staverman, A.J. and van Santen, J.H., Rec. Trav. Chim., 160, 76 and 640, (1941).

22. Flory, P.J., J. Chem. Phys., 10, 51, (1942).

23. Huggins, M.L., Ann. N.Y. Acad. Sci., 43, 1, (1942).

24. van der Waals, J.D. and Kohnstamm, Ph., Lehrbuch der Thermodynamik, Leipzig 1912, Vol. 11.

25. E.g. collected in: Lanbolt-Börnstein, 'Zahlenwerte und Funktionen'.

26. Kleintjens, L.A. and Koningsveld, R., Coll. Pol. Sci., 258, 711 (1980).

27. Kleintjens, L.A. and Koningsveld, R., J. Electrochem. Soc., 127, 2352 (1980).

28. See e.g. Schneider, G.M., Pure & Appl. Chem., 47, 277 (1976).

29. Welsch, H., PhD Thesis, Karlsruhe, 1973.

30. Sultanov, R.G., Skripka, V.G. and Namiot, A.Yu., Russian J. Phys. Chem., 46, 1238 (1972).

31. Schneider, G.M., 'Extraction with supercritical gases', Verlag Chemie, Weinheim (1980).

32. Orwoll, R.A. and Flory, P.J., J. Am. Chem. Soc., 89, 6814, 6822, (1967).

33. Nakajima, A. and Hamada, F., Kolloid Z. & Z. Pol., 205, 55, (1965).

34. Kodama, Y. and Swinton, F.L., Brit. Pol. J., 10(3), 191, (1978).

35. Koningsveld, R., Kleintjens, L.A., and Onclin, M.H., J. Macromol. Sci., Phys., 18, 363 (1980),.

CHAPTER 13

BINARY PHASE DIAGRAMS FROM
A CUBIC EQUATION OF STATE

Glenn T. Hong and Michael Modell
MODAR, Inc., 14 Tech Circle, Natick, Massachusetts 01760.

Jefferson W. Tester
Department of Chemical Engineering
Massachusetts Institute of Technology
Cambridge, Massachusetts 02139

Abstract

The Peng-Robinson equation of state is used with a constant interaction parameter to generate P-T-x diagrams over the entire fluid range for four binary systems. The systems considered are n-hexadecane- carbon dioxide, naphthalene-carbon dioxide, naphthalene-ethylene, and benzene-water. These systems exhibit various types of liquid-liquid immiscibilities and critical phenomena. In addition, the naphthalene systems involve the presence of a solid phase. Predictions include two-phase and three-phase equilibrium, azeotropic points, critical lines, critical end points, and spinodal curves. Fugacity-composition plots are shown to be a valuable tool in phase equilibrium calculations, and for defining stable, metastable, and unstable equilibria. Failure to consider metastable and unstable equilibria can result in erroneous predictions. The method is useful in the evaluation and design of phase equilibrium processes, including those involving supercritical fluids. It is also of value in experimental studies, since it points out regions of the P-T-x space which merit detailed attention.

INTRODUCTION

In 1876, van der Waals first proposed his now famous equation of state. Over the next 30 years, work with this equation showed that it possessed many intriguing qualities. It was able to describe both gaseous and liquid states, and could be used to predict critical and spinodal points, as well as stable, metastable, and unstable phase equilibria. While the lengthy numerical calculations were quite limiting, analysis of the equations proved of substantial benefit to van der Waals' more experimentally oriented colleagues at Leiden and Amsterdam. The equation aided greatly in the interpretation of both expected and unexpected results (see e.g. Smits, 1903; 1905).

In 1969, van Konynenburg completed a study on the effect of the parameters in the van der Waals equation on the predicted phase behavior. It was found that the equation could describe most, if not all, of the various types of phase behavior found in binary fluid systems. Van Konynenburg's predictions included both two- and three-phase equilibrium, as well as critical lines and azeotropic lines. However, calculations and comparisons were not carried out for specific binary systems. In light of van Konynenburg's results, it seems that a modern cubic equation of state might provide a good representation of overall phase behavior in many binary systems. Thus, a small amount of experimental data may be used to obtain information over an extensive range of temperatures and pressures.

This paper illustrates the application of the Peng-Robinson equation of state for prediction of overall P-T-x diagrams for four binary systems: n-hexadecane – carbon dioxide, naphthalene – carbon dioxide, naphthalene – ethylene, and benzene – water. These systems were chosen because of their relevance to supercritical fluid processes. A single constant interaction parameter is used in each case. Relevant constants for the pure components are given in Table I.

METHOD

The Peng-Robinson (P-R) equation is:

$$P = \frac{RT}{V-b} - \frac{a}{V(V+b) + b(V-b)} \quad (1)$$

(Peng and Robinson, 1976). It is a modified form of the van der Waals equation. The parameter b is associated with the volume occupied by the molecules, and the temperature dependent parameter $a = a(T)$ with the attractive forces between

Table 1

Constants of Pure Components

Substance	T_c (K)	P_c (atm)	actual v_c (m³/kgmol)	predicted v_c (m³/kgmol)	actual ρ_c (kg/m³)	predicted ρ_c (kg/m³)	actual Z_c	ω
Ethylene (C_2H_4)	282.4	49.7	0.129	0.143	217	196	0.276	0.085
Carbon dioxide (CO_2)	304.2	72.8	0.094	0.105	468	419	0.274	0.225
Benzene (C_6H_6)	562.1	48.3	0.259	0.293	301	266	0.271	0.212
Water (H_2O)	647.3	217.6	0.056	0.075	321	240	0.229	0.344
n-Hexadecane ($C_{16}H_{34}$)	717.0	14.0	-	1.290	-	175	-	0.742
Naphthalene ($C_{10}H_8$)	748.4	40.0	0.410	0.471	312	272	0.267	0.302

Note - The P-R equation predicts a universal Z_c of 0.307.

molecules. The equation is cubic in volume, so that volume may be analytically determined if pressure and temperature are specified. As with all cubic equations, the P-R equation cannot accurately model volume close to the critical point. It is one of the most accurate cubic equations available for use in this region, however, and gives a reasonably satisfactory qualitative picture. Figure 1 illustrates this point for the PVT diagram of pure water. The constants a and b are chosen according to water's critical temperature and pressure, and to satisfy the critical point criteria $(\partial P/\partial V)_T = 0$ and $(\partial^2 P/\partial V^2)_T = 0$. At the critical point, the predicted critical volume is about 34% too large. In the liquid phase region, the isotherms rise less steeply than they should.

For mixtures, the constants in the P-R equation are determined by combining the pure component constants (see Appendix). An expression for the fugacity of a component in a mixture is derived from the equation of state, and this expression may then be used to calculate multicomponent phase equilibrium. To obtain good agreement with data, it is often necessary to supply binary interaction parameters, δ_{ij}. A δ_{ij} expresses the deviation of the energy parameter a from the geometric mean of a_i (= a for component i) and a_j (= a for component j). Interaction parameters are arbitrarily chosen to fit available data, and are frequently taken as independent of temperature and pressure. For any given temperature, composition and interaction parameter(s), volumetric discrepancies similar to those encountered with pure components are likely to be found. For present purposes, we are more concerned with modelling of the critical region in the P-T-x space. Figure 2 depicts two P-x sections for the binary system isobutane-carbon dioxide (Peng and Robinson, 1976). At both temperatures, the vapor and liquid compositions are predicted reasonably well. Near the critical point, however, the inaccuracies become greater.

In order to describe fluid-solid equilibria, the Peng-Robinson equation is used in conjunction with an expression for the solid phase fugacity. For the systems treated here, it is valid to neglect the solubility of the light component in the solid phase and to take the molar volume of the solid as constant. With these assumptions, the solid fugacity is given as:

$$f^s = P^{sub} \exp(V^s (P-P^{sub})/RT) \qquad (2)$$

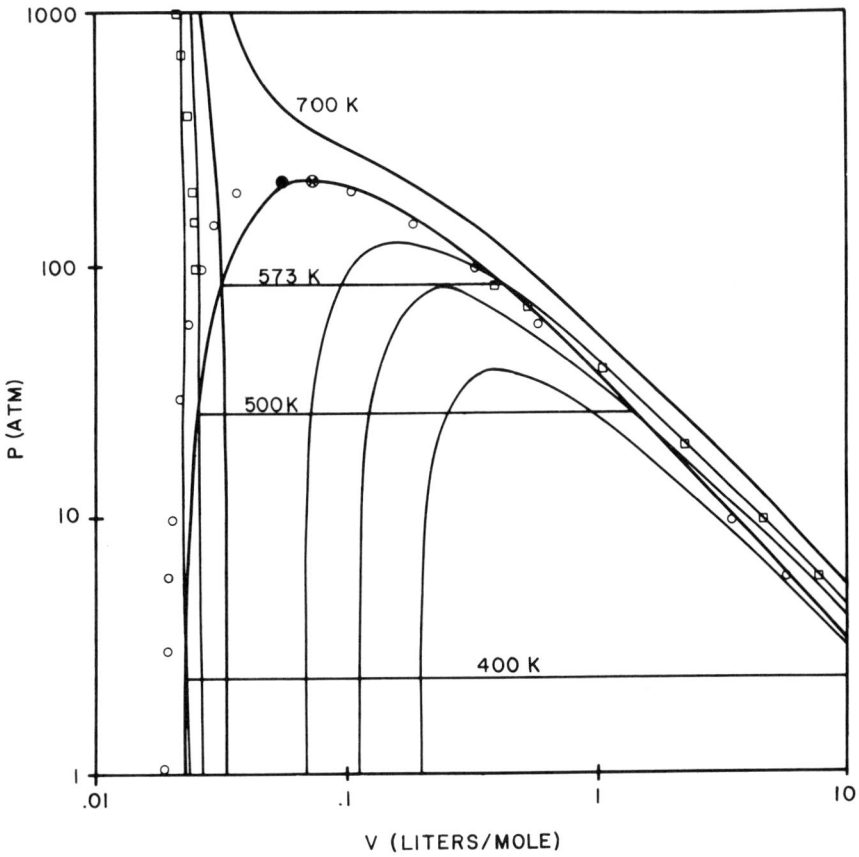

Figure 1. P-V diagram for water.
   ○ - Vapor-liquid equilibrium (Keenan, et al. 1978)
   □ - 573 K isotherm (Keenan, et al. 1978)
   ● - Experimental critical point
   ⊗ - Predicted critical point

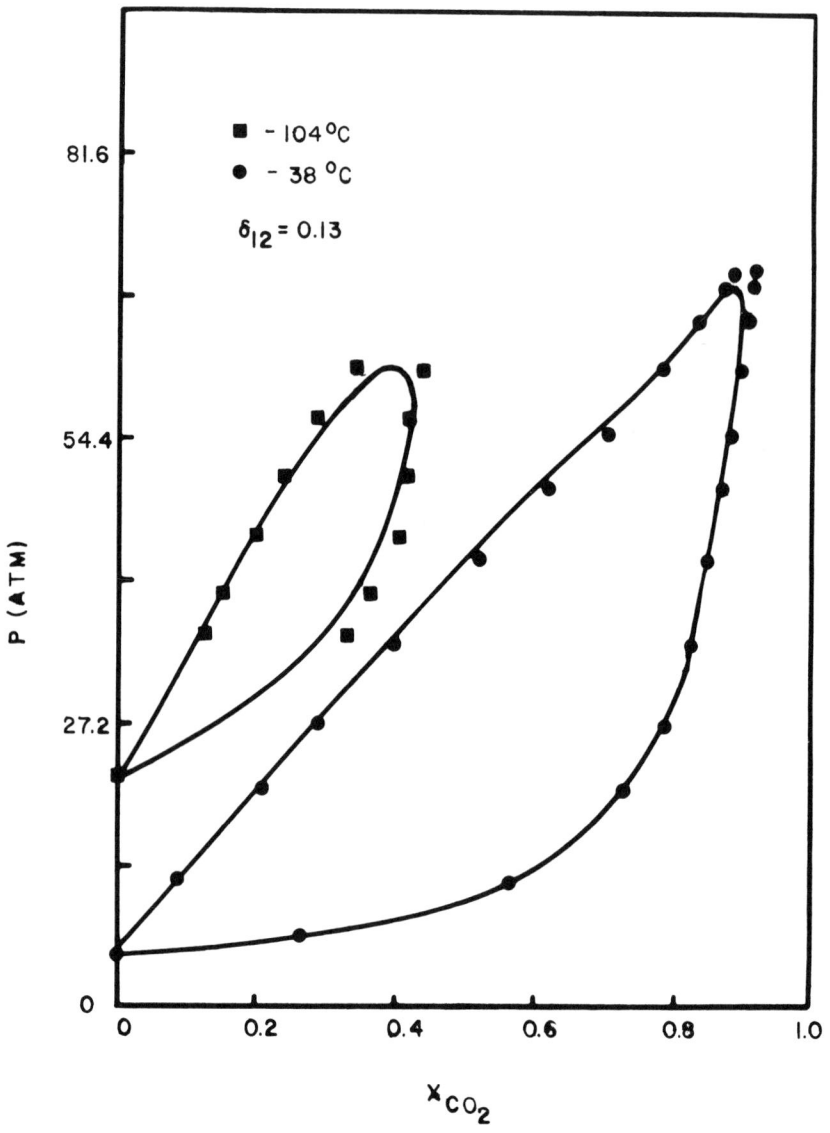

Figure 2. P-x diagram for isobutane-carbon dioxide. (Peng and Robinson, 1976).

where P^{sub} is the sublimation pressure and V^s the molar volume of the solid naphthalene. The sublimation pressure was obtained from the measurements of Ambrose, et al., (1975).

Equilibrium of phases requires that the phases be of equal temperature and pressure, and that the fugacity (or chemical potential) of a given component be the same in every phase. It has been found that plots of fugacity versus composition, at constant temperature and pressure, provide an excellent means of visualizing the criterion of equal fugacities. Consider Figure 3, which is a log-log plot of n-hexadecane (subscript H) fugacity versus n-hexadecane mole fraction for the n-hexadecane – carbon dioxide system, at a temperature of 300 K and a pressure of 0.1 atm. The systems represented by this diagram are a result of mixing liquid n-C_{16} with gaseous CO_2, and vapor-liquid equilibrium is to be expected if the two components are present in a certain range of proportions. The most striking feature of the plot is that there are actually two separate curves. Segment G is essentially linear over the entire range of composition, with a slope of unity and an intercept at the pure C_{16} axis of 0.1 atm. This is the equivalent of ideal gas behavior, since

$$f_H = x_H P = p_H \qquad (3)$$

where p_H is the partial pressure of C_{16}. The second curve appears when the cubic equation has three real roots. In the composition range from the infinity in df_H/dx_H to $x_H = 1$, segment G corresponds to the high volume (gaseous) root, segment L to the low volume (liquid) root, and segment U to the middle volume root. Between the slope infinity and x_{min}, df_H/dx_H for segment L is negative, indicating that the liquid phase is materially unstable in this composition range. For x_H greater than x_{min}, $df_H/dx_H > 0$, and segment L represents a stable or metastable liquid phase. Segment U appears to be materially stable in this plot. As with the middle volume root for pure component systems, however, it violates the criterion of mechanical stability (i.e., $(\partial P/\partial V)_T > 0$) and thus does not represent stable conditions. As indicated by the crossing of segments G and U in the graph, the middle volume root does not always give the middle fugacity.

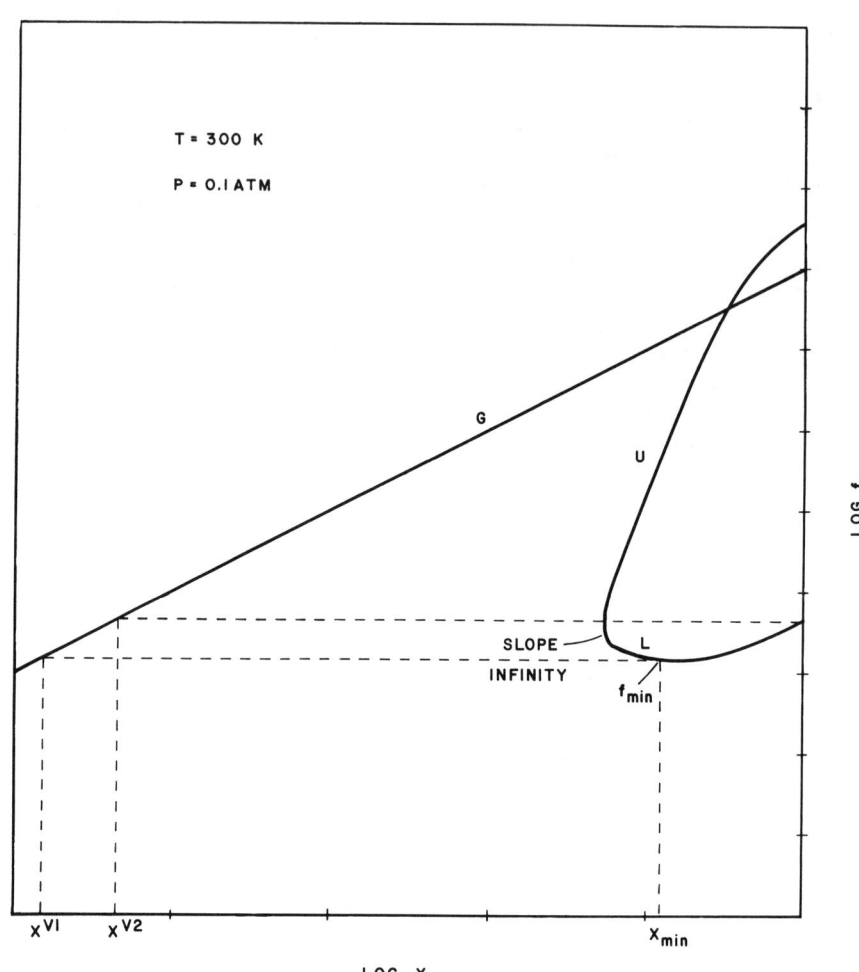

Figure 3. An f_H-x_H plot for n-hexadecane-carbon dioxide. Segment G is the gaseous root, segment U is the mechanically unstable root, and segment L is the liquid root. Segments U and L start at the slope infinity and continue to the pure hexadecane axis.

Suppose now that suitable quantities of C_{16} and CO_2 are present in the system so that two phases exist with compositions x^V and x^L, and fugacity f_H^E. The f-x plot shows that x^L must lie between x_{min} and $x_H = 1$, and furthermore, that f_H^E is between $f_H(x_{min})$ and $f_H(x_H = 1)$. For the coexisting vapor phase, which must lie on the straight line segment, equality of fugacities requires that:

$$x^{v1} < x^v < x^{v2} \qquad (4)$$

This limiting procedure illustrates the usefulness of the f-x plots in determining two-phase equilibrium. To find the actual compositions for vapor-liquid equilibrium, it is of course necessary to match the partial fugacities of carbon dioxide as well. Metastable or unstable equilibria are found when a metastable or unstable portion of the f-x curve is used in satisfying the criteria of equilibrium. Fugacity-composition plots have many shapes other than that shown in Figure 3. Hong (1981) has given this topic a more extensive treatment.

In addition to stable, metastable, and unstable two-phase equilibria, a number of other features of binary P-T-x diagrams are presented below. These features and the criteria applied for their determination are listed in Table II.

RESULTS

N-hexadecane - carbon dioxide

The n-$C_{16}H_{34}$ - CO_2 binary is a fluid-fluid system in the temperature range between the two pure component critical points and up to very high pressures. The P-T coordinates of the critical line beginning at the hexadecane critical point have been measured by Schneider, et al. (1967), and are plotted in Figure 4a along with P-R predictions for various values of the interaction parameter. A δ_{12} of 0.081 was chosen as being optimal for phase equilibrium predictions over the entire range, although it is clear that no single value will be completely satisfactory. Using this interaction

Table II.

Criteria Used in the Calculation of P-T-x Diagrams

Binodal points	Equality of component fugacities
Spinodal points	L1 = 0 (see Appendix)
Critical points	L1 = 0 and M1 = 0 (see Appendix)
Critical stability	Stability of the binodal locus which terminates at the given critical point.
Azeotropic points	Shape of the binodal locus. Zero length tieline, but not a critical point.
Three-phase equilibrium points	Intersection of three binodal regions at a single temperature and pressure.
Critical end points	Intersection of three phase line and critical line, or point of horizontal inflection in the binodal locus not terminated at the critical end point.

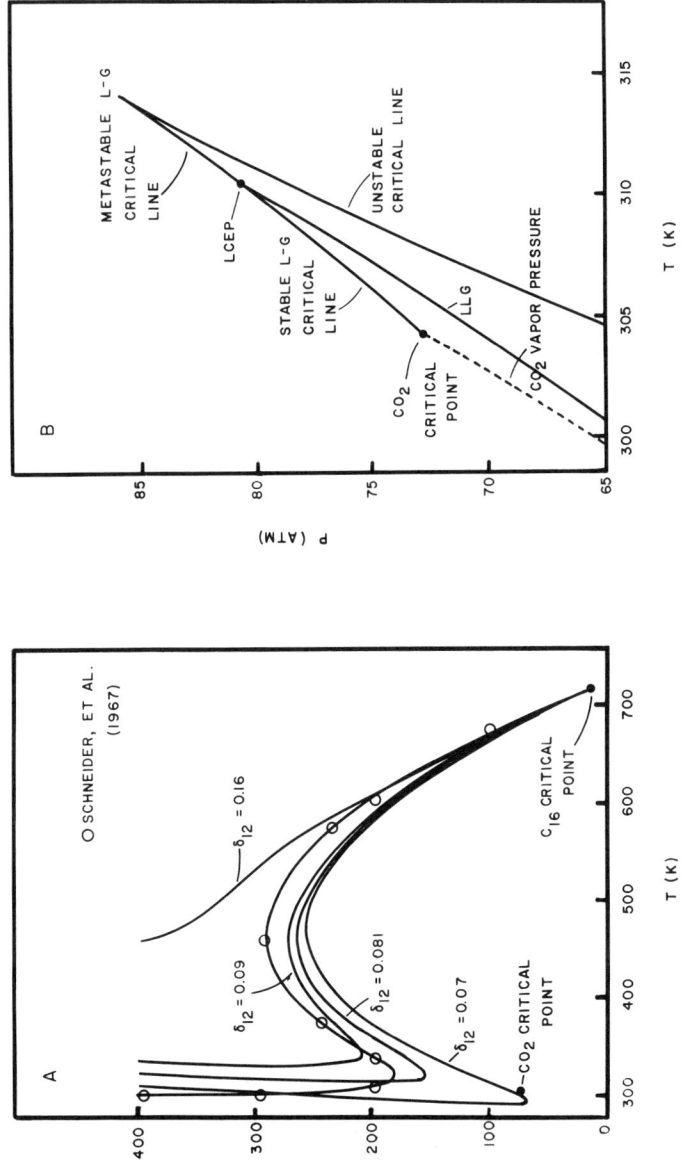

Figure 4. Diagram A is the P-T projection for the C_{16}-CO_2 system, and shows the effect of δ_{12} on the predicted upper critical line. The pure CO_2 critical point is not a part of this line. Diagram B gives the P-T projection of the lower critical locus for C_{16}-CO_2. The stable critical line is metastable between LCEP and the cusp.

parameter, the P-R equation also predicts the critical line beginning at the pure CO_2 critical point, which has not been experimentally determined. This lower critical line is depicted in Figure 4b. Its stable portion is terminated at the lower critical end point (LCEP) where it intersects a three-phase LLG line. The lower critical line continues as a metastable segment until it is terminated by intersection with an unstable critical line at a cusp. The unstable critical line represents the terminus of an unstable binodal surface, as will be seen shortly.

Figure 5 gives a series of P-x diagrams at different temperatures. At 300 K, Figure 5a, neither component is supercritical. Thus, a $G+L_1$ binodal region terminates at carbon dioxide's vapor pressure along the pure CO_2 axis (not visible on the logarithmic concentration axis) and a $G+L_2$ binodal region terminates at hexadecane's vapor pressure along the pure C_{16} axis. At low pressures, the gaseous branch of the $G+L_2$ binodal has a very low concentration of hexadecane, and does not appear on the plot. The L_1+L_2 region extends to very high pressures without reaching a critical point, in agreement with Figure 4a. The $G+L_1$ and $G+L_2$ regions mutually intersect with the L_1+L_2 region to give a three-phase line at about 64 atm, as indicated by the dashed line in the figure. The pressure and temperature of the three-phase equilibrium are seen to be on the LLG line in Figure 4b. The predicted phase equilibria near the three-phase line are fairly complex, so an enlargement has been given in Figure 6a. Each stable binodal curve continues past the three-phase line as a metastable segment. The metastable binodals in turn reach a maximum or minimum pressure, and then continue as unstable binodal curves. Note that two of these unstable binodal curves meet to form an unstable critical point. This point satisfies the first and second criteria of criticality but not the critical stability criterion (see Appendix). The $G+L_1$ and $G+L_2$ binodals are both part of the same continuous curve, while the L_1+L_2 binodal is separate, terminat-

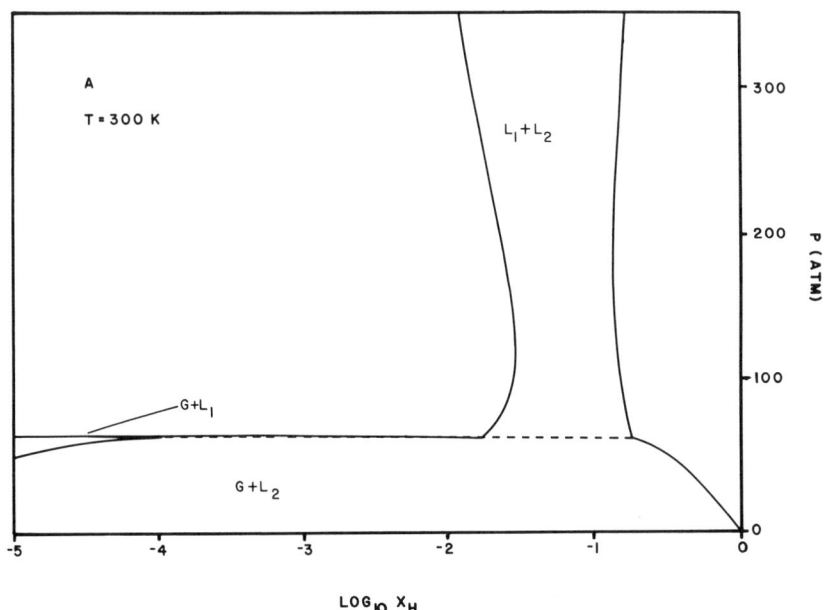

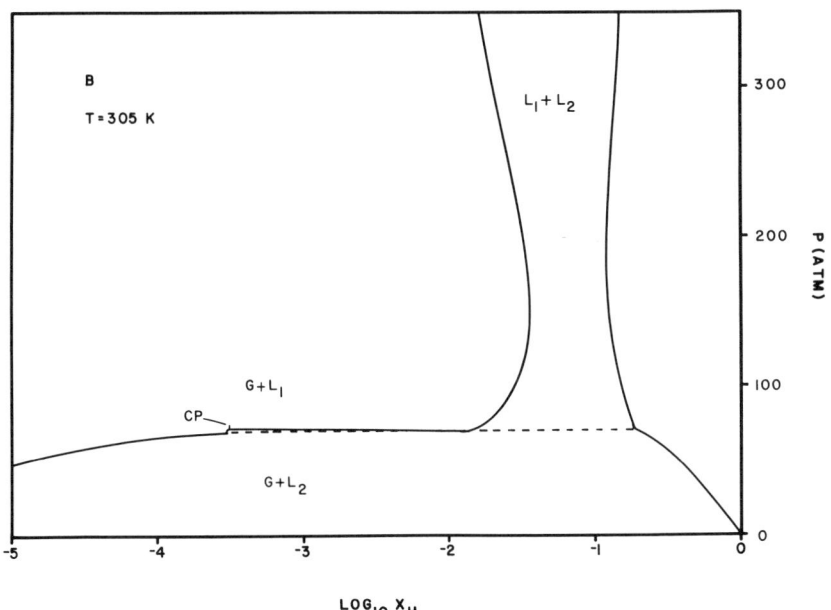

Figure 5. P-x sections for n-hexadecane-carbon dioxide. (First of three pages.)

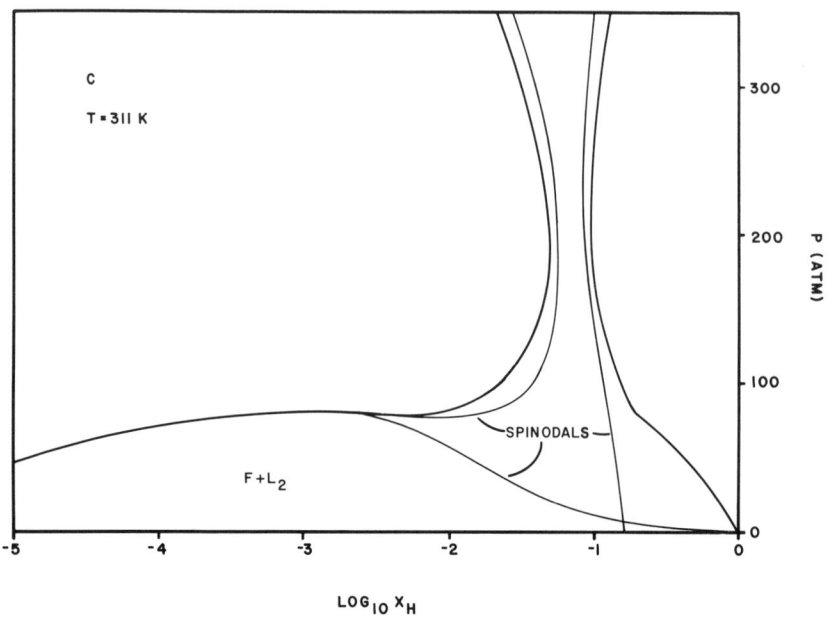
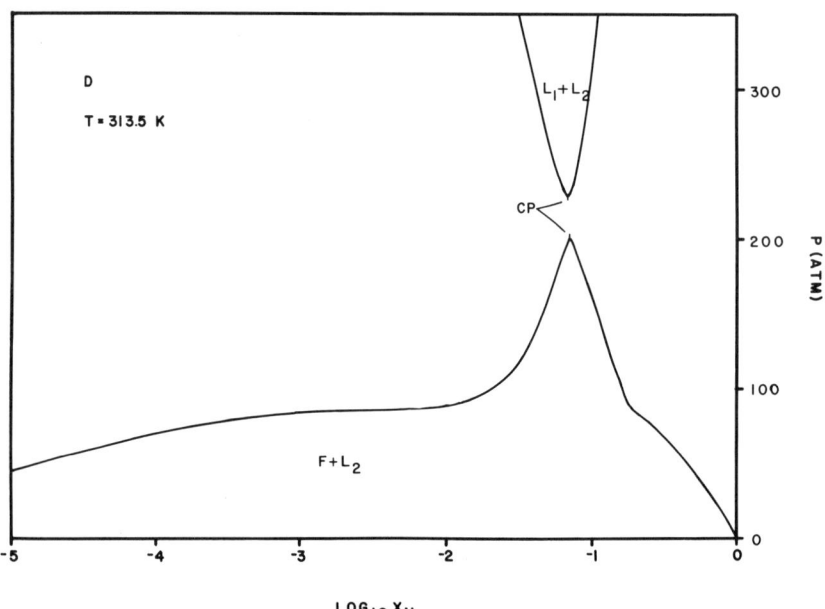

Figure 5. P-x sections for n-hexadecane-carbon dioxide. (Second of three pages.)

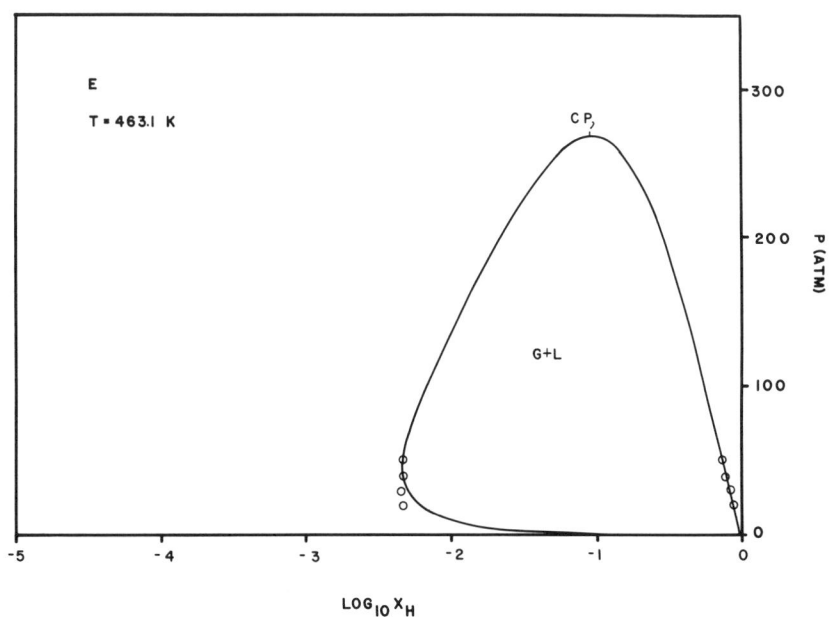

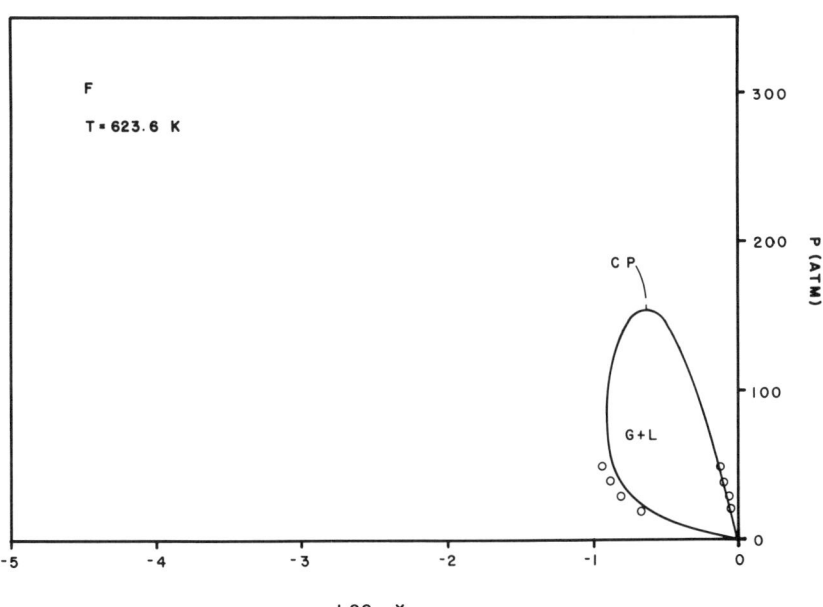

Figure 5. P-x sections for n-hexadecane-carbon dioxide. Data points are from Sebastian, et al. (1980). (Third of three pages.)

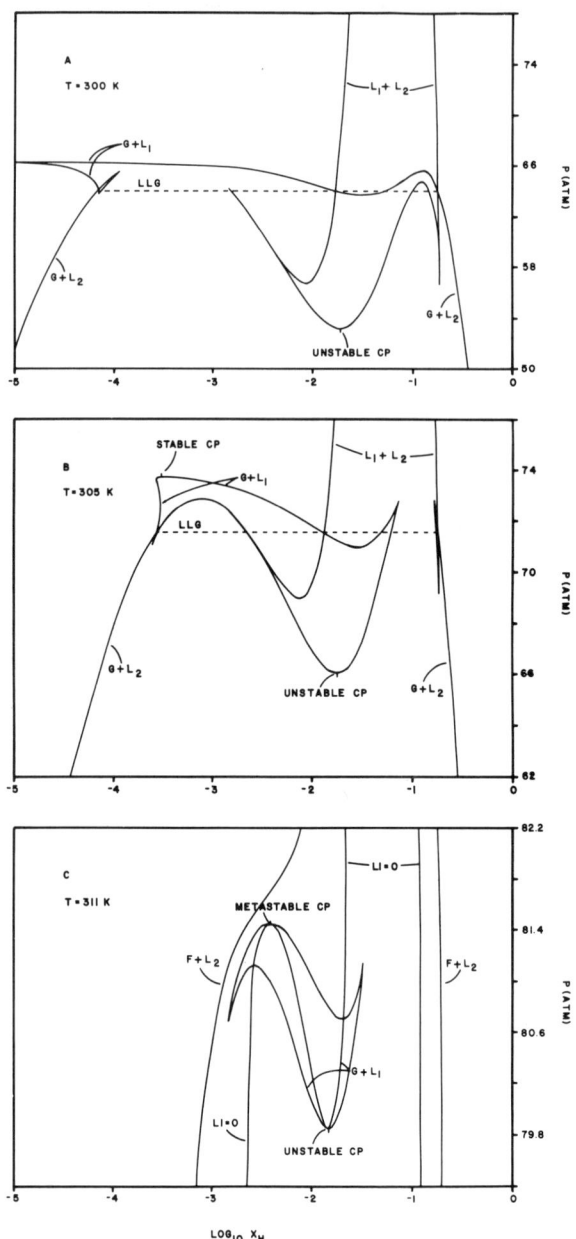

Figure 6. Enlarged P-x sections for n-hexadecane-carbon dioxide. The dashed line denotes liquid-liquid-gas three-phase equilibrium. Metastable and unstable binodal curves are included. In diagram A, the L_1+L_2 binodal forms a closed curve, while in diagrams B and C, the $G+L_1$ binodal forms a closed curve.

ing at low pressure at the unstable critical point. Although not shown, a small region of solid-liquid equilibrium should appear near the pure hexadecane axis above about 200 atm.

Figure 5b gives the P-x section at 305 K, a temperature above the critical point of CO_2. The $G+L_1$ region has now detached from the pure CO_2 axis, a fact which is more evident in the enlargement, Figure 6b. The inclusion of the metastable and unstable phase equilibrium solutions shows that the $G+L_1$ binodal is now a separate curve. This binodal exhibits both a stable and an unstable critical point, in accordance with Figure 4b.

With reference to Figure 4b, the presence of a metastable critical line between the LCEP and the cusp is noted. The meaning of this metastability is clarified upon generating a P-x section at a temperature in this range. Figure 5c shows a P-x section at 311 K, including stable binodal curves and spinodal curves. At about 80 atm, the binodal curve at low x_H displays a nearly horizontal segment. This might be regarded as the influence of the lower critical end point, which occurs a few tenths of a degree below 311 K. At the LCEP, the binodal curve would exhibit a horizontal point of inflection. Enlarging this region, the diagram of Figure 6c results. With the three-phase line no longer present, the $G+L_2$ and L_1+L_2 regions are not clearly distinguishable. These regions have thus collectively been labelled the $F+L_2$ region. As at 305 K, the $G+L_1$ binodal exhibits both a stable and an unstable critical point. Now, however, the entire curve lies within the $F+L_2$ region. Under these conditions, the stable $G+L_1$ segments represent a metastable equilibrium. At the LCEP, the $G+L$ critical point would be in a state of incipient metastability, and would just touch the $F+L_2$ binodal at its horizontal inflection point. At 311 K, the $G+L_1$ critical points lie much closer together than they did at 305 K. This circumstance is also indicated in Figure 4b. It is now evident that the cusp in that figure forms as the $G+L_1$ region shrinks to a point where the metastable and unstable critical points coincide.

Figure 5d shows the P-x section predicted at 313.5 K. The necked shape of Figures 5a through 5c has given way to a splitting of the $F+L_2$ region. The occurrence of the split is necessitated by the temperature minimum in the upper critical line, as shown in Figure 4a. The existence of a temperature minimum has been experimentally confirmed by Schneider, et al. (1967).

At 463.1 K, Figure 5e, the VLE region no longer extends to very low hexadecane concentrations, and is beginning to look more like a normal VLE envelope in a completely miscible system. This trend continues in Figure 5f, at a temperature of 623.6 K. At these temperatures, some low pressure composition data are available as shown in the diagrams. On the basis of the concentration of the lean component, the liquid phase predictions are within fifteen percent and the vapor phase predictions are within twenty-five percent of the measured values. This agreement is considered quite good in view of the fact that no compositional data was used in the determination of the interaction parameter.

Naphthalene - carbon dioxide

The naphthalene (N) - carbon dioxide system adds the complication of pure solid formation to the P-T-x space diagram. The triple point of naphthalene is nearly 50 K higher than the critical temperature of CO_2, a circumstance which often leads to the interruption of the critical lines by three-phase S-L-F lines.

The predictions in this section are carried out using an interaction parameter of $\delta_{12} = 0.11$, a value selected on the basis of data available in the supercritical fluid region. With this interaction parameter, it is possible to generate a fluid phase P-T-x space, as was done with the n-hexadecane-carbon dioxide system. In other words, the Peng-Robinson equation represents only the gaseous and liquid phases, and does not indicate the possibility of solid formation. Introduction of solid-fluid equilibria to the P-T-x space requires the use of the previously given equation for the fugacity of the pure solid phase.

The first features of interest in the $N-CO_2$ P-T-x diagram are the critical loci. These involve only fluid phases, and are thus generated from the equation of state alone. Figure 7 shows the critical curves for $N-CO_2$. As

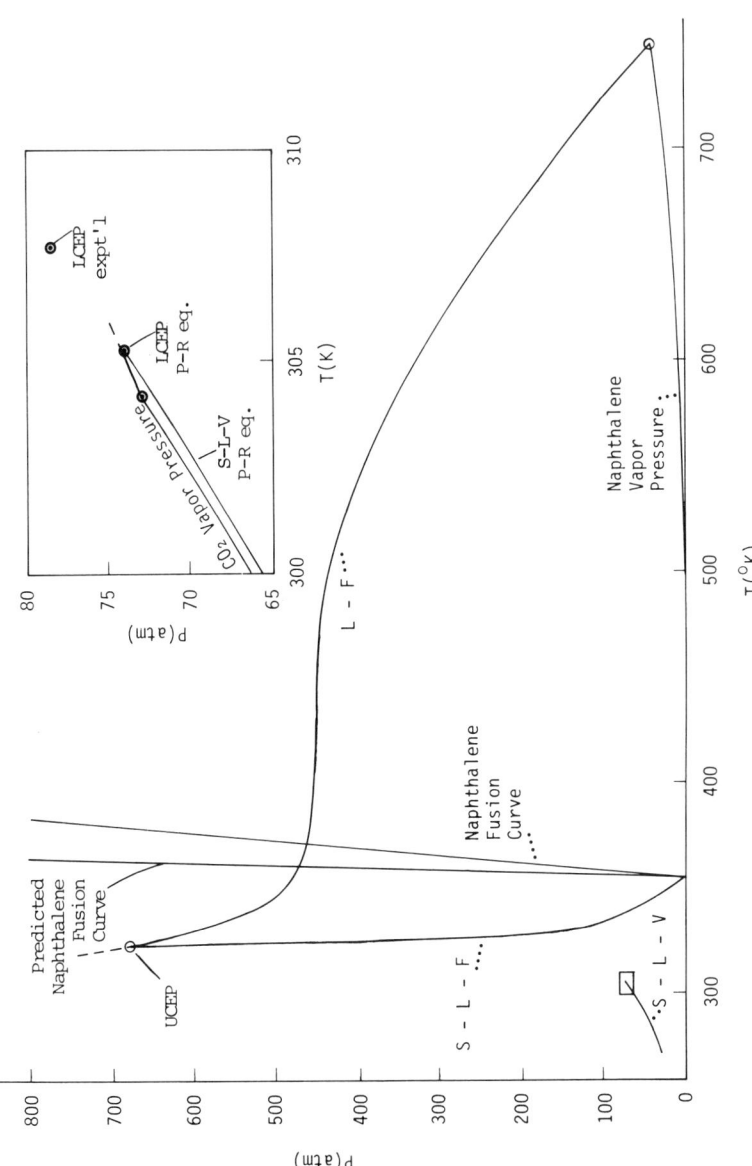

Figure 7. The P-T projection for the naphthalene-carbon dioxide system. The range of temperatures between the lower critical end point (LCEP) and the upper critical end point (UCEP) is the super-critical fluid region.

with the hexadecane-CO_2 system, there are two separate critical lines, each beginning at a pure component critical point. The hexadecane-CO_2 system, however, had only one critical end point, where liquid and vapor phases became identical in the presence of a third, liquid phase. There are now two critical end points, where the liquid-vapor or liquid-fluid critical phenomenon occurs in the presence of a solid phase. Another way of stating this is that, while the critical lines in general indicate the critical phenomenon for unsaturated solutions, the critical end points represent the critical phenomenon for saturated solutions.

Figure 7 also indicates that the prediction of critical lines does not stop at the critical end points. The upper critical locus continues indefinitely while the lower critical locus would continue until it met with an unstable critical locus in a cusp. A continuation of the three-phase lines after the CEPs is not possible, on the other hand, since beyond the critical line the equation of state predicts only one fluid. In addition to the critical and three-phase lines, the P-T projection in Figure 7 depicts the pure component vaporization curves and the naphthalene fusion curve. The vaporization curves predicted by the P-R equation in pure component form are at this scale indistinguishable from the experimentally measured curves. The naphthalene fusion curve is predicted by matching the solid phase fugacity with the pure liquid phase fugacity. It shows a significant deviation as a result of the poor modelling of liquid phase densities by the P-R equation.

A sequence of P-x sections calculated from the P-R equation is presented in Figure 8. Figure 8a shows the P-x section at 300 K. Except for a small region at about 70 atm, the S+F binodal represents the stable equilibrium state. The fluid-fluid binodal falls largely within the S+F region, where it represents metastable equilibrium. A metastable L_1L_2G tieline has been shown in the figure. The SLG tieline is not easily drawn until the 70 atm region is enlarged. Figure 9a gives this enlargement, and shows both three-phase tielines and spinodal curves and the labelling of the fluid-fluid binodals. The L_1+G binodals, which were not visible in Figure 8a, form a closed loop. Since 300 K is below CO_2's critical temperature, the stable L_1+G binodal extends to the

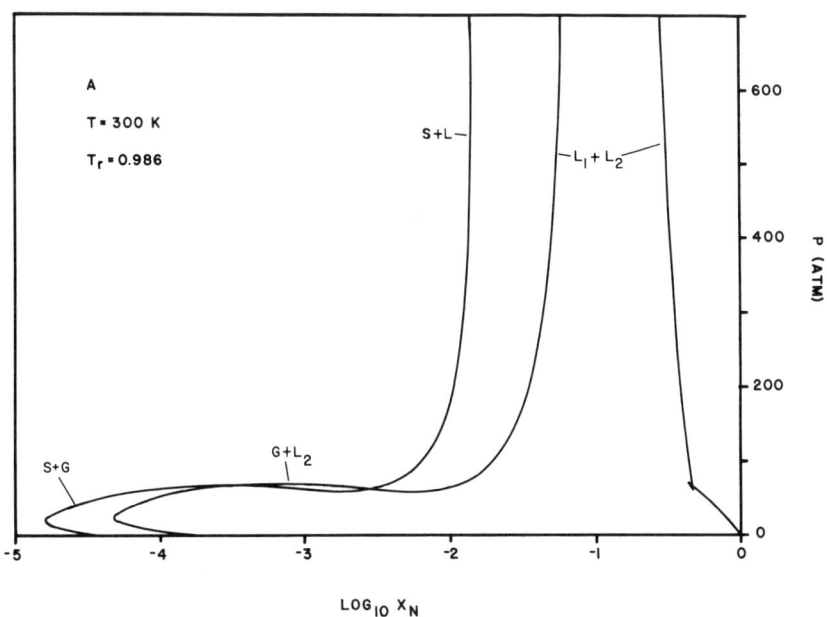

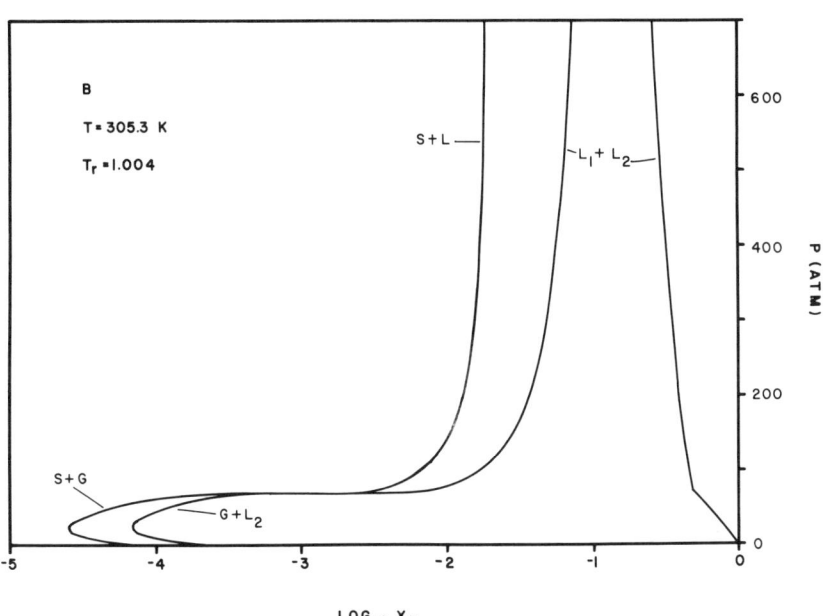

Figure 8. P-x sections for naphthalene-carbon dioxide. (First of three pages.)

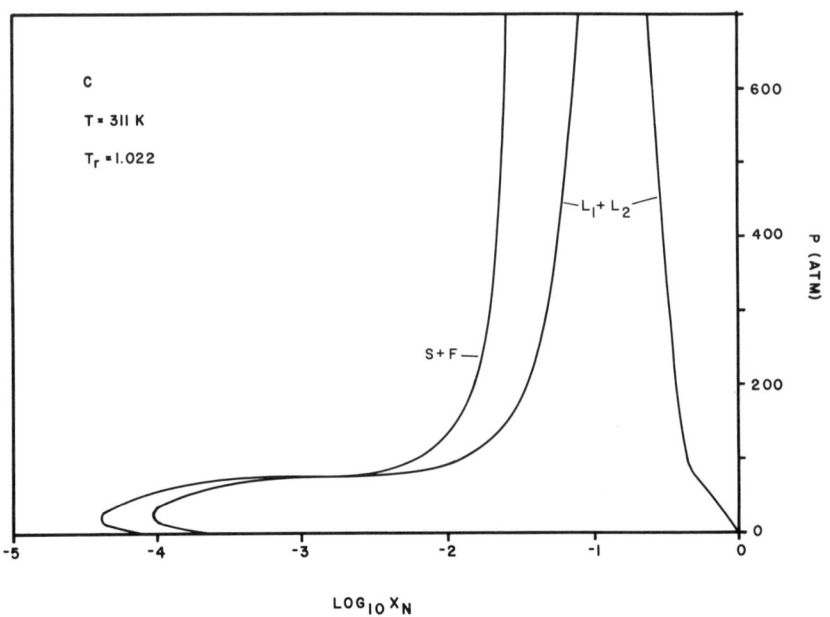

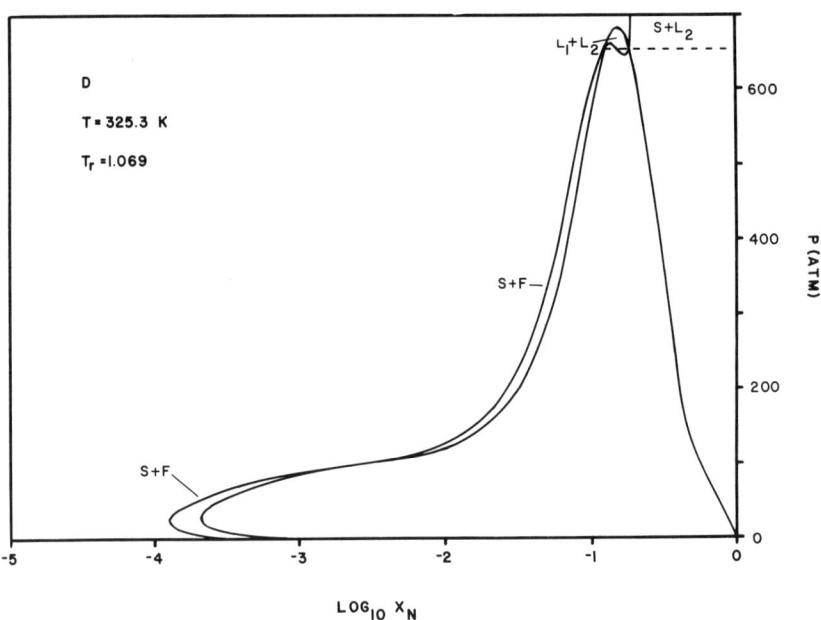

Figure 8. P-x sections for naphthalene-carbon dioxide. (Second of three pages.)

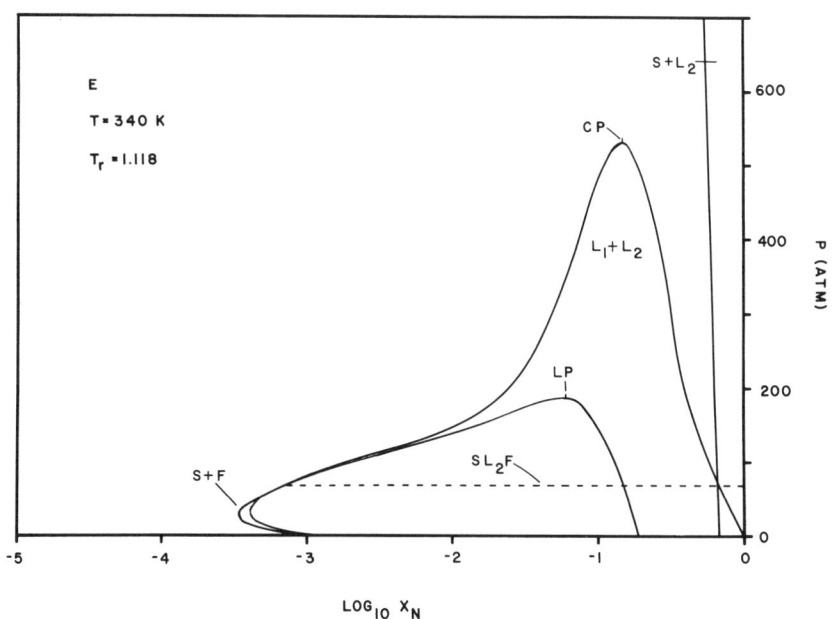

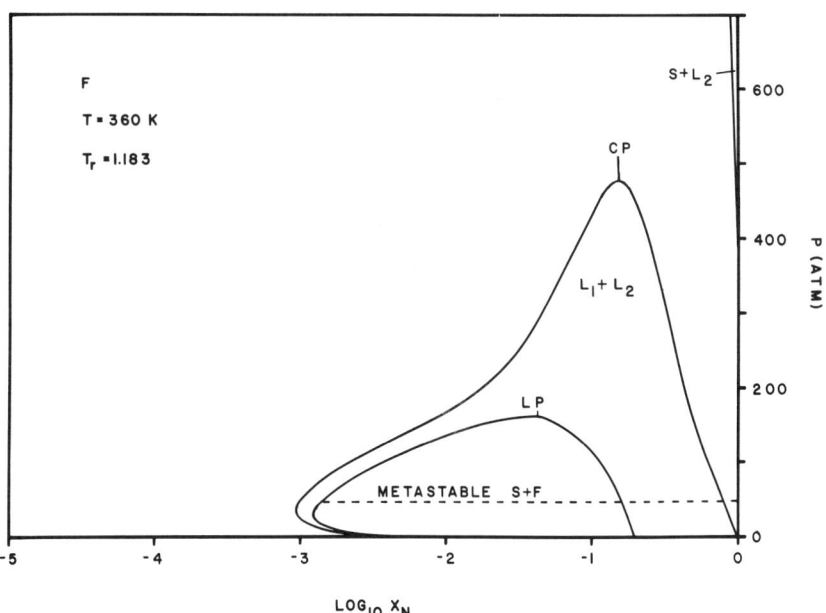

Figure 8. P-x sections for naphthalene-carbon dioxide. (Third of three pages.)

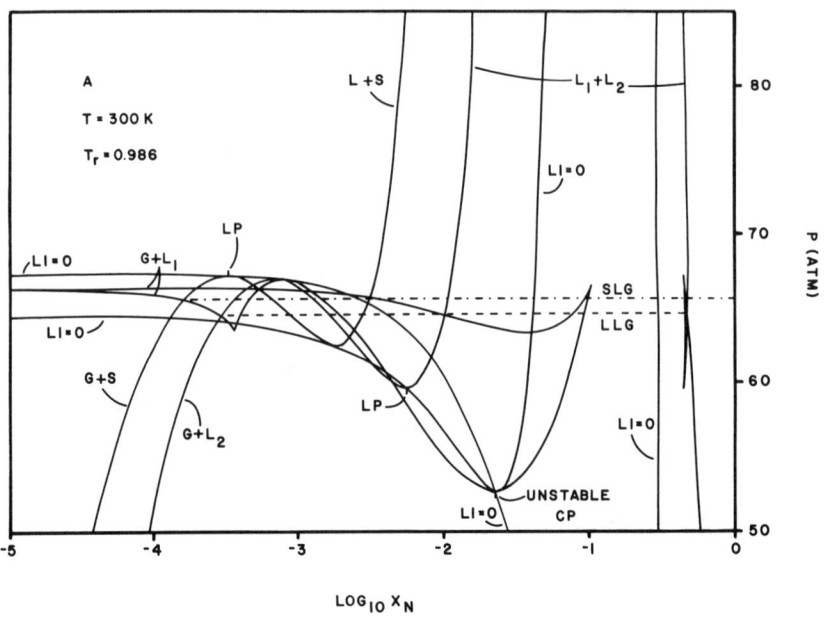
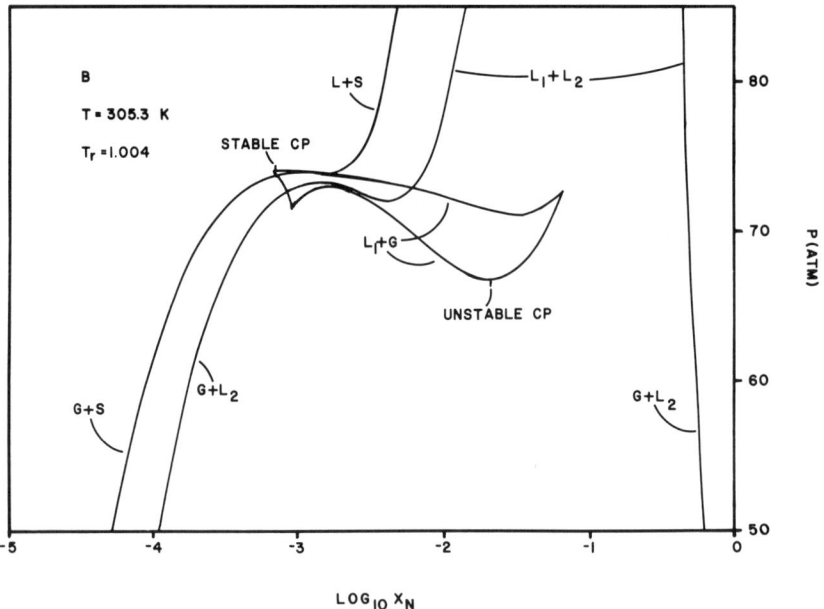

Figure 9. Enlarged P-x sections for naphthalene-carbon dioxide. Metastable and unstable binodal curves are included.

pure CO_2 axis. The metastable L_2+G binodals are joined to the metastable L_1+L_2 binodals by unstable segments.

Solid-fluid equilibria are represented in Figure 9a by a single curve. The S+G binodal starts at low pressures and rises to a smooth maximum in pressure. This point is designated as a limit point, LP, in the figure. The curve then enters an unstable region, reaches a smooth pressure minimum, and finally proceeds to higher pressures along the S+L binodal. The pressure minimum is also designated as a limit point. An interesting feature is noted upon following the S+G binodal more carefully. Starting at low pressure, the three-phase S-L-G pressure is reached at about 66.6 atm. In continuing to higher pressures, the metastable S+G curve passes entirely through the G+L_1 region before becoming intrinsically unstable at the limit point. It is well known that a single phase may exist in a metastable state when a two-phase system represents the true stable equilibrium. The present case demonstrates that the opposite can also occur, i.e., that two phases may exist in a metastable state in a stable one-phase region.

Figure 8b depicts the P-x section at 305.3 K, which is very nearly the LCEP temperature. The section is similar to that at 300 K, except that the fluid-fluid binodal no longer crosses the solid-fluid binodal. The L_1+G binodal still intersects the solid-fluid binodal, though, as shown in the enlargement of Figure 9b. Having detached from the pure CO_2 axis, the L_1+G region now terminates in a stable critical point. At a temperature slightly above 305.3 K, the stable critical point will remain as the last point of contact between the L_1+G binodal and the solid-fluid binodal. This is the lower critical end point. At this point, the solid-fluid binodal exhibits a horizontal point of inflection. It is also evident from the figure that, as temperature is raised further, a metastable LCEP will be predicted as the metastable critical point meets the metastable L_1L_2G line.

Figure 8c gives a P-x section in the supercritical fluid region, at 311 K. The solid-fluid binodal is now the stable two-phase system throughout the entire pressure range. The solubility increase between about 40 and 80 atm is actually not as dramatic as it was at lower temperatures, although at higher pressures the solubility is somewhat greater than previously. The solubility increase associated with the upper critical end point is not visible in this diagram since it occurs above 700 atm.

Increasing temperature to 325.3 K, the P-x section of Figure 8d is found. This is slightly above the upper critical end point temperature. A region of liquid-liquid equilibrium now exists between 650 and 680 atm. It intersects the solid-fluid binodal in two places to form the three-phase SLF line. Note that the solid-fluid binodal is again a single curve with two extrema. As the diagram suggests, at the UCEP temperature the L-L critical point intersects the solid-fluid binodal at a horizontal inflection point.

The P-x section at 340 K is given in Figure 8e. A three-phase S-L-F line is predicted at about 65 atm. Above this pressure, depending upon overall system composition, liquid-liquid or solid-liquid is the stable two-phase system. Below this pressure, the S+F binodal represents the stable two-phase system. Consequently, the low pressure portion of the L+F envelope is metastable. The connection of the S+F and S+L binodals in this case is discernable only at very low pressures.

Figure 8f gives the predicted P-x section at 360 K, somewhat above naphthalene's triple point temperature. At sufficiently high pressures, it is still possible to have solid present in the system, in equilibrium with a very concentrated solution. This two-phase region begins at a point along the naphthalene fusion curve. A metastable S+F tieline has been included in the diagram to emphasize that the limit point is not a critical point. Although two S+F binodals meet at the limit point, the two phases in equilibrium (solid and fluid) do not become identical. At low pressure, the S+F loci cross, with the metastable segment intersecting the pure naphthalene axis at a metastable sublimation pressure.

A limited amount of phase equilibrium data is available for the naphthalene-CO_2 system. The most frequently cited measurements are those of Tsekhanskaya, et al. (1964), whose observations have recently been confirmed by the work of McHugh and Paulaitis (1980) and Kurnik, et al. (1981). These data, which fall within the supercritical fluid region, are typically plotted as shown in Figure 10. Data of Najour and King (1966) in the compressed gas region is included as well. The metastable S+F solution matches the data quite well, even at the highest pressure of 320 atm. It is plausible, given the limited accuracy of the P-R equation of state, that the experimental points do in actuality represent metastable,

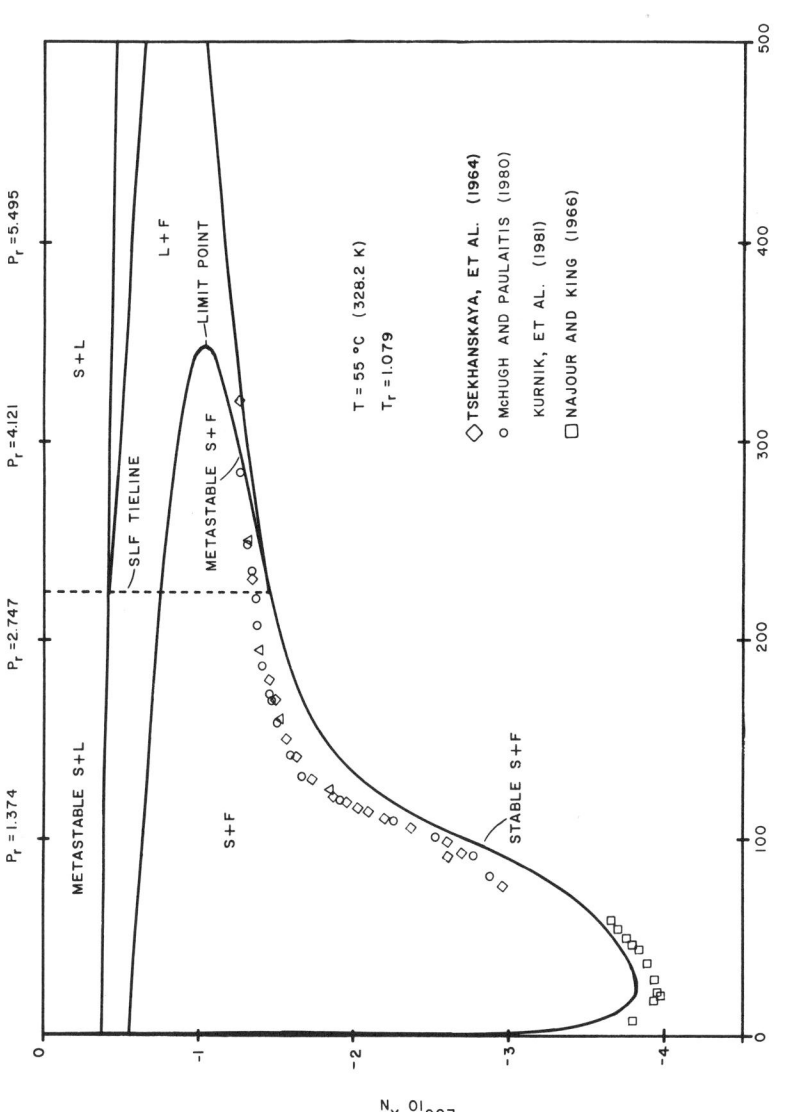

Figure 10. The solubility of naphthalene in carbon dioxide at 55°C. All predicted solid-fluid binodals, and the stable liquid-fluid binodals, are shown.

subsaturated S+F equilibrium. The existence of metastable phases in this system could be readily ascertained by approaching the SLF line from pressures above the limit point (i.e., 400 atm or more).

Measurements or estimates of the critical end points are also available for the naphthalene-CO_2 system. These are compared with the predicted values in Table III.

Naphthalene – ethylene

The naphthalene – ethylene system has been more completely characterized experimentally than the naphthalene – carbon dioxide system. In order to allow a comparison between the predictions for these two systems, only data in the supercritical fluid regime were used in the determination of the interaction parameter. A value of $\delta_{12} = 0$ was found to give reasonably good fit.

With δ_{12} set equal to zero, the P-T projections of Figure 11 are generated. Qualitative agreement with the experimental phase diagram is quite good over the entire range of temperatures and pressures. The calculated pure component vapor pressure curves for ethylene and naphthalene are indistinguishable from the experimental curves. The calculated fusion line, however, has a steeper slope than it should. Figure 11 differs from Figure 7, the naphthalene-CO_2 P-T projection, in that a maximum is predicted in the L-F critical line. The L-F critical line is once again shown continuing past its intersection with the S-L-F line at the UCEP. This metastable portion forms a CEP with the metastable L-L-G locus, as depicted in the enlargement, Figure 12. There is a second CEP formed by the intersection of the metastable L-L-G locus and the metastable L-G critical line. The figure also shows the generally excellent agreement between calculated and measured S-L-G loci. Note that the S-L-G line is not terminated by the metastable L-F critical line.

A comparison of the stable critical end points for this system is included in Table III. As with naphthalene-CO_2, the temperature and pressure coordinates for the LCEP agree quite closely, while the predicted mole fraction is too low, in this instance by a factor of 6 or 7. At the UCEP the predicted temperature is about 20 percent too low, and the predicted naphthalene mole fraction about 35 percent too low. This is similar to what was found for naphthalene-CO_2. The

TABLE III.
Critical End Points for Naphthalene-Ethylene
and Naphthalene-Carbon Dioxide

System	CEP-1 T(°C)	CEP-1 P(atm)	CEP-1 x	CEP-2 T(°C)	CEP-2 P(atm)	CEP-2 x
Naphthalene-ethylene, experimental	10.7[a]	51.2[a]	.002[a]	52.1[b]	174[b]	.17[b]
Naphthalene-ethylene, Peng-Robinson $\delta_{12} = 0$	10.0	50.4	.0003	40.7	161.0	.11
Naphthalene-CO_2, experimental	34.5[c]	78.5[d]	.0031[e]	63[f]	240[f]	.16[f]
Naphthalene-CO_2, Peng-Robinson $\delta_{12} = 0.11$	32.5	74.5	.0012	52.1	682	.15

a - van Gunst, et al. (1953)

b - van Welie and Diepen (1961)

c - Büchner (1906) and Tsekanskaya, et al. (1962)

d - Estimated from data of Tsekhanskaya, et al. (1966)

e - Tsekhanskaya, et al. (1962)

f - Estimate of McHugh and Paulaitis (1980)

Most of the predicted CEP coordinates have been revised from the values given by Modell, et al. (1979).

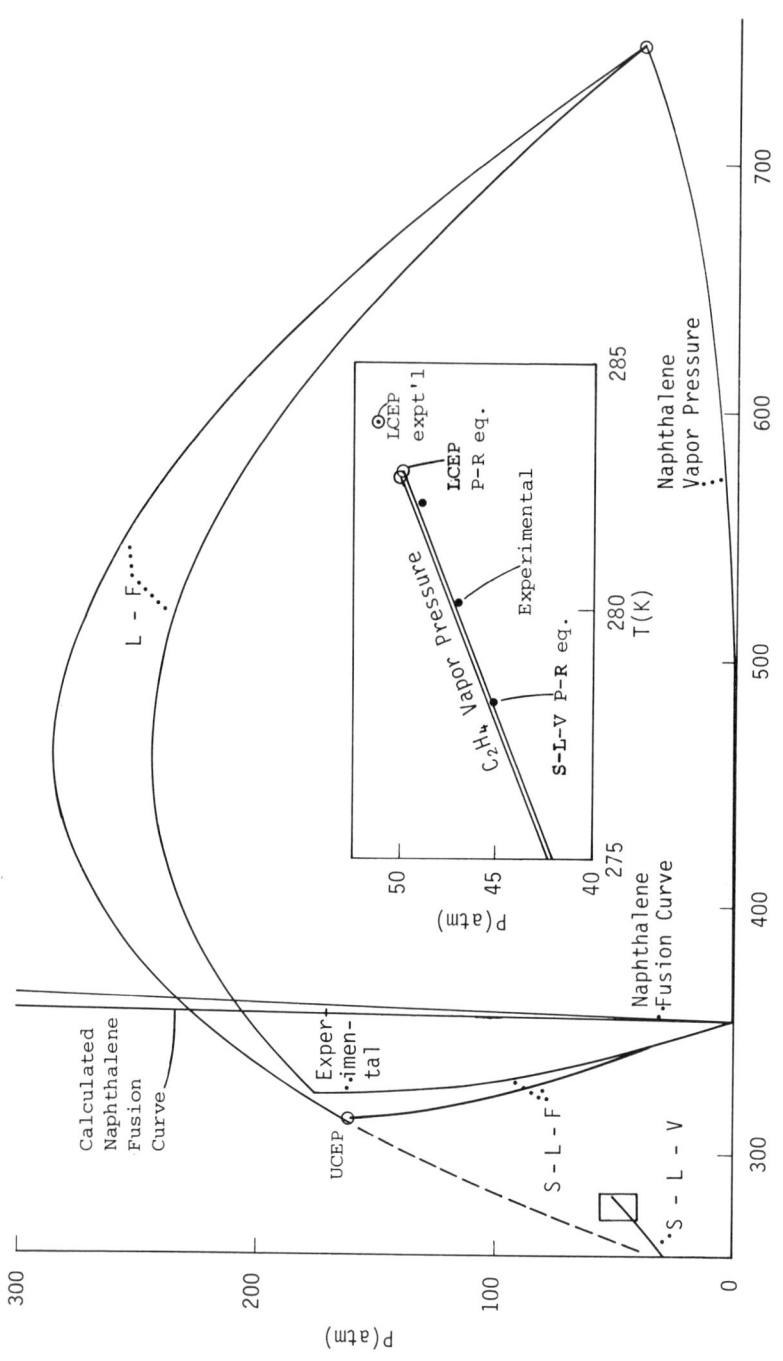

Figure 11. The P-T projection of the naphthalene-ethylene system. The inset shows the region near ethylene's critical point. (See also Figure 12.)

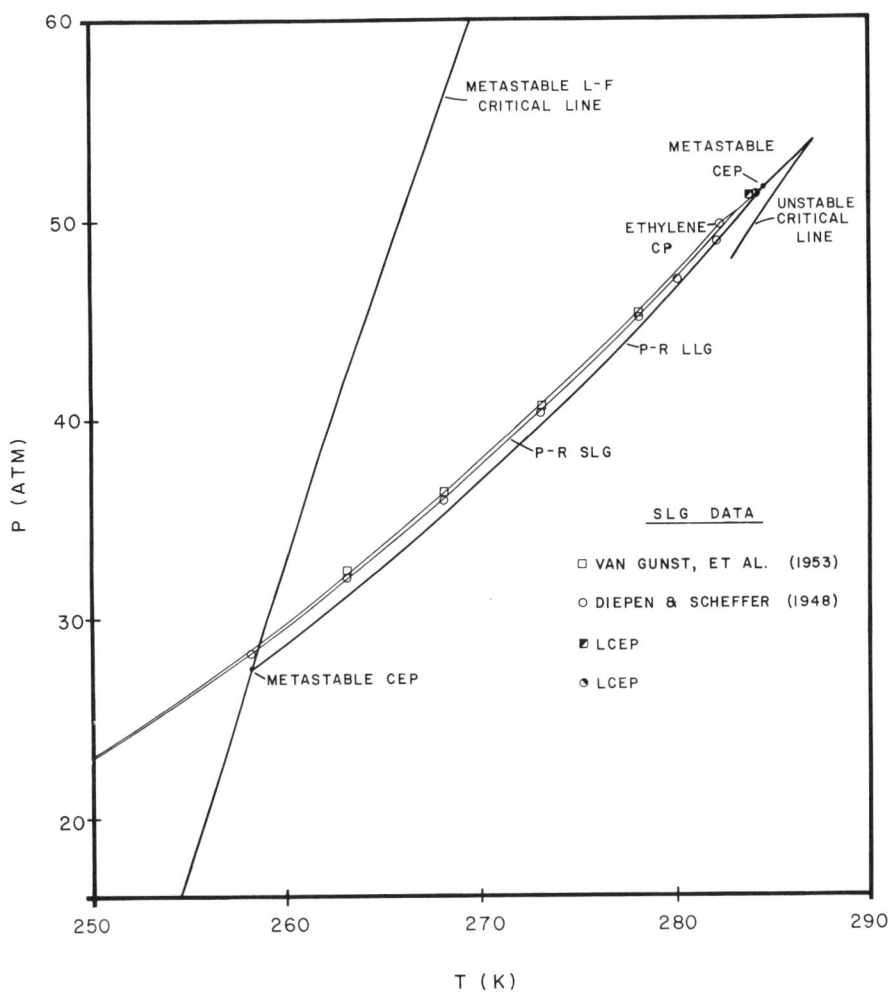

Figure 12. The P-T projection of the naphthalene-ethylene system in the vicinity of ethylene's critical point.

predicted UCEP pressure is only slightly low, which might speak for the reliability of the calculated UCEP pressure for naphthalene-CO_2. In that case, however, the calculated UCEP pressure was almost three times the estimated experimental value. This is perhaps due to the steep slopes of the L-F critical line and the S-L-F line in the naphthalene-CO_2 P-T projection at temperatures near the UCEP. From a calculational standpoint, an inaccuracy of a few degrees in the predicted UCEP could change the predicted pressure by almost one hundred atmospheres. From an experimental point of view, assuming that a region of steep slopes does indeed exist, it will be detected only by measurements at carefully selected temperatures, pressures, and compositions.

P-x sections at six different temperatures have been drawn up in Figure 13. Some of the sections have been chosen at the same reduced temperature, with respect to the solvent, as their naphthalene-CO_2 counterparts. Experimental data is not available for these diagrams.

Figure 13a gives the P-x section at 278.5 K, a temperature below ethylene's critical point. The metastable $G+L_2$ and L_1+L_2 regions are of a familiar shape, and completely enclosed by the S+F region. The stable $G+L_1$ region only shows up on an expanded scale. Figure 13a may be compared with Figure 8a which has been drawn at the same reduced temperature with respect to the solvent. In Figure 8a, the solubility in the S+G region is considerably higher than in Figure 13a. This is due to the lower absolute temperature in Figure 13a, which leads to a lower naphthalene sublimation pressure, and hence a lower solubility.

At 283.2 K, Figure 13b, the LCEP temperature has been reached. The S+F equilibrium curve exhibits a point of horizontal inflection. Coincident with this point is the stable critical point of the metastable $G+L_1$ region which is, however, too small to appear in the figure. The shape of this region has previously been given in Figure 9b.

Increasing temperature further, the supercritical fluid region is entered. Figure 13c is drawn at 288.7 K, the same reduced temperature with respect to the solvent as Figure 8c.

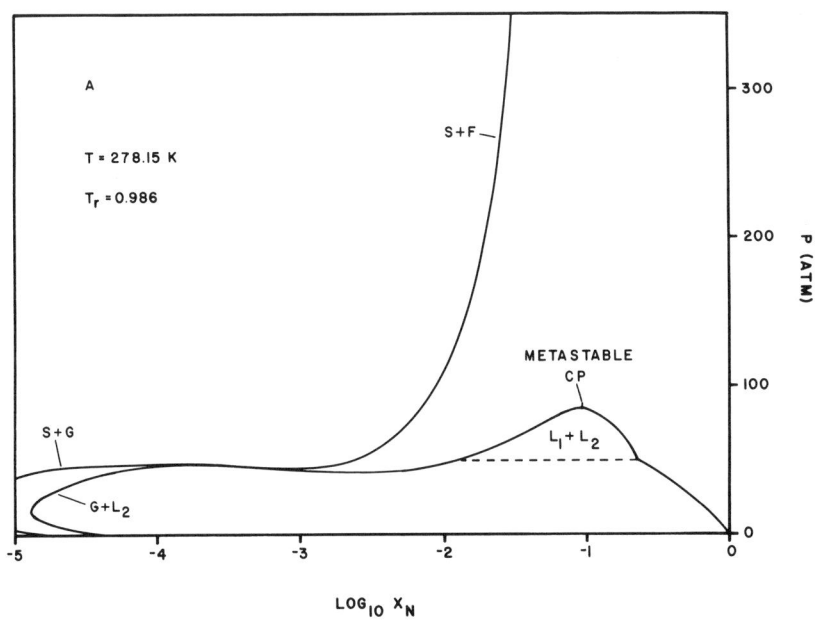

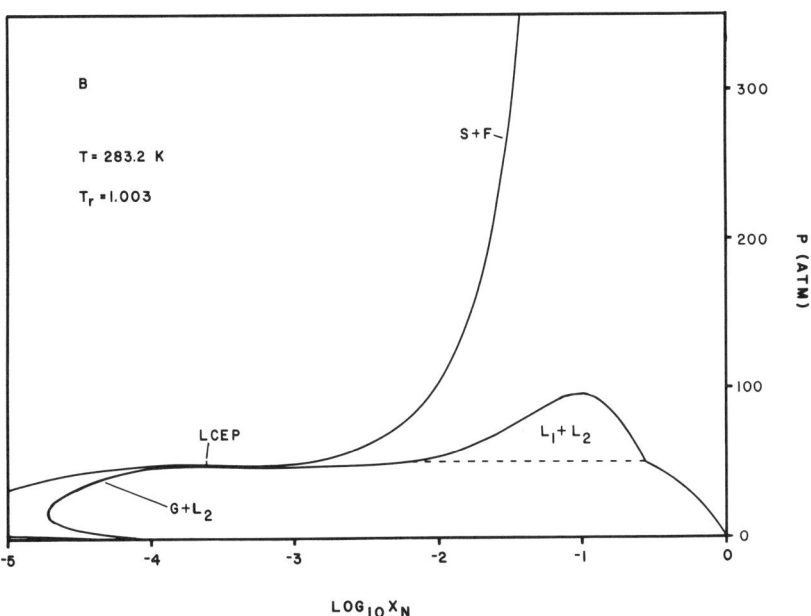

Figure 13. P-x sections for naphthalene-ethylene. (First of three pages.)

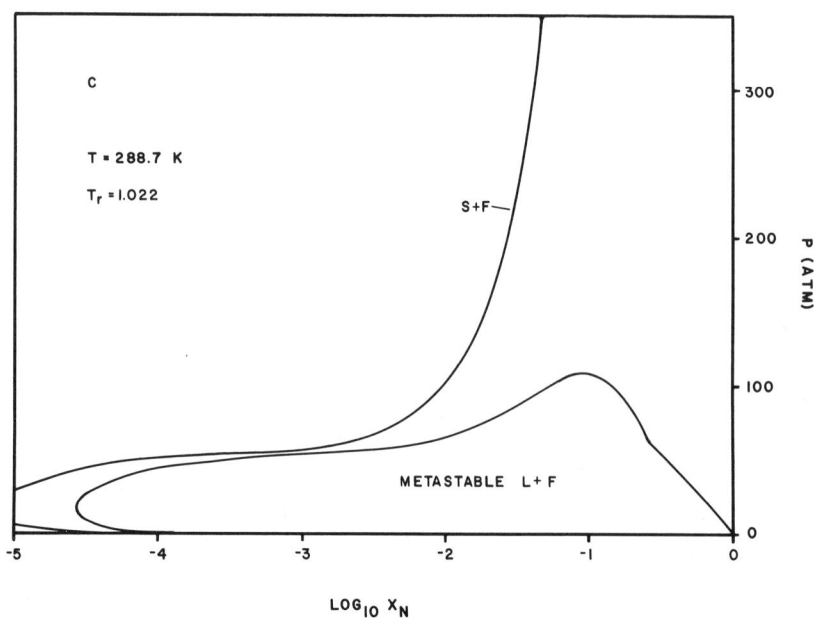

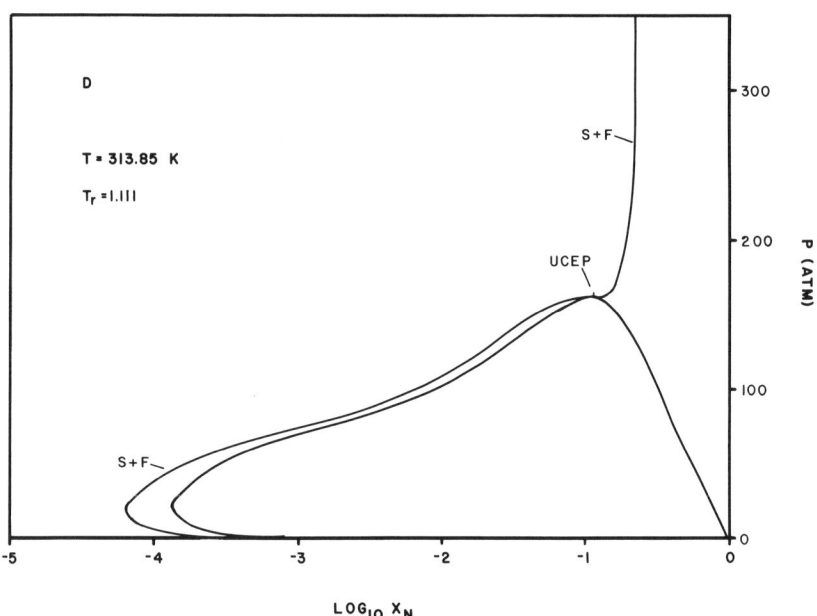

Figure 13. P-x sections for naphthalene-ethylene. (Second of three pages.)

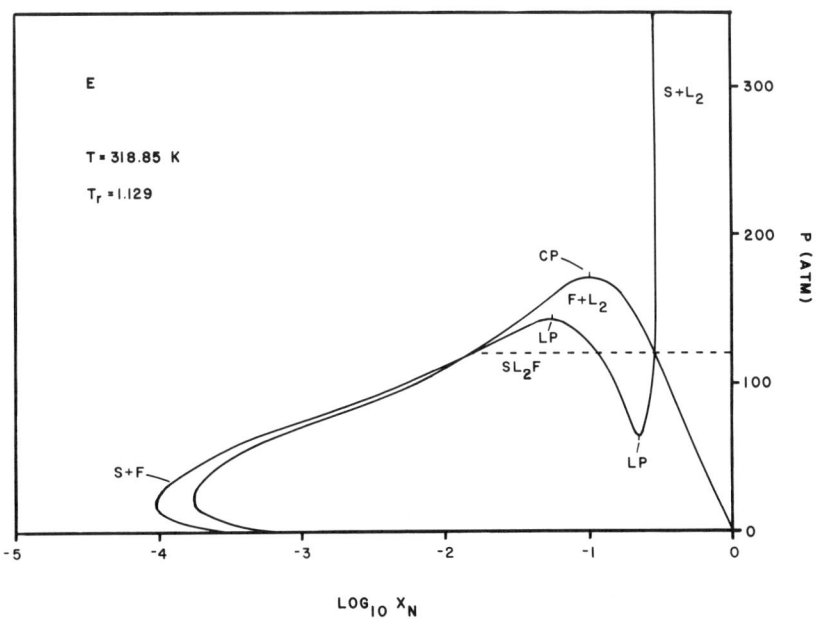

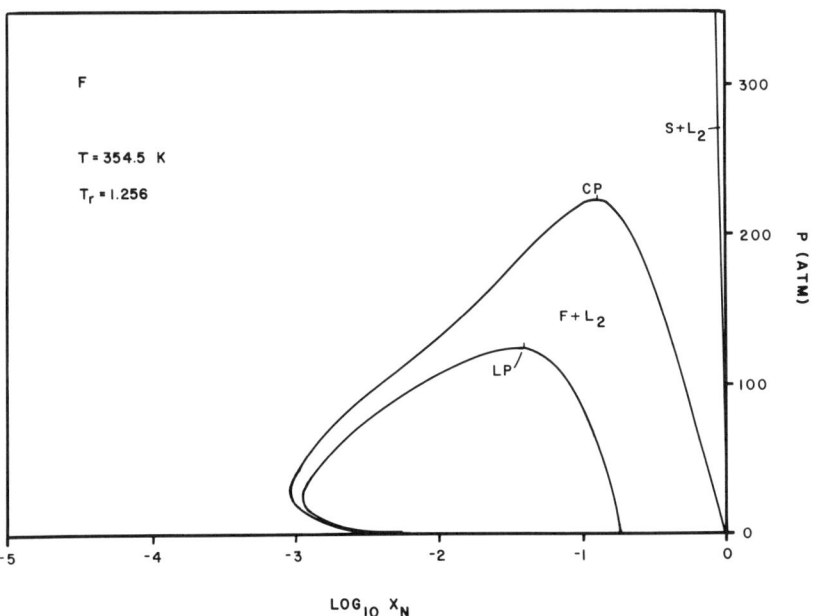

Figure 13. P-x sections for naphthalene-ethylene. (Third of three pages.)

The stable two-phase system for the entire pressure range is S+F. This temperature is above the upper metastable CEP in Figure 12, so that separate G+L_2 and L_1+L_2 regions no longer exist. The G+L_1 binodals, though too small to appear in the figure, are still present. They fall entirely within the L+F region, and exhibit the usual stable and unstable critical points. These critical points persist up to the cusp in Figure 12.

Figure 13d is drawn at 313.85 K, which is essentially the upper critical end point temperature. The solid-fluid binodal now exhibits a horizontal inflection point, which is coincident with the stable critical point of the metastable L+F region. This diagram is to be compared with Figure 8d.

Figure 13e gives the P-x section at 318.85 K, somewhat above the UCEP temperature. As in Figure 8e, a stable region of L+F is now present, along with the associated SLF tieline. The solid-fluid binodal now possesses two limit points, where the metastable and unstable segments are joined. Below the SLF tieline, S+F is the stable two-phase system. At pressures above the tieline, depending on the overall system composition, the stable two-phase system will be L+F or S+L. At naphthalene's triple point, the P-x section of Figure 13f is found. The S+L and L+F binodals meet at the triple point pressure.

Benzene - water

The preceding discussion has demonstrated the utility of the P-R equation in modelling nonpolar systems. In these systems the components were considerably different in molecular size, leading to nonidealities manifested as liquid-liquid immiscibilities. In the present case, nonidealities are a result of both size and polarity differences. The polar compound will have a strong self-affinity, in this instance the hydrogen bonding of water. Peng and Robinson (1976b) have used their equation to predict three-phase equilibrium in multicomponent hydrocarbon - water systems with some degree of success. Figure 1 also indicated that the P-R equation gives a reasonable picture of the phase behavior of pure water.

At 455.2 K, the vapor pressures of benzene and water are equal. Below this temperature, benzene is the more volatile component, while at higher temperatures water is more volatile. The crossing of the vapor pressure curves requires the occurrence of an azeotrope in the system, although not necessarily a stable one. A fairly extensive amount of experimental data has been gathered for the benzene-water system, by a number of investigators. No single study has covered the entire fluid region, but between the work of Rebert and Kay (1959) at low to moderate pressures, the work of Connolly (1966) at moderate to high pressures, and the mapping of the upper critical locus by Alwani and Schneider (1967), a good overall picture of the phase behavior is obtained.

An interaction parameter for this system was selected by matching a single point along the P-T projection of the upper critical locus. Figure 14 shows the experimental upper critical locus along with two predicted loci. The value of δ_{12} = 0.052 is reasonably optimal in reproducing the course of the critical line, and was used for all further predictions in the benzene-water system.

The critical curves in Figure 14 have been depicted as smooth and continuous. It is necessary to examine the region near the benzene critical point more closely to elucidate this behavior. Figure 15a illustrates the phase relationships as deduced from experimental measurements. The lower critical locus begins at the benzene critical point and tends toward lower temperatures and higher pressures. It reaches a temperature minimum, turns toward higher temperatures, and shortly thereafter is terminated at the lower critical end point. Rebert and Kay's measurements indicate that the three-phase LLG line has a metastable continuation beyond the lower critical end point. It terminates at a critical solution end point, simultaneously intersecting the L-L and L-G critical lines. The transition between the L-L and L-G critical lines has been drawn as smooth, although experimental measurements indicate that this is not actually true. No measurements of the L-L critical line are available in this region to indicate its correct shape, however.

Figure 15b gives the calculated P-T projection in the vicinity of benzene's critical point. The critical curve starting from the pure water critical point does not meet the liquid-liquid critical line as in Figure 15a, but rather reaches a minimum in temperature and then returns to the benzene critical point. The liquid-liquid critical line on the other hand reaches a minimum in pressure shortly before being terminated by an unstable critical line. Figure 15b also gives the three-phase LLG line which terminates at a critical end point on the LL critical line. The short segment of the LL critical line between the CEP and the point of the cusp where the unstable critical line is joined is metastable.

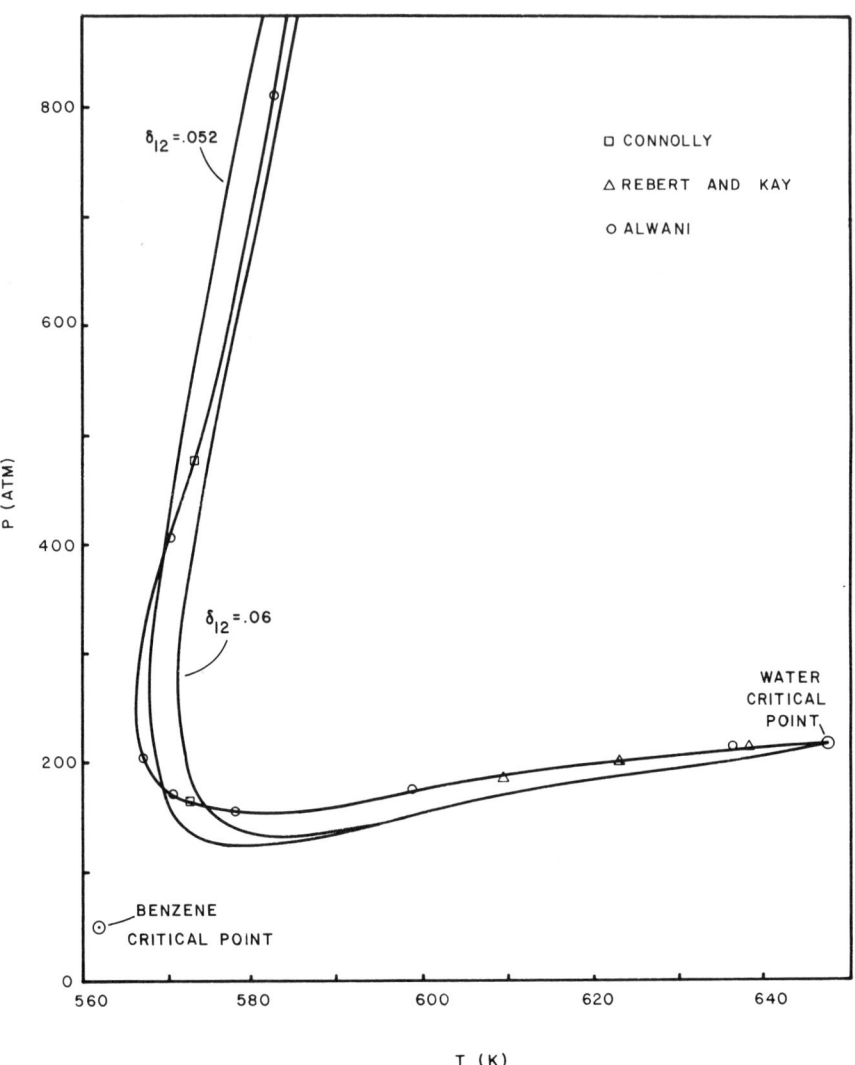

Figure 14. The P-T projection of the upper critical locus for benzene-water.

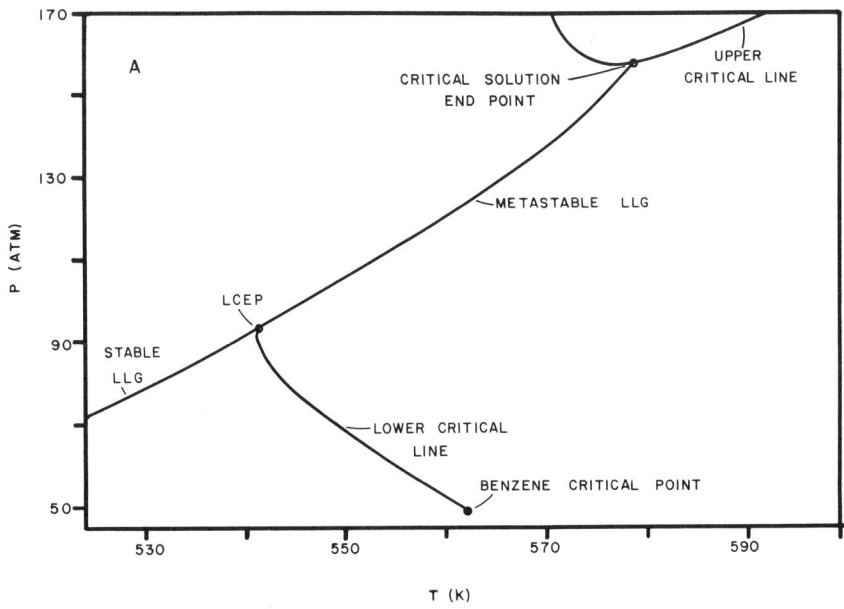

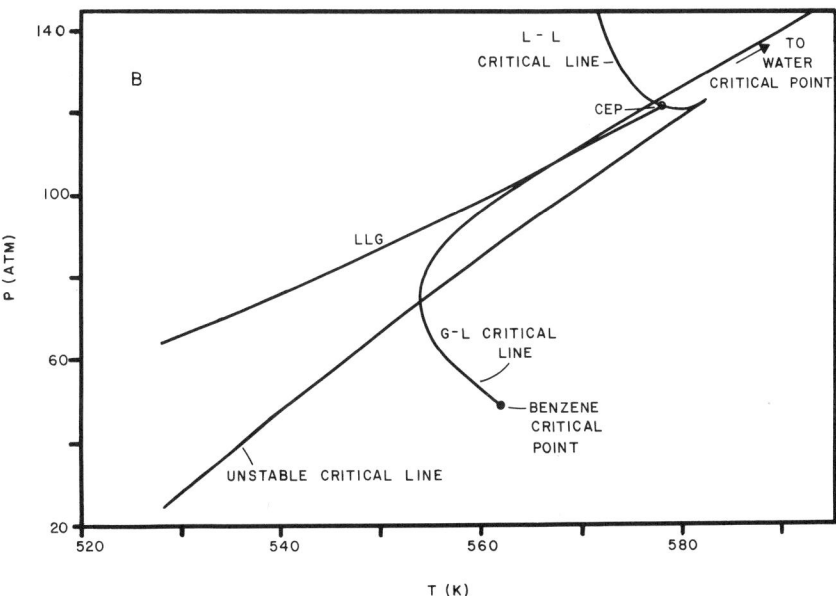

Figure 15. The benzene-water P-T projection in the vicinity of the pure benzene critical point. Diagram A gives the experimental results. Diagram B shows the predicted loci.

The experimental and predicted P-T projections of Figure 15 differ mainly in that the former depicts a metastable portion of the LLG line. Rebert and Kay assert that this metastable line corresponds to the intersection of two binodal regions. However, data were taken on a fairly coarse grid, and a third binodal region of limited extent may well have gone undetected. As consideration of the P-x sections will show, the equation of state predictions indicate that a third binodal region does exist, and that very careful measurements will be necessary to determine the stability or metastability of the LLG line up to the critical solution end point.

A fairly complete set of P-x diagrams has been published by Rebert and Kay (1959). Some of these have been reproduced in Figure 16. High pressure data from Connolly (1966) has been included in Figures 16d and 16e. Figure 16a shows the P-x section at 528.15 K, a temperature below the critical end point in Figure 15a. The diagram is typical for systems in which the three-phase LLG line lies at a higher pressure than the vapor pressure of either component. Systems of this type are known as heteroazeotropes. In this instance the azeotropic composition is given by x.

Figure 16b is drawn at 541.45 K, a temperature slightly above the critical end point. The benzene-rich $G+L_1$ region has separated from the water-rich $G+L_2$ and L_1+L_2 regions. This separation first occurs at the temperature minimum in the lower critical locus in Figure 15a. In the few tenths of a degree between the lower critical locus minimum and the critical end point, there is a small region of benzene-rich VLE attached to the three-phase line in addition to the separate "island" stemming from the vapor pressure of pure benzene. This must be so, since at any temperature just above the minimum in the critical locus there must be two critical points.

The L_1+L_2 and $G+L_2$ regions in Figure 16b are shown to intersect along a horizontal tieline (dashed) which corresponds to a point on the metastable LLG tieline in Figure 15a. Thus, this three-phase line represents the intersection of two, rather than three binodal regions, as is normally the case. At the metastable three-phase pressure, a discontinuity in the slope of the binodal locus has been indicated. Rebert and Kay have drawn the section in this manner, although their data seem insufficient to prove such a discontinuity. The equation of state does not yield slope discontinuities when only two binodal regions merge.

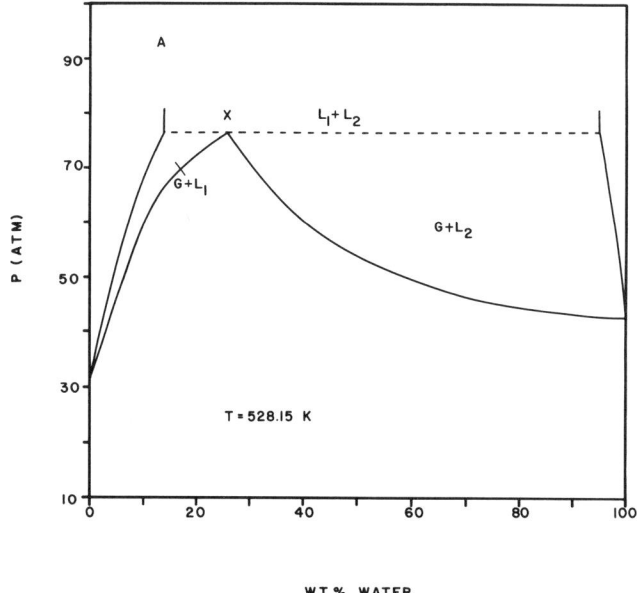

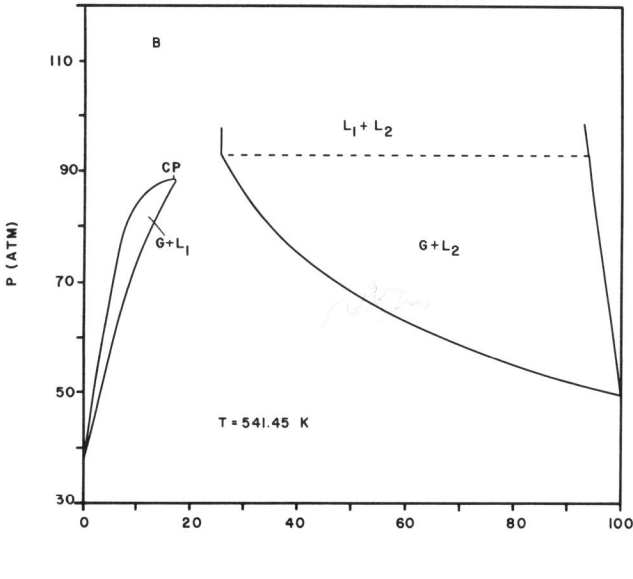

Figure 16. Smoothed experimental P-x sections for benzene-water (Rebert and Kay, 1959). (First of three pages.)

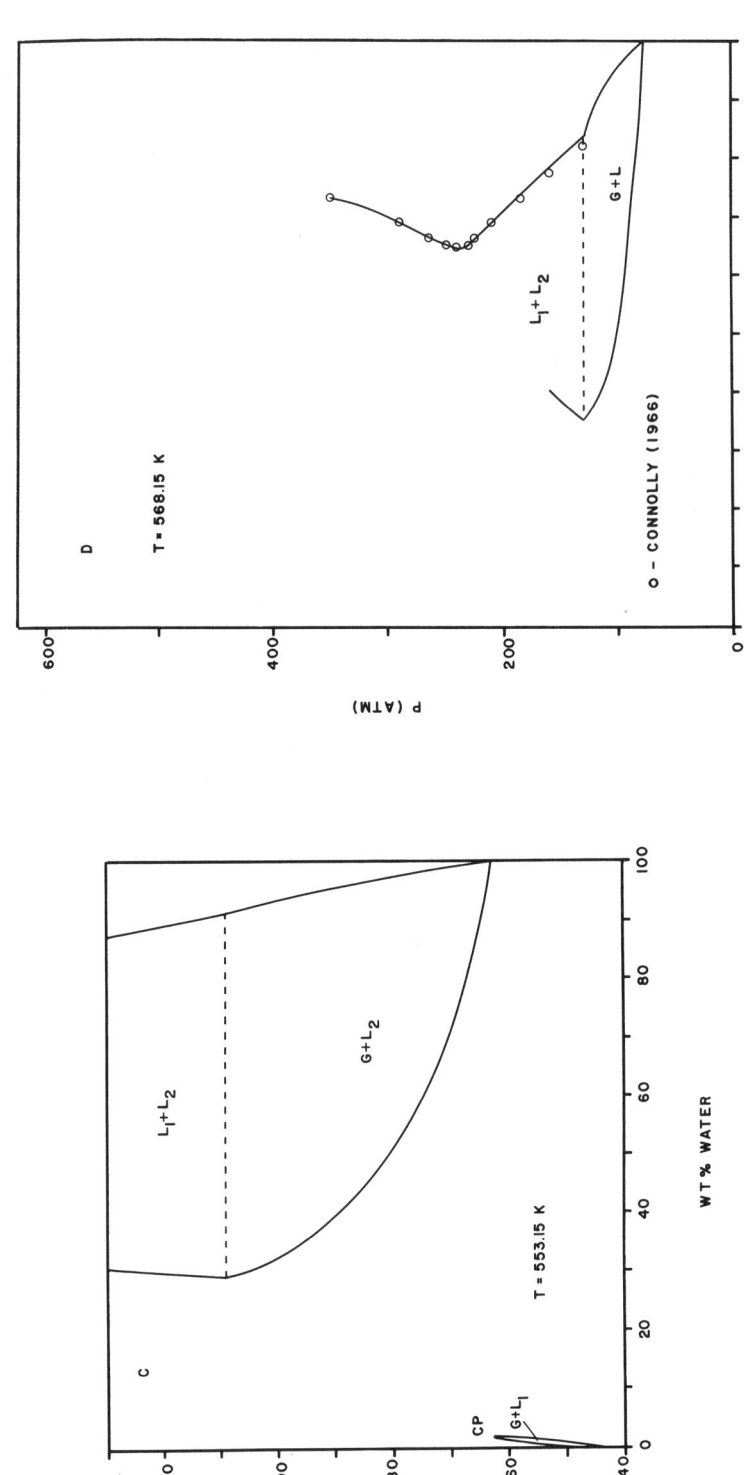

Figure 16. Smoothed experimental P-x sections for benzene-water (Rebert and Kay, 1959). Data due to Connolly has been included in diagram D. (Second of three pages.)

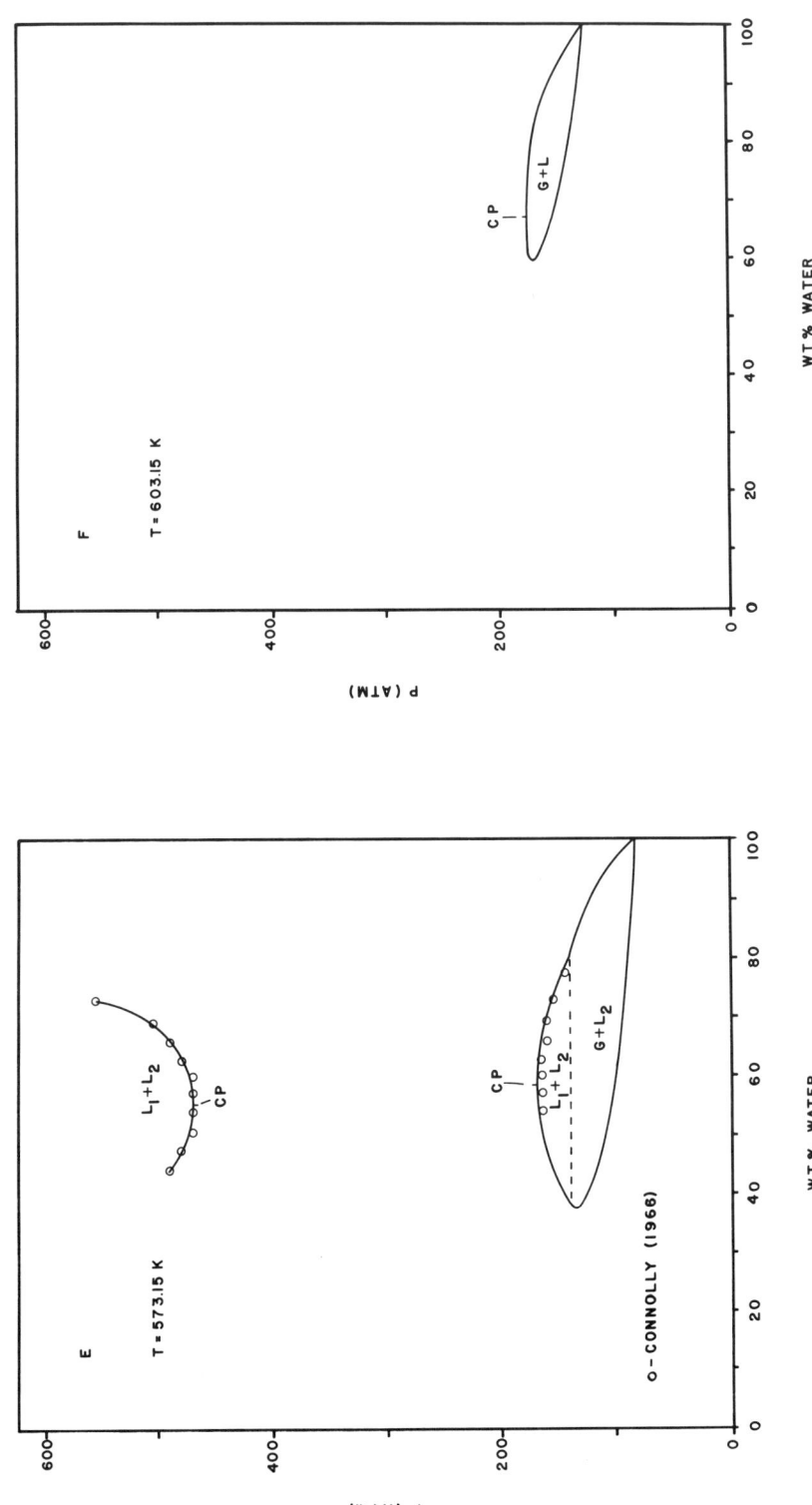

Figure 16. Smoothed experimental P-x sections for benzene-water (Rebert and Kay, 1959). Data due to Connolly has been included in diagram E. (Third of three pages.)

Increasing temperature to 553.15 K, the section of Figure 16c is found. The benzene-rich island is now quite small. Figure 16d is drawn at 568.15 K. This temperature is above benzene's critical point, so that the benzene-rich island no longer exists. The figure also includes some higher pressure data along the L_1+L_2 envelope. Only the water-rich leg of the curve has been measured. It exhibits a pronounced minimum in water composition. The benzene leg would have a corresponding maximum in water composition. At 573.15 K, as shown in Figure 16e, the pinching effect created by these extrema has been completed, yielding two distinct L_1+L_2 regions separated by a miscibility gap. As temperature is further increased, the lower L_1+L_2 region shrinks and finally disappears at the critical solution end point. At temperatures above this end point, the VLE envelopes are of the usual shape, as illustrated in Figure 16f.

Figure 17a gives the calculated P-x section at 528.15 K. Three stable binodal regions are correctly predicted. Figure 17b shows the P-x section for 553.15 K. The benzene-rich VLE envelope has narrowed. At this temperature, a previously metastable normal azeotropic point has moved into the stable region. Thus, a normal, minimum boiling azeotrope is now predicted rather than a heteroazeotrope. Experimental measurements have not been detailed enough to verify the presence of a normal azeotrope in this small region. A normal azeotrope must exist, however, unless by chance the lower critical locus is coincident with the heteroazeotrope-normal azeotrope transition point.

Increasing temperature slightly, to 554 K, the temperature minimum in the vapor-liquid critical locus has just been exceeded. From Figure 15b, it is evident that there must now be a separate benzene-rich region. Figure 17c illustrates these features. As previously mentioned, there must be a similar break with the formation of two critical points just above the temperature minimum in Figure 15a.

Figure 17d is the P-x section at 568.15 K. The azeotropic region has diminished in size considerably. The necking effect in the L_1+L_2 region, due to the proximity of the critical solution temperature, is evident. As temperature is further increased, the benzene-rich branch of the azeotrope continues to shrink. Eventually, the azeotrope critical point will meet the azeotrope point in a cusp. Figure 17e, at

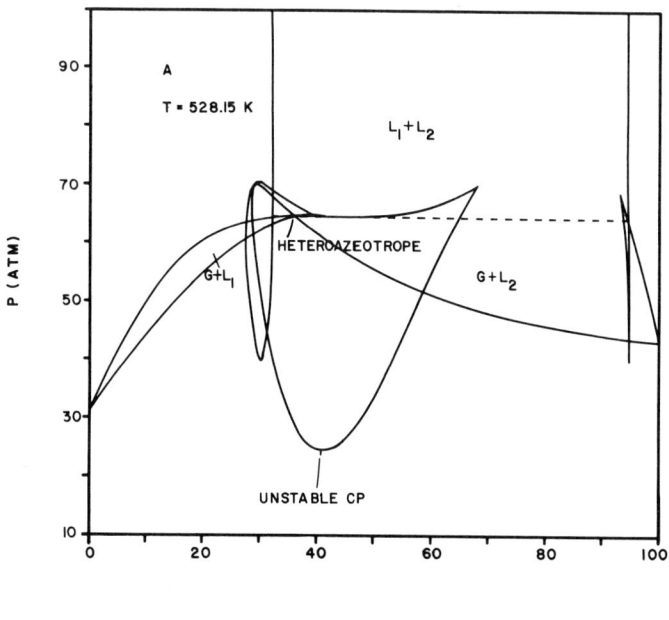

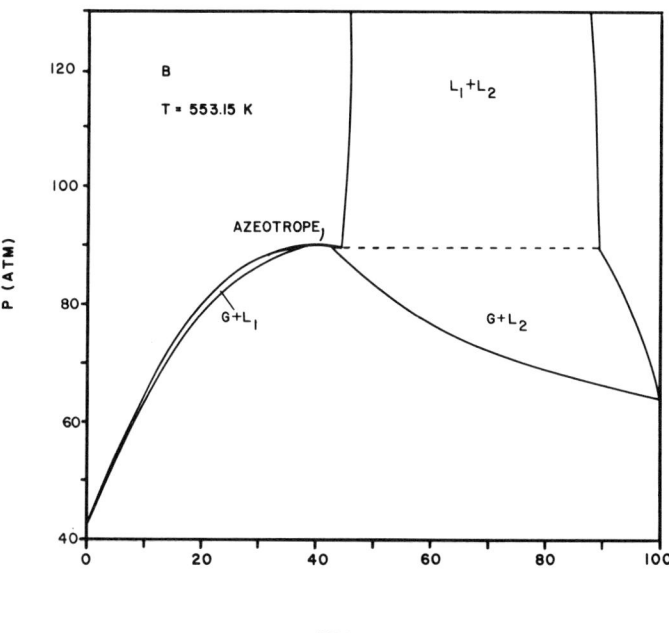

Figure 17. Predicted P-x sections for benzene-water. (First of three pages.)

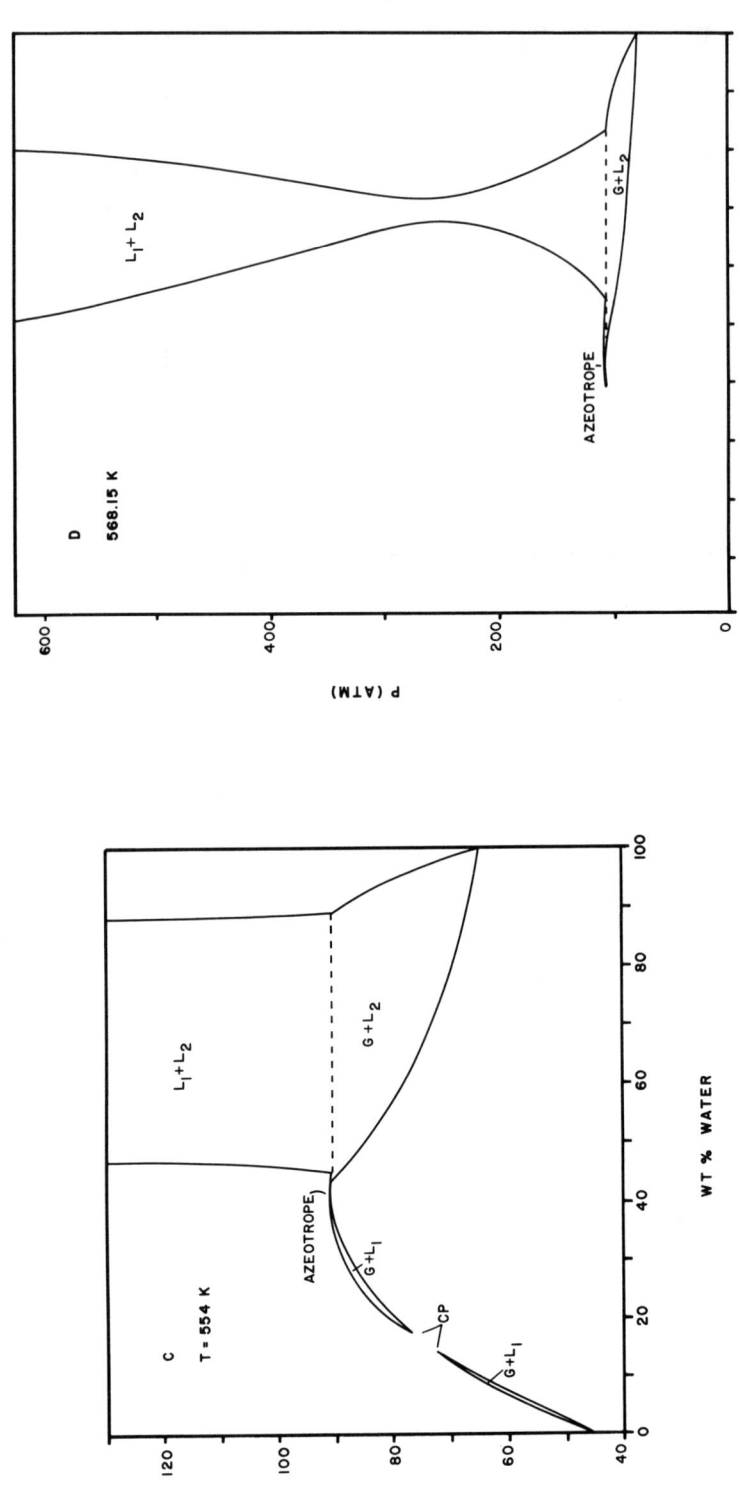

Figure 17. Predicted P-x sections for benzene-water. (Second of three pages.)

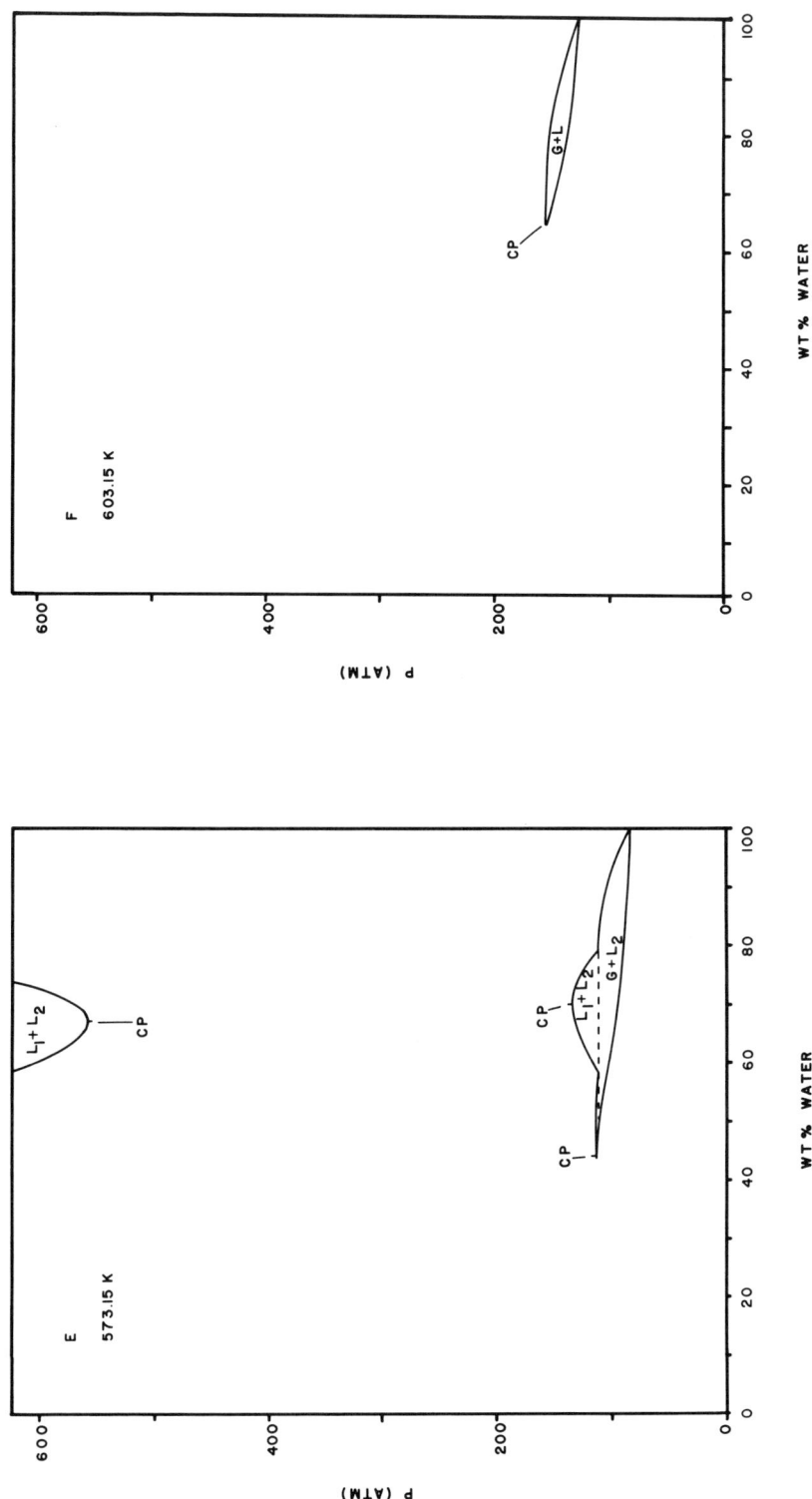

Figure 17. predicted P-x sections for benzene-water. (Third of three pages.)

573.15 K, is slightly above the cusp temperature. A critical point appears very near the tip of the $G+L_2$ region. The miscibility gap has been correctly predicted. Figure 17f, drawn at 603.15 K, is at a temperature above the L_1+L_2 critical line termination. The VLE envelope is now of the usual type.

CONCLUSIONS

The Peng-Robinson equation with a single, constant interaction parameter has been shown to give a qualitatively correct representation of phase behavior over the entire fluid range for several nonpolar binary systems. These systems were nonideal with respect to molecular size of the components. For the benzene-water system, some qualitative discrepancies may exist, although further experimental work is required to ascertain this fact. In all cases, the equation of state has pointed out features of the phase diagrams which were not readily apparent from the existing data. The qualitative correctness of the predictions over extensive temperature and pressure ranges indicates that semiquantitative predictions should be attainable in any fluid region through the use of a nonconstant interaction parameter.

These considerations indicate that modern cubic equations of state are well suited to an orienting or preliminary study of binary phase behavior. The minimal computing time required by these equations makes them an excellent choice for such an application. The preliminary study maps out regions where more detailed investigation, in both the theoretical and experimental sense, is warranted. From the theoretical standpoint, once these regions have been defined, it may in some cases be desirable to switch to a more accurate equation of state. In other cases the cubic equation will provide sufficient accuracy with only a change in the interaction parameter. From the experimental standpoint, the preliminary study indicates the regions where further data should be taken, and serves as a check on the consistency of results.

Finally, the occurrence of multiple phase equilibrium solutions associated with the three-phase region should be emphasized. All equations of state will exhibit this type of behavior, and care must be taken that the proper solution is chosen.

Appendix   The Equation of State and Derived Expressions

The Peng-Robinson equation of state is

$$P = \frac{RT}{v-b} - \frac{a(T)}{v(v+b) + b(v-b)} \tag{A.1}$$

where $a(T)$ is a temperature dependent attraction energy parameter. This equation may be rewritten in cubic form as

$$Z^3 - (1-B)Z^2 + (A-3B^2-2B)Z - (AB-B^2-B^3) = 0 \tag{A.2}$$

where

$$A = \frac{aP}{R^2T^2} \tag{A.3}$$

$$B = \frac{bP}{RT} \tag{A.4}$$

$$Z = \frac{Pv}{RT} \tag{A.5}$$

By requiring equation (A.1) to satisfy the pure component critical point criteria,

$$a(T_c) = 0.45724 \frac{R^2 T_c^2}{P_c} \tag{A.6}$$

$$b = 0.07780 \frac{RT_c}{P_c} \qquad (A.7)$$

$$Z_c = 0.307 \qquad (A.8)$$

For temperatures other than T_c, let

$$a(T) = a(T_c) \cdot \alpha(T_r, \omega) \qquad (A.9)$$

with

$$\alpha^{\frac{1}{2}} = 1 + \kappa(1-T_r^{\frac{1}{2}}) \qquad (A.10)$$

$$\kappa = 0.37464 + 1.54226\omega - 0.26992\omega^2 \qquad (A.11)$$

The fugacity of a pure component is derived from the relation

$$\ln\frac{f}{P} = \int_0^P \left(\frac{v}{RT} - \frac{1}{P}\right) dP \qquad (A.12)$$

Using equation (A.1), the integration yields

$$f = P\exp\left[Z - 1 - \ln(Z-B) - \frac{A}{2\sqrt{2}\,B} \ln\left(\frac{Z+2.414B}{Z-0.414B}\right)\right] \qquad (A.13)$$

For a mixture,

$$a = \sum_i \sum_j x_i x_j a_{ij} \qquad (A.14)$$

$$b = \sum_i x_i b_i \qquad (A.15)$$

$$a_{ij} = (1-\delta_{ij})(a_i a_j)^{\frac{1}{2}} \qquad (A.16)$$

where δ_{ij} is an empirically determined interaction parameter.

The fugacity of a component in a mixture is found from the relation

$$RT\ln \frac{f_i}{x_i P} = \int_{\underline{V} \to \infty}^{\underline{V}} \left(\frac{RT}{\underline{V}} - \left(\frac{\partial P}{\partial N_i} \right)_{T,\underline{V},\{N_i\}} \right) d\underline{V} - RT\ln Z \qquad (A.17)$$

Substituting equation (A.1), the result for component 1 in a binary system is

$$f_1 = x_1 P \exp \left[\frac{b_1}{b}(Z-1) - \ln(Z-B) - \frac{A}{2\sqrt{2}B} \left(\frac{2(a_1 x_1 + a_{12} x_2)}{a} - \frac{b_1}{b} \right) \right.$$
$$\left. \cdot \ln \left(\frac{Z + (\sqrt{2}+1)B}{Z - (\sqrt{2}-1)B} \right) \right] \qquad (A.18)$$

Use of the above fugacity expressions allows the calculation of phase equilibrium.

The mechanical stability of a phase is found from equation (A.1). For a stable system,

$$\left(\frac{\partial P}{\partial V} \right)_{T,x} = -\frac{RT}{(v-b)^2} - \frac{2a(v+b)}{v^2+2bv-b^2} < 0 \qquad (A.19)$$

Determination of material stability spinodal curves was carried out using the L1 determinant of Beegle, et al. (1974b). This is given in intensive form for a binary system as

$$L1 = \begin{vmatrix} A_{vv} & A_{vx} \\ A_{vx} & A_{xx} \end{vmatrix} = A_{vv}A_{xx} - A_{vx}^2 \qquad (A.20)$$

Compositional derivatives (subscript x) are taken with respect to component 2.

This quantity is zero at a point on the stability limit. The M1 determinant of Reid and Beegle (1977) was used as the second criterion of criticality. This is expressed in intensive form for a binary system as

$$M1 = \begin{vmatrix} A_{vv} & A_{vx} \\ \left(\dfrac{\partial L1}{\partial V}\right)_{T,x} & \left(\dfrac{\partial L1}{\partial x}\right)_{T,V} \end{vmatrix} = A_{vv}A_{xx}A_{vvx} + A_{vv}^2 A_{xxx} -$$

$$- 3A_{vv}A_{vx}A_{vxx} - A_{xx}A_{vx}A_{vvv} + 2A_{vx}^2 A_{vvx} \qquad (A.21)$$

At a critical point, this quantity is zero. Evaluation of L1 and M1 requires an expression for the Helmholtz free energy $A = A(T, V, x)$. Using their equation, Peng and Robinson (1977) have formulated this expression as

$$A = RT \sum x_i \ln \frac{x_i RT}{v-b} - \frac{a}{2\sqrt{2}\,b} \ln\left(\frac{v+(\sqrt{2}+1)B}{v-(\sqrt{2}-1)B}\right) +$$

$$\int_{T_o}^{T} \sum x_i C_{Pi}^{*} dT - T \int_{T_o}^{T} \sum x_i \frac{C_{Pi}^{*}}{T} dT - RT \qquad (A.22)$$

where the asterisk denotes an ideal gas state.

Using equation (A.22), the derivatives in equations (A.20) and (A.21) are as follows:

$$a_x = 2(-a_1 x_1 + a_{12}(1-2x_2) + a_2 x_2)$$

$$a_{xx} = 2(a_1 - 2a_{12} + a_2)$$

$$b_x = b_2 - b_1$$

$$r = a/b$$

$$r_x = \frac{a_x b - a b_x}{b^2}$$

$$r_{xx} = \frac{a_{xx} - 2 r_x b_x}{b}$$

$$r_{xxx} = \frac{-3 r_{xx} b_x}{b}$$

$$W = v^2 + 2bv - b^2$$

$$W_x = 2 b_x (v-b)$$

$$W_{xx} = -2 b_x^2$$

$$X = v + (1+\sqrt{2})b$$

$$X_x = (1+\sqrt{2})b_x$$

$$Y = v + (1-\sqrt{2})b$$

$$Y_x = (1-\sqrt{2})b_x$$

$$A_{vv} = \frac{RT}{(v-b)^2} - \frac{2a(v+b)}{W^2}$$

$$A_{vx} = \frac{-RTb_x}{(v-b)^2} + \frac{a_x}{W} - \frac{aW_x}{W^2}$$

$$A_{xx} = RT\left(\frac{1}{x_1 x_2} + \left(\frac{b_x}{v-b}\right)^2\right) - \left(r_{xx}\ln\frac{X}{Y} + 2r_x\left(\frac{X_x}{X} - \frac{Y_x}{Y}\right)\right.$$
$$\left. - r\left(\left(\frac{X_x}{X}\right)^2 - \left(\frac{Y_x}{Y}\right)^2\right)\right) \Big/ \sqrt{8}$$

$$A_{vvv} = \frac{-2RT}{(v-b)^3} + \frac{2a}{W^2}\left(\frac{4(v+b)^2}{W} - 1\right)$$

$$A_{vvx} = \frac{2RTb_x}{(v-b)^3} - \frac{2(a_x(v+b) + ab_x)}{W^2} + \frac{4aW_x(v+b)}{W^3}$$

$$A_{vxx} = \frac{-2RTb_x^2}{(v-b)^3} + \frac{a_{xx}W - 2a_x W_x - aW_{xx}}{W^2} + \frac{2aW_x^2}{W^3}$$

$$A_{xxx} = RT\left(\left(\frac{2x_2-1}{x_1 x_2}\right)^2 + 2\left(\frac{b_x}{v-b}\right)^3\right) -$$

$$\left(r_{xxx}\ln\frac{X}{Y} + 3r_{xx}\left(\frac{X_x}{X} - \frac{Y_x}{Y}\right) - 3r_x\left(\left(\frac{X_x}{X}\right)^2 - \left(\frac{Y_x}{Y}\right)^2\right)\right.$$

$$\left. + 2r\left(\left(\frac{X_x}{X}\right)^3 - \left(\frac{Y_x}{Y}\right)^3\right)\right)\bigg/\sqrt{8}$$

NOTATION

A - Helmholtz free energy; parameter in cubic equation of state.
B - Parameter in cubic equation of state.
CEP - Critical end point.
CP - Critical point.
F - Fluid.
G - Gas (or vapor).
H - Hexadecane.
L - Liquid
L1 - Spinodal (first criticality criterion) determinant.
LCEP - Lower critical end point.
LP - Limit point.
M1 - Critical (second criticality criterion) determinant.
N - Mole number; Naphthalene.
P - Pressure.
R - Gas constant.
S - Solid.
T - Temperature.
TP - Triple point.
UCEP - Upper critical end point.
V - Vapor (or gas).
Z - Compressibility factor.

a,b - Parameters in cubic equation of state.
f - Fugacity.
v - Molar volume.
x - Mole fraction.

Subscripts
c - Critical property.
i, j, 1, 2 - Component indices.
r - Reduced property.

Superscripts
g - Gas.
s - Solid.

Greek Letters

δ_{12} - interaction parameter

ω - acentric factor

REFERENCES

Alwani, Z. and G.M. Schneider. "Phasengleichgewichte, kritische Erscheinungen und PVT - Daten in binaren Mischungen von Wasser mit aromtischen Kohlenwasserstoffen bis 420 C und 2200 bar." Ber. Bunsenges., 73, 294 (1969).

Ambrose, D., I.J. Lawrenson and C.H.S. Sprake. "The Vapour Pressure of Naphthalene." J. Chem. Thermo., 1, 1173 (1975).

Beegle, B.L., M. Modell and R.C. Reid. "Thermodynamic Stability Criterion for Pure Substances and Mixtures." AIChE J., 20, 1200 (1974b).

Büchner, E.H. "Flüssige Kohlensaure als Losungsmittel." Z. Phys. Chem., 54, 665 (1906a).

Connolly, J.F. "Solubility of Hydrocarbons in Water Near the Critical Solution Temperature." J. Chem. Eng. Data, 11, 13 (1966).

Hong, G.T., "Binary Phase Diagrams From a Cubic Equation of State", ScD. Thesis, Massachusetts Institute of Technology, August, 1981.

Kurnik, R.T., S.J. Holla and R.C. Reid. "Solubility of Solids in Supercritical Carbon Dioxide and Ethylene." J. Chem. Eng. Data., 26, 47 (1981).

McHugh, M. and M.E. Paulaitis. "Solid Solubilities of Naphthalene and Biphenyl in Supercritical Carbon Dioxide." J. Chem. Eng. Data., 25, 326 (1980).

Modell, M., G.T. Hong and A. Heiba. "The Use of the Peng-Robinson Equation in Determining Supercritical Behavior." Paper presented at the 72nd Annual Meeting, AIChE, San Francisco, California, November 26, 1979.

Najour, G.C. and A.O. King, Jr. "Solubility of Naphthalene in Compressed Methane, Ethylene, and Carbon Dioxide. Evidence for a Gas-Phase Complex Between Naphthalene and Carbon Dioxide." J. Chem. Phys., 45, 1915 (1966).

Peng, D.Y. and D.B. Robinson. "A New Two-Constant Equation of State." Ind. Eng. Chem. Fundam., 15, 59 (1976).

Peng, D.Y. and D.B. Robinson. "A Rigorous Method for Predicting the Critical Properties of Multicomponent

Systems from an Equation of State." <u>AIChE J.</u>, <u>23</u>, 137 (1977).

Rebert, C.J. and W.B. Kay. "The Phase Behavior and Solubility Relations of the Benzene-Water System." <u>AIChE J.</u>, <u>5</u>, 285 (1959).

Reid, R.C. and B.L. Beegle. "Critical Point Criteria in Legendre Transform Notation." <u>AIChE J.</u>, <u>23</u>, 726 (1977).

Schneider, G.M., Z. Alwani, W. Heim, E. Horvath and E.U. Franck. "Phasengleichgewichte und kritische Erscheinungen in binaren Mischsystemen bis 1500 bar." <u>Chem. Ing. Tech.</u>, <u>39</u>, 649 (1967).

Sebastian, H.M., J.J. Simnock, H. Lin and K. Chao. "Vapor-Liquid Equilibrium in Binary Mixtures of Carbon Dioxide and n-Decane and Carbon Dioxide and N-Hexadecane." <u>J. Chem. Eng. Data.</u>, <u>25</u>, 138 (1980).

Smits, A. "The Course of the Solubility Curve in the Region of Critical Temperatures of Binary Mixtures I." <u>Proc. Roy. Acad. Sci. Amsterdam</u>, <u>6</u>, 171 (1903a).

Smits, A. "The Course of the Solubility Curve in the Region of Critical Temperatures of Binary Mixtures II." <u>Proc. Roy. Acad. Sci. Amsterdam</u>, <u>6</u>, 484 (1903b).

Smits, A. "On the Hidden Equilibria in the P-x Diagram of a Binary System in Consequence of the Appearance of Solid Substances." <u>Proc. Roy. Acad. Sci. Amsterdam</u>, <u>8</u>, 196 (1905a).

Smits, A. "On the Hidden Equilibria in the P-x Sections Below the Eutectic Point." <u>Proc. Roy. Acad. Sci. Amsterdam</u>, <u>8</u>, 568, (1905b).

Tsekhanskaya, Y.V., M.B. Iomtev and E.V. Mushkina. "Solubility of Diphenylamine and Naphthalene in Carbon Dioxide Under Pressure." <u>Russ. J. Phys. Chem.</u>, <u>36</u>, 117 (1962).

Tsekhanskaya, Y.V., M.B. Iomtev and E.V. Mushkina. "Solubility of Naphthalene in Ethylene and Carbon Dioxide Under Pressure." <u>Russ. J. Phys. Chem.</u>, <u>38</u>, 1173 (1964).

Tsekhanskaya, Y.V., N.G. Roginskaya and E.V. Mushkina. "Volume Changes in Naphthalene Solutions in Compressed

Carbon Dioxide." <u>Russ. J. Phys. Chem.</u>, <u>40</u>, 1152 (1966).

van der Waals, J.D. Thesis. A.W. Sigthoff, Leiden (Dutch ed.); Barth, Leipzig (German ed.), 1873.

Van Gunst, C.A., F.E.C. Scheffer and G.A.M. Diepen. "On Critical Phenomena of Saturated Solutions in Binary Systems II." <u>J. Phys. Chem.</u>, <u>57</u>, 578 (1953).

Van Konynenburg, P.H. "Critical Lines and Phase Equilibria in Binary Mixtures." PhD. Dissertation, U.C.L.A., (1968).

Van Welie, G.S.A. and G.A.M. Diepen. "The P-T-x Space Model of the System Ethylene-Naphthalene (III)." <u>Rec. Trav. Chim.</u>, <u>80</u>, 673 (1961).

CHAPTER 14

PHASE EQUILIBRIA OF HIGH-BOILING ORGANIC
SOLUTES IN COMPRESSED SUPERCRITICAL FLUIDS.
EQUATION OF STATE WITH NEW MIXING RULE.

K. W. Won
Fluor Engineers and Constructors, Inc.
3333 Michelson Drive
Irvine, California 92730

ABSTRACT

Rational design of supercritical fluid extraction and the solvent regeneration process requires accurate calculation of phase equilibria and other thermophysical properties of the compressed fluid mixtures at wide temperature and pressure ranges.

Cubic equations of state with a simple mixing rule have played a very important role in the calculation of phase equilibria and enthalpies of fluid mixtures of small molecular sizes. When the mixtures contain molecules of much larger sizes, the phase equilibria of larger molecules, are, in general, less accurately calculated. This paper presents our progress to date on the development of new mixing rules for simple equations of state which calculate the phase equilibria and enthalpies of supercritical fluid containing high-boiling hydrocarbons as well as polar chemical solutes.

INTRODUCTION

Fluids slightly above their critical temperatures and pressures exhibit high solvencies towards high-molecular weight solutes. This solvency is very sensitive to the system pressure and temperature changes. It is this high sensitivity of the supercritical fluid's solvency to the system pressure and also to temperature that makes supercritical fluid extraction a potentially economic separation process, despite the expensive high-pressure equipment and power requirement for compressors and pumps.

Supercritical carbon dioxide has been investigated as a means of removing adsorbed materials from activated carbon (Modell et al, 1979 and 1978) and as an extractant for caffeins (Vitzthum, 1975), for ethanol from water (McHugh et al, 1981) and for the recovery of primary solvent dissolved in the treated water (Won, 1981), while liquid carbon dioxide has been used for selective aroma and essence extraction from fruit juices (Shulz and Randall, 1970; Randall et al, 1971; Shultz et al, 1974). Supercritical fluids, probably toluene and other similar hydrocarbon mixtures, have been reported as an effective solvent for coal-derived oil deashing (Corbett et al, 1981). Supercritical propane was reported as a de-asphalting agent for heavy residual oils (Solomon, 1971; Zhuse, 1960). Supercritical pentane is the solvent for ROSE (Residuum Oil Supercritical Extraction) process (Kerr-McGee Refining Corp., 1981).

For the rational design of such processes utilizing supercritical fluids, a calculation method is needed for phase equilibria and other thermophysical properties of supercritical fluid mixtures at wide temperature, pressure and concentration ranges. It is also important that the method be predictive with reasonable accuracies for the solvencies of the related solutes.

For this purpose, two cubic equations of state were selected rather than more complicated equations of state containing a larger number of mixture parameters. A modified Soave Redlich-Kwong equation of state was selected because of its simplicity and a modified Van der Waals equation of state was chosen because this equation may not need critical constants of pure components. Critical constants of close-boiling fractions of petroleum or coal-derived crudes are, in general, not accurately known at the present. With only one adjustable binary constant, these two equations of state could not accurately calculate the sensitive pressure dependency of the solvent capability of supercritical fluids.

We report in this paper that a modified Soave R-K and a modified Van der Waals equation of state could quantitatively calculate the phase equilibria of a high-boiling naphthalene solute at wide temperature and pressure ranges, with a new mixing rule for the energy parameter.

Thermodynamic Framework

At equilibrium, the fugacity of component i in α phase, f^{α} is equal to that in the supercritical fluid phase, f^{β};

$$f_i^{\alpha} = f_i^{\beta} \tag{1}$$

For liquid phase, α,

$$f_i^{\alpha} = \gamma_i X_i f_i^S(T) \exp[\frac{\bar{V}_i(P-P^S)}{RT}] \tag{2}$$

where γ is activity coefficient at constant reference pressure set at zero, $f^S(T)$ is standard state fugacity corrected to zero pressure and $\bar{V}$ is the partial molar volume of component i.

When phase α is pure solid,

$$f_i^\alpha = P_i^S \phi_i^S \exp\left[\frac{V_i(P-P_i^S)}{RT}\right] \quad (3)$$

where P^S is the sublimation pressure, ϕ^S is the fugacity coefficient of solute i in the saturated gas state, and V is the molar volume, which is approximated incompressible. For the supercritical fluid phase,

$$f_i^\beta = (Y_i P)\phi_i \quad (4)$$

where ϕ is the fugacity coefficient of solute i in the supercritical fluid solvent.

Substitution of equations (3) and (4) into equation (1) yields the solubility of high-boiling solid, Y as a function of temperature and pressure;

$$Y_i = \frac{P_i^S}{P} \frac{\phi_i^S}{\phi_i} \exp\left[\frac{V_i(P-P_i^S)}{RT}\right] \quad (5)$$

The first term in the right-hand side of equation (5) is the ideal solubility as a function of temperature and pressure, the second term stands for the effect of gas phase non ideality and the last exponential term represents the effect of pressure on the condensed phase fugacity, commonly known as the Poynting correction.

When the sublimation pressure, P^S is sufficiently low, in the order of 0.01 atmosphere, the fugacity coefficient of saturated gas ϕ^S can be taken as unity as long as the solute does not associate in the gas phase.

When accurate solid solubility data, Y are available, we could calculate accurate fugacity coefficients, ϕ without the extensive gas phase volumetric data (P - V - T - Y) of the mixture.

$$\phi_i = \frac{P_i^S}{P} \frac{\phi_i^S}{Y_i} \exp\left[\frac{V_i(P-P_i^S)}{RT}\right] \quad (5\text{-a})$$

The validity of any equation of state and its mixing rule could be tested by comparing the calculated fugacity coefficients with those of equation (5-a).

Cubic Equations of State

Cubic equations of state have been widely accepted in gas, petroleum and chemical processing industries due to their simplicity. The pressure, P in the cubic equations of state consists of two parts; one representing the molecular repulsion, P^{rep} and the second representing the molecular attraction, P^{attr}.

$$P = P^{rep} + P^{attr} \quad (6)$$

Most of the cubic equations of state use the same form

for the pressure due to molecular repulsion, P^{rep}.

$$P^{rep} = \frac{RT}{V-b} \tag{7}$$

where the constant b stands for a measure of the molecular size.

The difference among cubic equations of state stems from the difference in the pressure due to attraction forces.

The Redlich-Kwong equation of state is

$$P^{attr} = -\frac{a(T)}{V(V+b)} \tag{8}$$

where $a(T)$ is a measure of molecular attraction.

The term, $a(T)$ of the modified Soave-Redlich-Kwong (MSRK) is

$$a(T) = (a_c - a_p)[1 + m_i(1 - \sqrt{Tr_i})]^2 + \frac{a_p}{Tr_i^3} \tag{9}$$

where Tr is reduced temperature.

The constants a and b are calculated from the critical properties of the pure fluid and the constants, a_p and m_i are specific to the polar components. For several polar components, the constants a_p and m_i are published elsewhere (Won and Walker, 1979).

Substitution of equations (6), (7) and (8) into the thermodynamic definition of the fugacity coefficient, ϕ, yields:

$$\ln\phi_i = \ln\frac{RT}{P(V-b)} + \frac{\partial(nb)}{\partial n_i}(\frac{PV}{RT} - 1)\frac{1}{b} - \ln(1 + \frac{b}{V})$$

$$X \; (\frac{1}{nb}\frac{\partial(n^2a)}{\partial n_i} - \frac{a}{b^2}\frac{\partial(nb)}{\partial n_i})/RT \tag{10}$$

The modified Van der Waals equation of state is

$$P^{attr} = -\frac{a(T) \; S^2}{(V+S-b)^2} \tag{11}$$

where the parameter S is an effective volume which is larger than, but approaches the parameter b at high temperature.

$$S = b \; \exp(\frac{C_0}{Tr'}) \tag{12}$$

and $a(T) = \frac{RT}{S}(C_1 + \frac{C_2}{Tr'} + \frac{C_3}{Tr'^2} + \frac{C_4}{Tr'^3}) \tag{13}$

where the pseudo-reduced temperature, Tr' defined and used in this work is

$$Tr' = \frac{T}{300} \tag{14}$$

Pseudo-reduced temperature, Tr' is defined in this work so that the equation of state parameters, b, S and a(T) can be determined by vapor pressure and latent heat of vaporization of undefined fractions.

Critical properties, which are needed to calculate the true reduced temperature, are measured and reported for most of the low-molecular weight components. However, high-boiling undefined fractions of petroleum and coal-derived oils are thermally unstable and decompose before reaching the critical temperatures. Therefore, there is a need to determine the equation of state constants from readily measureable data such as vapor-pressure and latent heat of vaporization in place of critical properties.

The modified Van der Waals equation of state is similar to that proposed by Wilson and Weiner (1977), where true reduced temperature was used in equations (12) and (13).

Table 1 presents the equation of state constants, b, a, and S determined by fitting the vapor pressure and enthalpy departure of carbon dioxide, ethene and naphthalene at 3 temperatures.

TABLE 1
EQUATION OF STATE PARAMETERS FOR PURE COMPONENTS

Component	b, cm^3/g mol			S, cm^3/g mol	$a^{(4)}$	$aS^{2(4)}$
	MSRK	MVdW		MVdW	MSRK	MVdW
CO_2	29.7	26.6	(1)	41.6	3.72	3.62
			(2)	39.1	3.10	3.13
			(3)	37.2	2.57	2.77
C_2H_4	40	36.6	(1)	55.6	4.41	4.34
			(2)	52.3	3.99	3.78
			(3)	50.0	3.61	3.35
Naphthalene	134.5	87.5	(1)	899	73.7	383
			(2)	645	68.6	202
			(3)	502	63.9	156

(1) At 300 K
(2) At 350 K
(3) At 400 K
(4) Unit is (liter/g mole)^2·atm

Substitution of equations (6), (7) and (11) into the classic thermodynamic relation of the fugacity coefficient gives:

$$\ln \phi_i = \ln \frac{RT}{P(V-b)} + \frac{P}{RT} \frac{\partial (nb)}{\partial n_i} - \left(\frac{\partial (n^2 S^2 a)}{\partial n_i} \frac{1}{V+S-b} - \frac{\partial (nS)}{\partial n_i} \frac{S^2 a}{(V+S-b)^2} \right) \frac{1}{RT} \quad (15)$$

Mixing Rules of Equation of State Parameters

For pure components, the accuracy of equations of state can be judged by calculating the vapor-pressure as well as other thermophysical properties and comparing them with experimental data. Therefore, much attention has been paid to the temperature dependence of the equation of state parameters and also to the volume dependence of the pressure as shown in the denominator of p^{attr} and p^{rep}. Until fairly recently, less attention has been paid to the composition dependence of the equation of state parameters of mixtures.

The most widely accepted mixing rules are;

$$b = \Sigma Y_i b_i \quad (16)$$

$$a = \Sigma Y_i Y_j a_{ij} \quad (17)$$

and
$$a_{ij} = \sqrt{(a_i a_j)_n} (1-k_{ij}) + \sqrt{(a_i a_j)_p} \quad (18)$$

The subscripts n and p in equation (18) represent non-polar and polar contributions to equation of state energy parameter, a.

In general, the quadratic mole fraction average of energy parameter, a, and arithmetic mole fraction average of size parameter, b, have been adequate for non-polar hydrocarbon mixtures. Arithmetic mole fraction average is also used for the effective size parameter, S, of the modified Van der Waals equation of state.

$$S = \Sigma Y_i S_i \quad (19)$$

For non-hydrocarbon gases such as N_2, CO_2 and H_2S, in hydrocarbon mixtures, small corrections from geometric mean of energy-parameter cross term represented by (1-k) in equation (18) have been used with remarkable success.

For polar components, a new temperature-dependent energy parameter was used as shown in equation (9), but the conventional quadratic mixing rule was still adequate in the dilute region. When the constituent molecules of a mixture differ much in their chemical properties or molecular sizes, the simple mixing rule outlined in equations (16) to (18) is not adequate. In that case, a new mixing rule is necessary. The new mixing rule uses the excess Gibbs energy, G^e, of the fluid mixture similar to the liquid state activity coefficient model.

The excess Gibbs energy, G^e, is related to the fugacity coefficients:

$$G^e = RT \Sigma Y_i (\ln \phi_i - \ln \phi_i^*) \tag{20}$$

where the fugacity coefficient, ϕ, of component i in mixture is defined in equation (10) or (15) and the fugacity coefficient, ϕ^*, is that of pure component calculated at the same temperature and pressure. At the hypothetical infinite pressure, the volume, V, will approach the size parameter, b. At this condition,

$$G^e_{MSRK} = (\Sigma \frac{a_i}{b_i} Y_i - \frac{a}{b}) \ln 2 \tag{21}$$

$$G^e_{MVDW} = \Sigma Y_i S_i a_i - Sa \tag{22}$$

The subscript MSRK in equation (21) indicates that the fugacity coefficient derived from the modified Soave Redlich-Kwong equation of state was used, while the subscript MVDW stands for the result calculated from the modified Van der Waals equation of state. Equations (20) and (21) were published by Huron and Vidal (1979).

Calculation Procedure

To calculate the solubility of solid naphthalene in compressed gases, equation (5) was used with the reported vapor pressure, p^S, solid molar volume for $\overline{V}$ and the fugacity coefficient, ϕ, calculated by three different methods.

(a) Modified Soave Redlich-Kwong equation of state with conventional mixing rule.

The equation of state is defined by equations (6), (7), (8), and (9). The conventional mixing rule is defined by equations (16), (17) and (18). Then, equation (10) yields the fugacity coefficient, ϕ,

$$\ln \phi_i = \ln \frac{RT}{p(V-b)} + \frac{b_i}{b} (\frac{PV}{RT} - 1) - \ln(1 + \frac{b}{V})$$
$$\times \left(2 \frac{\Sigma Y_i a_{ij}}{RTb} - \frac{ab_i}{RTb^2} \right) \tag{23}$$

To start the calculation, the vapor pressure, P^S and the solid molar volume are calculated at the experimental temperature and pressure. At a given binary constant, k, the equation of state parameters, a and b are calculated from equations (16) to (18). Then the cubic equation of state defined by equations (6), (7) and (8) is solved. The largest root is always taken as the compressibility factor of the supercritical fluid mixture. The molar volume, V, and the compressibility factor, PV/RT, were substituted into equation (23) to calculate the fugacity coefficient, ϕ. This fugacity coefficient, ϕ, is used in equation (5) in order to calculate the solid solubility, Y.

(b) <u>Modified Soave Redlich-Kwong equation of state with new mixing rule.</u>

The modified Soave Redlich-Kwong equation of state used in this method is the same as the first method. The equation of state parameter, b, of the mixture was calculated as the arithmetic mole fraction average of those of the constituent components. However, the equation of state parameter, a, was calculated by the new mixing rule defined in equation (21).

For the excess Gibbs Energy, G^e_{MSRK}, one of the local compositions models or Non-Random Two Liquid (NRTL) was used to calculate the attractive contribution to the pressure,

$$P^{attr} = -\frac{b \Sigma Y_i}{V(V+b)} \times \left[\frac{a_i}{b_i} - \frac{1}{\ln 2} \frac{\Sigma Y_j (g_{ji}-g_{ii}) \exp[-\frac{\alpha(g_{ji}-g_{ii})}{RT}]}{\Sigma Y_j \exp[-\frac{\alpha(g_{ji}-g_{ii})}{RT}]} \right] \quad (24)$$

Then, equation (10) gives,

$$\ln \phi_i = \ln \frac{RT}{P(V-b)} + \frac{b_i}{b}(\frac{PV}{RT} - 1) - \ln(1+\frac{b}{V})[\frac{a_i}{RTb_i} - \frac{\bar{g}^e_i}{RT\ln 2}] \quad (25)$$

where $\bar{g}^e_i = \Sigma Y_i^2 [\tau_{ji}(\frac{G_{ji}}{AG_{ji}})^2 + \tau_{ij}\frac{G_{ij}}{AG_{ij}^2}]$ (26)

where $\tau_{ji} = (g_{ji} - g_{ii})/RT$ (27)

$G_{ji} = \exp(-\alpha \tau_{ji})$ (28)

$AG_{ji} = y_j + Y_i G_{ij}$ (29)

(c) <u>Modified Van der Waals equation of state</u>

The equation of state parameters, b and S for mixtures are calculated by the arithmetic mole fraction averages of those of pure constituent molecules as shown in equations (16) and (19). The equation of state parameter, a, was calculated by the new mixing rule defined in equation (22). For the excess Gibbs energy, G^e_{MVDW}, Van Laar equation was used. Then, the attractive contribution to the pressure is given by:

$$P^{attr} = -\frac{S[\Sigma Y_i(S_i a_i - \Sigma RTY_i Y_j A_i A_j / \Sigma Y_i A_i)]}{(V+S-b)^2} \quad (30)$$

Substitution of equation (30) to equation (15) yields the fugacity coefficient. For binary mixtures,

$$\ln \phi_1 = \ln \frac{RT}{P(V-b)} + \frac{Pb_1}{RT} + [\frac{S^2 a}{(V+S-b)^2} - \frac{S(a+a_1)}{V+S-b}] S_i +$$

$$\frac{RT}{V+S-b} \frac{A_1}{(1 + \frac{Y_1 A_1}{Y_2 A_2})^2} \quad (31)$$

RESULTS

The solubility of naphthalene in carbon dioxide system was used to determine the suitability of modified Soave Redlich-Kwong equation of state with the conventional mixing rule.

The solubility measurements were reported by Tsekhanskaya et al (1964) and also by McHugh and Paulaitis (1980).

Figure 1 presents the solubilities of naphthalene in the compressed supercritical carbon dioxide as a function of pressure at 55°C. The three lines represent the result calculated by equation (5) using the fugacity coefficient, ϕ calculated from equation (23) with three different binary parameters, k. In this method, the fugacity coefficient was calculated by equation (23) from the experimental solubility data.

The result calculated by the modified SRK equation of state with conventional mixing rule using one binary parameter k = 0.1 over-estimates the solid solubility at high pressure, where the supercritical fluid extraction may operate, while at low pressures the calculation under-estimates the solid solubilities where the solvent regeneration process may operate.

Therefore, if the modified SRK equation of state is used with the conventional one binary mixture parameter for the process design calculation of the supercritical fluid extraction, it will lead to the under-estimation of the supercritical solvent circulation.

When the temperature of the system is slightly above the upper critical end point temperature, the solubility of the solid in the supercritical fluid, probably liquid, abruptly increases. This abrupt increase of the solubility data at 64.9°C and at about 230 atmospheres is equivalent to the abrupt decrease of the fugacity coefficient at that condition.

Figure 2 shows that the SRK equation of state with the conventional quadratic mixing rule under-estimates the fugacity coefficient of naphthalene by an order of magnitude at pressures above 230 atmospheres. Even at the lower pressures, the effect of the pressure on the fugacity coefficient (or the solubility of naphthalene) cannot be quantitatively calculated with a single binary constant of the conventional quadratic mixing rule for the SRK energy parameter. The result shown in Figure 2 leads us to investigate a new mixing rule for the modified SRK equation of state.

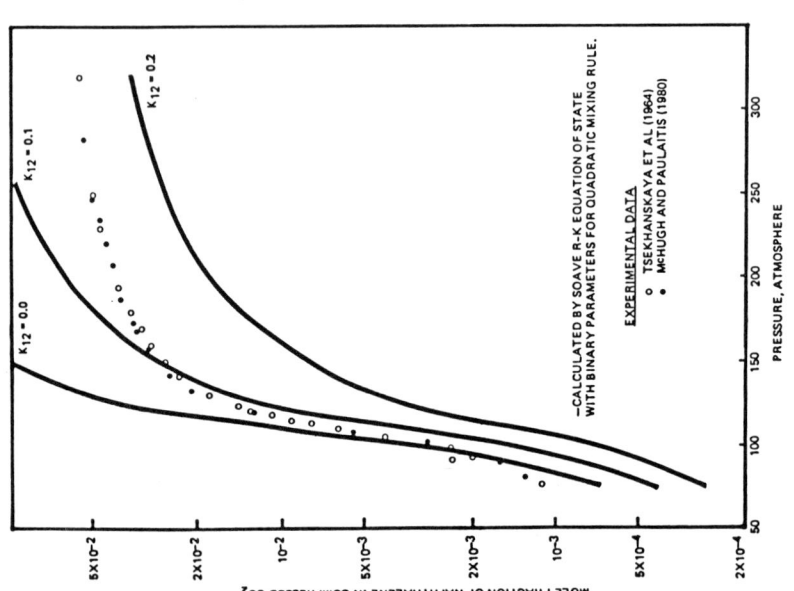

FIGURE 1.—EFFECT OF PRESSURE ON THE NAPHTHALENE SOLUBILITY IN COMPRESSED CARBON DIOXIDE.

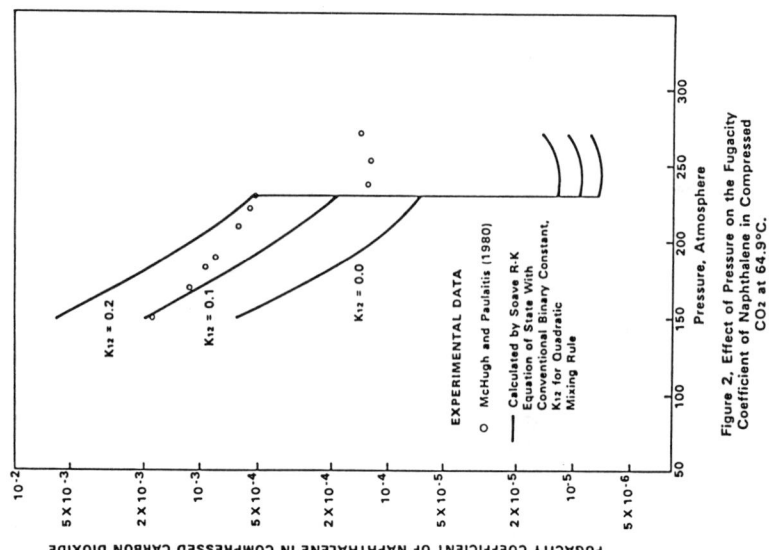

Figure 2. Effect of Pressure on the Fugacity Coefficient of Naphthalene in Compressed CO_2 at 64.9°C.

Figure 3 presents the calculated as well as the experimental solubilities of naphthalene in carbon dioxide at three temperatures; 35, 55 and 64.9°C. The solid lines represent the result calculated by the SRK equation of state with a new mixing rule in which NRTL solution model was used. The broken lines in Figure 3 represent the result calculated by a modified Van der Waals equation of state with another new mixing rule in which Van Laar solution model was utilized.

The result presented in Figure 3 indicates that the new mixing rules enabled the two simple cubic equations of state to quantitatively calculate the discontinuous solubility jump at 64.9°C as well as the continuous solubility change as a function of pressure at lower temperatures.

The accuracy of the result calculated by the modified Van der waals equation of state is comparable to that of the SRK in most of the temperature and pressure domain. At the pressures above about 230 atmospheres, the modified Van der Waals equation of state over-estimates the naphthalene solubilities at 64.9°C by about a factor of 1.5 and it underestimates slightly the solubilities at 35°C.

Figure 4 presents the calculated as well as the experimental naphthalene solubilities in the supercritical ethylene at 25, 45 and 50°C. At 50°C, this system exhibits a marked naphthalene solubility increase with pressure starting from about 150 atmospheres. Contrary to the naphthalene solubility discontinuity observed in carbon dioxide, the solubility increase at 50°C in ethylene solvent is continuous, probably because the temperature is slightly below the upper critical end point temperature of the mixture, which was reported to be 52.1°C. This continuous increase of the naphthalene solubility with pressure is quantitatively predicted both by the modified SRK and by the modified Van der Waals equation of state with the new mixing rules.

DISCUSSION AND CONCLUSION

The phase equilibria involved in the supercritical solvent systems are complex. Simple cubic equations of state with the conventional mixing rule appears to provide at the best the semi-quantitative result for binary mixture when applied at wide composition and pressure ranges.

Small errors in the binary coefficient, k, may sometimes result in the erroneous prediction of two fluid phases. As an example, Table 2 presents the SRK equation of state parameter, a, the fugacity coefficient, ϕ, and the fugacity of the naphthalene solute in the supercritical fluid CO_2 at 50°C and 150 atmospheres. At this condition, naphthalene exists as pure solid and its fugacity in CO_2 must be equal to that of the solid, which is about 0.002 atmosphere. The fugacities of naphthalene calculated by SRK equation of state with a binary coefficient k (0.1) indicates that the gas phase mole fraction of the solute is about 0.03. It is noteworthy

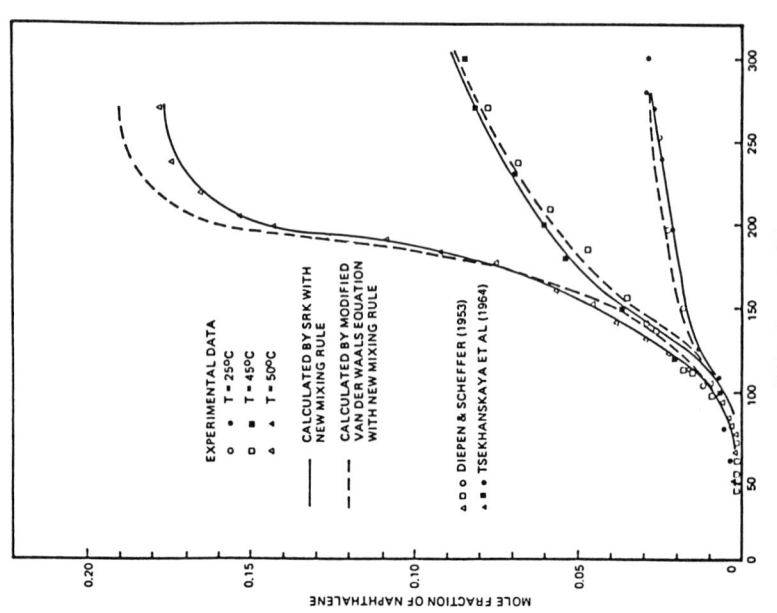

FIG. 4 – EXPERIMENTAL AND CALCULATED SOLUBILITIES OF NAPHTHALENE IN SUPERCRITICAL ETHYLENE AS A FUNCTION OF PRESSURE AT THREE TEMPERATURES.

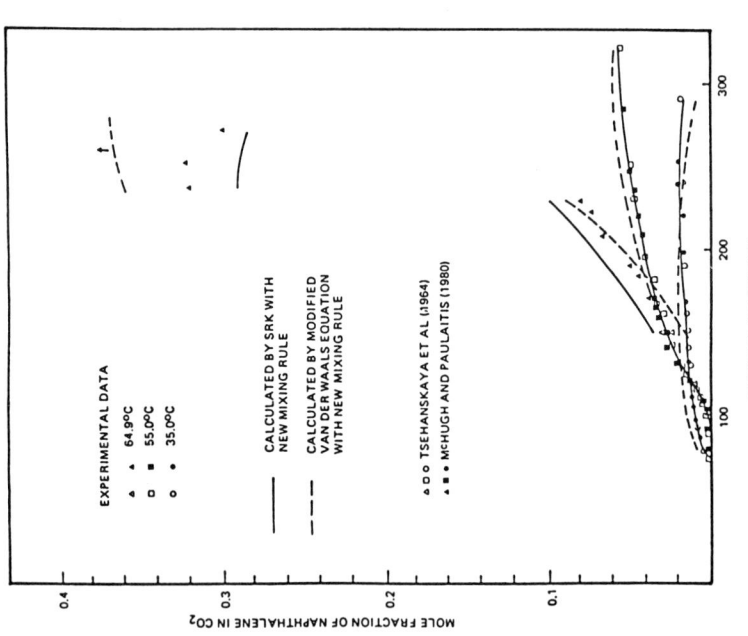

FIG. 3 – EXPERIMENTAL AND CALCULATED SOLUBILITIES OF NAPHTHALENE IN SUPERCRITICAL CARBON DIOXIDE AS A FUNCTION OF PRESSURE AT THREE TEMPERATURES.

that SRK equation of state with binary coefficient calculates a sharp maximum naphthalene fugacity at this condition. If we have chosen binary coefficient k somewhat larger than 0.1, we would have found two solubilities of naphthalene; one smaller than and the other larger than 0.03 mole fraction. The fugacities of naphthalene shown in Table 2 as a function of the solute mole fraction suggests that the result calculated by SRK equation of state is very sensitive to the small errors in the binary coefficient k.

TABLE 2. THE EFFECT OF COMPOSITION ON THE SRK EQUATION OF STATE ENERGY PARAMETER, a, AND THE FUGACITY COEFFICIENT, ϕ_N OF NAPHTHALENE IN SUPERCRITICAL CO_2 AT 150 ATM AND 50°C

Y_N	$a, (\frac{liter}{mole})$atm		$\phi_N(f_N, atm)$	
	Conventional	New	Conventional[3]	New[4]
0.0001	3.43	3.43	$1.7 \times 10^{-3} (2.6 \times 10^{-5})$	$9 \times 10^{-4} (1.4 \times 10^{-5})$
0.01	3.68	3.66	$1.1 \times 10^{-3} (1.7 \times 10^{-3})$	$8.7 \times 10^{-4} (1.3 \times 10^{-3})$
0.03[1]	4.20	4.15	$4.3 \times 10^{-4} (1.9 \times 10^{-3})$	$4.6 \times 10^{-4} (2.1 \times 10^{-3})$
0.10	6.43	6.03	$4.2 \times 10^{-5} (6.3 \times 10^{-4})$	$2.2 \times 10^{-4} (3.3 \times 10^{-3})$
0.20	10.7	9.4	$7.0 \times 10^{-6} (2.1 \times 10^{-4})$	$1.4 \times 10^{-4} (4.2 \times 10^{-3})$
0.30[2]	16.2	13.6	$2.7 \times 10^{-6} (1.2 \times 10^{-4})$	$9.3 \times 10^{-5} (4.2 \times 10^{-3})$
0.40	23.	18.7	$1.7 \times 10^{-6} (1.0 \times 10^{-4})$	$6.3 \times 10^{-5} (3.8 \times 10^{-3})$
0.50	31.	24.8	$1.3 \times 10^{-6} (9.8 \times 10^{-5})$	$4.1 \times 10^{-5} (3.1 \times 10^{-3})$

(1) Maximum fugacity with the conventional mixing rule

(2) Maximum fugacity with the new mixing rule

(3) Binary constant k_{12} was set to 0.1.

(4) $g_{12} - g_{22}$ was 3500 and 3150, $g_{21} - g_{11}$ was 550 and 200 Cal/g-mole at 35 and 55°C, respectively, and α was -0.40. All the parameters in this work were considered empirical.

The SRK equation of state with the new mixing rule calculates very mild plateau of the solute fugacities as a function of the solute mole fraction. On the other hand, the new mixing rule contains at least two parameters for Van Laar model and three for NRTL model. To determine these parameters of the new mixing rule, experimental data of sufficient accuracy are needed.

When the composition changes much, particularly in the condensed phase(s) such as liquid 1-liquid 2-vapor equilibria, it is expected that the new mixing rule will provide quantitatively satisfactory results.

Further work will be needed to validate this capability of the new mixing rule.

The new mixing rule for the cubic equations of state does not predict the composition dependence of the second virial coefficient of mixture. It is our intention to sacrifice this theoretical rigor for the sake of the simplicity of the calculations. From industrial applications viewpoint, small deviations at low pressures does not appear to present significant problems.

ACKNOWLEDGEMENT

The author is grateful to the management of Fluor Corporation for permission to publish this work.

LITERATURE CITED

Ambrose D., Lawrenson I.J. and Sprake C.H.S. "The Vapor Pressure of Naphthalene," J. Chem. Thermodynamics 7:1173 (1975).

Corbett R.W., Janka R.C. and Gir S. "Development in Critical Solvent Deashing," Paper presented at the AIChE Spring National meeting, Houston Texas, April (1981).

Huron M.J. and Vidal J. "New Mixing Rules in Simple Equations of State for Representing Vapor-Liquid Equilibria of Strongly Non-Ideal Mixtures," J. Fluid Phase Equilibria 3:255 (1979).

Kerr-McGee Refining Corporation "Novel Solvent Recovery Enhances Residum Upgrading," Chem. Eng. 88(24):69 (1981).

Modell M., de Fillippi R., Krukonis V. Paper presented at the ACS meeting, Miami, Florida, September (1978).

Modell M., Hong G. and Heiba A. "The Use of the Peng-Robinson Equation in Determining Supercritical Behavior" Paper presented at the 72nd Annual Meeting of AIChE, San Francisco, California, November (1979).

McHugh M. and Paulaitis M.E. "Solid Solubilities of Naphthalene and Biphenyl in Supercritical Carbon Dioxide," J. Chem. Eng. Data 25(4):326 (1980).

McHugh M.A., Mallett M.W. and Kohn J.P. "High Pressure Fluid Phase Equilibria of Alcohol-Water-Supercritical Solvent Mixtures," Paper presented at the AIChE Annual Meeting, New Orleans, Louisiana (1981).

Randall J.M., Schultz W.G. and Morgan A.I. "Extraction of Fruit Juices and Concentrated Essences with Liquid Carbon Dioxide," Confructa 16(1):10 (1971).

Schultz W.G. and Randall J.M. "Liquid Carbon Dioxide for Selective Aroma Extraction," Food Technology 24(11):94 (1970).

Schultz W.G., Schultz T.H., Carlson R.A. and Hudson J.S. "Pilot-Plant Extraction with Liquid CO_2," Food Technology 32: June (1974).

Solomon H.J. Paper presented at the ACS meeting, Washington D.C., September (1971).

Tsekhanskaya Y.V., Iomtev M.B. and Mushkina E.V. "Solubility of Naphthalene in Ethylene and Carbon Dioxide Under Pressure" Russ.J.Phys.Chem. 38:1173 (1964).

Vitzthum O. and Hubert P. German Patent 2357590 (1975).

Whiting W.B. and Prausnitz J.M. "Equations of State for Strongly Non-Ideal Fluid Mixtures; Application of the Local-Composition Concept" Paper presented at the 1981 Spring

AIChE Meeting, Houston, Texas, April (1981).

Wilson G.M. and Weiner "A Computer Model for Calculating Physical and Thermodynamic Properties of Synthetic Gas Process Streams," Chapter 14, "Phase Equilibria and Fluid Properties in the Chemical Industry" ACS Symposium Series 60. Edited by T.S. Storvick and S.I. Sandler, published by Am. Chem. Soc. (1977).

Won, K.W. Preliminary Patent Disclosure. "Supercritical Extraction of the Extractant for Acetic Acid and Other Organic Chemicals Recovery" October 1981.

Won, K.W. and Walker C.K. "An Equation of State for Polar Mixtures; Calculation of High-Pressure Vapor-Liquid Equilibria of Polar Solutes in Hydrocarbon Mixtures," Chapter 13 of Equations of State in Engineering and Research, Advances in Chemistry Series, 182

Zhuse T.P. *Petroleum* (London)23:298 (1960).

NOTATION

A	= Van Laar type coefficient defined in equation (30)
a	= energy parameter for equation of state
b	= size parameter for equation of state
f	= fugacity, atmosphere
S	= effective size parameter for equation of state
C_n	= coefficients defined in equation (13)
k	= conventional binary coefficient defined in equation (18)
G	= Gibbs free energy
$\bar{g}$	= partial molar Gibbs free energy
g_{ij}, G_{ij} & ΔG_{ij}	= parameters for NRTL type activity coefficient equation
R	= gas constant, $0.08206 \dfrac{\text{liter} \cdot \text{atmosphere}}{\text{g-mol} \cdot \text{Kelvin}}$
P	= absolute pressure, atmosphere
V	= molar volume, liter/g-mol
T	= absolute temperature, Kelvin
X	= mole fraction in liquid
Y	= mole fraction in gas

Greek Symbols

γ	= activity coefficient
ϕ	= fugacity coefficient
τ	= parameter for activity coefficient equation defined in equation (27)
α	= a parameter for activity coefficient equation used in equation (24)

Subscripts

i	= component
p	= polar nature
n	= non polar nature
c	= at critical state
r	= reduced variable
MSRK	= modified Soave Redlich-Kwong equation of state
MVDW	= Modified Van der Waals equation of state

Superscripts

α, β	= phase
s	= saturation state
rep	= repulsive interaction
attr	= attractive interaction
$*$	= pure state
e	= excess property

CHAPTER 15

THE CORRELATION AND PREDICTION
OF CRITICAL STATES OF MIXTURES
USING A CORRESPONDING STATES
PRINCIPLE

<u>Amyn S. Teja</u> **and**
<u>Richard L. Smith</u>
School of Chemical
Engineering
Georgia Institute of
Technology
Atlanta, GA 30332

Abstract

The calculation of critical states using a Corresponding States Principle is described. It is shown how a correlation of binary critical $T^c - P^c$ behavior can lead to predictions of high pressure as well as low pressure phase equilibria and PVT properties. Critical states (including $V^c - x^c$ behavior) of hydrocarbon systems as well as systems containing polar components have been successfully correlated using this method. The applicability of the method to asymmetric mixtures which are typical of supercritical extraction processes is examined.

Introduction

The PVT behavior of mixtures in the critical region is of practical importance in the petroleum and energy industries and also in processes involving supercritical gas extraction. Indeed, a knowledge of phase equilibria in the critical region is necessary for assessing the feasibility of supercritical gas extraction processes.

The calculation of critical lines of mixtures is also of theoretical interest, since the topology of these lines in the PVTx space is complex and it may be argued that their calculation provides a most severe test of any theory.

In this work, we outline an extended Corresponding States method which has been used with considerable success for the calculation of densities [1-3], vapor-liquid equilibria [3-5] and critical and azeotropic states [6,7] of mixtures when the components of the mixture do not differ too greatly in size. The applicability of the method to asymmetric mixtures (such as ethylene + napthalene, which are typical of supercritical extraction processes) is examined.

The Extended Corresponding States Method

A large number of methods for the calculation of thermodynamic properties make use of extensions of the (two parameter) Corresponding States Principle. These extended Corresponding States methods may be divided into two broad categories. The first, and by far the largest, category involves a perturbation of a thermodynamic (configurational) property about that of a spherical reference fluid. The theoretical justification for this perturbation was presented by Pitzer [8] who showed that the compressibility Z_{ii} of fluid i at reduced temperature T_R and reduced pressure P_R is given by the sum of the compressibility $Z^{(0)}$ of a spherical reference fluid at the same reduced temperature and pressure and a deviation $\omega_{ii} Z^{(1)}$ which is proportional to the acentric factor ω_{ii} of substance i. Thus

$$Z_{ii}[T_R, P_R] = Z^{(0)}[T_R, P_R] + \omega_{ii} Z^{(1)}[T_R, P_R] \qquad (1)$$

Teja et. al. [9] later showed that the perturbation may be carried out about the property of a nonspherical reference fluid.

The second category of extended Corresponding States methods involves a perturbation of the variables T_R and V_R (or P_R) and has been termed the shape factor approach by Leland [10]. It is this approach that is described below. For convenience, the shape factor approach will be described in terms of the configurational Helmholtz energy rather than compressibility, as the former is more suited to the calculations performed in this study.

Two pure substances i and o are defined to be in corresponding states if the configurational Helmholtz energy of substance i at temperature T and volume V may be obtained from the configurational Helmholtz energy of substance O at temperature $T/f_{ii,o}$ and volume $V/h_{ii,o}$ as follows:

$$A_{ii}[T,V] = f_{ii,o} A_o[T/f_{ii,o}, V/h_{ii,o}] - RT \ln h_{ii,o} \qquad (2)$$

where the subscripts ii,o signify a property of i relative to substance o. The parameters $f_{ii,o}$ and $h_{ii,o}$ are related to the critical constants of substances i and o by

$$f_{ii,o} = (T_{ii}^c / T_o^c) \, \theta_{ii,o} \qquad (3)$$

$$h_{ii,o} = (V_{ii}^c/V_o^c)\, \phi_{ii,o} \qquad (4)$$

where the shape factors $\theta_{ii,o}$ and $\phi_{ii,o}$ are unity for simple spherical fluids and are slowly varying functions of reduced temperature T_{Ri} and reduced volume V_{Ri} in the general case. They may be thought of as small perturbations of the reduced temperature and volume. Analytical expressions for the shape factors have been reported by Leland and Chappelear [10] and by Ely and Hanley [11]. They are of the form:

$$\theta_{ii,o} = 1 + (\omega_{ii} - \omega_o)\, F_1[T_{Ri}, V_{Ri}]$$

$$\phi_{ii,o} = (Z_o^c/Z_{ii}^c) \{1 + (\omega_{ii} - \omega_o)\, F_2[T_{Ri}, V_{Ri}]\}$$

where F_1 and F_2 are functions of the reduced temperature and volume. Although the Leland shape factor correlations were originally obtained for the n-alkanes (up to about n-decane) relative to methane, the utility of this method lies in the fact that they have been used successfully for other hydrocarbons and for polar and quantum fluids [3,7]. Deviations from experiment are small for the n-alkanes and somewhat larger for polar and quantum fluids, as expected [7]. It should be noted that the deviations of the shape factors from unity are proportional to the differences in the acentric factors as well as to the ratio of the critical compressibility factors of the two substances i and o. For normal fluids, the critical compressibility factor Z^c is a linear function of the acentric factor and a three-parameter form of the Corresponding States Principle results. For fluids for which Z^c is no longer a linear function of ω, a four-parameter Corresponding States Principle is obtained.

Given the properties of a reference substance (these may, for example, be obtained from an accurate equation of state or from tabulated PVT properties) and the shape factor correlations, then the calculation of the thermodynamic properties of any substance i requires a knowledge only of its critical constants T_{ii}^c, V_{ii}^c, Z_{ii}^c and ω_{ii}.

Equation (2) may be extended to mixtures by assuming that the configurational Helmholtz energy of a mixture, after subtraction of a combinatorial term, is equal to that of a single hypothetical equivalent substance. This is the so-called one-fluid model for mixtures.

$$A_{mixture}[V,T,x] = A[V,T,x] + A_{comb} \qquad (7)$$

The configurational Helmholtz energy of the equivalent substance $A[V,T,x]$ may then be obtained from that of the reference substance using eqn. (2):

$$A[V,T,x] = f\, A_o[T/f,\, V/h] - RT \ln h \tag{8}$$

where the parameters f and h depend on the composition of the mixture. Various prescriptions may be written for these parameters. A convenient form of these prescriptions is as follows:

$$f\, h^\delta = \Sigma\Sigma\, x_i x_j f_{ij} h_{ij}^\delta \tag{9}$$

$$h = \Sigma\Sigma\, x_i x_j h_{ij} \tag{10}$$

When $\delta=0$, these equations reduce essentially to the modified Kay's model proposed by Prausnitz and Gunn [12]; setting $\delta=1$ leads to the van der Waals one fluid model [10], and setting $\delta=0.25$ leads to the model proposed by Plöcker et. al. [13] for mixtures in which the molecular sizes of the components differ by a factor of 3 or more. It should be added that the critical temperatures and volumes used in the original one-fluid models described above are "weighted" by the shape factors in equations (9) and (10). Moreover, to obtain the shape factors in a rigorous manner, an iterative solution of equations (9) and (10) is necessary.

In using eqns. (9) and (10), the like terms (i=j) may be obtained from pure component properties, but the unlike terms (i≠j) require mixture data for their evaluation. The usual procedure is to transfer the problem of evaluating f_{ij}, h_{ij} to the problem of evaluating the binary interaction coefficients ξ_{ij} and η_{ij}, where:

$$f_{ij} = \xi_{ij}(f_{ii}f_{jj})^{1/2} \tag{11}$$

$$h_{ij} = \eta_{ij}\{(h_{ii}^{1/3} + h_{jj}^{1/3})/2\}^3 \tag{12}$$

Values of ξ_{ij} and η_{ij} are generally close to unity and no further information is required to predict the properties of ternary and higher mixtures [7].

For nonpolar mixtures in which the molecular sizes of the components differ by a factor of 3 or less, it is usually sufficient to use one adjustable coefficient to characterize each binary system and it is common to assume $\eta_{ij}=1.0$, with ξ_{ij} being calculated from experimental data. Values of ξ_{ij} estimated from the correlation of various thermodynamic properties and phase equilibria agree very well with each other [7], but depend on the particular prescription (eqns. 9-10) chosen.

The combinatorial contribution to the Helmholtz energy for an ideal mixture is given by:

$$A_{comb} = RT \sum_i x_i \ln x_i \tag{13}$$

When the molecules differ appreciably in size, this equation is known to be inadequate. The Flory-Huggins equation was derived for monmer-polymer mixtures (i.e. for molecules made up of like segments), but it has been applied to many different asymmetric mixtures. This equation is given by:

$$A_{comb} = RT\Sigma \, x_i \ln \psi_i \tag{14}$$

where the volume fraction ψ_i is given by:

$$\psi_i = x_i V_{ii} / \Sigma_j x_j V_{jj} \tag{15}$$

In the binary case, we may write the two volume fractions by:

$$\psi_1 = x_1/(x_1 + r\,x_2) \tag{16}$$

$$\psi_2 = rx_2/(x_1 + rx_2) \tag{17}$$

where

$$r = V_{11}/V_{22} \tag{18}$$

is the ratio of molar volumes (or the ratio of semiarbitrarily defined segments). Since the systems considered in this study are not polymer solutions, we have treated r as an empricial parameter in this study. Previous studies have shown [7,14] that r = 1 when the components of a mixture do not differ appreciably in size.

Critical States of Binary Mixtures

Critical states of binary mixtures satisfy the following equations [6]:

$$A_{2x} \cdot A_{2v} - A_{xv}^2 = 0 \tag{19}$$

$$A_{3x} \cdot A_{2v}^3 - 3\,A_{2xv}A_{xv}A_{2v}^2 + 3\,A_{x2v}A_{xv}^2 A_{2v} - A_{3v}A_{xv}^3 = 0 \tag{20}$$

where the subscripts denote derivatives e.g.

$$A_{2xv} = \partial^3 A/\partial x^2\, \partial V$$

It should be noted that these equations contain second and third derivatives of the Helmholtz energy (and, therefore, of the one-fluid model-eqns. 9, 10 and 14) with respect to composition. It may therefore be argued that the calculation of critical states represents a severe test of the mixture model chosen, since errors will tend to increase with successive differentiation of the function.

Solution of the two equations at a given critical composition gives the critical temperature and volume (and hence the critical pressure) of the system. Details of the procedure are given elsewhere [6].

Results

Methane was chosen as the reference substance, since an accurate analytical equation of state is available for this substance and shape factor correlations relative to methane are also available. The critical properties and acentric factors of the pure components were taken from the compilation by Ambrose [15]. Results of the calculation of critical curves of binary mixtures are discussed below.

Figures 1-3 show typical results for a nonpolar hydrocarbon mixture (methane + propane) in which the components do not differ appreciably in size.

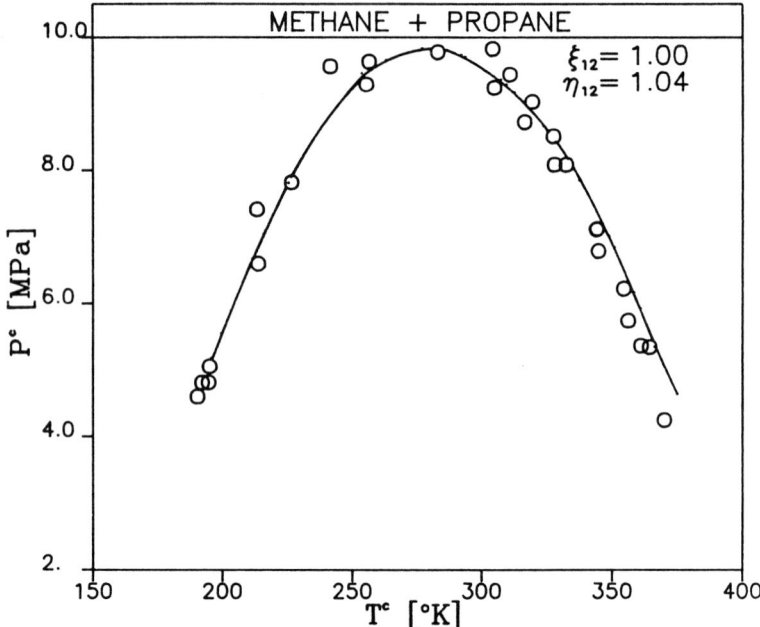

Figure 1. P^c vs T^c behavior of the methane + propane system. The predicted critical curve is shown by the full line whereas the experimental points are shown by circles.

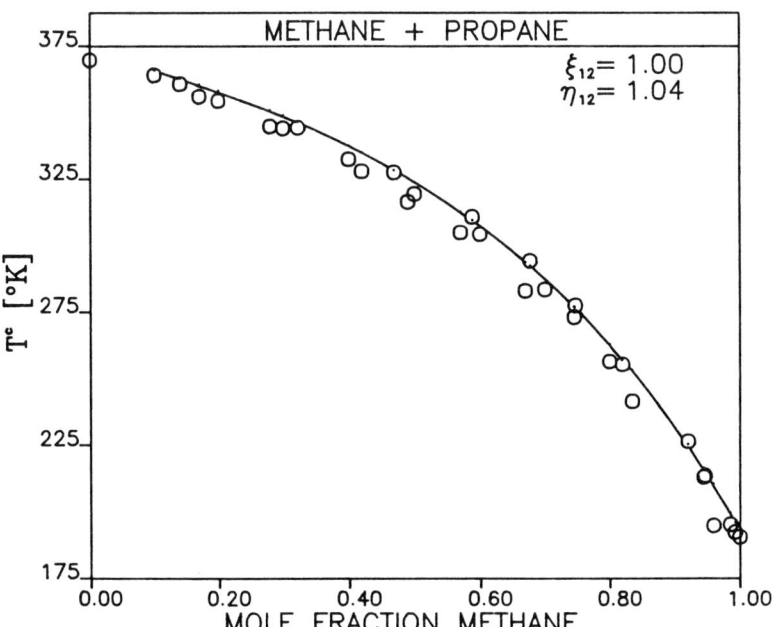

Figure 2. The predicted and experimental T^c vs x^c behavior of the methane + propane system.

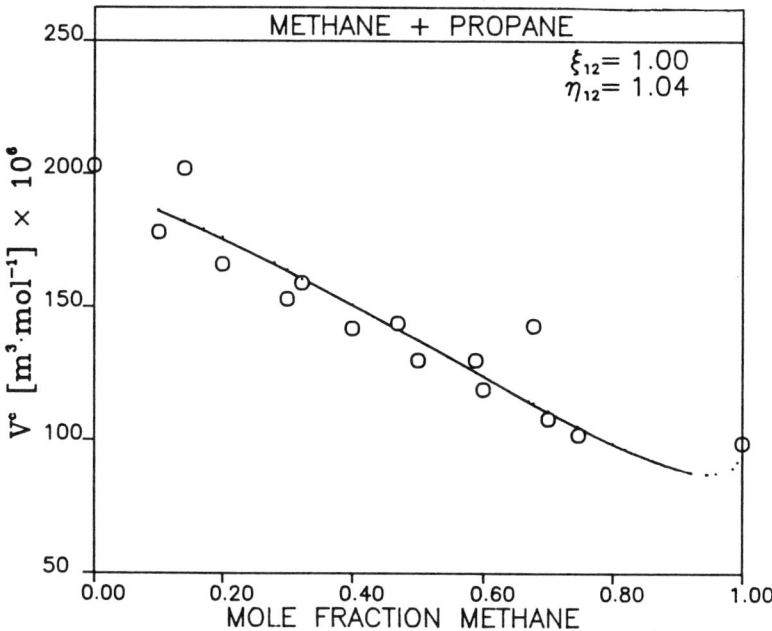

Figure 3. The predicted and experimental V^c vs x^c behavior of the methane + propane system with $\xi_{12} = 1.0$ and $\eta_{12} = 1.04$.

It is seen that P^c vs T^c vs. x^c curves can be correlated with $\xi_{12} = 1.0$ and $\eta_{12} = 1.04$ when the van der Waals one-fluid prescription ($\xi=1$) and the ideal combinatorial term (r=1) are used for this system. In general, only one binary interaction coefficient - close to unity - is required for such systems [6,7]. Moreover, Barber et. al. [16] have shown that the single binary interaction coefficient calculated from the critical states of mixtures can be used to predict vapor-liquid equilibria and second virial coefficients of the system. This is shown in Figs. 4 and 5 which show the azeotropic locus and second virial coefficients of propane + perfluorocyclobutane mixtures calculated with $\xi_{12} = 0.89$ which was obtained from the correlation of the critical states of the system [16]. A number of authors [1,3,5] have used two binary interaction coefficients to characterize each binary mixture. Our studies have shown that the use of two adjustable coefficients leads to values which are not unique and which do not necessarily show a regular trend for a series of binary mixtures with a common component. The latter is usually true when only one adjustable coefficient is used in the calculations.

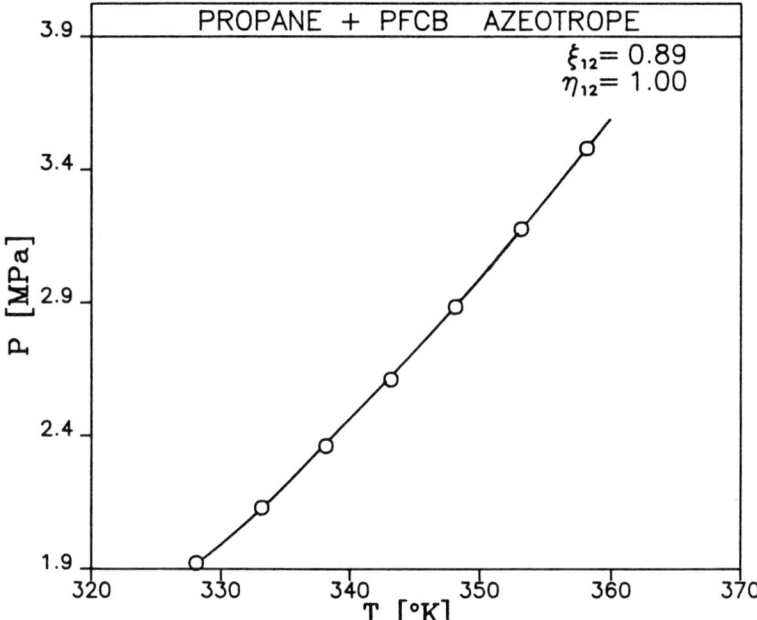

Figure 4. The azeotropic locus of propane + perfluorocyclobutane mixtures predicted with $\xi_{12} = 0.89$, $\eta_{12} = 1.0$.

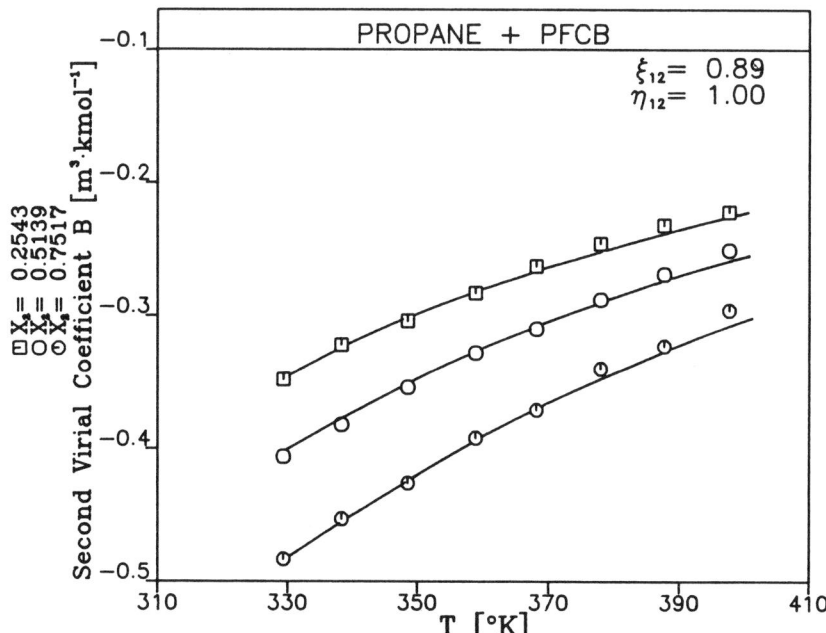

Figure 5. Second virial coefficients of propane + perfluorocyclobutane mixtures predicted with $\xi_{12} = 0.89$, $\eta_{12} = 1.0$

Figures 6-12 show the results of our calculations for three asymmetric mixtures (methane + n-decane, CO_2 + n-decane and ethylene + napthalene) in which the components differ in size (as measured by the ratio of their critical volumes) by a factor of 3 or more. These systems are typical of those found in supercritical extraction. In general, there is a great deal of uncertainty in the available values of the critical volumes of large molecules (indeed, the critical volumes of many single and multi-ring aromatic compounds are not known). For asymmetric mixtures, we have therefore replaced eqn. (4) with

$$h_{ii,o} = \{(T^c_{ii}/P^c_{ii})/(T^c_o/P^c_o)\} \phi_{ii,o} \tag{21}$$

This equation has the advantage that critical volumes are not required and is equivalent to a slight modification of the mixture model (i.e. of multiplying $h_{ii,o}$ in eqns. (9) and (10) by Z^c_{ii}/Z^c_o). We have found that this modification leads to better predictions than the original model (there is of course little change for mixtures of small molecules since $Z^c_{ii} \approx Z^c_o$). Figs. 6-12 show that one adjustable binary interaction coefficient is no longer adequate for the correlation of the critical properties of asymmetric mixtures.

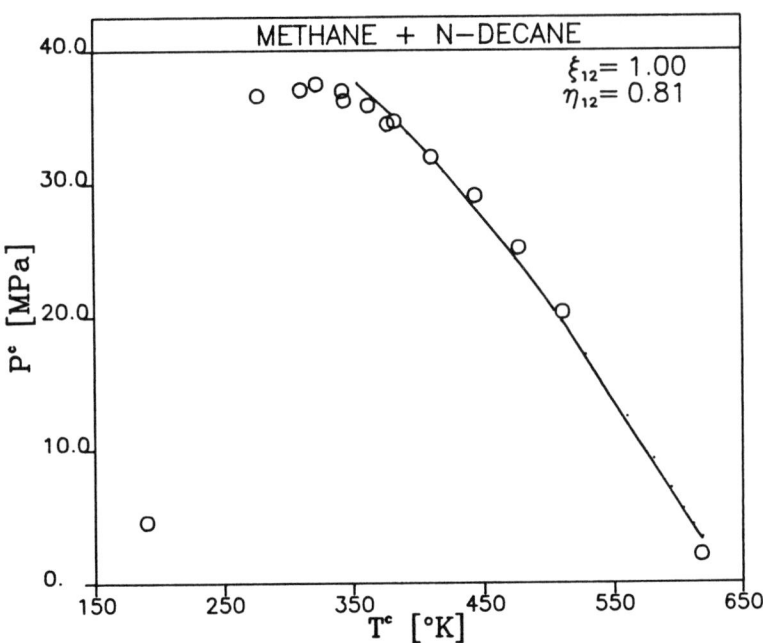

Figure 6. P^c vs T^c behavior of the methane + n-decane system

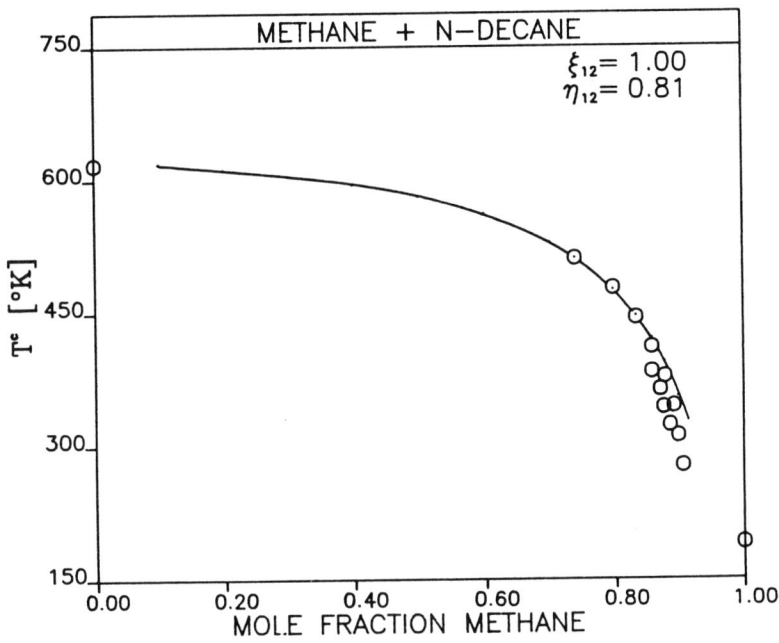

Figure 7. The predicted and experimental T^c vs x^c behavior of the methane + n-decane system

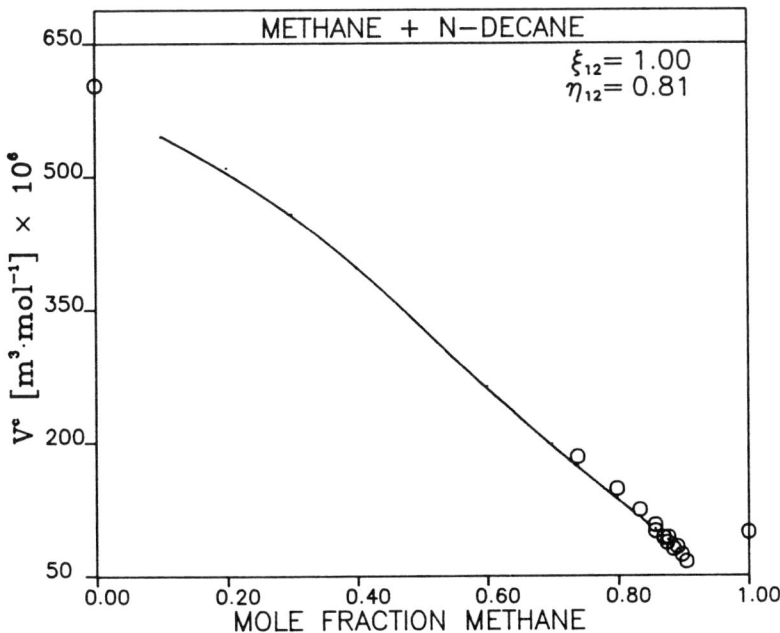

Figure 8. V^c vs x^c behavior of the methane + n-decane system

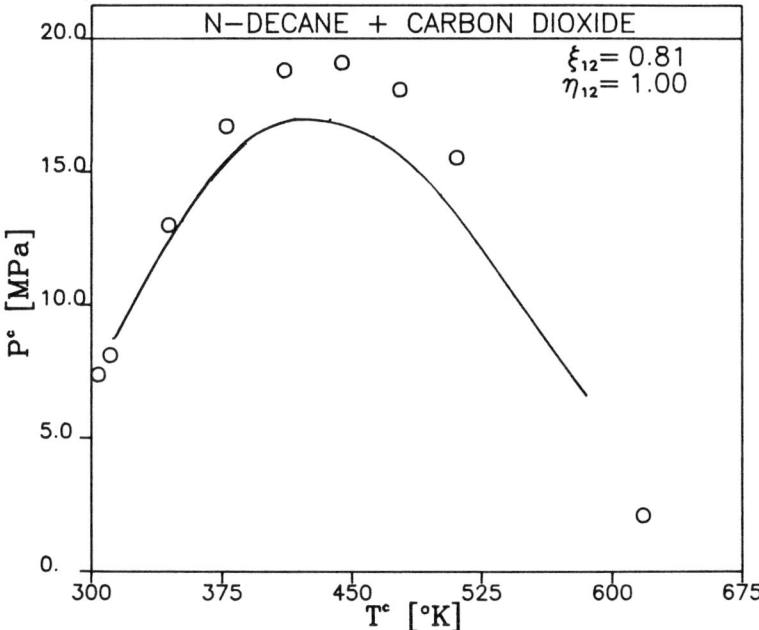

Figure 9. The predicted T^c vs x^c behavior of the n-decane + CO_2 system

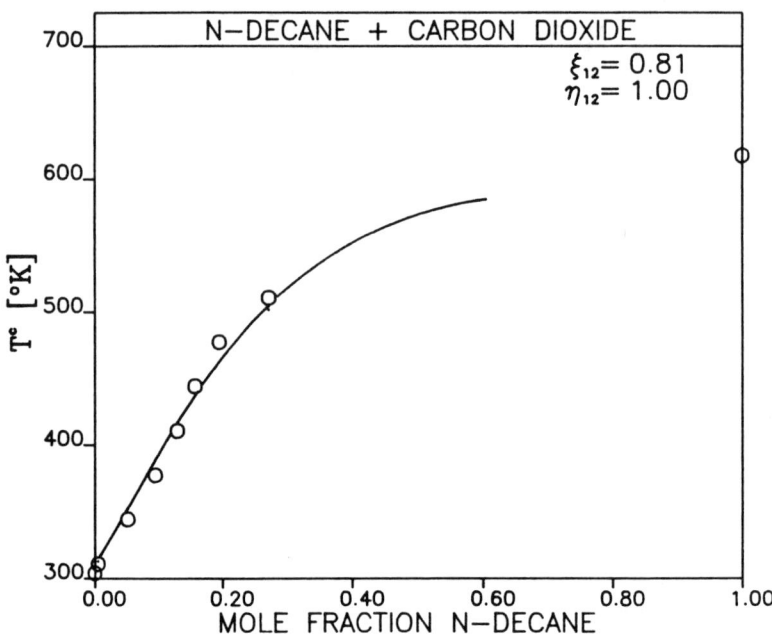

Figure 10. The predicted and experimental T^c vs x^c behavior of the n-decane + CO_2 system

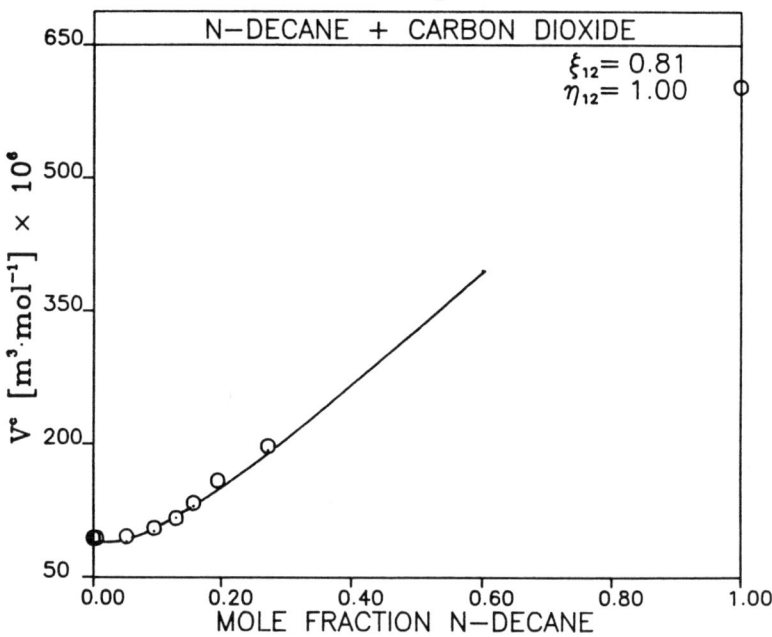

Figure 11. The predicted and experimental V^c vs x^c behavior of the n-decane + CO_2 system

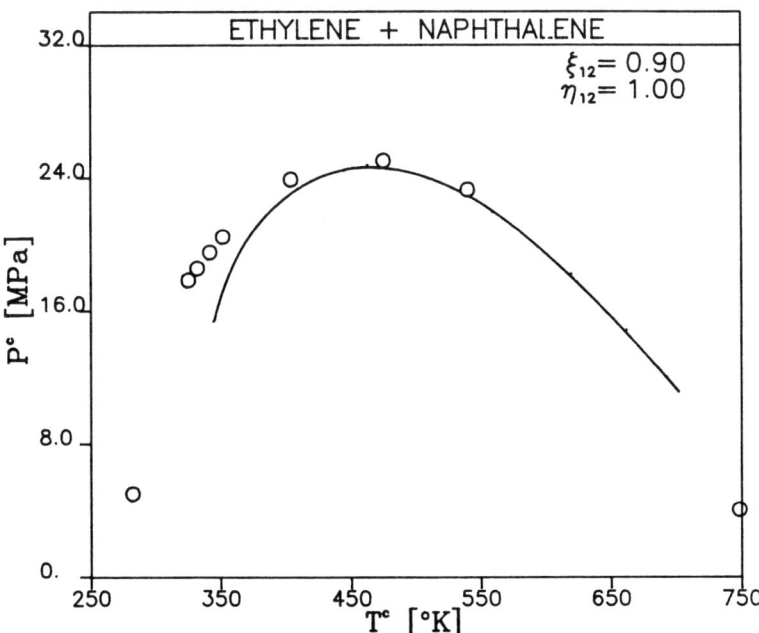

Figure 12. The predicted and experimental P^c vs T^c behavior of the ethylene + napthalene system

The use of the Flory-Huggins combinational term did not lead to a reduction in the number of adjustable coefficients required. It would therefore appear that only a small contribution to the Helmholtz energy can be attributed to the combinatorial term. The major effect of varying r appears to be on the critical pressure of the system, with critical temperature remaining practically unchanged. However, the accuracy of critical data in practice does not warrant the use of three adjustable constants in any method and r was set to unity in all other systems reported here. Use of the Plocker et. al. [13] one-fluid model ($\delta=0.25$) did not also lead to a reduction in the number of adjustable constants required. Hence the van der Waals one-fluid model ($\delta=1$) was used in all calculations presented in this work. Our calculations lead us to believe that only slight modifications of the van der Waals one-fluid model are required for the treatment of asymmetric mixtures.

It may be argued that part of the errors that arise in the treatment of asymmetric mixtures may result from the fact that the properties of large molecules (such as n-decane or n-hexadecane) cannot be obtained from the properties of methane using the shape factor correlations. That this is not so is demonstrated conclusively in Figures 13 and 14, which shown T^c vs x^c curves for n-hexane + n-tridecane and n-hexane + cis-decalin mixtures. These systems can be adequately correlated with the extended corresponding states method, although the components of the mixture are large. The relative size differences in these systems are, of course, less than 3.

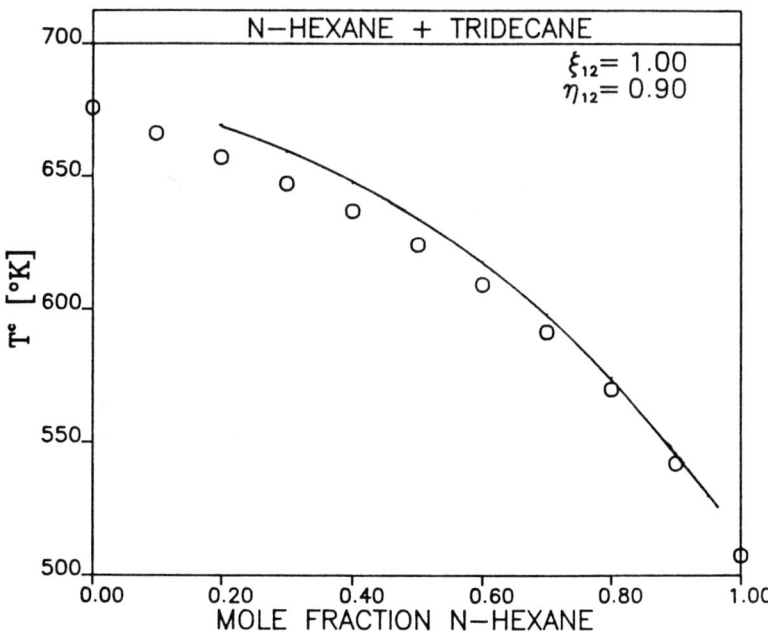

Figure 13. Experimental and calculated T^c vs x^c behavior of the n-hexane + n-tridecane system

A final observation that should be mentioned is that T^c vs x^c behavior of many systems (including asymmetric mixtures) can often be correlated equally well by various mixture models (e.g. van der Waals, Plöcker et. al. etc.). It is obvious that T^c vs x^c behavior alone is not sufficient to distinguish between the various models and we therefore suggest that a simultaneous fit of P^c-T^c-V^c-x^c data be used to obtain any adjustable constants. In general, we have found the van der Waals one-fluid model to be the best among the mixture models tested - although there is room for improvement for asymmetric mixtures.

Conclusions

We have examined the applicability of an extended Corresponding States Method using shape factors to the calculation of critical states of asymmetric mixtures which are typical of systems used in supercritical extraction. Various modifications of the method have been evaluated. In general, the van der Waals one-fluid model appears to yield results

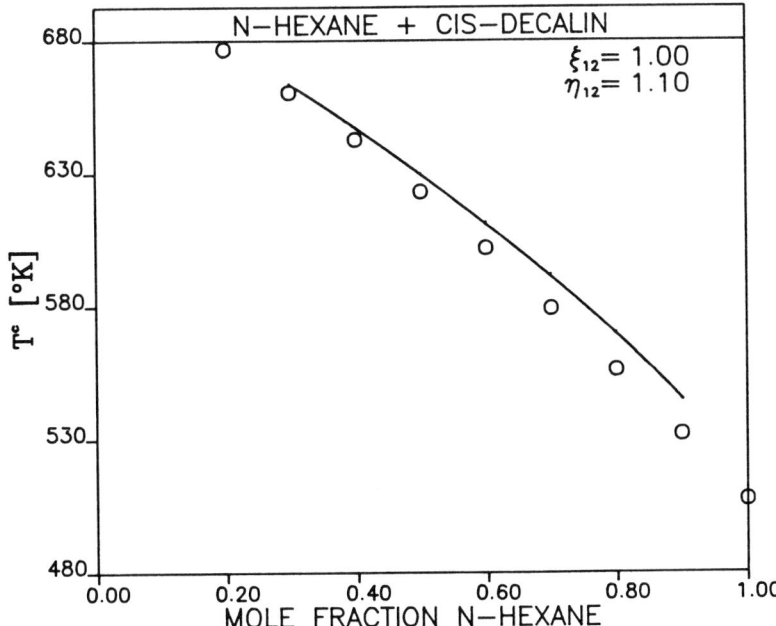

Figure 14. Experimental and calculated T^c vs x^c behavior of the n-hexane + cis-decalin system

which compare well with experiment, although there remains room for improvement. The method provides a means for the quantitative prediction of the behavior of interest in supercritical extraction.

Acknowledgment

Part of this work was carried out under a National Science Foundation Grant no. CPE-8104201 to the Georgia Tech Research Institute. Support was also provided by a Scholarship from the Tenneco Oil Company.

Nomenclature

A   Helmholtz energy
F   functions in eqns. (5) and (6)
f   Corresponding States energy parameter
h   Corresponding States size parameter
P   pressure
R   gas constant
r   adjustable constant
T   temperature
V   volume
x   mol fraction
Z   compressibility factor

Greek Symbols

δ exponent in mixture model (eqn. 9)
ξ binary interaction coefficient
η binary interaction coefficient
ω acentric factor
θ shape factor for energy (temperature)
ϕ shape factor for size (volume)
ψ volume (or segment) fraction

Subscripts

comb   combinatorial
i,j   components i.j
o   reference fluid
1,2   components 1,2

References

1. Mollerup J. and Rowlinson J. S., "The prediction of the densities of liquefied natural gas and of lower molecular weight hydrocarbons", Chem. Eng. Sci., 29, 1373 (1974).

2. Teja A. S. and Rice P., "Densities of benzene-n-alkane mixtures", J. Chem. Eng. Data, 21, 173 (1976).

3. Mentzer R. A., Greenkorn R. A. and Chao K. C., "The principle of corresponding states and prediction of gas-liquid separation factors and thermodynamic properties: a review", Sep. Sci. and Tech., 15, 1613 (1980).

4. Watson I. D. and Rowlinson J. S., "The prediction of the thermodynamic properties of fluids and fluid mixtures. II Applications". Chem. Eng. Sci., 24, 1575 (1969).

5. Gunning A. J. and Rowlinson J. S., "The prediction of the thermodynamic properties of fluids and fluid mixtures. III Applications", Chem. Eng. Sci., 28, 521 (1973).

6. Teja A. S. and Rowlinson J. S., "The prediction of the thermodynamic properties of fluids and fluid mixtures. IV Critical and azeotropic states", Chem. Eng. Sci., 28, 529 (1973).

7. Genco J. M., Teja A. S. and Kay W. B., "Study of the critical and azeotropic behavior of binary mixtures", J. Chem. Eng. Data, 25, 350, 355 (1980).

8. Pitzer K. S., Lippman D. Z., Curl R. F. Jr., Huggins C.M. and Petersen D. E., "The volumetric and thermodynamic properties of fluids II compressibility factor, vapor pressure and entropy of vaporization". J. Am. Chem. Soc., 77, 3433 (1955).

9. Teja, A. S., Sandler S. I. and Patel N. C., " A generalization of the corresponding states principle using two nonspherical reference fluids", Chem. Eng. J., 21, 21 (1981).

10. Leland T. W. and Chappelear P. S., "The principle of corresponding states", Ind. Eng. Chem., 60 (7), 15 (1968).

11. Ely J. F. and Hanley H. J. M., "Prediction of transport properties", Ind. Eng. Chem. Fundam., 20, 323 (1981).

12. Prausnitz J. M. and Gunn R. D., "Pseudocritical constants from volumetric data for gas mixtures", AIChE J., 4, 494 (1958).

13. Plöcker U., Knapp H. and Prausnitz J. M., "Calculation of high pressure equilibria from a corresponding states correlation with emphasis on asymmetric mixtures", Ind. Eng. Chem. Process Des. Dev., 17, 324 (1978).

14. Teja A. S., "The use of the corresponding states principle for mixtures containing polar components", Ind. Eng. Chem. Fundam., 18, 435 (1979).

15. Ambrose D., "Vapour-liquid critical properties", National Physical Laboratory, Teddington, England, Report (Oct. 1979).

16. Barber J. R., Kay W. B. and Teja A. S., "A Study of the volumetric and phase behavior of binary systems", AIChE J., 28, 142 (1982).

CHAPTER 16

CORRESPONDING STATES THEORIES
FOR CHAIN MOLECULES

Carol K. Hall and Barbara A. Hacker
Department of Chemical Engineering
Princeton University
Princeton, New Jersey 08544

ABSTRACT

A new corresponding states correlation for asymmetric and chain molecules is described. The correlation is similar to the corresponding states correlation for simple fluids with the exception that the reduced temperature variable T/T_c is replaced by $B_2(T)/B_2(T_c)$ where $B_2(T)$ is the second virial coefficient. The correlation has been tested on thirty compounds using the perturbed-hard-chain equation and found to be successful in the supercritical region for reduced volumes greater than 0.5. An empirical equation of state based on this correlation is described.

INTRODUCTION

Efficient design of separations processes requires reliable thermodynamic and phase equilibria properties of fluids and fluid mixtures. Since experimental data are often unavailable for systems at the conditions of interest it is important to develop methods for the estimation of thermodynamic properties. The principle of corresponding states has been useful in this regard because it allows the prediction of the unkown properties of many fluids from the known properties of a few fluids. In this paper, a new corresponding-states theory is described which correlates equation of state data on assymetric and chain molecules.

REVIEW OF CORRESPONDING STATES THEORIES

The corresponding states idea was introduced in the 1880's by van der Waals [1] who wrote his equation of state in the reduced form

$$P/P_c = f(V/V_c, T/T_c)$$
$$Z = f(V/V_c, T/T_c) \tag{1}$$

where P,V, and T are the pressure, molal volume and temperature, Z = PV/RT is the compressibility factor, R is the universal gas constant, the subscript c indicates the value at the critical point and the function f is a universal function. (In this paper the symbol f will be used to denote a universal function of the variables enclosed in parentheses.) Although first derived only for fluids satisfying van der Waal's eqn., corresponding states is actually quite general and can be derived on the basis of a simple set of assumptions using statistical mechanics [2]. The corresponding-states theory given in equation (1), which is sometimes referred to as simple corresponding-states theory, was found to give accurate results only for classical monatomic fluids such as Ar, Kr, Xe, etc.

To extend the principle of corresponding states to more complex fluids, two different approaches have been taken. In the first approach a third parameter has been added so that

$$Z = f\left(\frac{V}{V_c}, \frac{T}{T_c}, X\right) \tag{2}$$

Thus a group of molecules having the same third parameter will then conform to the same reduced equation of state. Numerous third parameters have been developed, the best known of which is the Pitzer acentric factor, ω. Some workers have also introduced a fourth parameter in order to include polar fluids in a corresponding states-development.

In the second approach, called the shape-factors approach [3,4,5], the temperature and volume are modified not only by the associated critical values but also by the so-called shape factors ϕ and θ which are themselves functions of temperature. In this approach the compressibility factor Z_i for a fluid i at temperature T_i and specific volume V_i is

$$Z_i = f(T_i/T_{ci}\theta_i, V_i/V_{ci}\phi_i) \tag{3}$$

where the universal function f is defined such that for some reference fluid, designated fluid 1, usually taken to be methane, the compressibility factor Z_1 is given by

$$Z_1 = f(T_1/T_{c1}, V_1/V_{c1}).$$

Thus in this approach one only need know Z_1 accurately as a function of its variables and the values for the shape factors of substance i referenced to substance 1.

The shape factors approach can be derived from statistical mechanics by first assuming that the intermolecular potential between asymmetric molecules can be angle-averaged and then using the resulting temperature-dependent pair potential in a corresponding-states type argument to arrive at temperature-dependent shape factors. More often the shape factors θ_i and ϕ_i are found empirically by fitting data on substance i with the reference data on substance 1. Shape factors found empirically will generally have a weak density dependence as well as temperature dependence.

A number of authors [3,4,5] have developed shape factors with respect to methane for hydrocarbons and related compounds and for mixtures of these substances. These have been obtained by fitting data on such thermodynamic properties as compressibility, saturation liquid volumes, enthalpies and P-V-T data. The correlations are most successful for the lower molecular weight substances which are similar in size and shape to the reference substance, methane. Thus the usefulness of the shape-factors approach is dependent on the similarity between the reference substance and the other substances to be correlated.

POLYMER SCALING THEORIES

It is of interest to develop corresponding-states theories for asymmetric and chain molecules which are independent of the choice of reference substance. Such corresponding states theories have been developed recently for polymer solutions and have come to be known as polymer scaling theories [6,7,8]. In this section the polymer scaling theories are described and it is shown that while there is no theoretical basis for assuming that they would be valid for pure fluids containing short chains, they are nevertheless suggestive. In the following section it will be shown that an extension of the polymer scaling theories to pure fluids containing short chains yields, in practice, a successful corresponding-states theory for pure fluids of asymmetric or chain molecules.

Polymer scaling theory states that at temperature T and total volume V_T, the osmotic pressure Π of a polymer solution consisting of n very long chains of length ℓN (where ℓ is the length of a flexible link and N is the number of links) is given by

$$\frac{\Pi V}{kT} = f \left\{ \frac{\ell^3 N^{\frac{3}{2}}}{V}, \frac{v(T)}{\ell^3} N^{\frac{1}{2}} \right\} \tag{4}$$

where $V = V_T/n$, k is Boltzmann's constant, $v(T)$ is the excluded volume parameter and f is a universal function of its variables. The excluded volume parameter $v(T)$ is a measure of the volume excluded to one chain by the presence of another chain and is given by

$$v(T) = \int [1 - \exp(W(r_{ij})/kT)] d^3 r_{ij} \qquad (5)$$

where $W(r_{ij})$ is the pair potential between two links. It can be shown that the second virial coefficient $B_2(T)$ is a function of $N^p v(T)$ where p is a positive number.

In order to determine the significance of polymer scaling theories for pure fluids of very short chains which have relatively inflexible links, it is necessary to extend the theory to pure polymer fluids (i.e. bulk polymer) and to develop corrections to polymer scaling theories for chain shortness and link flexibility. To extend the theory to pure polymer fluids one need only take the solvent to be a vacuum and replace the osmotic pressure, Π, by the pressure, P. Attempts to calculate corrections to polymer scaling for chain shortness and link inflexibility indicate however, that the corrections to scaling are large and that there is no theoretical basis for assuming that polymer scaling should be valid for short chains of inflexible links. Nevertheless it is of interest to explore whether the theories might be useful in practice. Accordingly, if one assumes that

$$\frac{PV}{kT} = f\left(\frac{\ell^3 N^{\frac{3}{2}}}{V}, \frac{B_2(T) N^p}{\ell^3}\right) \qquad (6)$$

(where the excluded volume parameter has been replaced by the second virial coefficient and p is a simple fraction) is valid for short chains and calculates the critical properties as is done in the usual statistical mechanical derivation of corresponding states theory [2], the following is obtained

$$\frac{Z}{Z_c} = f\left(\frac{V}{V_c}, \frac{B_2(T)}{B_2(T_c)}\right). \qquad (7)$$

Since the dimensionless second virial coefficient for a square-well potential may be approximated by

$$\frac{B_2(T)}{b_o} \cong \frac{(T - T_b)}{T}$$

where b_o is the hard-sphere second virial coefficient, and T_b is the Boyle temperature, the postulated corresponding-states theory becomes

$$\frac{Z}{Z_c} = f\left(\frac{V}{V_c}, X_r(T)\right) \qquad (8)$$

where

$$X_r(T) = \frac{T_c}{T} \frac{(T - T_b)}{(T_c - T_b)} . \tag{9}$$

CORRESPONDING-STATES THEORY USING $X_r(T)$ AS A TEMPERATURE VARIABLE

In order to investigate the supposition that Equation 8 would make a good corresponding-states correlation we have studied the scaling properties of the perturbed-hard-chain equation [9] developed by Donohue, Prausnitz and co-workers. When the perturbed-hard-chain equation (which is a three parameter equation) was plotted for different substances* on a graph of P_r/T_r versus $X_r(T)$ for fixed V_r where $P_r = P/P_c$, $T_r = T/T_c$ and $V_r = V/V_c$) a single curve was obtained whereas when the data was plotted as P_r/T_r versus $1/T_r$ for fixed V_r, a different curve was obtained for each substance. Figures 1, 2, and 3 show these graphs for values of $V_r = 0.5$, 2.0 and 6.0 respectively for over thirty compounds including such diverse compounds as methyl chloride, ammonia and eicosane. The outliers on these graphs are eicosane and hexadecane. As can be seen from Figures 1 to 3 the correlation is quite good for values of $X_r(T)$ between -1.5 and 1.0 which corresponds to a range in T_r of between 0.9 and 20.0. The correlation is not well satisfied for values of $V_r < 0.5$. Even though these plots are not of real data but rather of fits of real data to an equation of state, they illustrate the potential usefulness and also the simplicity of this type of approach.

It is of interest to ask why the second virial coefficient should be a better corresponding-states temperature variable than the reduced temperature. A possible explanation is that $B_2(T)$, which for asymmetric molecules is given by

$$B_2(T) = \frac{-1}{2\pi} \int d^3 r_{ij} \int d\Omega_j \int d\Omega_{jj} \int d\phi_{ij} \left| \overline{\exp(-U(r_{ij}, \Omega_i, \Omega_j, \phi_{ij})/kT) - 1} \right|$$

(where r_{ij} is the distance between molecules i and j and Ω_i, Ω_j and ϕ_{ij} represent the angles between asymmetric molecules i and j and the vector $\underset{\sim}{r}_{ij}$), is a good measure of how the pair potential and temperature enter the partition function. Because the temperature enters the partition function only when coupled to the pair potential in the form $\exp(-U/kT)$,

---

*The parameters used were those parameters fitted by Donohue and Prausnitz to a wide variety of data. The values of T_c, P_c and V_c used were those calculated using the perturbed-hard-chain equation.

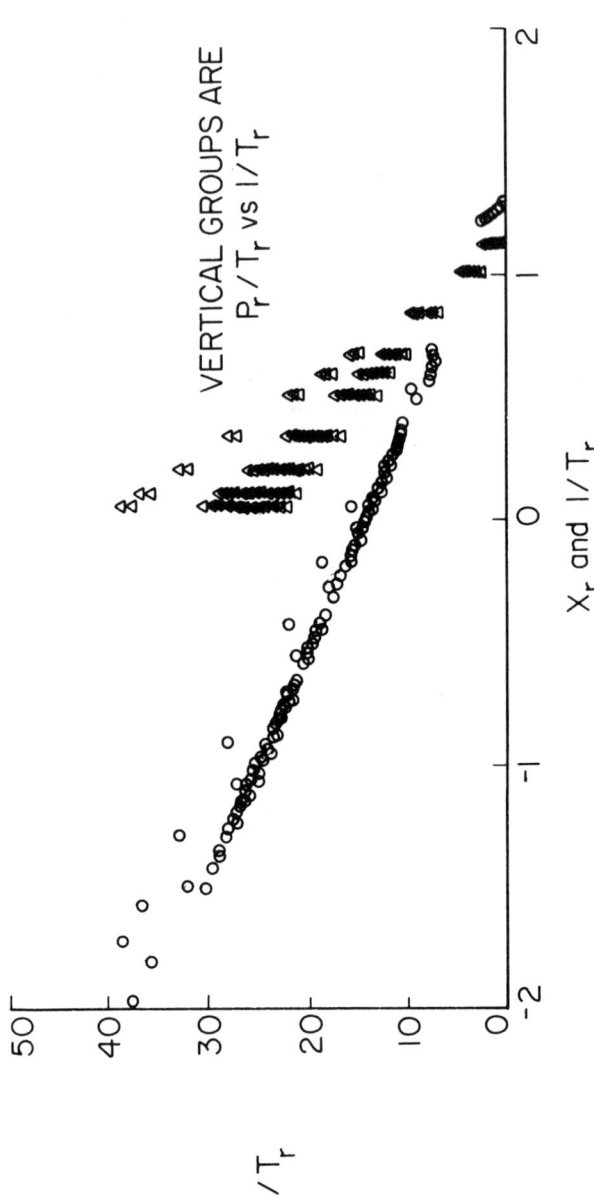

Figure 1. Comparison of simple corresponding states theory and the proposed corresponding states theory for thirty compounds at $V_r = 0.5$. The thirty compounds are Methane, Ethane, Propane, Butane, Pentane, Hexane, Heptane, Octane, Nonane, Decane, Dodecane, Hexadecane, Eicosane, Isobutane, Isoheptane, 2,2,5 - Trimethylhexane, Cyclohexane, Benzene, Toluene, m-Xylene, Naphthalene, Methyl Chloride, Ethylene, Propylene, Water, Sulfur dioxide, Ammonia, Carbon monoxide, Carbon dioxide, Nitrogen.

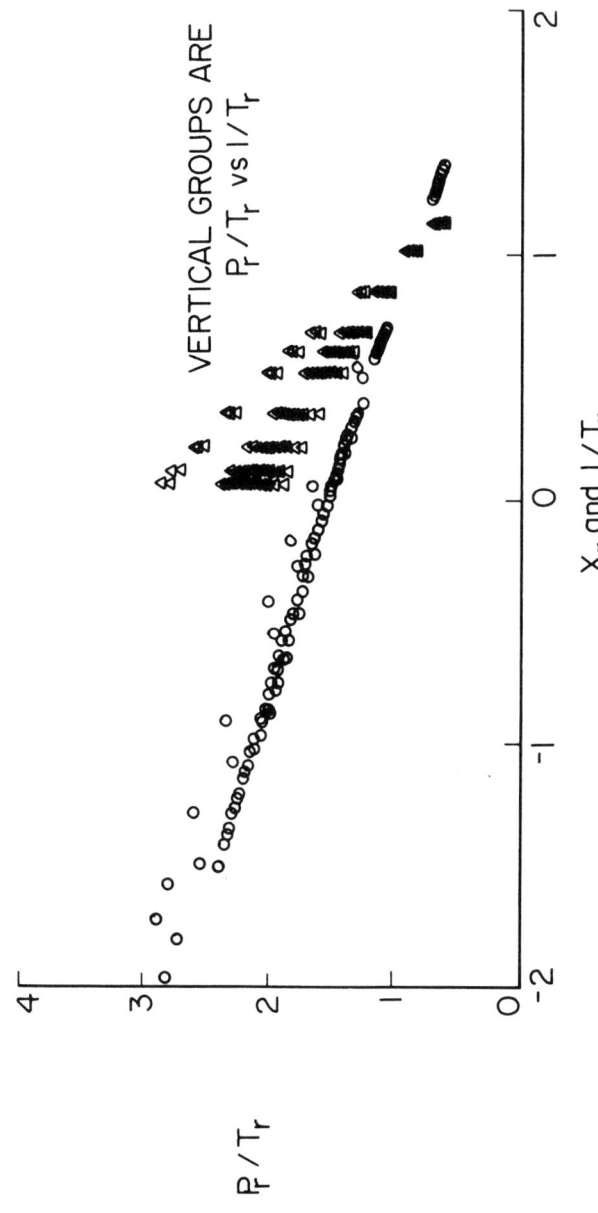

Figure 2. Comparison of simple corresponding states theory and the proposed corresponding states theory for thirty compounds at $V_r = 2.0$.

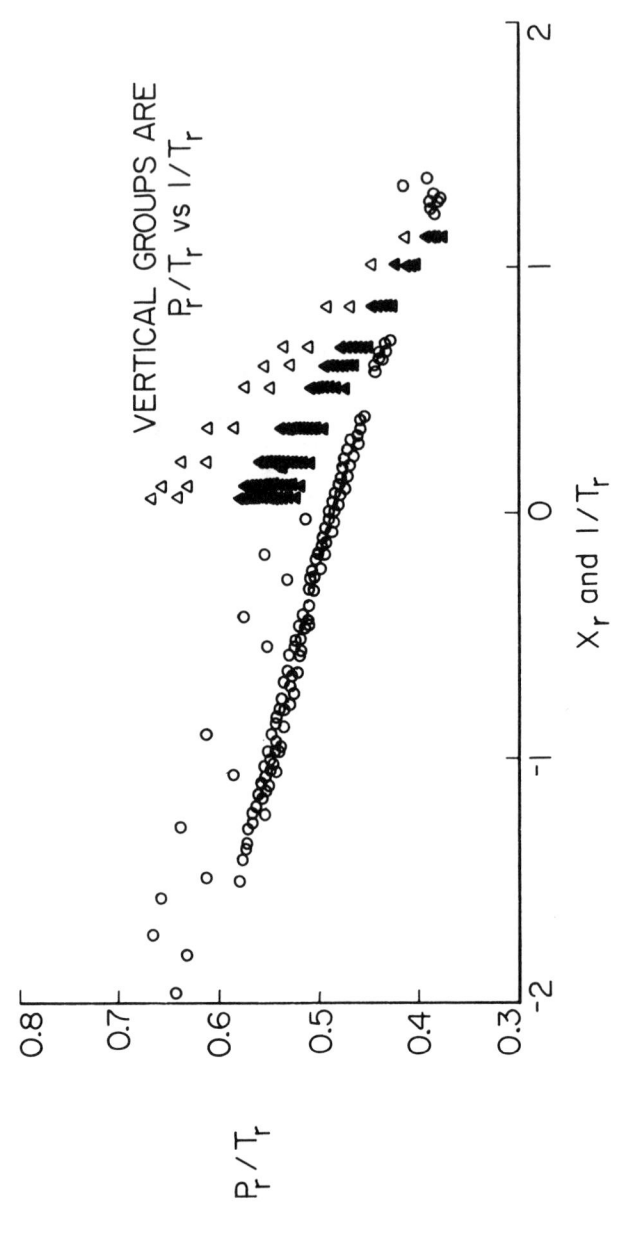

Figure 3. Comparison of simple corresponding states theory and the proposed corresponding states theory for thirty compounds at $V_r = 6.0$.

any function of exp(-U/kT) should weight the important parts of the pair potential appropriately. The second virial coefficient is the simplest function of exp(-U/kT) which is measurable thermodynamically. Notice also that the second virial coefficient contains information about the angle-averaged potential. Finally, in statistical mechanical arguments leading to simple corresponding states theory [2]. i.e. Equation 1, it is necessary to make the assumption that all molecules have pair potentials of the form $U(r_{ij}) = \varepsilon \phi(r_{ij}/\sigma)$ where the function ϕ is the same for all molecules. This rather restrictive assumption is unlikely to be obeyed by a wide class of molecules. No such restriction need be imposed in the corresponding-states theory in which $B_2(T)$ is the temperature variable. In summary, it appears that the proposed corresponding-states theory is an improvement over simple corresponding-states theory because it contains more information about how the temperature couples to the pair potential in the partition function.

EMPIRICAL EQUATION OF STATE

An examination of the corresponding states correlation described in the previous section in which P_r/T_r is a universal function of $X_r(T)$ for fixed V_r where $V_r \gtrsim 0.5$ suggests that an empirical equation of state which is universal may be developed. Since the graphs of P_r/T_r versus $X_r(T)$ are approximately linear with only slight curvature for all values of V_r, the following equation of state was tried

$$\frac{P_r}{T_r} = A + B\, X_r(T) + C\, X_r^2(T) \tag{10}$$

where A, B and C are functions of V_r. The functional forms of $A(V_r)$, $B(V_r)$ and $C(V_r)$ were obtained by fitting the "data" (which is taken to be the values predicted by the perturbed-hard-chain equation) at various values of V_r for six compounds which were thought to be representative, namely, methane, octane, 2-2-5 trimethyl hexane, benzene, water and CO_2.

It was found that $A(V_r)$ which measures the intercept of the curve and $B(V_r)$ which measures its slope are indeed universal functions of V_r but that $C(V_r)$ which measures the (slight) curvature is not a universal function of V_r. Instead, one finds that C is a function of the variable V_r' which is defined to be

$$V_r' = \frac{V_r^*}{(V_r - V_r^*)}$$

where V_r^* is the ratio of the hard core volume (obtained from the perturbed-hard-chain equation of state) to the critical volume. Thus the empirical equation of state satisfied by these six compounds and presumably by the other twenty-two compounds* considered above is given by †

$$\frac{P_r}{T_r} = A(V_r) + B(V_r)X_r(T) + C(V_r')X_r^2(T) . \qquad (11)$$

Figures 4, 5 and 6 show graphs of A versus V_r, B versus V_r and C versus V_r' for the six compounds. The functional forms of these curves were found to be

$$A(V_r) = 2.67 \left(\frac{1}{V_r}\right) + 1.07 \left(\frac{1}{V_r}\right)^2 - 0.949 \left(\frac{1}{V_r}\right)^3$$
$$+ 0.746 \left(\frac{1}{V_r}\right)^4 \qquad (12a)$$

$$B(V_r) = -2.63 \left(\frac{1}{V_r}\right)^2 \qquad (12b)$$

$$C(V_r') = 0.035 - 2.07 \left(\frac{1}{V_r'}\right) + 13.0 \left(\frac{1}{V_r'}\right)^2$$
$$- 16.6 \left(\frac{1}{V_r'}\right)^3 \qquad (12c)$$

Figure 7 contains a graph of P_r/T_r versus $X_r(T)$ for the six compounds. The points are the data calculated according to the perturbed-hard-chain equation and the curves are for the empirical equation given in Equations (11) and (12). The agreement is reasonable with the exception of octane at $V_r = 0.5$.

Although the coefficients in Equation (12) were determined by fitting to the perturbed-hard-chain theory, the results of this section suggest that an equation of state of the form given in Equation (11) can be used to fit experimental equation-of-state data for a wide variety of compounds. It is of interest to ask whether an equation of state of the form given in Equation (11) which contains the second virial coefficient, yet is valid for liquid volumes, can be derived using statistical mechanics. A possible answer to this question may lie in the work of Barboy and Gelbart [10]. Barboy and Gelbart showed that when the equation of state for hard particles is expanded in V_r', the coefficients are related to the virial coefficients and that a truncation of the series to include only three terms gives excellent agreement with

---

*Since eicosane and hexadecane did not satisfy the corresponding-states theory as indicated in Figures 1, 2, and 3 it was not expected that they would satisfy the empirical equation of state.
†Note that since C is not a universal function of V_r an inconsistency will occur in equation (11) at the critical point, i.e. when $X_r = 1$.

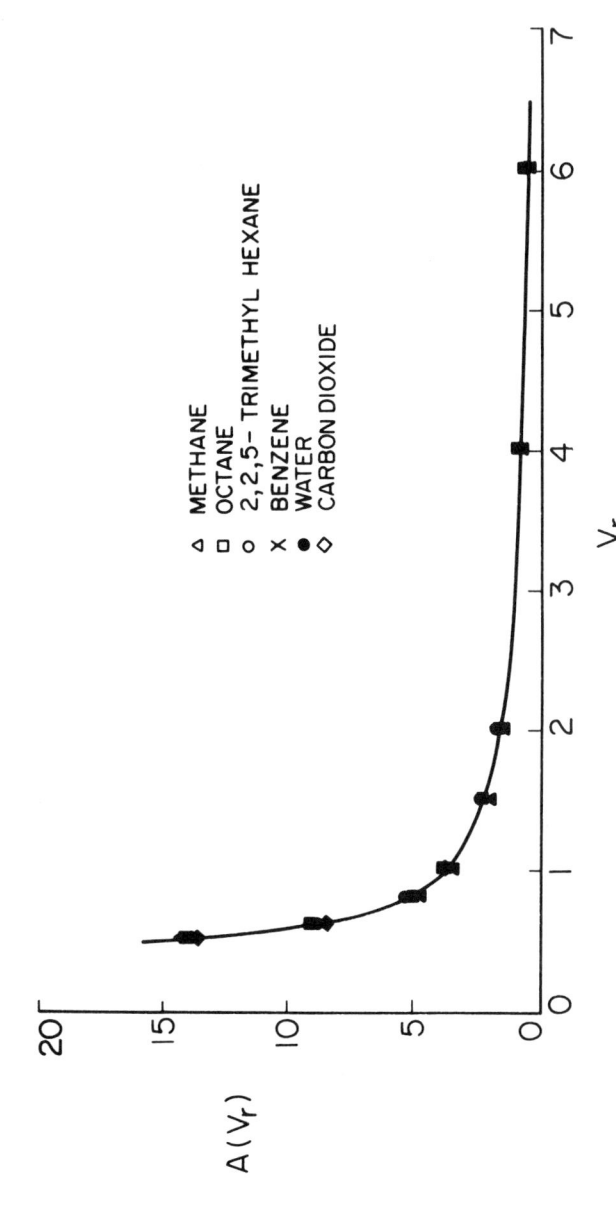

Figure 4. Graph of $A(V_r)$ versus V_r. The points are the data and the curve is given by Eqn. 12a.

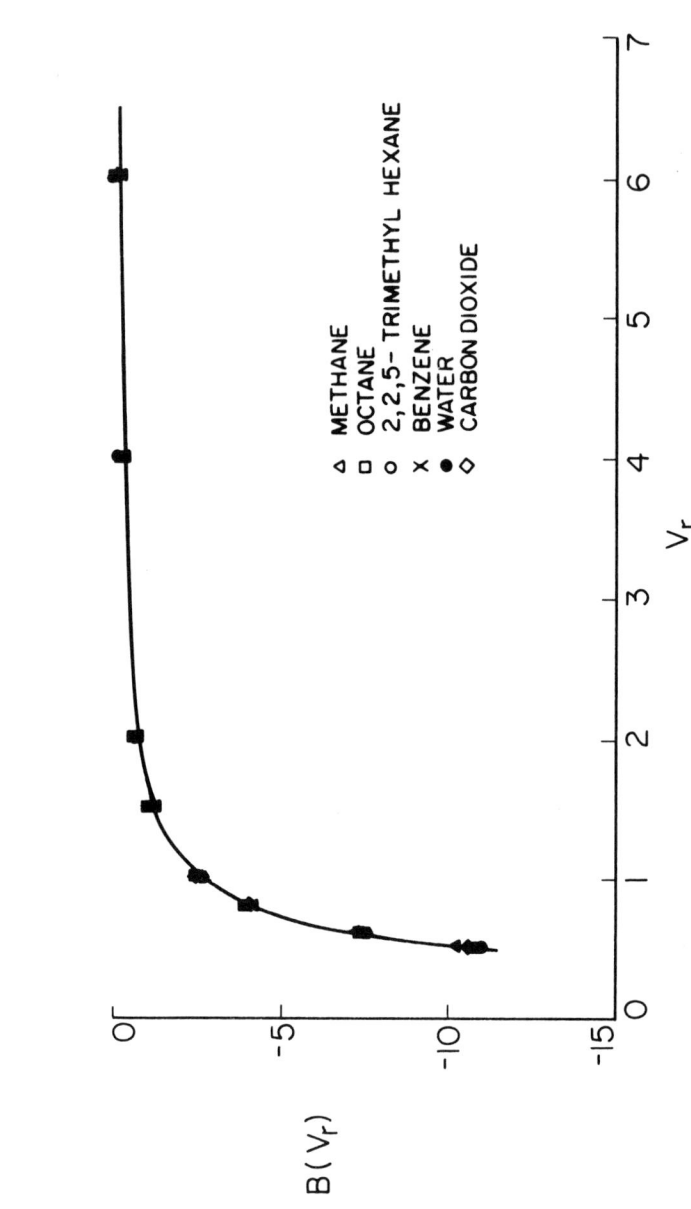

Figure 5. Graph of $B(V_r)$ versus V_r. The points are the data and the curve is given by Eqn. 12b.

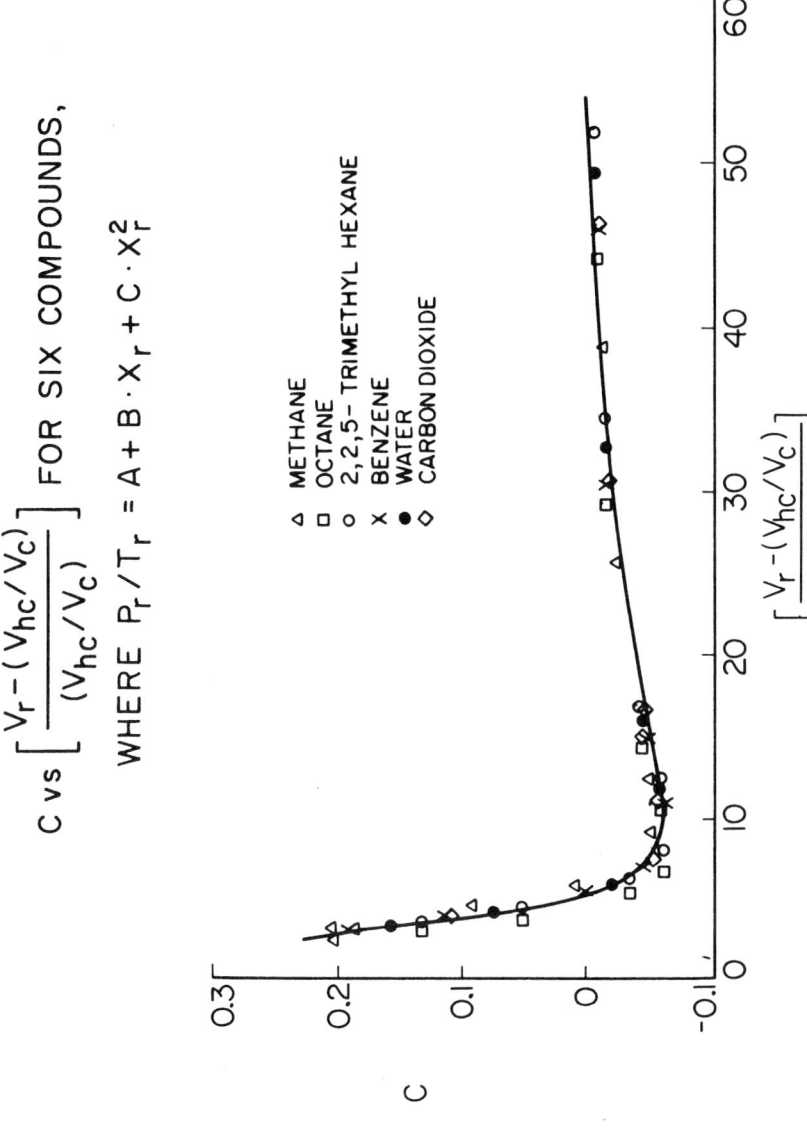

Figure 6. Graph of $C(V_r')$ versus V_r'. The points are the data and the curve is given by Eqn. 12c.

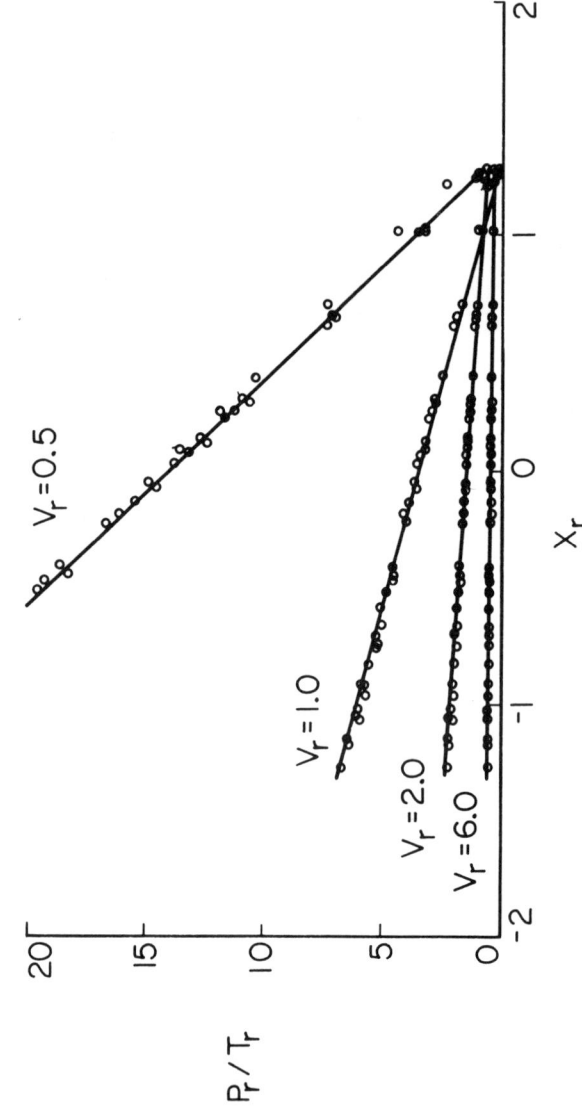

Figure 7. Graph of P_r/T_r versus $X_r(T)$ for V_r = 0.5, 1.0, 2.0 and 6.0. The points are the data and the curves are given by Eqn. 11.

molecular dynamics calculations. The relationship between the empirical equation of state obtained in this paper and the work of Barboy and Gelbart is currently under invertigation.

REFERENCES

1. J. D. van der Waals, thesis (Leiden 1873); Amsterdam Verslagen 10, 321, 337 (1876) and other papers.
2. T. M. Reed and K. E. Gubbins, Applied Statistical Mechanics, McGraw Hill (1973).
3. T. W. Leland and P. S. Chappelaer, Ind. & Eng. Chem. 60, 15 (1968).
4. R. A. Mentzer, R. A. Greenkorn and K. C. Chao, Separ. Sci. & Tech. 15, 1613 (1980).
5. J. Mollerup, Fluid Phase Equilibria 4, 11 (1980).
6. P. G. de Gennes, Scaling Concepts in Polymer Physics, Cornell University Press, Ithaca, New York (1979).
7. M. Kosmas and K. Freed, J. Chem. Phys. 69, 3647 (1978).
8. C. K. Hall, J. Chem. Phys. 73, 1446 (1980).
9. M. D. Donohue and J. M. Prausnitz, AIChE J. 24, 849 (1978).
10. B. Barboy and W. M. Gelbart, J. Chem. Phys. 71, 3053

OVERVIEW

PART III. APPLICATIONS

J.M.L. Penninger
  Akzo Zout Chemie
  The Netherlands

Chapters 17 through 26 deal with a wide range of applications for supercritical fluids. Supercritical fluid technology has found its way into the field of synfuels research, and several chapters of this book focus on the production of liquid fuels from coal by treatment with supercritical fluid (SCF) solvents.

In Chapter 17, Fong and coworkers demonstrate an empirical correlation between the yield of extracted oil and second virial coefficients of SCF solvents in the extraction of Illinois #6 coal. They conclude that mixed SCF solvents will be more effective than the individual solvent components, as demonstrated by their experimental results for mixtures of toluene and methanol. A similar observation is made by Olcay and coworkers in Chapter 19 for the extraction of lignites with a supercritical fluid mixture consisting of dioxane and toluene. Oil yields from SCF solvent extraction, however, are found to be primarily a function of extraction temperature, and the highest yields have been observed in the range of 350°-400°C, where coal pyrolysis normally occurs. Thus chemical activation of the coal seems to be necessary for improved extraction yields. This is illustrated by the work of Scarrah in Chapter 18 where SCF extraction yields of only a few percent are obtained with supercritical methanol and with supercritical hexane at 265 C, but higher yields of 11% are obtained with supercritical water at 400°C. One potential advantage of SCF extraction over liquid extraction of coal appears to be the ease of separating the oils from the extraction residue. This is emphasized in Chapter 21 by Brunner and coworkers, who demonstrate the enhanced rate of sedimentation of micron-size solids in SCF solvents, and provide a conceptual description of a process for the separation of fine solids from viscous oils.

Supercritical fluid technology applied to carbon resources other than coal are presented in Chapters 20 and 25. Eisenbach and coworkers report in Chapter 20 that oil from tar sands can be recovered more selectively by extraction with SCF solvents. Their experimental results demonstrate that different qualities of oil extracts can be obtained from the same Athabasca sample, depending on the nature of the SCF solvent. The treatment of biomass with SCF solvents also appears to be promising. Köll and coworkers show in Chapter 25 that thermal decomposition of cellulose in supercritical acetone yields almost entirely soluble products, whereas pyrolysis in inert atmospheres generates substantial amounts of char. Liquefaction of wood, peat, and chitin by pyrolysis in supercritical fluids are also described in this chapter.

Specialized applications of SCF solvents involving the preparation of aerogels, regeneration of activated carbon, and dense gas or SCF chromatography are described in Chapters 22 through 24, respectively. These studies demonstrate the significance of enhanced mass transfer in supercritical fluids. The use of supercritical fluids as reaction solvents as well as reactants is described in Chapter 26, where thermal organic reactions are studied at elevated temperatures and pressures. The experimental work shows that high yields of addition products can be obtained, which are enhanced by the high density of the reactants.

CHAPTER 17

EXPERIMENTAL OBSERVATIONS ON A SYSTEMATIC APPROACH TO SUPER-CRITICAL EXTRACTION OF COAL

W. S. Fong, P. C. F. Chan, P. Pichaichanarong, and W. H. Corcoran, Chemical Engineering Laboratory, California Institute of Technology, Pasadena, California

D. D. Lawson, Jet Propulsion Laboratory Pasadena, California

ABSTRACT

Extraction of coal by supercritical fluids was studied in a semibatch reactor. A bituminous coal was extracted and the effects of temperature, pressure, and solvent types investigated. Larger molecules appeared to have greater solubilizing effects but showed mass-transfer limitations because of the small pore structure of the coal. A suggestion is made that the use of a supercritical-fluid mixture having both large and small molecules would be a practical way to improve extraction. For example, use of a small amount of methanol in toluene was found to increase the extraction, and surface-area measurements of the extracted coal supported the concept of improved extraction using the binary system.

INTRODUCTION

Extraction of coal by fluids under supercritical conditions continues to be of interest as one facet in coal liquefaction. Much work has been done nationally and internationally in the area, and specific experimental results are reported here.

The supercritical extraction process is based on improved mass transfer and increased distribution coefficients for soluble material between the solid phase and the fluid phase in the supercritical region. There is a useful equilibrium distribution of soluble material into the fluid phase, and the fluid phase in turn has sufficiently low density and viscosity so that it moves reasonably well through the pore structure of the coal.

The supercritical approach was first applied to coal conversion in the last decade by the National Coal Board in the United Kingdom [1]. The extractables can be used as liquid

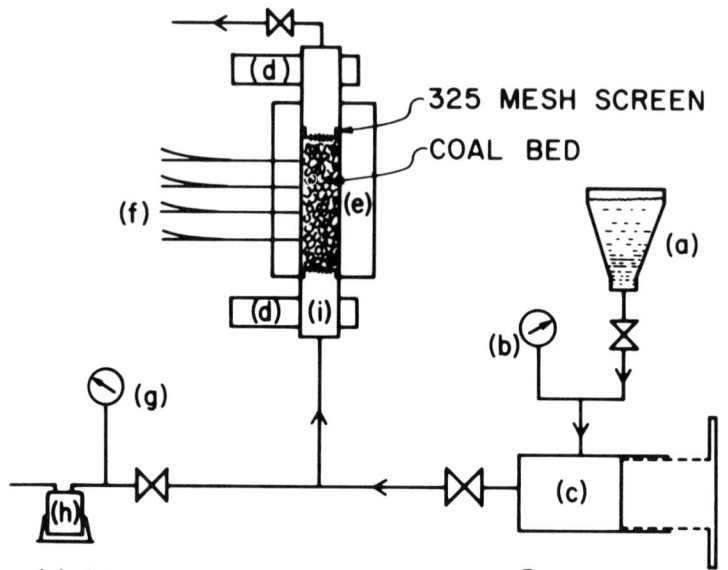

(a) SOLVENT RESERVOIR (250 cm^3)
(b) HIGH-PRESSURE GAUGE [0-13.79×10^6 Pa (2,000 PSIG)]
(c) HYDRAULIC PLUNGER
(d) COLD-WATER CIRCULATION
(e) HEATING ELEMENT
(f) THERMOCOUPLES
(g) VACUUM GAUGE
(h) VACUUM PUMP
(i) EXTRACTION TUBE

Figure 1. Schematic Diagram of Experimental Apparatus.

fuels upon hydrocracking or can be used as chemical feedstocks.

Theoretical aspects of supercritical extraction have been outlined by Paul and Wise [2], Gangoli and Thodos [3], and Bartle et al. [4] discussed structural analysis of supercritical solvent extraction of coal.

In the present work, temperature, pressure, and second virial coefficients, as empirical measures of cohesive energy, were used as parameters in investigating the extracting characteristics of various solvents. In addition, care was given to consideration of the general physical structure of the coal by observing the change in the apparent surface area of coal particles as a result of extraction [5] and by examining the extracted particles using the scanning-electron microscope. Other techniques have been used by investigators in the past such as the internal pressure of the solvent [6] and the non-polar solubility parameter [7]. The present experimental work using temperature, pressure, and the second virial coefficient is an additional attack to understanding the nature of the extraction process and provides data both supplementary and complementary to work by previous investigators.

EXPERIMENTAL PROCEDURE

Figure 1 is a schematic diagram of the experimental apparatus used in the extraction studies. The extraction tube was made of 304 stainless steel with an O.D. of 0.955 cm (0.376 in.) and an I.D. of 0.724 cm (0.285 in.). It was heated on the outside by an 800-w heater. Temperature was monitored by chromel-alumel thermocouples. In order to displace fluid in the extraction tube, a hydraulic plunger with a volumetric displacement of 15 cm^3 was used.

In the work, an Illinois Number 6, high-volatile, bituminous coal designated as PSOC 190 was used. Its mesh size was 16-24. In Table I the ultimate analysis is given on a dry basis. In the extraction, a sample of 5.5 g of the PSOC 190 coal was supported in the tube between two 325-mesh, stainless-steel screens.

Figure 2 shows the procedure used in the extraction. Experiments were conducted with 16 different solvents. Reagent-grade solvents were used and were outgassed to remove air. Coal was outgassed before adding solvent. A heating time of 15 min was used, and the cooling time was 5 min for a total of 90 min for each experiment. The extraction time was established by experiment, and it was found that the 90-min run provided reasonable equilibrium in the extraction yields. In the extraction, a total of 45 cm^3 of solvent were introduced. An increase in the solvent-coal ratio by a factor of two did not yield any significant changes in extraction.

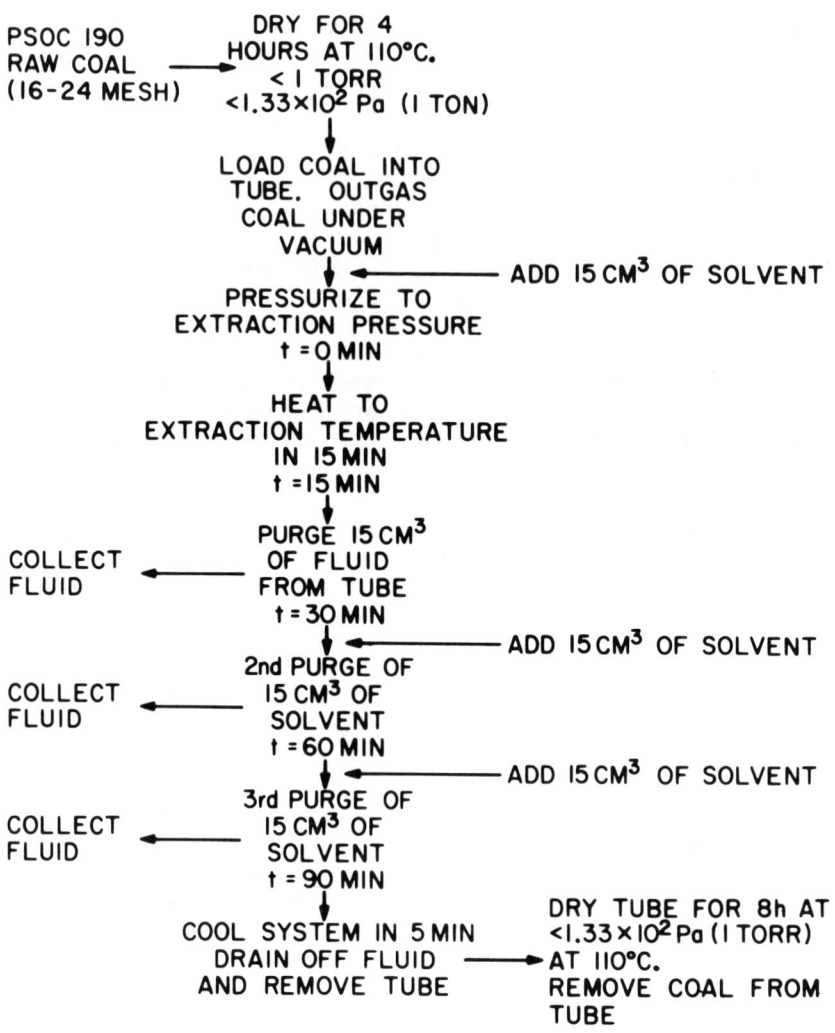

Figure 2. Procedure for Extraction.

Table I. Analysis of PSOC 190 Coal on a Dry Basis
Illinois Number 6, Bituminous

Component	wt %
C	70.4
H	4.99
N	1.36
O	7.75
S	2.86
Ash	12.6
Moisture	0.17
Volatile	36.7
Fixed C	55.3

At the end of an experiment the whole tube was removed from the system and dried in a vacuum oven. The wt % extraction was defined as:

$$\text{wt \%} = \left[1 - \frac{\text{weight of extracted coal (dried)}}{\text{weight of raw coal (dried)}} \right] \times 100$$

Residence time for the solvent beyond a certain critical level was not significant. In one set of experiments with PSOC coal, extraction at 350°C and 13.79×10^6 Pa (2000 psig) with toluene, for example, gave an extraction of 16.8 wt % for a 4-hr run and 16.4 wt % for a 2-hr run. Temperature and pressure, however, were major factors in the per cent extraction that was observed.

EXPERIMENTAL RESULTS

Effect of Temperature

The role of temperature in the extraction was examined by using toluene at a pressure of 13.79×10^6 Pa (2000 psig) and at temperatures between 274°C and 386°C. Table II is a tabulation of the results which are also plotted in Figure 3. In going from 320°C to 380°C, the amount of extraction doubled. Note should be made, however, that when the toluene was taken from a subcritical temperature past the critical temperature (318.6°C) into the supercritical region, only gradual changes in extractability were observed. That

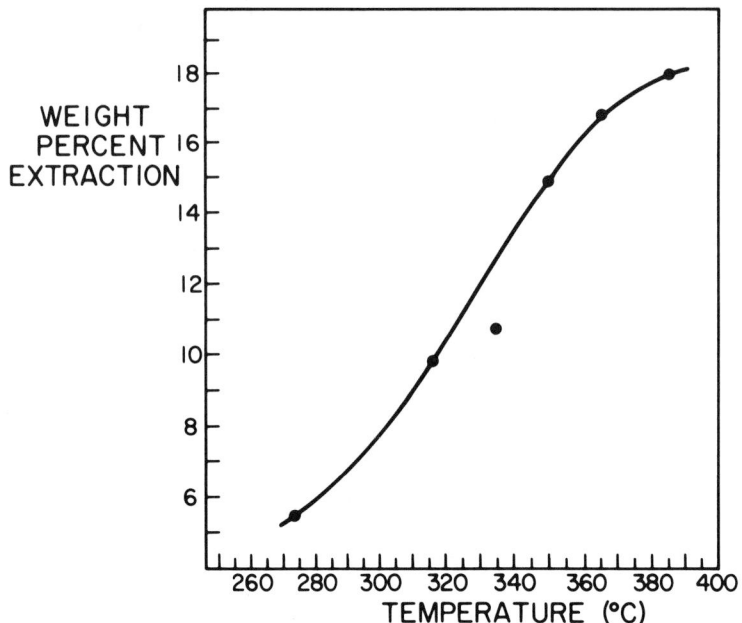

Figure 3. Extraction with Toluene at 13.79×10^6 Pa (2000 PSIG) and a Temperature Range of 274–386°C.

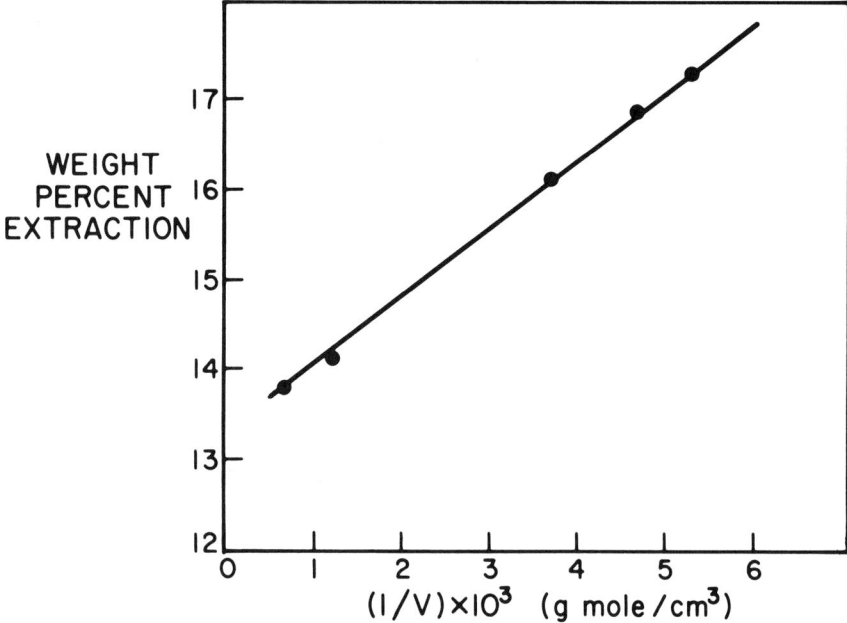

Figure 4. Extraction with Toluene at 365°C and Different Pressures as Represented by the Inverse of Molal Volumes of Solvent.

suggested the increase in extraction might not be due to changes in density or viscosity of the supercritical fluid but rather to the thermal depolymerization accompanied by increased distribution of soluble compounds into the fluid phase.

Table II. Extraction of PSOC 190 Coal by Toluene at 13.79 x 10^6 Pa (2000 psig) and Different Temperatures

Temperature (°C)	Weight Loss of Coal on Dry Basis (%)
274	5.5
316	9.8
335	10.6
350	14.8
366	16.7
386	17.8

Effect of Pressure

Toluene was also used to examine the effects of pressure in the extraction. Table III shows the results at 365°C.

Table III. Extraction of PSOC Coal by Toluene at 365°C and at Different Pressures

Pressure		Weight Loss of Coal on Dry Basis (%)
Pa x 10^{-6}	psig	
2.758	400	13.8
4.137	600	14.1
6.894	1000	16.1
9.653	1400	16.8
13.790	2000	17.2

The critical pressure of toluene is 4.104 x 10^6 Pa (595.4 psi), and so the pressure was varied from below the critical region up to 13.79 x 10^6 Pa (2000 psig). As the pressure was increased, there was a slight increase in extraction. Figure 4 gives a plot of the data with the inverse of the molal volume of the solvent, $\frac{1}{V}$, as the measure of pressure.

At 365°C and 13.79 x 10^6 Pa (2000 psig) m-cresol extracted 26.8% of the PSOC 190 coal. Toluene for the same conditions extracted 17.2% and methanol 8.7%. With the assumption that the depolymerization of the coal itself is basically a temperature function, then the experiments with temperature and pressure show that a major consideration in extraction is the nature of the distribution coefficient of a given compound between coal and the extracting fluid. The next step, then, in the sequence of study was to consider the thermodynamics of the phase equilibria between coal and the extracting fluid for given temperatures, pressures, and extracting fluids.

Phase Equilibria

A reasonable treatment of phase equilibria for coal relative to an extracting fluid is possible if an assumption is made that coal may be treated as one single compound and if critical properties can be assigned to the vapor of that compound. Just for purposes of analysis and thought, anthracene was chosen as a model coal compound, and its solubility in toluene was calculated. At 13.79 x 10^6 Pa (2000 psig) and 335°C, the method by Prausnitz [8] with third virial coefficients taken into account gave a value of 5.4% for the solubility of anthracene in toluene compared to a 10.6% solubility for coal in toluene as noted in Table II.

In order to correlate the data in the present analysis an approximate method was used. By equating the chemical potential of a pure solid designated as component 2 to the chemical potential of its vapor in the presence of a compressed fluid designated as component 1, a simple expression results if the virial equation of state of the supercritical fluid is approximated through the second term of the virial equation and if the vapor concentration of component 2 is small as discussed by Rowlinson and Richardson [9]. The resulting expression is given as follows:

$$\ln \frac{C_2}{C_{2_o}} = \frac{V_s - 2B_{12}}{V} \qquad (1)$$

Here $\frac{C_2}{C_{2_o}}$ is a ratio of concentration C_2 of the vapor of component 2 in the presence of the compressed extracting fluid, and C_{2_o} is its concentration in the absence of compressed fluid, all at the same temperature. V_s is the molal volume of solid. B_{12} is the interaction second virial coefficient, and V is the molal volume of the mixture. Although the resulting analysis is only approximate, it provides an opportunity to correlate extraction of a substance like coal through interaction with a solvent molecule. B_{12} is a measure of the Lennard-Jones potential, Γ_{12}, between molecules 1 and 2. For spherical molecules, the interaction second virial coefficient is:

$$B_{ij} = -2/3 \, \pi \, N_o \left(\frac{\sigma_i + \sigma_j}{2} \right) \qquad (2)$$

where N_o is the Avogadro number, and σ_i and σ_j are Lennard-Jones' parameters. A larger molecule will have a deeper potential well and hence will interact more strongly with other molecules. It will also then show a more negative value of B_{ij}. In reference back to Equation 1 it may be seen that a more negative value of B_{12} implies a higher value of $\frac{C_2}{C_{2_o}}$. That result agrees with the molecular relationships.

If another approximation is made that

$$B_{12} = 1/2 \, (B_{11} + B_{22}) \qquad (3)$$

then Equation 1 would have the form:

$$\ln \left(\frac{C_2}{C_{2_o}} \right) = \frac{(V_s - B_{22}) - B_{11}}{V} \qquad (4)$$

Use of Equation 4 allows two observations to be made relative to extraction of coal. First, the extraction by any supercritical solvent will increase if V decreases which would be the same as increasing the pressure. The second point is that for the same value of V, a solvent with a more negative value of B_{11}, can extract more material. Figure 4 shows the extraction by toluene as a function of $\frac{1}{V}$ at 365°C. Because the concentration of the vapor of component 2 is small, the molal volume of the total fluid, V, is equal to V_1, the molal volume of the solvent. For the range of pressures studied, the correlation gives a straight line.

The discussion above indicates that correlations can be made between the second virial coefficients of solvents with the amount of material extracted. Table IV gives extraction results for various solvents at 250°C, 350°C, and 400°C, respectively. In establishing the virial coefficients for comparison with experiment, the method by Tarakad and Danner [10] was used. For some solvents, excluding those in Table IV, their correlation of the acentric factor to find the radius of gyration was used.

Table IV. Extraction of PSOC 190 Coal at 11.03 x 10^6 Pa (1600 psig) with Different Solvents

Figures in table shown are per-cent weight loss of coal on a dry basis.

Solvent	250°C	350°C	400°C
methanol	3.9	9.2	16.8
ethanol	4.1	9.7	19.7
1-propanol	3.6	13.9	20.3
1-butanol		16.6	
1-hexanol	1.8	22.7	25.9
1-octanol	0		22.3
benzene	5.1		21.6
toluene	5.3	16.2	20.8
xylene	2.7		20.7
tetralin	2.0		82 (+)
pyridine	6.1		21.5
acetone		6.6	
pentane		11.4	18.5
hexane		11.4	20.2
heptane			
decane		12.2	18.6

(+) Extensive degradation of coal. Small particles of coal escaped from trap.

Figure 5 shows the plot of per cent extraction as a function of $-B_{11}$ calculated by the techniques of Tarakad and Danner [10]. For lower values of $-B_{11}$ the curves are more or less straight lines. A maximum is reached, and then the extraction decreases at higher values of $-B_{11}$. A possible explanation is that for solvents with large values of $-B_{11}$, they could, in principle, dissolve larger amounts of coal, but their larger size prevents them from moving through the pore structure of the coal. In a continuation of that point, the process of supercritical extraction can be broken down into four steps as noted by Gangoli and Thodos [3]. Their four steps were:

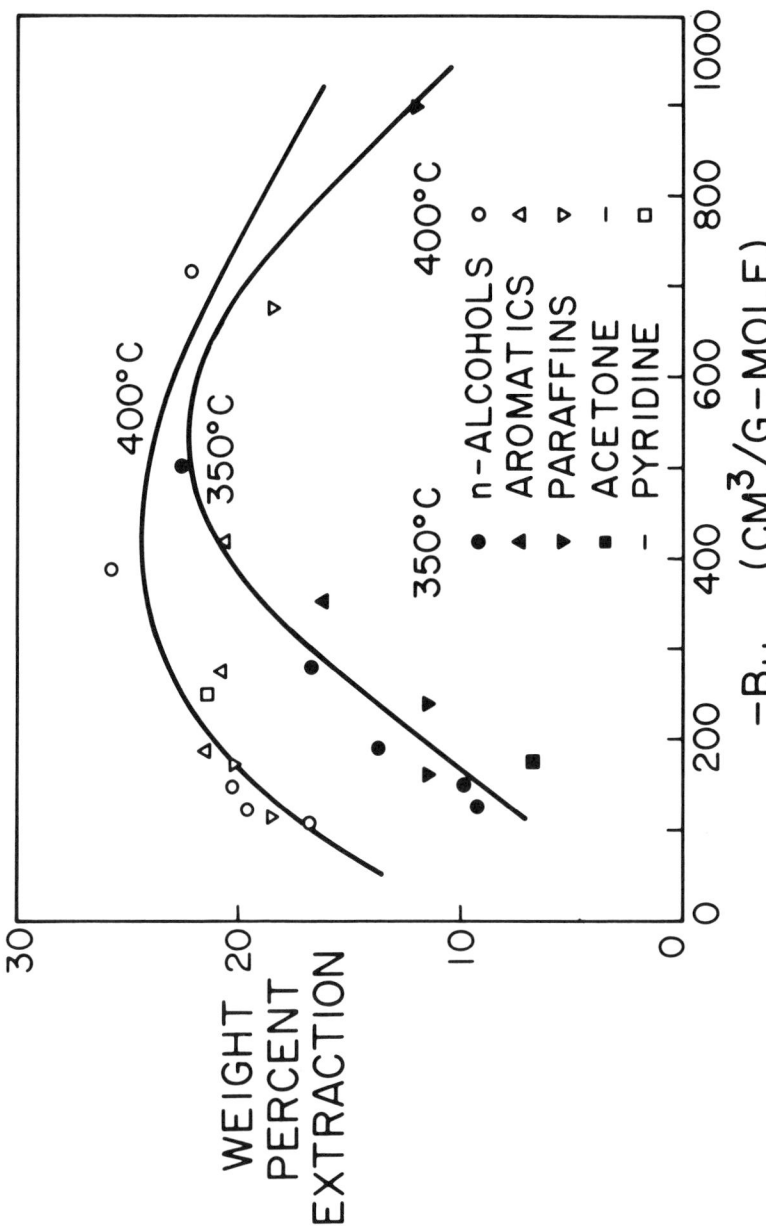

Figure 5. Extraction of Coal at 11.03×10^6 Pa (1600 PSIG) as a Function of the Second Virial Coefficients of Solvents.

1. Penetration of the coal-micropore structure by the solvent molecules.
2. Depolymerization of large molecular aggregates and the dissolution of the resulting products in the solvent.
3. Breaking up of the resulting molecular bonds between the coal structure and the molecular species to be extracted.
4. Diffusion of the extract and the fluid from the pore structure of the coal.

Consistent with the above, Table IV shows that 1-octanol, for example, extracts less than 1-hexanol probably because of the relative size of the molecules. A similar result was developed in a comparison between hexane and decane. For very large molecules, step 1 in the four steps from Gangoli and Thodos [3] cannot be effected. As part of the present analysis it is worthwhile to examine the data of Kiebler [6]. His extraction results have been correlated with $-B_{11}$ and are shown in Figure 6. Napthalene and biphenyl are much larger molecules than phenol, m-cresol, or aniline. Ostensibly they are thus not able to move easily through the pore structure of the coal.

The concept of molecular size and shape needs to be examined relative to pore-size distribution. For the PSOC 190 coal that was used, the pore-size distribution is shown in Figure 7. The average pore diameter was found to be 31 Å [11]. In measurements on another Illinois Number 6, high-volatile, bituminous coal it was observed that 30.2% of the pore volume was smaller than 12 Å; 52.6% was between 12 Å and 300 Å; and 17.2% was above 300 Å [12]. Examination of the longest dimension of each solvent molecule is worthwhile. The length of n-heptane is 9.5 Å, that of 1 octanol about 11 Å, and that of naphthalene about 7 Å. These dimensions are comparable to the average pore diameter. Benzene, methanol, and ethanol are not larger than 4.5 Å. It is apparent that penetration of the micropore structure of the coal can be achieved more easily by the smaller molecules.

Surface Area of Extracted Coal

Table V shows the resulting apparent surface area of coal after extraction by various solvents in the present experiments. The areas were established by the single-point-BET method using nitrogen adsorption. Note is made that the surface area was decreased by about 97% for aromatic solvents to 99.5% for alcohols. It was shown that coal extracted with smaller molecules showed smaller apparent surface areas at the end of extraction and that coal extracted at higher temperatures showed larger areas. The large decrease in apparent surface areas probably results from the retention of extract in the

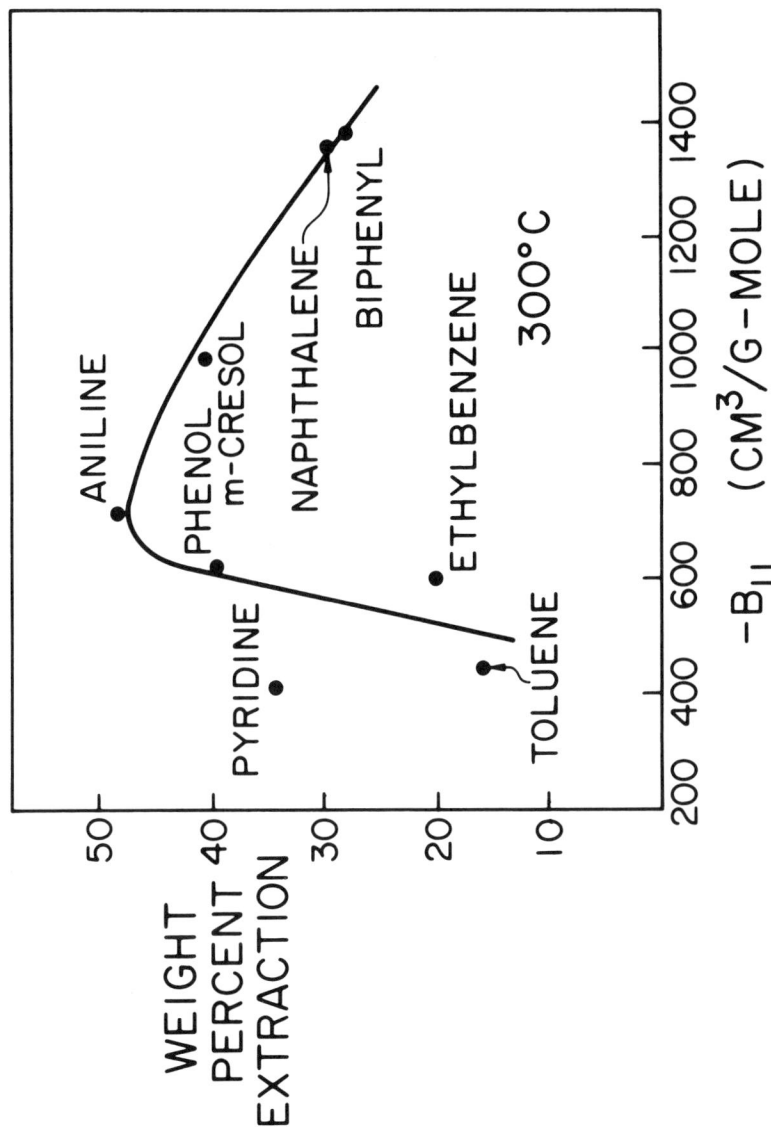

Figure 6. Correlation of Kiebler's Work [6] at Various Bomb Pressures Using Second Virial Coefficients.

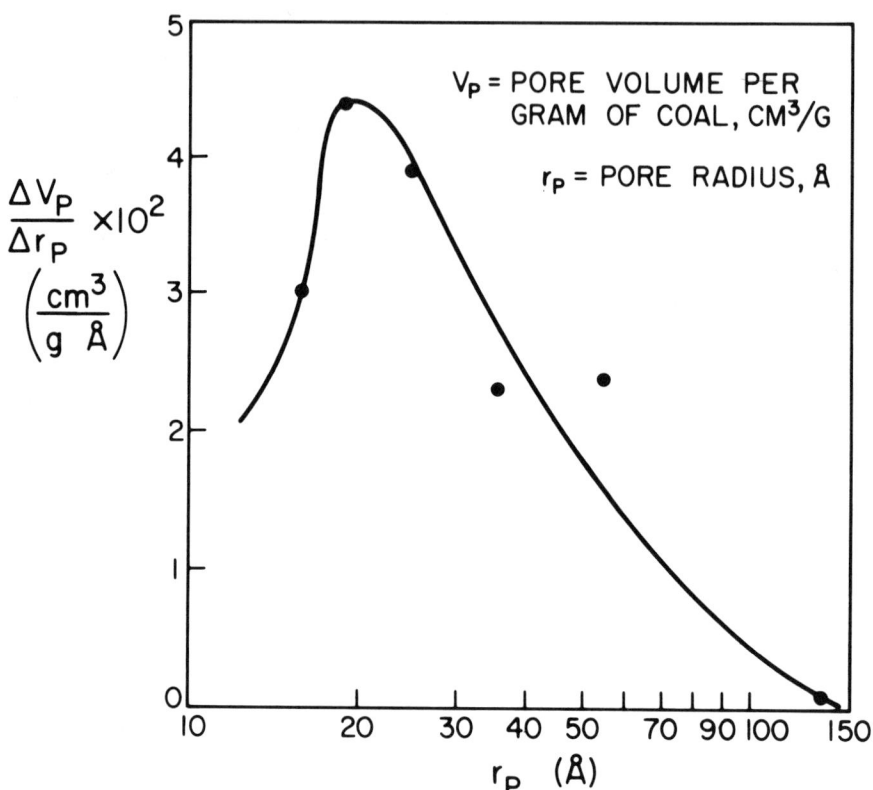

Figure 7. Pore Size Distribution of PSOC 190 Coal.

Table V. Surface Area (m^2/g) of Residual PSOC 190 Coal After Extraction at 11.03×10^6 Pa (1600 psig) (Measured by H_2 Adsorption)

Solvent	350°C	400°C
Untreated raw coal	45.0	
methanol	0.219	0.228
ethanol	0.202	0.459
1-propanol	0.207	
1-hexanol	0.482	0.542
1-octanol		0.658
acetone	0.273	
toluene	1.16	1.46
benzene		1.03
xylene		1.12
pentane	0.445	

pores. Upon removal of the fluid at the end of an extraction cycle, the relative volatility of the light alcohols, for example, is greater than that of heavy fluids, and so they could more readily leave the pores than the heavy fluids. They would leave condensed extract in the pores with resulting blockage. At higher temperatures, a heavier fluid in the pores could volatilize at a reasonable rate and leave larger surface areas as observed.

Use of Two-Component Solvents

The thoughts expressed above about the relative size of fluid molecules and of pores in the coal structure suggest that larger solvent molecules could be mixed with smaller molecules in order to optimize extraction yield. This thought suggests, for example, that a light molecule like carbon dioxide could be mixed with, say, naphthalene to achieve higher extraction. This suggestion has not been verified specifically, but in earlier work there was increased extraction in a methanol-toluene solvent pair as shown in Figure 8 and tabulated in Table VI. A peak in the wt % extraction is shown at about 5 wt % methanol in the toluene-methanol system.

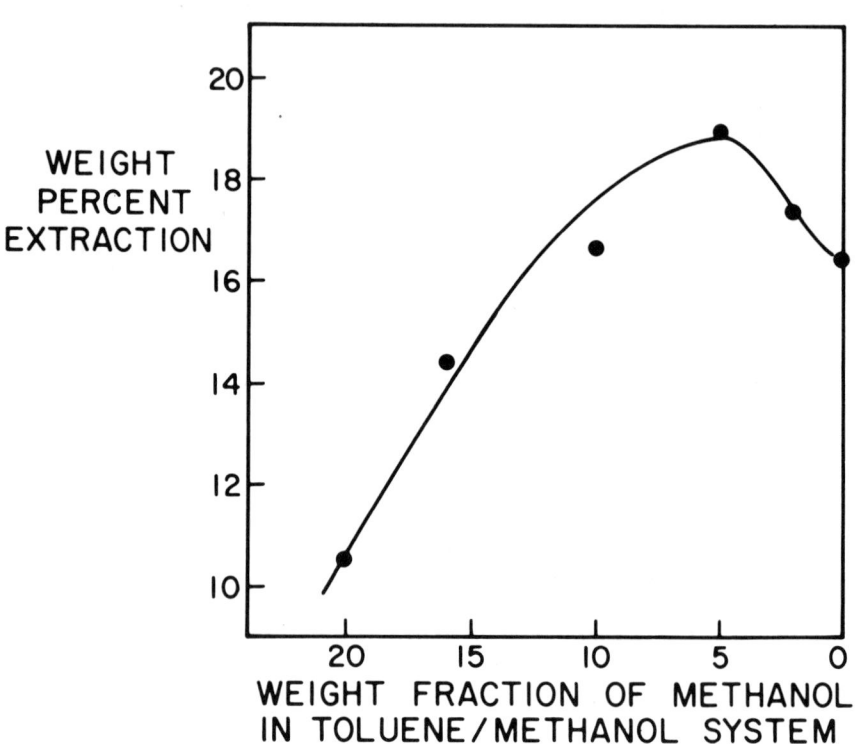

Figure 8. Mixed Solvents in Extraction of a Coal at 13.79 x 10^6 Pa (2000 PSIG).

Table VI. Extraction with a Toluene/Methanol Mixture at 360°C and 13.79 x 10^6 Pa (2000 psig)

Volume Ratio, Methanol/Toluene (at room temperature)	Weight Loss of Coal (%)
0/100	16.4
2/98	17.4
5/95	19.0
10/90	16.6
16/84	14.4
20/80	10.5

CONCLUSIONS

A systematic approach to solvent extraction of coal is possible by consideration of empirical correlations using second virial coefficients as has been displayed in Figures 5 and 6 of this paper. The resulting maxima in extraction at a given pressure for various temperatures and values of the second virial coefficient are explained by solvent size. Further experiments using binary and multicomponent systems to provide optimum diffusion into and out of the coal structure during the course of the extraction process would appear to be useful. The optimum performance has already been displayed by the methanol-toluene system as shown in Figure 8. Further experiments with binary and multicomponent systems to examine the relationships among the second virial coefficients, solvent size, temperature, and pressure appear in order.

ACKNOWLEDGMENT

Portions of the work reported here were supported by the Bechtel Corporation and other portions by the Jet Propulsion Laboratory of the California Institute of Technology under NASA Contract NAS 7-100. The support of the two groups is gratefully acknowledged.

REFERENCES

1. Whitehead, J. C. "Development of a Process for the Supercritical Gas Extraction of Coal," presented at 88th National Meeting of the American Institute of Chemical Engineers, Philadelphia, June, 1980.

2. Paul, P. F. M., and W. S. Wise *The Principles of Gas Extraction* (London: Mills & Boon, Monographs CE/5, 1971).

3. Gangoli, N., and G. Thodos *Ind. Eng. Chem. Process Des. Dev.* 16:208 (1977).

4. Bartle, K. D., W. R. Ladner, T. G. Martin, C. E. Snape, and D. F. Williams *Fuel* 58:413 (1979).

5. Harris, E. C., Jr., and E. E. Petersen *Fuel* 58:599 (1979).

6. Kiebler, M. W. *Ind. Eng. Chem.* 32:1389 (1940).

7. Angelovich, J. M., G. R. Pastor, and H. F. Silver, *Ind. Eng. Chem. Process Des. Dev.* 9:106 (1970).

8. Prausnitz, J. M. *Molecular Thermodynamics of Fluid-Phase Equilibria* (New Jersey: Prentice-Hall, 1969).

9. Rowlinson, J. S., and M. J. Richardson *Adv. Chem. Phys.* 2:85 (1959).

10. Tarakad, R. R., and R. P. Danner *AIChE J.* 23:685 (1977).

11. Wang, J. K. "Supercritical Extraction of Coal," Master's Thesis, California Institute of Technology, Pasadena, CA (1980).

12. Gan, H., S. P. Nandi, and P. L. Walker *Fuel* 51:272 (1972).

CHAPTER 18

LIQUEFACTION OF LIGNITE USING LOW COST SUPERCRITICAL SOLVENTS

Warren P. Scarrah
Department of Chemical
Engineering
Montana State University
Bozeman, Montana

INTRODUCTION

The separation of coal components using supercritical fluid extraction (SCFE) is attractive because it avoids the use of reactants and catalysts associated with most coal liquefaction processes. This study was conducted to investigate the use of inexpensive solvents.

The raw materials used for this experiment included Beulah, North Dakota lignite whose properties are shown in Table I.

Table I. Properties of Beulah 3, North Dakota Lignite

Proximate Analysis (as received)		wt %
Moisture		28.22
Volatile Matter		29.18
Fixed Carbon		30.84
Ash		11.76
	Total	100.00
Ultimate Analysis (moisture free)		wt %
Hydrogen		3.62
Carbon		57.81
Nitrogen		0.85
Oxygen (by difference)		19.05
Sulfur		2.29
Ash		16.38
	Total	100.00

The distinguishing characteristics of this coal were its high moisture (28.2 wt %), ash (11.8 wt %), and oxygen (19.1 wt %, H_2O-free) contents and low sulfur (2.3 wt %, H_2O-free). The coal was ground to a minus 60-mesh (U.S. Sieve Series). The solvents included carbon dioxide, methanol, and water--the primary consideration was cost and all three are inexpensive. As explained below they are most effective at a temperature close to their critical temperature which is 31 °C for carbon dioxide, 240 °C for methanol, and 374 °C for water. Critical pressures are 72.8, 79.9, and 217.6 atmospheres respectively. Because 31 °C is too low for coal extraction, carbon dioxide would be useful primarily in combination with water; however, at any appreciable carbon dioxide level, the critical pressure of the carbon dioxide-water mixture is so large that supercritical conditions cannot be attained at reasonable pressures. Therefore, the only mixture investigated contained 4 mol % carbon dioxide. Methanol and water have specific hydrogen bonding properties so it was decided to test hexane for comparison purposes. Hexane has a critical temperature of 234 °C and a critical pressure of 29.3 atmospheres; its solvent power is based on nonspecific London dispersion forces. Thus the validity of treating SCFE as a physical rather than chemical phenomenon could be explored.

The principal process variables were temperature and pressure. As in many rapidly growing fields, SCFE applications are developing much faster than the accompanying theory. Earlier interpretations explained the extremely high solvent power of supercritical solvents for solid solutes in purely physical terms [1]. The enhanced solubilities were partially due to the effect of pressure (Poynting correction); however, it was shown that a larger effect could be due to the second cross-virial coefficient for the gaseous mixture. When empirical estimates of the cross-virial coefficient were made considering only dispersion forces between molecules, it was shown that the solubility enhancement was exponentially proportional to the density of the gaseous mixture. For pure components the density of the gaseous phase is greatest at the critical temperature as long as the pressure is greater than the critical pressure. This state is commonly referred to as a supercritical fluid and is characterized by a density similar to a liquid, a viscosity similar to a gas, and a diffusivity between that of a gas and a liquid. It was also shown that the temperature should be increased to raise the vapor pressure and thus the concentration of the solid solute in the supercritical fluid but conversely it should be decreased to raise the density of the supercritical fluid. The guidelines developed were followed in this investigation and stated that a solvent should be used slightly above its critical temperature and selected so that the solute is relatively volatile at this temperature. The pressure should be as high as possible but certainly greater than the critical pressure. With a feed material as heterogeneous as lignite, some components should be relatively volatile at any reasonable temperature.

This investigation was conducted in two stages: (1) batch exploratory experiments and (2) semi-continuous variable screening experiments. The purpose of the former was to evaluate the process potential of SCFE for this application with little effort devoted towards product characterization. The latter set of experiments had the objective of exploring the effects of process variables.

BATCH EXPLORATORY EXPERIMENTS

Equipment

The equipment used for this series of experiments consisted primarily of a 1-liter stirred-tank reactor fabricated by Autoclave Engineers. The lignite was placed in an alundum thimble which in turn was held in position using a sample holder (Figure 1). The groove in the top of the sample holder was to allow for the expansion of the alundum thimble. The advantages of this equipment were simplicity and ease of analysis--the total material that was extracted could be readily determined (see Analyses below). The disadvantages included poor pressure control, the inability to obtain high pressures, and the possibility of solvent saturation.

Procedure

Approximately 60 grams of lignite were placed in the alundum thimble and the thimble and any liquid solvent placed inside the reactor. After the reactor was sealed, any gases to be added were introduced at that time. The heat-up period was approximately 1 hour for 265 °C experiments and 2 hours for 400 °C experiments. After attaining the desired temperature the reactor was held for 1 hour; the vapors were then removed from the reactor and bled through an ice water trap, a dry ice-isopropanol trap, and a wet test meter.

Analyses

The feed lignite for each experiment was analyzed for moisture and ash. Product yields from the cold traps were

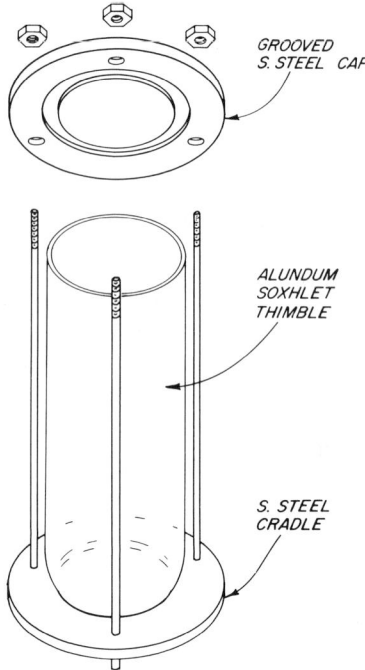

Figure 1. Sample Holder for Batch Supercritical Fluid Extraction.

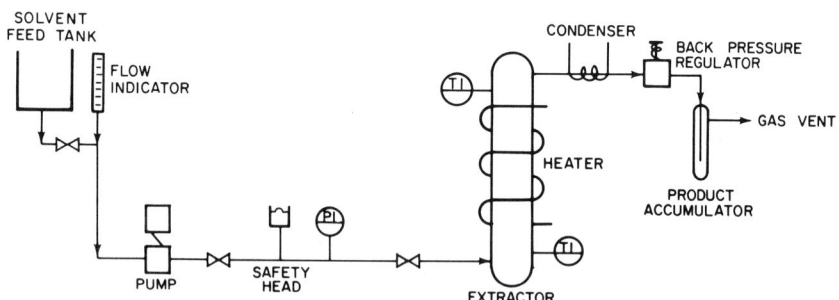

Figure 2. Flowsheet for Semicontinuous Supercritical Fluid Extraction.

combined with any organic materials washed from the reactor walls and evaporated. If the solvent had been water, the sample was first extracted with methylene chloride and then the methylene chloride evaporated in a rotary vacuum evaporator. Organic solvents were evaporated directly in the rotary vacuum evaporator. Yield was defined as the organic material remaining in the evaporator. The lignite residue remaining in the thimble was extracted using a Soxhlet apparatus with tetrahydrofuran. The thimble was then oven dried at 110 °C and weighed. An ash analysis was made on a portion of the dried residue to put the extraction data on a moisture-and-ash free basis. Extraction was defined as the difference between the organic material charged and that remaining after supercritical fluid extraction. For selected experiments, elemental analyses were made on product yields and residue prior to ashing.

Results and Conclusions

Water Runs

In addition to investigating the use of water as a supercritical solvent, this set of runs was also used to develop procedures and examine the effect of carbon dioxide addition. It is obvious from Table II that there was no apparent difference between placing the coal loose in the reactor or retaining it in the alundum thimble. Also small amounts of carbon dioxide did not seem to improve extraction. Finally, it was apparent there were large differences between the total amount of organics extracted and the yields attained.

Table II. Batch Supercritical Fluid Extraction of Lignite Using Water

Temp = 400 °C (T_r = 1.04)
Pres = 4100 psig (P_r = 1.30)

	Extraction (wt % MAF)	Yield (wt % MAF)
Lignite Inside Thimble	39.7	10.7
Lignite Loose	33.6	11.4
96 mol % H_2O, 4 mol % CO_2	40.3	10.2

Methanol Runs

Because of its lower critical temperature, the methanol experiments were made at 265 °C compared to 400 °C for water. In addition to investigating the use of methanol, a comparison was also made of the effect of drying the lignite prior to supercritical fluid extraction rather than using the as-received lignite. It is apparent from Table III that the yields were approximately the same regardless of whether dry or wet lignite was used, but the extractions varied considerably.

Table III. Batch Supercritical Fluid Extraction of Lignite Using Methanol

	Extraction (wt % MAF)	Yield (wt % MAF)
Dry Lignite	5.4	3.1
As Received Lignite	38.6	3.5

This constancy of yields is in agreement with data showing little effect of moisture on yields varying between 2 and 15 wt % when supercritical toluene was used to extract a Wyodak coal [2].

Hexane Runs

Hexane has approximately the same critical temperature as methanol so reaction temperatures were similar. This series of experiments was also used to gain an appreciation for the reproducibility of the experimental procedures. It is seen from Table IV that yields were consistent but one total extraction was quite high (~ 55 wt %) compared to those of the other two runs (~ 8 wt %). Because the 55-wt % total extraction had been obtained from the initial hexane run and was significantly higher than the 40-wt % total extractions attained using water and methanol, the two replicate hexane runs were made to verify the results of initial run. Obviously an error had been made during the initial run - it probably occurred during a weighing. In addition to emphasizing the need for experimental replication, this experience also promoted confidence in using a recoverable product (yield) rather than a disappearing reactant (total extraction) as a basis for evaluating experimental results.

Table IV. Batch Supercritical Fluid Extraction
of Lignite Using Hexane

Temp = 265 °C (T_r =1.06)
Pres = 3050 psig (P_r = 6.97)

Extraction (wt % MAF)	Yield (wt % MAF)
54.9	3.4
7.7	3.5
8.7	3.2

Solvent Comparison

The three solvents, water (including water-carbon dioxide), methanol, and hexane are compared in Table V.

Table V. Supercritical Solvent Comparison for the
Batch Extraction of Lignite

Solvent	Avg. Yield (wt % MAF)	Temp (°C)
H_2O & $H_2O\text{-}CO_2$	10.8	400
Methanol	3.3	265
Hexane	3.4	265

At first glance it appears the yields with water (~11 wt % MAF) were much higher than with the two organic solvents (~3 wt % MAF). However, this superiority is misleading because of the high temperature (400 °C) required with water. At 300-350 °C lignite pyrolyzes to form oils and tars. Therefore, part of the higher yield shown with the water experiments was due to pyrolysis rather than supercritical fluid extraction.

Product Characterization

Elemental analyses of the starting lignite and two representative experiments were used to gain some appreciation for the type of product that was being extracted. From Table VI, it is obvious that the lighter hydrocarbons were removed from lignite by supercritical fluid extraction. The hydrogen:carbon ratio for the products using both water and hexane was about 1.55:1. This ratio was considerably higher than the 0.75:1 hydrogen:carbon ratio in the starting lignite. Also, the residue had a lower hydrogen:carbon ratio (0.52-0.66:1) which showed that lighter material had been removed. The lower carbon and hydrogen contents in the product yield using water compared to the product yield using hexane indicated that much less oxygen was present in the product when hexane was used.

Table VI. Product Characterization from Batch Supercritical Fluid Extraction of Lignite

(MAF basis)

	C (wt %)	H (wt %)	H:C (mol ratio)
Starting Lignite	69.13	4.33	0.75:1
Supercritical H_2O:			
Yield	54.59	7.08	1.56:1
Residue	83.93	3.61	0.52:1
Supercritical Hexane			
Yield	79.92	10.28	1.54:1
Residue	73.41	4.04	0.66:1

Overall Conclusion

The batch supercritical fluid extraction experiments are suitable for exploratory process evaluations but yields and product characterizations are questionable. There was a possibility that the solvent could have become saturated due to the limited amount of solvent that could be contained in the reactor; also, during the pressure let-down there existed a dynamic condition that could have allowed some of the solute that had already been extracted to recondense as the pressure was reduced.

SEMI-CONTINUOUS VARIABLE SCREENING EXPERIMENTS

Equipment

Figure 2 is a flowsheet showing the semi-continuous extraction apparatus. Approximately 30 grams of lignite were placed in a 1-inch diameter, stainless steel extraction tube. The solvent was continuously pumped in with a high pressure pump. One undesirable component was the cooler on the discharge stream between the extractor and the back pressure regulating valve. Because the valve contained plastic and rubber, it was necessary to reduce the temperature of any material flowing through it to prevent malfunction of the valve. Preferably, the temperature would not have been reduced until after the pressure was reduced. The advantages of this semi-continuous unit were good pressure control, no concern about solvent saturation, and a product collection method allowing reliable determination of yields. The disadvantage was that the equipment was moderately complex compared to batch experiment apparatus. It required continual operator attention, longer experimental run times, and contained more components that could malfunction.

Procedure

The system was initially filled with room-temperature solvent while running the pump until the back pressure regulating valve had been adjusted to the desired pressure. The solvent pump was shut off and the extractor heated to the desired temperature which took approximately 1½-2 hours. The pump was then started at a feed rate of approximately 2 milliliters per minute and the product collected in a separable vacuum trap placed in an ice water bath. Receivers on the accumulator were replaced whenever 100 milliliters of discharge material had been collected. Crystals of product were seen to settle to the bottom of the receiver and discolorization of the solvent indicated that some solute remained in solution.

Analyses

Moisture determinations were made on all lignite charged to the extractor and ash analyses were made about once a week. Two methods were used to evaporate the solvent from the product that had been collected in the accumulator. Initially the material was subjected to oven drying at 110 °C. It was observed that the product deteriorated (probably due to oxidation) and some of the more volatile material was lost. Subsequently, all product was subjected to rotary vacuum evaporation to retain the characteristics of the product. If enough product yield was present, it was extracted with hexane in a micro-Soxhlet apparatus to estimate the amount of oils present. Product was retained for characterization by nuclear magnetic resonance (NMR) spectroscopy; this portion of the investigation has just begun. The Department of Energy's Grand Forks Energy Technology Center will be using ^{1}H-NMR and Montana State University using ^{13}C-NMR to examine the samples.

Results and Conclusions

Methanol Runs

Three temperatures and two pressures were examined (Figure 3). Because of the relatively high critical pressure of methanol our apparatus only allowed a reduced pressure range of 1.5-2.5 to be examined. It is apparent that at reduced pressures greater than 2.5 an increase in temperature resulted in an increased product yield. At reduced pressures less than 1.5 an increase in the temperature caused a decreased product yield. This reversal in the temperature effect on product yields with increasing pressure agrees with a previous discussion of the supercritical extraction of napthalene with ethylene [1]. Also, as several of the experiments were repeated, the variation in yields showed that replicates should be made for each experimental condition.

Hexane Runs

Two temperatures and two pressures were examined during these tests. Because the critical pressure of hexane is considerably lower than that of methanol, a much larger reduced-pressure range could be examined (2.0-5.0). As shown in Figure 4, at reduced pressures less than 2.0 an increase in temperature resulted in a decreased yield.

Methanol-H_2O Runs

It was decided to combine two solvents with somewhat different specific solubility properties. Methanol is both a strong proton donor and proton acceptor while water is a strong proton donor but a somewhat weak proton acceptor. As shown in Figure 5, water concentrations less than about 13.5 wt % did not seem to affect yields. The primary effect of the water addition appeared to be a reduction in the critical temperature in the solvent mixture. This statement is based on a comparison of Figures 3 and 5. At all water contents it is obvious that the yield is greater at the higher pressure than at the lower pressure (Figure 5). However, in Figure 3 it is obvious that at 250 °C there was very little difference in yields at the two pressures. The 250 °C temperature was very close to the critical temperature of the pure methanol while it was probably above the critical temperature of the methanol-H_2O mixture. This reduction in product yield as the temperature was raised above the critical temperature while maintaining low reduced pressures (< 1.5) is also in agreement with the effect of

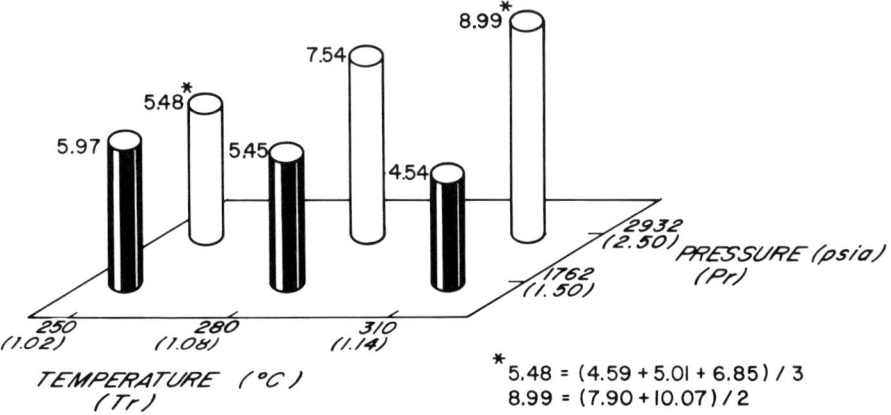

Figure 3. Yields Using Supercritical Methanol (wt % MAF lignite).

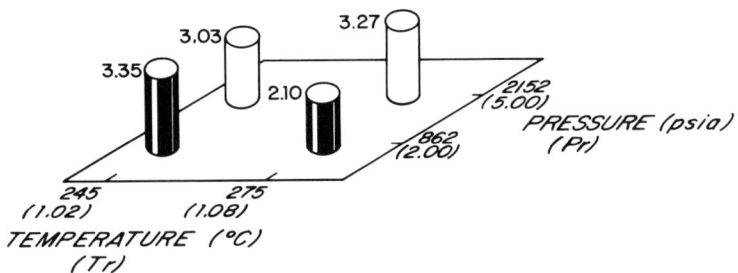

Figure 4. Yields Using Supercritical Hexane (wt % MAF lignite).

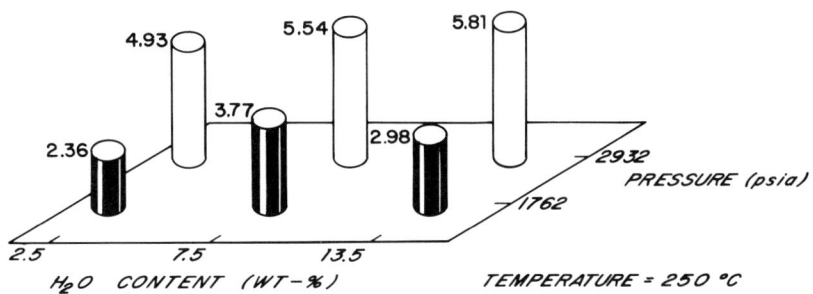

Figure 5. Yields Using Supercritical Mixtures of Methanol and Water (wt % MAF lignite).

temperature on the previously-mentioned napthalene-ethylene system [1]. Finally, the yields obtained using oven drying seemed to be larger than those using vacuum evaporation. At 250 °C, the data presented in Figure 3 (oven drying) shows that the yields were close to 5 wt % for both pressures while in Figure 5 (vacuum evaporation) the yields were only about 2 wt % for the lower pressure and 5 wt % for the higher pressure. It was probable that oven drying led to oxidation of the product and thus to higher yields.

Overall Conclusions

Significant hydrocarbon yields can be obtained from lignite using inexpensive supercritical solvents at temperatures well below those required for pyrolysis and solvent refining. Also, the reduced pressure at which the temperature effect on yields is reversed is probably unique for each supercritical solvent.

FUTURE

Continued work at Montana State University will be directed as follows:

1. Significant effort will be expended to develop product characterization capabilities.
2. Supercritical fluid mixtures will be investigated.
3. Product fractionation using sequential pressure changes will be explored.
4. Semicontinuous process equipment and procedures will be modified to reduce experimental variations.

ACKNOWLEDGEMENT

The support of these projects by the U.S. Department of Energy's Grand Forks Energy Technology Center (GFETC) is gratefully acknowledged. The cooperative arrangement between the Associated Western Universities and GFETC provided the opportunity for the investigation to be initiated. The laboratory assistance of Theodore Poppke, Muthuswamy Rameswaran, and James Heath was outstanding.

REFERENCES

1. Paul, P. F. M., and Wise, W. S. *The Principles of Gas Extraction*, (London: Mills and Boon Ltd., 1971), pp. 23-25.

2. Maddocks, R R., J. Gibson and D. F. Williams. "Supercritical Extraction of Coal," Chem. Eng. Prog. 75(6): 49-55 (1979).

CHAPTER 19
THE SUPERCRITICAL GAS EXTRACTION
OF LIGNITES AND WOOD

A. Olcay

T. Tuğrul

A. Çalımlı

Faculty of Science

Ankara University, Ankara, Turkey

ABSTRACT

The chemical nature and yield of supercritical gas (SCG) extracts from Turkish coking coal, lignites and softwood obtained with different solvents at various pressures, were investigated with a view to assessing the potential of SCG extraction as an alternative to pyrolysis for chemicals and liquid fuels production. SCG extracts were subjected to solvent extraction and subsequently to preliminary separation by column chromatography. Low molecular weight compounds were characterized by GC/MS combined system.
The results are discussed in terms of the prevailing views on the composition of coal, lignite and wood.

INTRODUCTION

Over the past 60 years a wide range of aromatic compounds has been obtained by processing petrochemicals; the energy demand has been met by petroleum due to the ease of transportation and handling. However, in recent years because of limited resources of petroleum, there has been a growing interest to explore new energy sources as an alternative to crude oil and substantial research is carried out for better utilization of resources.
Coal, the most important of the fossil fuels with relative-

ly large reserves when compared to oil, will certainly not only be the main source for power generation but also will furnish the basic raw materials for many essential industries.

On the other hand, wood provides a renewable resource for both energy and chemicals. Today, the most important use of wood is the production of pulp. In the pulping process lignin, which is present in wood in an amount of 19-23 % in intimate association with structural polysaccharide (1), is removed with the waste water. The annual waste of lignin amounts to 40 million tons which is equivalent to 24 % of the carbon used in the chemical industry. So in the short term, lignin being a complex polymer of substituted phenols, could be regarded as a renewable source of phenols. In the long term, wood can be regarded as a potential source of petro-chemicals.

Chemicals from both coal and wood were initially and mostly obtained by destructive distillation. However, since the oil crisis of 1973, many different processes for the conversion of coal to liquid fuels are being developed (2). One of the promising processes is the extraction of coal and wood with supercritical gases. The objective of this work was to study the influence of solvent, extraction pressure and the coal rank on the nature and the yields of the coal extract and on the yield of phenols obtained from wood. Recently, we have shown that supercritical gas extraction of wood and coal with different solvents produces 23 % of soluble material (3, 4) which is only slightly cracked. Destructive distillation of wood and coal on the other hand yields less distillate but more gas and char or coke.

EXPERIMENTAL

Supercritical extraction was carried out in a one liter autoclave. The weight ratio of the solvent to wood, lignite or coal was 10 to 1. The temperature of the autoclave was gradually increased and the final temperature was maintained for 30 minutes. According to the theory of supercritical gas extraction, the volatility enhancement is greatest when the extraction temperature is near the critical temperature of the solvent (5). Therefore, the extraction was carried out at a temperature 15 to 20°C higher than the critical temperature of the solvent. The time required to reach the extraction temperature was about an hour. The gas phase containing volatile compounds was withdrawn through a valve into a condenser and collected in a flask at atmospheric pressure. The solvent in the extract was removed by evaporation in a rotary evaporator at 60-80°C.

To investigate the reactions which occur during the heating period, lignite and solvent were heated to 10°C below the critical temperature of the solvent under exactly the same conditions as those used for the supercritical extraction, but not kept at the final temperature. The amount of distillate was very small indicating that during heating period little pyrolysis of lignite occurs.

The extract of both coal and wood contains a wide range of compounds with different molecular weights, including different functional groups and accordingly with different polarities. For complete resolution of all coal and wood extract, efficient separation methods should be used in connection with mass spectrometry. In this work the extracts were fractionated either chemically into acidic, basic and neutral compounds or were partitioned by solvent extraction to yield n-pentane solubles, n-pentane insoluble/benzene solubles and n-pentane/benzene insolubles. Each fraction was then subjected to column chromatography on silicagel and by successive elution with n-pentane, benzene and methanol three fractions were obtained.

The low-molecular-weight compounds present in each fraction were analyzed by GC/MS. A Packard 427 gas chromatograph was fitted with a 1-m column composed of 3-5 % SE-30 on chromosorb W-AW (80-100 mesh) and coupled through a capillary glass jet separator to a Finnigan 3000E quadrupole mass spectrometer, helium was used as the carrier gas. Ionization was carried out at 70 eV electrons and at a gas pressure of 1.06 mPa.

RESULTS

Table I records the analysis of the three lignites, the coking coal and spruce wood used in the supercritical gas extraction.

Table I. Analysis of coal, lignites and spruce wood.

	Zonguldak coking coal	Tunç-bilek lignite	Seyit Ömer lignite	Elbistan Afşin lignite	Spruce wood
C(wt % daf)	85.7	70.7	68.5	64.2	50.1
H(wt % daf)	5.2	4.7	5.0	5.4	6.2
S(wt % daf)	0.6	2.1	2.6	6.5	-
N,O(wt % daf)	8.5	22.5	23.9	23.9	43.7
Ash (wt %)	14.5	13.7	15.2	23.8	-
Moisture(wt %)	0.5	4.9	21.9	14.3	-
Volatile matter (wt % daf)	30.5	43.4	49.8	45.4	-
Fixed Carbon (wt % daf)	69.5	56.6	50.2	54.6	-

In Figure 1, the yields obtained with supercritical dioxane from Tunçbilek lignite (at 330°C) and with supercritical toluene from Zonguldak coking coal (at 360°C) are plotted against pressure. The results show that the yield of extracted oil increases considerably with the pressure from 8 to 15 MPa. This is in accordance with the well known fact that the solvent power of the compressed gas increases with increasing density (5).

Increase of the temperature from 350°C to 400°C results in a yield increase of almost 50 % when Tunçbilek lignite is extracted with toluene. This might be due to the mild pyrolysis of the lignite at 400°C. Hence if the objective of the supercritical gas extraction of coal is to obtain liquid fuels, then the extraction should be carried out at temperatures at which pyrolysis of coals is favored. In supercritical gas extraction, the pyrolysis products can be recovered as they formed and the undesirable polymerization of primary compounds can be avoided. In pyrolysis of coal at low temperatures, it is not possible to recover the primary compounds because the temperature is too low for complete volatilization. An increase in temperature on the other hand enhances volatility but also promotes polymerization and degradation of the primary compounds.

Figure 2 represents the effect of extraction time on the yield obtained with supercritical dioxane and Tunçbilek lignite at 330°C and 8 MPa. The extract yield goes up from 14 % to 28 % as the extraction time increases from 30 minutes to 150 minutes. However, a similar increase in yield can also be obtained by increasing the pressure to i.e. 15 MPa. For structure elucidation of coal, supercritical gas extraction should be carried out at higher pressures and at temperatures close to the critical temperature of the lixiviant. The extraction time should be kept short as to avoid thermal degradation of coal so that the extracts contain compounds which were originally retained in the pores of coal (6,7) instead of decomposition products of coal.

As was stated earlier, the results indicated that the percentage of extractable material with the same solvent did not depend on the coal rank appreciably (8). However, the ratio of n-pentane solubles, n-pentane insolubles/benzene solubles and n-pentane/benzene insolubles fractions does change with coal rank. In Figure 3, the yields of these fractions are plotted against the carbon content of different coals. The yield of n-pentane solubles decreases with increasing carbon content. The yield of benzene solubles shows a maximum with the carbon content and is low for Zonguldak coking coal; n-pentane/benzene insolubles show a minimum with the carbon content.

Although in supercritical gas extraction, physical rather than the chemical properties of the lixiviant are significant(5); the results obtained with different solvents but with the same coal indicated that the composition of the extract depend on

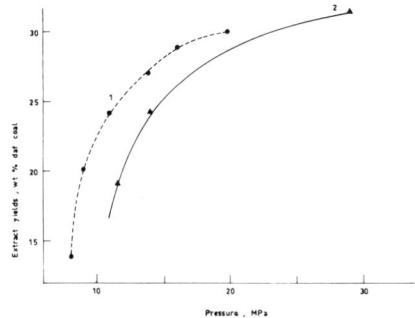

Figure 1. Effect of pressure on yields of extracts. 1, Tucbilek lignite extracted with dioxane at 350°C; 2, Zonguldak coking coal extracted with toluene at 360°C.

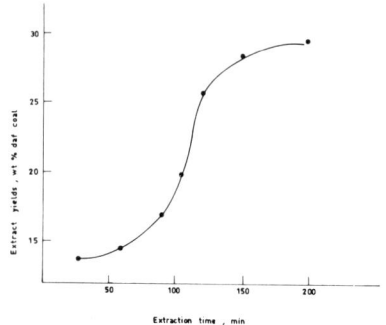

Figure 2. Effect of extraction time on yields of extract (Tuncbilek lignite ex-racted with dioxane at 330°C and 8 MPa).

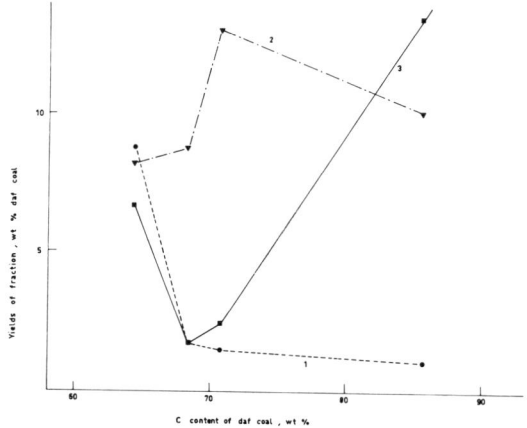

Figure 3. Effect of coal rank on composition of extracts. 1, n-pentane solubles; 2, n-pentane insoluble/ benzene soluble; 3, n-pentane/benzene insolubles (Supercritical extracts obtained at 400°C and 17 MPa).

the particular solvent.

This may be due to the differences in the rate of diffusion of the solvent through the micropore structure of coal.

The GC/MS analysis has indicated that both extracts of lignites and coking coal contained similar compounds (3,9). The yield of each compound depends on the type of the coal and the extraction conditions such as solvent, temperature and pressure (3,10). It is also found that lignites contain a large amount of compounds with molecular weights less than 1000 and these compounds can be classified as H-compounds and O-compounds as Wahrman suggested for bituminous coals (7). The H-compounds present in lignite extracts are mainly aromatic; smaller amounts of aliphatic hydrocarbons and of aromatic compounds which contain heterocyclic oxygen are also present.

The O-compounds comprise of a wide range of phenolic compounds and smaller amounts of components with ether bonds.

Table II emphasises the results obtained by extraction with supercritical toluene and with dioxane of Tunçbilek lignite. While the yield of the O-compounds in the super-

Table II. Yields of H-compounds and O-compounds in the extracts of Tunçbilek lignite.

Condition of SCG extraction		Yield of extract	H-compounds	O-compounds
Solvent	Press. (MPa) / Temp. (°C)	(wt % daf coal)	(wt % daf coal)	
Toluene	12.5 / 350	12.2	2.5	1.4
Dioxane	8.0 / 330	14.2	1.4	4.4

critical dioxane extract is higher, the yield of the H-compounds are higher in the toluene extract. Since the yield of the O-compounds is higher in the dioxane extract and the H-compounds is higher in the toluene and the critical temperatures of both solvents are close (T_c toluene, 320.8°C; T_c dioxane, 314.8°C) (11), it is reasonable to expect higher overall yields when the extraction is carried out with a mixture of supercritical toluene and dioxane. In fact, experiments do confirm this expectation. As can be seen from Table III similar amounts of benzene solubles were obtained from the supercritical dioxane and toluene extractions of Tunçbilek lignite. The amount of n-pentane solubles and n-pentane/ benzene insolubles were higher in the dioxane extract.

Table III. The yields of extracts and composition obtained with supercritical gas extraction

	Tunçbilek lignite			Spruce wood
	Toluene	Dioxane	Toluene-Dioxane (1/1)	Dioxane
Pressure (MPa)	12.5	8.0	20.0	9.0
Temperature (°C)	350	330	350	330
Yields of extract (wt % daf)	12.2	14.2	25.0	23.6
n-pentane solubles (wt % daf)	1.6	3.4	3.1	2.7
n-pentane insoluble/ benzene solubles (wt % daf)	8.1	7.4	13.8	17.3
n-pentane/benzene insolubles (wt % daf)	2.5	3.4	8.0	3.6

The yield of the n-pentane soluble fraction and of the n-pentane/benzene insolubles is higher when a 1:1 dioxane-toluene mixture is used instead of pure toluene to extract Tunçbilek lignite. Especially the insoluble fraction is favored by the mixed solvent.

To explain the differences in the nature of the extracts obtained with different kind of solvents, further studies are required with respect to the change in micropore structure of coal upon supercritical extraction.

The results of supercritical gas extraction of wood with dioxane is summarized in Table III. It shows that 23.6 % yield of liquid products was obtained from soft wood. Gas formation is considerable (30-40 %). It is noteworthy that lignin under the same conditions yields a larger liquid fraction and virtually no gas (12).

Previous work has established that in fact dioxane gives the largest yield of liquid products and these yields are significantly higher than those obtained by straightforward pyrolysis (13).

Inspection of the solid residues remaining after supercritical gas extractions indicated that some carbonization had occurred.

Supercritical dioxane and acetone extraction of lignin model compounds under the same conditions indicated that no chemical reaction took place between the model compounds and the supercritical fluids (14).

One of the objectives of this investigation was to study the effect of the solvent on the yield of phenols.

One observes that the yield of phenols by extraction of wood with supercritical acetone, tetrahydrofurane (THF), toluene and dioxane increases in that same order (Table IV). The lower

Table IV. Phenols contents of supercritical gas extracts of spruce wood.

Condition of SCG extraction			Yield of extract	Phenols content of extract
Solvent	Press. (MPa)	Temp. (°C)	(wt % dry wood)	(wt % dry wood)
Acetone	8.0	250	11.95	3.8
THF	8.0	290	21.19	6.9
Toluene	8.0	340	15.05	7.6
Dioxane	9.0	330	23.59	12.2

yields obtained with acetone and THF can be attributed to the lower extraction temperature. Increase of same tends to increase also the yield of the phenols.

CONCLUSIONS

Supercritical gas extraction should be regarded as a mild pyrolysis conducted in the presence of an inert diluent whose presence minimises secondary reactions.

The chemical composition of the supercritical extracts depends on the nature of the solvent as well as on the rank of the coal. The yield of the soluble material increases with increasing pressure, temperature and extraction time.

Supercritical gas extraction may become an important method of obtaining organic chemicals and liquid fuels from coal.

Under appropriate conditions considerable amounts of phenols can be obtained from the supercritical extraction of wood.

REFERENCES

1. Sarkanen, K. V. and C. H. Ludwig, Ed. Lignins, (New York: Wiley-Interscience, 1971).
2. Ellington, R. T. Ed. Liquid Fuels from Coal,(New York: Academic Press, 1977).
3. Tuğrul, T. and A. Olcay."Supercritical-Gas Extraction of Two Lignites", Fuel, 57: 415-420 (1978).
4. Çalımlı, A. and A. Olcay."Supercritical-Gas Extraction of Spruce Wood", Holzforschung, 32: 7-10 (1978).
5. Paul, P. F. M. and W. S. Wise. The Principle of Gas Extraction, (London: Mills and Boon Ltd., 1971).
6. Vahrman, M."The Smaller Molecules Drived From Coal and Their Significance", Fuel, 49: 5-16 (1970).
7. Vahrman, M. "The Smaller Molecules- An Overlooked Component of Coal", Chemistry in Britain, 8:19-22 (1972).
8. Bartle, K. D., A. Çalımlı, D. W. Jones, R. S. Matthews, A. Olcay, H. Pakdel and T. Tuğrul. "Aromatic Products of 340ºC Supercritical-Toluene Extraction of Two Turkish Lignites; an n.m.r. Study", Fuel, 58: 423-428 (1979).
9. Ceylan, R. and A. Olcay. "Supercritical-Gas Extraction of Turkish Coking Coal", Fuel, 60: 197-200 (1981).
10. Demirci, B."Supercritical-Dioxane Extraction of Tunçbilek Lignite", PhD Thesis, Atatürk University, Turkey (1981).
11. Kudchadker, A. D., G. H. Alani and B. J. Zwolinski."The Critical Constants of Organic Substances", Chem. Rew., 68: 659-735 (1968).
12. Çalımlı, A. "Supercritical Gas Extraction of Wood and of Lignin", PhD Thesis, Ankara University, Turkey (1978).
13. Olcay, A. and A. Çalımlı. "Supercritical Dioxane Extraction of Spruce Wood and of Dioxane Lignin and Comparison of the Extracts with Pyrolysis Products", Separation Science and Technology, The Special Topics Issue, 17: No.1 (1982).
14. Karahan, E. "Supercritical Gas Extraction of Lignin Model Compounds", MS Thesis, Ankara University, Turkey (1981).

CHAPTER 20

SUPERCRITICAL FLUID EXTRACTION
OF OIL SANDS AND RESIDUES FROM
OIL AND COAL HYDROGENATION

Wilhelm O. Eisenbach
Klaus Niemann
Peter J. Göttsch
  Max-Planck Institut für Kohlenforschung
  Kaiser-Wilhelm-Platz 1, Postfach 01 13 25
  D 4330 Mülheim a.d. Ruhr, West Germany

ABSTRACT

Supercritical fluids separate high boiling fractions from non-distillable materials. This procedure has been applied to the residues remaining after distillation of crude-oil and the products of the hydrogenation of lignite. In the latter case up to about 45 % of the vacuum residue could be separated. With oil sands a rapid separation of pure oil results.

INTRODUCTION

Rising consumption and diminishing crude-oil reserves have led to considerable activity in the search for alternative hydrocarbon sources. Two possible sources are coal and oil sand, which are both abundant. In addition an efficient treatment of the residues from crude-oil distillation would help to stretch our hydrocarbon reserves.
All these materials need some difficult operations for producing a sufficient feedstock for the petrochemical industries and there is a special lack for separation procedures in order to isolate high boiling products. What we want to present here is such a procedure, which was discovered and developed at the Max-Planck-Institut für Kohlenforschung in Mülheim-Ruhr (West Germany)

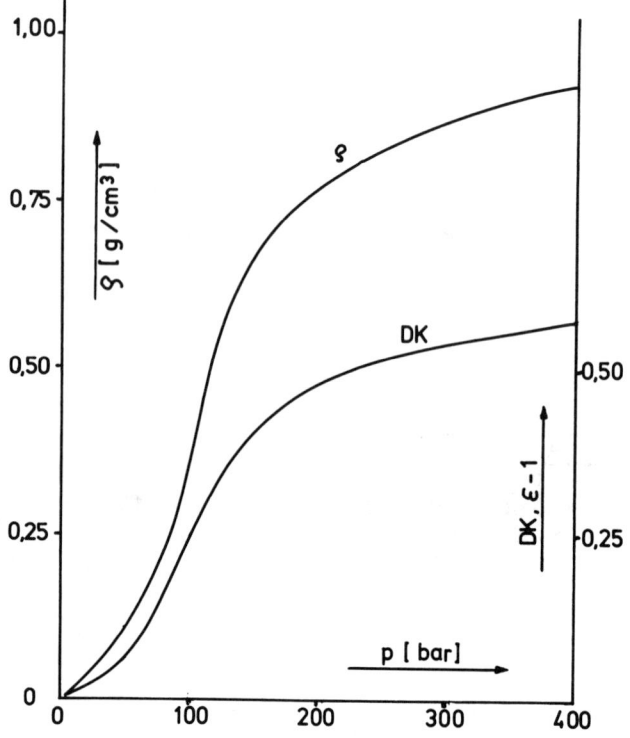

Figure 1. Density and Dielectric Constant of CO_2 as a Function of Pressure (50°C).

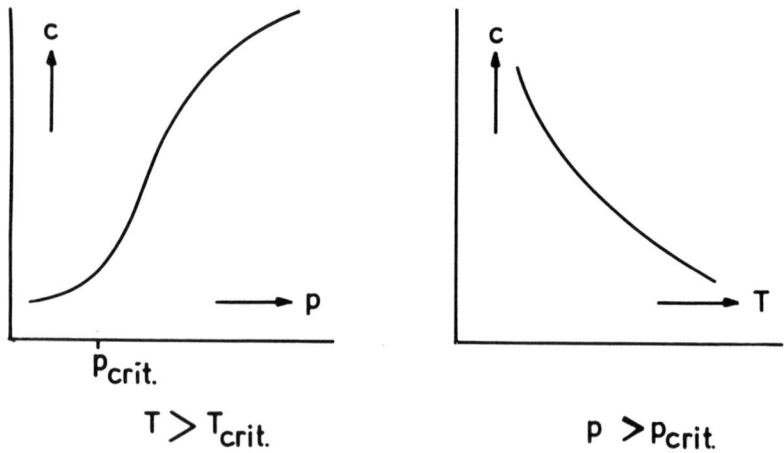

Figure 2. Dependence of Solubility on Pressure and Temperature.

some years ago by K. Zosel. Around 1960 K. Zosel[1] observed that fluids under supercritical conditions were potentially excellent solvents for high boiling substances and he realized that this could form the basis for a more general separation procedure.

The density and the dielectric constant of an organic compound are important factors which determine its solvent power. Figure 1 shows these parameters of carbon dioxide in relation to pressure at a constant temperature. At $50°C$ (which is above the critical temperature of $31°C$) both parameters increase with pressure starting from the critical pressure of 72.9 bars.

The density for example, is also dependent on the temperature and the pressure as well as on its ability to dissolve products. Figure 2 shows the relationship within the range of our purposes in a simplified manner.

As a result two possibilities present themselves for separating components dissolved in supercritical fluids. Both depend on a reduction of the density of the supercritical fluid; either one can increase the temperature at a constant pressure (isobaric method) or alternatively decrease the pressure at a constant temperature (isothermal method). Figure 3 shows a simplified flow scheme for the isobaric approach. The supercritical fluid in reactor (1) is loaded up with the product to be extracted while passing through the reactor. This solution is heated in heat exchanger (2) whereby the dissolved material precipitates and is then collected in separator (3). The hot supercritical fluid is cooled by the second heat exchanger (5) to reactor temperature and recycled to the reactor by circulating pump (4).

It turns out that the second approach, the isothermal method, is prefered for the isolation of thermally unstable products and for separating volatile components from solids. The flow scheme (Figure 4) is very similar to that of the isobaric method. The supercritical fluid is loaded in reactor (1) but contrary to the previous method is expanded into separator (3) where the dissolved material precipitates. The fluid (gas), pressure which may be either above or below the critical pressure, is raised to reactor pressure by membrane compressor (4) and recycled to the reactor.

After this short introduction we want to present a selection of applications of this extraction procedure. Two areas are of interest: on the one

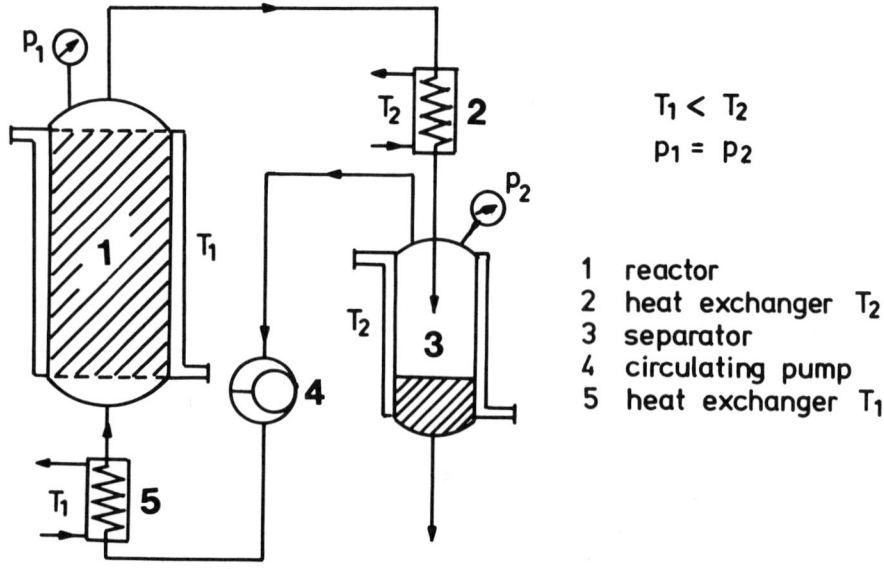

Figure 3. Flow Scheme of Isobaric SCF Extraction.

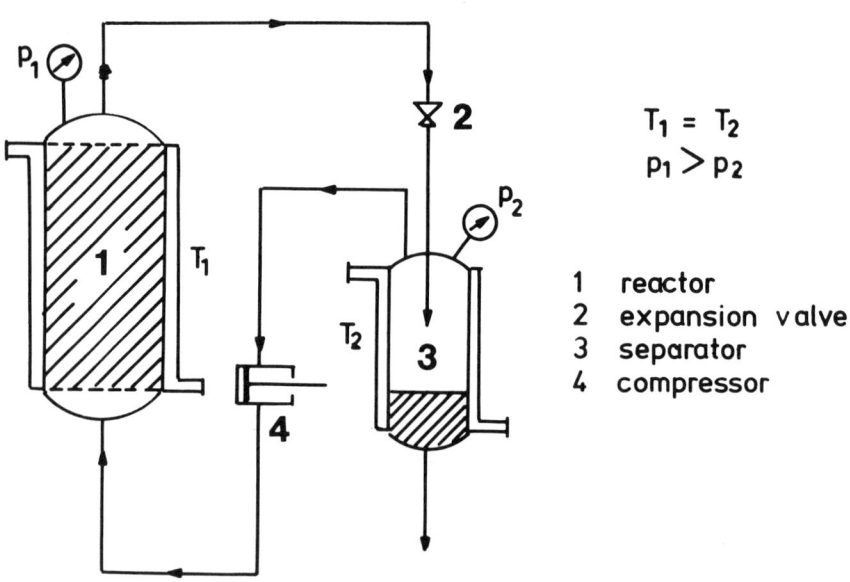

Figure 4. Flow Scheme of Isothermal SCF Extraction.

one hand the field of petrochemical products and chemicals and on the other hand the field of natural products. At our institute we have applied SCF extraction to a wide variety of separation problems, such as:

1. Applications to petrochemical products and chemicals

+ The deasphaltation of residues from crude-oil distillation [2]
+ The separation of paraffins from high temperature lignite tar [3]
+ The regeneration of waste lubricating oil
+ The separation of tall oil in resins and fatty acids

2. Applications to natural products

+ The decaffeination of green coffee beans, which is being used on an industrial scale in West Germany by "Kaffee HAG" [2], [4].
+ The separation of aromatics from spices, e.g. pepper, nutmeg and chilli [5], [7]
+ The separation of aromatics from waste materials such as tobacco dust [6]
+ The production of hop extract for the brewing industry [7], [8]
+ The isolation of oil from oil-containing seed, e.g. soy beans and jojoba beans [6]
+ The deodorization of vegetable and animal oil
+ The fractionation of high boiling mixtures such as cod-liver oil [2]

For this presentation we wish to highlight two applications which we have investigated in more detail, namely the separation of oil from oil sands and the extraction of lignite liquifaction residues.

SCF EXTRACTION OF OIL FROM OIL SANDS

Oil sand is found in immense deposits. Its estimated reserves exceed by fare current reserve figures of petroleum. The oil is isolated from the oil sand on a large industrial scale by treatment with hot water or steam. This method is very energy intensive and has its environmental problems. By using supercritical fluids, we are able to extract the oil directly; any further purification is not necessary.

In a pilot plant with a flow scheme similar to that of Figure 4, we have treated Athabasca oil sand with supercritical fluids under isothermal conditions. Initially, we worked with propane at $110°C$

and pressures of 200 bars in the reactor and 20 bars in the separator. The oil sand contained 13.5 % of organic material extractable with toluene. Under the conditions illustrated above 70 % of this material was obtained as a dark-red viscous oil. About 95 % of this amount was obtained in a relatively short time of one hour. The fraction which is obtained at longer contact times is more viscous -in other words it contains higher molecular components. The rate of this SCF extraction is shown in Fig. 5.

The oil produced consists mainly of paraffins which follows from the infrared spectrum in Fig. 6. The intensive CH-vibration bands at 2960-2850 cm^{-1}, the deformation bands of the methylene groups at 1420 cm^{-1} and of the methyl groups at 1375 cm^{-1} can be clearly seen.

A further 4 % of organic material was extracted with toluene at ambient conditions from the residue left after propane SCF extraction. This extract appeard as a black solid and consisted of the asphaltenes present in the oil sand.

A comparison between the propane SCF extract and the toluene extract obtained by soxhlet extraction of the oil sand reveals no significant difference in the element composition (Table I). On the other hand the asphaltene content and the Conradson number are very different. The absence of asphaltenes

Table I Oil from Athabasca Oil Sand
-Comparison between Propane SCF Extract and Toluene Extract-

		Propane	Toluene
Elementary Analysis	% C	83.80	81.61
	% H	11.03	10.22
	% N	0.48	0.76
	% O	1.17	1.31
	% S	3.07	4.36
	C/H	1.568	1.492
Yield (%)		9.5	13.5
Asphaltenes (%)	+)	0.0	16.9
Conradson Number	+)	3.3	15.9

+) related to the extract

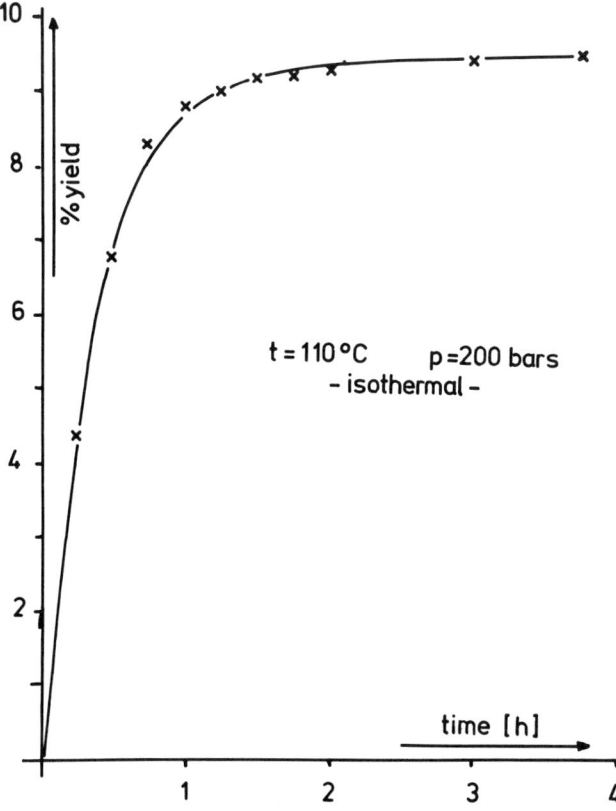

Figure 5. SCF Extraction of Oil Sand with Propane.

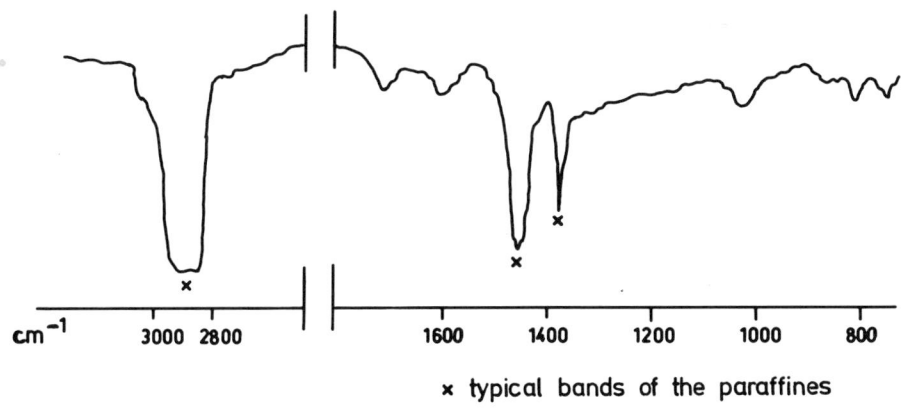

Figure 6. Infrared Spectrum of the Propane SCF Extract from Athabasca Oil Sand.

in the oil isolated by SCF extraction with propane shows the possibility of simultaneously isolating the oil and separating from the asphaltenes. Experience shows that this oil is more amenable to further refining operations such as hydrogenation, because the asphaltenes as coke precursors are absent.

The oil isolated by the two methods have a different distillation behaviour as shown by the boiling curves in Figure 7. Up to a temperature of $582°C$ only 37 % of the toluene extract could be distilled, whereas 76 % of the oil obtained with supercritical propane was distillable. Furthermore the initial boiling point of the toluene extract was $50°C$ higher.

SCF EXTRACTION OF LIGNITE HYDROGENATION RESIDUES

We have recently turned our attention to residues formed in the hydrogenation process of lignite with the intend to obtain useful products which cannot be isolated by conventional procedures. At present the hot separator bottoms from the hydrogenation plant are divided in a volatile oil and a residue (by vacuum distillation), but a substantial amount of useful organic material remains in the residue.

The objective of our investigations was to try to isolate these products by treatment with supercritical fluids. The isolated products contribute to the oil yield and thereby improve the economy of coal hydrogenation processes.

Two approaches are possible:
1. The separation of these useful products from the residue after vacuum distillation or
2. The treatment of the entire hot separator bottom without prior vacuum distillation.

Table II Critical Data

Fluid	Pressure (bar)	Temperature (°C)	Density (g/l)
Propane	42.0	97	224
i-Butane	36.0	136	222
n-Butane	37.5	152	225
n-Pentane	33.3	197	244

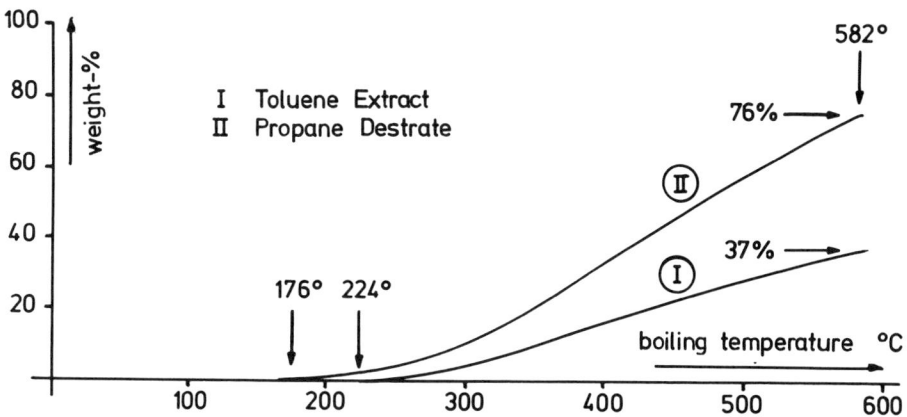

Figure 7. Boiling Curves of Products Isolated from Athabaca Oil Sand.

During these experiments both the extraction conditions and the nature of the supercritical fluids were varied. Table II summarizes the critical data of some hydrocarbons used. The experiments were carried out in the same apparatus as used for the treatment of the oil sand. The vacuum distillation residue was treated with fluids at a constant pressure of 200 bars in the reactor and temperatures of about 30°C above the particular critical temperature. Table III lists the conditions and results of these experiments.

Table III SCF Extraction of Vacuum Distillation Residue

Exp. Nr.	Fluid	Temp. (°C)	Pressure Reactor (bar)	Pressure Separator (bar)	Time (h)	Yield (%)
1	Propane	120	200	15	17.8	20.4
2	i-Butane	160	200	5	18.5	27.0
3	n-Butane	170	200	5	13.8	35.0
4	n-Pentane	210	200	5	7.0	46.8
5	Propane	210	200	20	10.0	30.0
6[+])	Propane	120	200	15	4.5	46.1

[+]) SCF extract from experiment 4

The yield of isolated hydrocarbons increases in the homologous series from propane to n-pentane; moreover the rate of SCF extraction increases. The experiments 4 and 5, which were carried out under the same temperature and pressure, show clearly the effect of the nature of the supercritical fluid. Although the temperature has been increased in the case of propane from 120°C to 210°C, the yield only increases from 20.4 % to 30.0 % (experiments 1 / 5) and still does not approach the 46.8 % obtained with n-pentane. This means that n-pentane extracts components in addition to those extracted with propane. This is confirmed by experiment 6 in which the SCF extract from experiment 4 is treated again with propane under the conditions of experiment 1; only 46.1 % of the n-pentane SCF extract could be solubilized in propane at 120°C and 200 bars.

The rate of SCF extraction for the different fluids are shown in Figure 8. Going from propane to

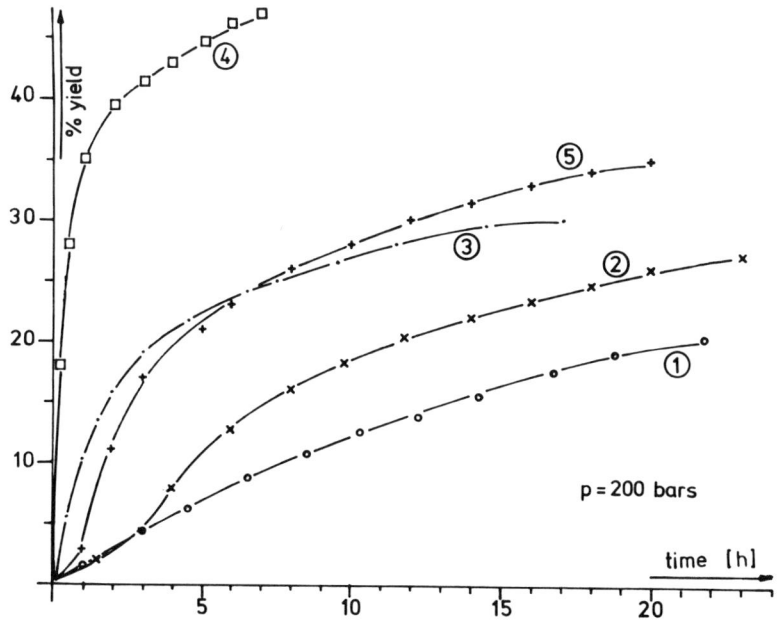

Figure 8. SCF Extraction of Vacuum Distillation Residues.

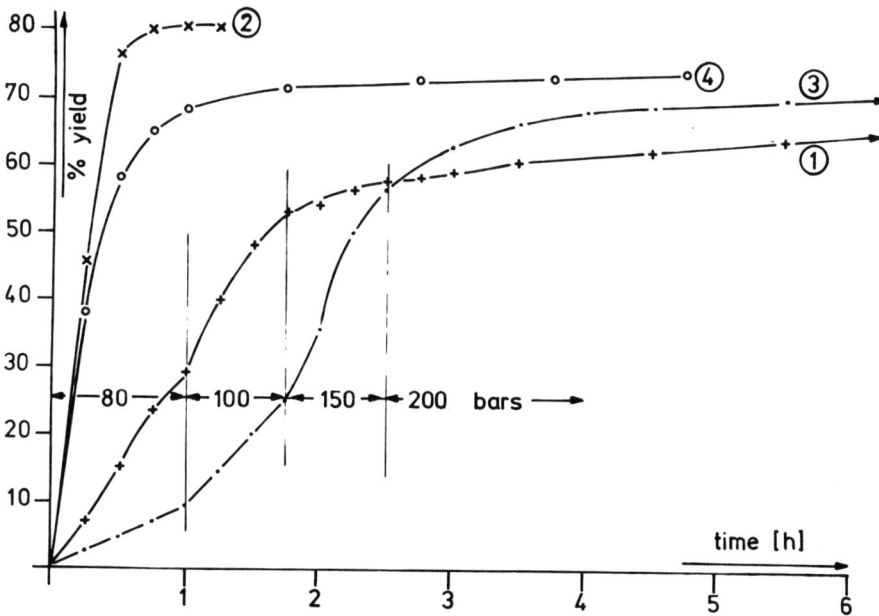

Figure 9. SCF Extraction of Hot Separator Bottoms.

n-pentane both the yield of the isolated product and the rate of SCF extraction increases. After one hour extraction time n-pentane at 210°C (curve 2) produces 20 times more product than propane at 120°C (curve 1).

A more reasonable approach would be to avoid the distillation step and instead to subject the whole of the hydrogenation residue to a supercritical fluid. Conditions and results of these experiments with only propane or n-pentane as the extraction medium are listed in Table IV.

Table IV SCF Extraction of Hot Separator Bottom

Exp. Nr.	Fluid	Temp. (°C)	Pressure Reactor (bar)	Pressure Separator (bar)	Time (h)	Yield (%)
1	Propane	135	80-200	20	10.0	68.0
2	n-Pentane	210	50-200	5	1.2	80.5
3	Propane	210	80-200	20	10.0	72.3
4	Propane	210	200	20	4.8	74.3

At temperatures 20 to 30°C above the critical temperature, the same trend is observed as in the previous experiments: with n-pentane the yield is higher than with propane. In these experiments the pressure has been initially raised in a stepwise manner, because the supercritical phase tends to become super-saturated and as a result some of the material in the reactor foams over into the separator. The difference in yield obtained with propane and n-pentane decreases when the temperature of propane is increased to 210°C. This shows clearly the positive influence of temperature even though the density of propane decreases.

In Figure 9 characteristic SCF extraction curves of the hot separator bottoms are combined. By raising the temperature of propane from 135°C to 210°C the yield after 10 hours increases by 4.3 %. However at the onset, as a result of the lower density of supercritical propane, the rate of SCF extraction is lower (curve 3).

At a temperature of 210°C propane at a pressure of 200 bars and n-pentane at a pressure of 50 bars have a similar density of about 300 g/l. A comparison between the experiments 2 and 4 shows that the rate of SCF extraction during the first hour is nearly equivalent, but the final yield is about 12 % higher in the case of n-pentane. Whereas no additional product could be extracted with n-pentane after this period, propane gave a further 6.3 % during the next 4 hours. Table V lists the material balance obtained in the experiments with propane and n-pentane.

Table V Material Balance

Fluid	Propane	n-Pentane
Temperature	210°C	210°C
Pressure	200 bar	50 bar
Density	303 g/l	300 g/l
Residue	25.7 %	19.8 %
Hot Separator Bottom Extract	74.3 %	80.2 %
Vacuum Residue Extract	11.0 %	17.4 %
Difference	63.3 %	62.8 %

A comparison is made between SCF extracts from hot separator bottoms and from vacuum distillation residues. It follows that the hot bottom gives a higher yield of SCF extract with either propane or n-pentane. The yield increase over the vacuum distillation residue is the same for either propane or n-pentane and corresponds to the quantity which can be separated from the hot separator bottom by vacuum distillation. This list shows again that supercritical n-pentane is able to extract higher molecular compounds which are insoluble in supercritical propane.

Analysis of the SCF extracts show clearly that the higher yield obtained with n-pentane consists mainly of asphaltenes. Table VI combines some of the analytical data. For the bottoms there are two values listed, namely the value (a) from samples taken at the beginning of the experiment and the value (b) from samples taken at the end. With propane the higher molecular compounds are extracted at the end of the treatment and this is mirrored in the higher values of the Conradson number and of the softening points. In the case of n-pentane no difference in the composition of the SCF extracts is observed throughout the treatment.

Table VI Analytical Data of SCF Extracts

Fluid		C/H	Asphaltenes (%)	Conradson Number	sp ($°C$)
Hot Separator Bottoms					
Propane ($135°C$)	a	1.383	0.0	0.2	28.0
	b	1.114	3.5	3.0	29.0
Propane ($210°C$)	a	1.380	4.4	0.2	26.0
	b	1.121	7.4	5.8	30.5
n-Pentane ($210°C$)	a	1.318	6.5	7.0	26.0
	b	1.224	7.5	6.5	29.0
Vacuum Residues					
Propane ($135°C$)		1.171	10.2	13.5	43.5
Propane ($210°C$)		1.196	11.0	17.0	45.2
n-Pentane ($210°C$)		1.100	37.6	20.5	61.5

sp = softening point (ring and ball, ASTM D 36-66 T)

The asphaltene content in the SCF extract isolated from the vacuum distillation residue is clearly higher. Even the propane extract includes a higher percentage of asphaltenes. This may be the result of the longer thermal treatment of the vacuum distillation, when presumable higher molecular compounds are formed, as well as of the longer SCF extraction period.

CONCLUSIONS

We have shown by two examples that SCF extraction is a promising procedure to separate valuable products from raw materials which cannot be isolated by conventional methods. In contrast to vacuum distillation of the hot separator bottom of coal liquifaction we are able to isolate a further amount of about 30 % heavy oil by this new method. So it seems to be economic to substitute vacuum distillation by SCF extraction.

Although we have still realized SCF extraction in a batchwise manner, we now turn our attention to developing a continuous procedure, which is a substantial condition for an industrial application.

REFERENCES

[1] K.Zosel US Patent 3 969 196 (13.09.1978)
Studiengesellschaft Kohle mbH.

[2] K.Zosel Ang.Chem.Int.Ed.Engl. $\underline{17}$, 702-9 (1978)

[3] W.Eisenbach, K.Niemann Erdöl & Kohle, Erdgas, Petrochemie $\underline{34}$, 296-300 (1981)

[4] K.Zosel US Patent 4 260 639 (7.04.1981)
Studiengesellschaft Kohle mbH.

[5] G.Vitzhum, P.Hubert US Patent 4 198 432
(15.04.1980) Studienges. Kohle mbH.

[6] W.Eisenbach 157th Meeting of the Electrochemical Society, St.Louis, Miss. May 1980

[7] G.Vitzhum, P.Hubert Ang.Chem.Int.Ed.Engl. $\underline{17}$, 710-5 (1978)

[8] G.Vitzhum, P.Hubert, R.Sirtel
US Patent 4 104 409 (1.08.1978)
Studiengesellschaft Kohle mbH.

CHAPTER 21

SEPARATION OF FINELY DISPERSED
SOLIDS FROM LOW-VOLATILE
VISCOUS MEDIA BY GAS EXTRACTION

D. Stützer. Lehrstuhl für Technische Chemie, Universität Erlangen-Nürnberg, Egerlandstrasse 3, D-8520 Erlangen, West Germany

G. Brunner. Kraftwerk Union AG, Hammerbacher Str. 12 + 14, D-8520 Erlangen, West Germany

S. Peter. Lehrstuhl für Technische Chemie, Universität Erlangen-Nürnberg, Egerlandstrasse 3, D-8520 Erlangen, West Germany

SUMMARY

The viscous media containing the µ-sized particles are processed with a compressed gas at high density and an entrainer. The conditions are chosen so that the ternary (or pseudoternary) system supercritical gas-entrainer-viscous substance (or mixture) is supercritical. In this way concentrations of 20 to 35 wt% of the viscous material in the gas phase can be achieved at conditions that permit rapid sedimentation of the suspended material. These results suggest that may provide the basis for the development of processes of economical interest.

INTRODUCTION

The removal of the µ-sized mineral particles found in bitumina from tar sands, in shale oil or in coal derived liquids is a rather difficult solid/liquid separation process. Clays and other minerals, which occur in the deposits find their way into the products. The mineral content of bitumina from tar sands can range from about 10 to 20 wt% for example.

The ash content of the coal derived liquids can range from about 4 to 20 wt% depending upon the liquefaction process. Ash levels must be reduced to about 0.4 wt% in the case of boiler fuel and less than 0.1 wt% for further refining purposes [1]. Suspensions of finely dispersed solids occur also as residues of catalysts in chemical products or as abrasion losses in used lubricating oil etc.

According to the SRC-process the oil is mixed with paraffinic solvents at a temperature of 300 to 450 $^\circ$C in a continuously operated thickener. Thus the rate of sedimentation is increased so that solid particles with a size of more than 40 µ separate quickly [2].

In the Lummus anti solvent deashing process a kerosine type promoter liquid is added to agglomerate micron-sized particles in solution. Separation occurs by gravity settling [3]. Furthermore flotation processes for separation of the finely dispersed solids from oil suspensions are discussed. Henry compares the expenditure of filtration, centrifugation and hydrocyclone processing [1]. According to that investigation the most favorable mechanical separation process is a rotating filter operated under elevated pressure. The products contain about 0.2 to 0.7 wt% of solid material. For mechanical processing the viscosity of the oil (or other viscous media) is usually reduced by adding a solvent of low viscosity.

In the critical-solvent deashing process of Kerr-Mc Gee Corporation [4] the oil is dissolved in light aromatic compounds at a temperature below the critical temperature and a pressure above the critical pressure of the aromatic solvent. The solid material is then separated from the solution by sedimentation or filtration. Subsequently the solution is heated to a temperature above the critical temperature of the solvent, say 400 to 450 $^\circ$C at constant pressure. Thereby two phases are formed. One is a liquid phase rich in oil and the other is a gaseous phase poor in oil. The aromatic solvents dissolved in the oil have to be recovered. As the critical pressure of benzene amounts to 48.6 atm (T_c = 288.9 $^\circ$C), of toluene to 41.6 atm (T_c = 320.8 $^\circ$C) and of xylene to 37.5 atm (T_c = 359 $^\circ$C) the process requires a pressure of about 50 bar.

Using compressed supercritical gases in combination with an entrainer as a solvent for the organic components it is possible to operate the separation of the suspended solids at a considerable lower temperature say 70 to 180 $^\circ$C. The pressure required for separation by supercritical gas extraction with an entrainer amounts to about 70 to 140 bar and is thus within a practicable range.

THERMODYNAMIC CONDITIONS OF THE SEPARATION PROCESS

The thermodynamic conditions for the separation of finely dispersed solids and liquids by gas extraction with the aid of an entrainer are demonstrated by means of Figure 1. The phase behaviour of the ternary system shale oil/entrainer/supercritical gas is represented in the triangular diagram. The pressure is assumed to be 1.1 to 4 times the critical pressure of the supercritical component. The temperature may be the 1.1 to 1.5 multiple of the critical temperature of the supercritical component. The solid forms an additional dispersed solid phase. The organic components are insoluble in the solid material and vice versa. The influence of the solid particles on the phase behaviour is therefore negligible.

Between shale oil and gas a miscibility gap exists. With the increasing content of an entrainer the miscibility gap diminishes and disappears at a high content of the entrainer. A mixture with a composition represented by a point within the vertical hatched area disintegrates in a liquid phase rich in shale oil and a gaseous phase rich in gas.

If the composition of a mixture is represented by a point out of the hatched area we have only one phase containing the oil, the entrainer and the supercritical component besides the solid particles. In that case no difference exists between liquid and gaseous phase.

For the separation of the finely dispersed solids from shale oil, temperature, pressure and composition are chosen so that the mixture of shale oil, entrainer and supercritical component form one fluid phase. This fluid phase is able to take up larger portions of the oil. Even at relatively high oil content the viscosity of the fluid phase is so low that the total sedimentation of the finely dispersed mineral matter occurs within a short time. In Figure 1 the suitable area of concentrations of the quasi ternary system of shale oil, entrainer and gas is cross hatched. The fluid containing the oil is withdrawn from the extraction vessel.

The oil can be separated from the withdrawn fluid phase by diminishing its density. Reduction of the density may be caused by increasing the temperature or by decreasing the pressure. The thermodynamic conditions suitable for the precipitation of the solid free oil are demonstrated by means of Figure 2. The composition of the mixture considered is marked by point I. The temperature T_2 at which precipitation occurs is about 30 to 50 °C higher than the temperature T_1 of the extraction process. The pressure remains unchanged.

By the increase of the temperature the two phase region is enlarged. The temperature is increased to such an extent that the enlarged two phase area comprises the mixture represented by point I. As the composition of the fluid phase

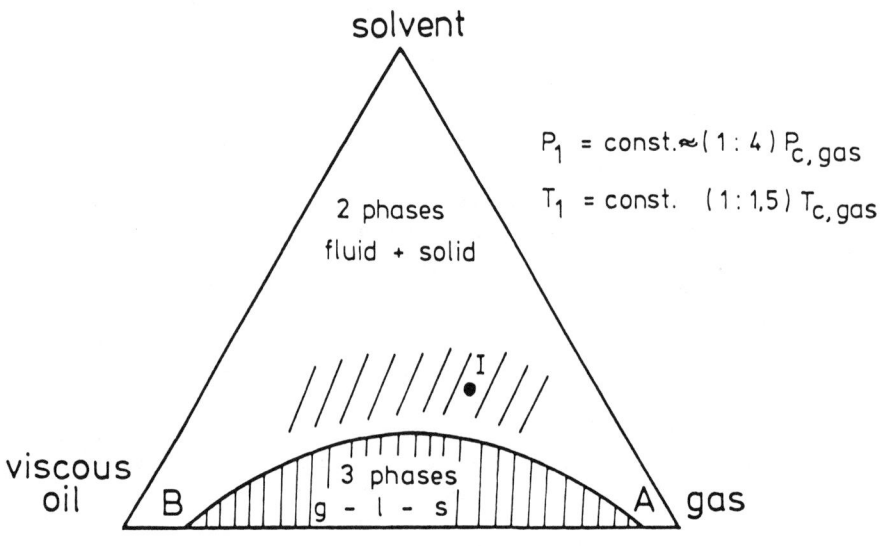

Figure 1. Charging the supercritical extracting agent.

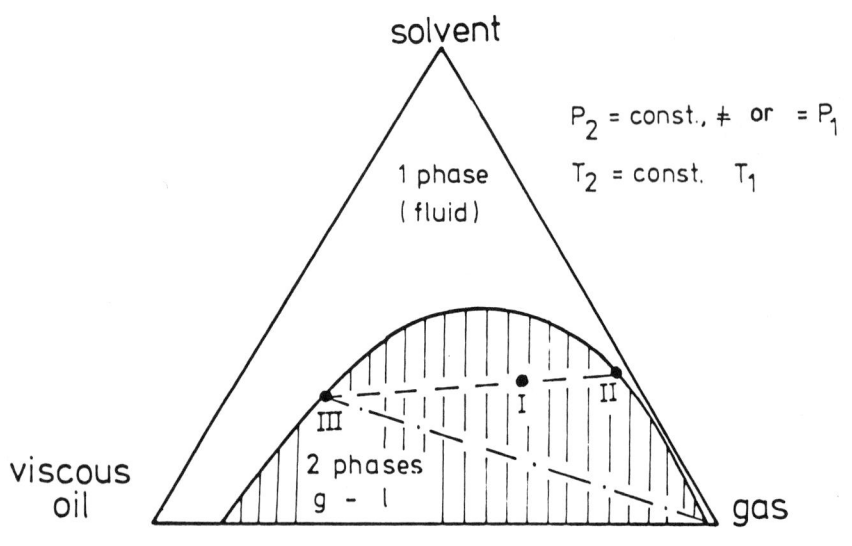

Figure 2. Precipitation of the dissolved oil.

lies within the two phase area, the fluid phase I disintegrates into the liquid phase III rich in oil and into the gaseous phase II rich in gas. Phase II contains only small amounts of oil.

The liquid phase III is withdrawn and relieved. The dissolved gas and part of the dissolved entrainer are removed. The entrainer remaining dissolved in the oil has to be recovered. The gas phase is cooled to its initial temperature and recirculated. The portion of entrainer and gas carried out with the oil has to be replaced.

APPARATUS AND EXPERIMENTAL RESULTS

In Figure 3 the apparatus for the separation of finely dispersed solids from viscous media is shown schematically. Feed, entrainer and gas are charged to the mixing vessel in the desired ratio. After adjusting temperature and pressure the content is thoroughly mixed. Oil, entrainer and gas form a uniform fluid phase. Next, the mixture is transported to the separation vessel where the μ-sized particles settle within a short time. Upon separation of solids and fluid phase the solid material is withdrawn from the bottom of the separation vessel. The fluid phase passes through a heat exchanger and into the regenerator. By adjusting of the temperature the two phase area is enlarged so, that the fluid phase disintegrates into a liquid phase (rich in oil) and a gas phase (poor in oil). The liquid phase is withdrawn from the bottom of the regenerator and expanded into the vessel for recovery of the released gas and entrainer. Released gas and entrainer are recompressed and recirculated via a heat exchanger.

The feasibility of separating finely dispersed solids from viscous liquids by gas extraction in the presence of an entrainer was demonstrated by means of separating μ-sized mineral matter from shale oil. Shale oil containing 25 wt% μ-sized mineral matter was charged to an autoclave. Subsequently 20 wt% toluene were added. The mixture was contacted with compressed ethylene at a temperature of 70 $^\circ$C and a pressure of 135 bar. After stirring for about half an hour the content of the autoclave was left undisturbed for 15 to 90 min, so that sedimentation of the suspended material could take place. Subsequently part of the gaseous phase was withdrawn until the pressure had dropped about 40 bar and ethylene was pumped again into the autoclave until the pressure reached the initial value. This procedure was repeated several times. After nine cycles about 60 wt% of the feed oil was withdrawn from the autoclave. The average concentration

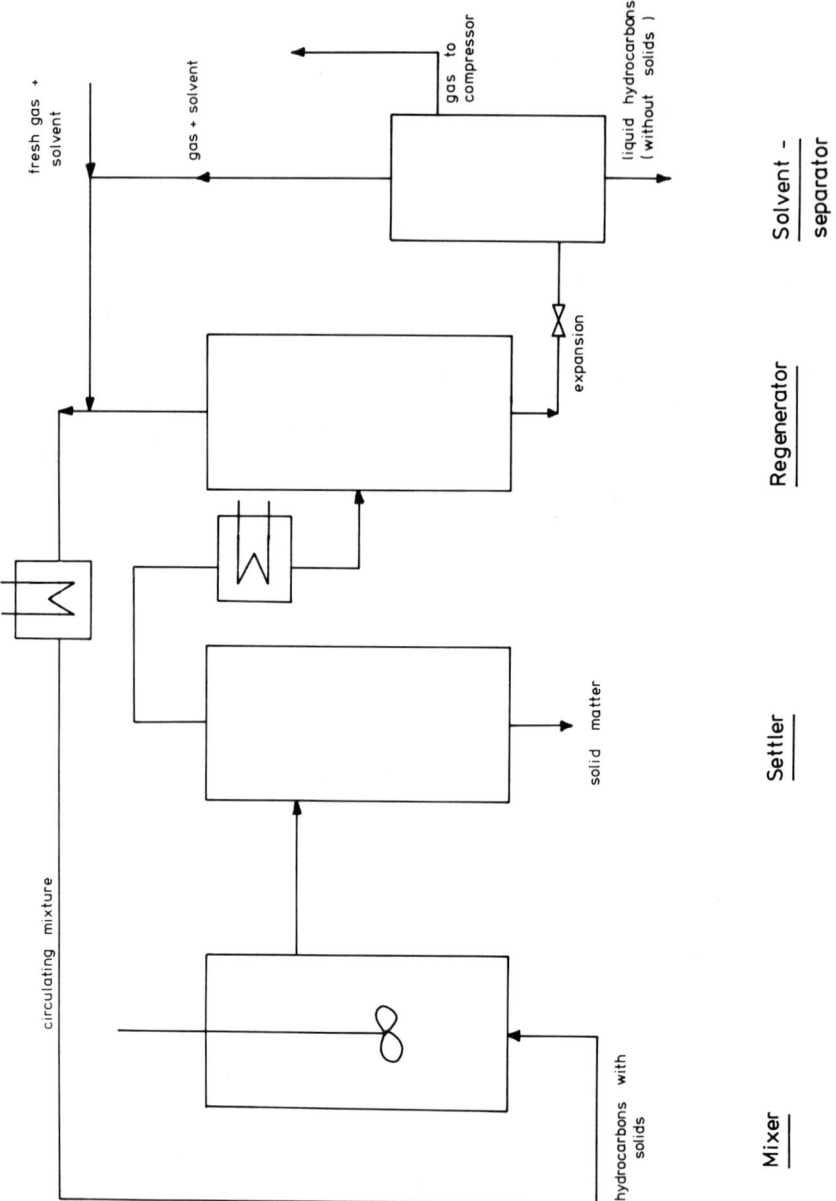

Figure 3. Apparatus for deashing of viscous oil.

of the oil in the gaseous phase amounted to about 30 wt%. The concentration in the first cycle amounted to about 50 wt% and diminished to about 5 wt% in the ninth cycle. The produced oil was substantially free of mineral matter.

As the density of the gaseous phase increases with the oil content the time needed for total sedimentation grows. This is demonstrated by the experimental results shown in Table I.

Table I. Sedimentation time as a function of the oil concentration in the system shale oil - toluene - ethylene at 70 °C and 135 bar

Oil concentration (wt%)	Sedimentation time (minutes)
20	2 - 4
40	10
50	30
> 50	rapid increase of time to more than several hours

The viscosity of a supercritical phase with a density of 0.3 to 0.5 g cm^{-3}, which is suitable for gas extraction is closer to the viscosity of the gas than to the viscosity of the liquid, as has been pointed out by Franck [5]. Also diffusion in supercritical phases of high density is very fast as shown by Schneider [6]. Further the density of a compressed gaseous phase suitable for gas extraction lies in the range of 0.2 to 0.5 g/cm^3. The difference between the density of the gaseous phase and the dispersed mineral matter is sufficient for rapid sedimentation. As the oil content increases the gaseous phase becomes more viscous, which consequently increases the settling time.

In further series of experiments the separation of micro sized solid particles from shale oil was investigated in the presence of toluene as an entrainer and propane as a supercritical component. 600 g shale oil with 22 wt% of suspended mineral matter was treated at a temperature of 119 °C and a pressure of 80 bar with a mixture of propane and toluene. This mixture was pumped with a rate 0.8 liter/hour through the extraction vessel so that the time of direct contact between gas and liquid phase amounted to about 3 min.

The extracting medium consisted of about 46 wt% of toluene and about 54 wt% of propane. The oil content of the supercritical solution transferred to the separator decreased with time as is shown in Figure 4. The decrease in oil con-

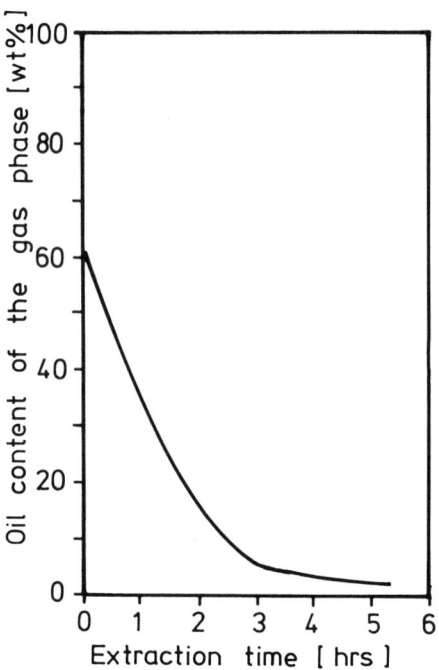

Figure 4. Oil content of the gas phase in the mixer vessel as a function of the extraction time.

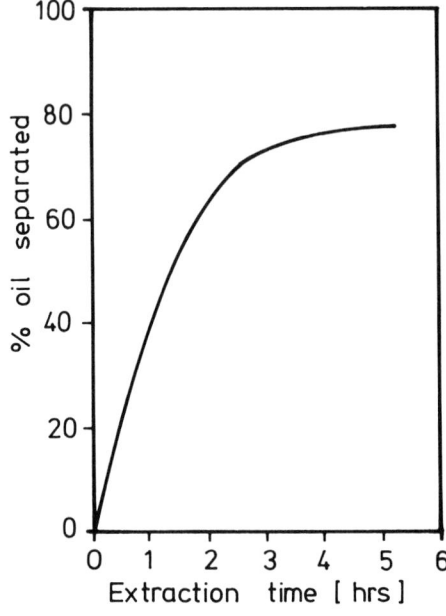

Figure 5. Separated ash free oil as a function of time.

tent of the gas phase in the course of the extraction is possibly due to the fact that compounds of low molecular weight are more soluble in compressed gases than compounds of high molecular weight. A partial recirculation of the more soluble components of the extracted oil would improve the process in that it would act as an entrainer.

By heating the supercritical solution to a temperature of 170 °C at a constant pressure, a liquid phase is precipitated. In Figure 5 the fraction of recovered ash-free oil is plotted against the extraction time. The oil accumulating during the experiment consisted of about 59 wt% of oil and 41 wt% of toluene with a small amount of dissolved propane. Within 5 hours about 75 wt% of the charged oil was removed in this way. The regenerated gas phase contained less than 1 wt% of shale oil.

Within the time of direct contact of 3 minutes an oil concentration of the gas phase could be achieved which amounted to 65 % of the equilibrium. Considering that dispersing of the gas in the liquid phase was not intensive the dissolving rate may be regarded as rather good. After more than 5 hours about 90 wt% of the charged oil could be recovered.

Apart from toluene, solvents as acetone, benzene, cyclohexane, hexane, xylene etc. are also appropriate as an entrainer. Carbon dioxide and hydrocarbons such as methane, ethane, butane etc. are suitable supercritical substitutes for ethylene and propane.

REFERENCES

[1] Henry, J. D., M. E. Prudrich, K. R. Vaidyanathan
Separation and Purification Methods, $\underline{8}$ (2) 81 - 118 (1979)
[2] Kleinpeter, J. A., D. C. Jones, P. J. Dudt, F. P. Burke
Ind. Eng. Chem. Process Des. Dev., $\underline{18}$ (3) 541 - 546 (1979)
[3] Burke, D. P.
Chem. Week (11. 9. 1974), 538 - 543, DOS 2355 606
[4] U.S. Patent 4,162,965 (1979)
[5] Franck, E. U.
Ber. Bunsen-Ges. Phys. Chem. $\underline{73}$, 135 - 142 (1969)
[6] Schneider, G. M.
Angew. Chemie Int. Ed. in English, $\underline{17}$, 716 - 727 (1978)

CHAPTER 22

THE PREPARATION OF ACID-
CATALYZED SILICA AEROGEL

W. J. Schmitt,
R. A. Grieger-Block, and
T. W. Chapman
  Chemical Engineering Dept.
    University of Wisconsin
      Madison, WI 53706

ABSTRACT

   Highly porous silica aerogel was synthesized and found to be both transparent and a good thermal insulator. The aerogel was created by the acid-catalyzed hydrolysis of ethyl silicate in an alcoholic medium and dried supercritically in an autoclave. The process of supercritical drying replaces the liquid permeating a gel with a non-condensable gas without causing the collapse of the solid matrix. Several physical properties of the aerogels were measured. These include the optical transmittance, strength, density, and thermal conductivity of selected samples. A description of the method of formation of this unusual material is presented along with a general discussion of supercritical fluid drying. An aerogel was formed between two panes of commercial flat glass thereby creating an "aerogel window". This composite was found to insulate better than a standard double-pane arrangement containing an air gap of similar geometry.

INTRODUCTION

   An aerogel is a highly porous synthetic material consisting of an expanded, small-particle-size solid matrix permeated by a non-condensable gas such as air. Such material was first produced by Kistler (1932) to demonstrate that a colloidal gel could be dried without the usual collapse of the solid skeleton [1]. The transformation requires an uncommon unit operation called supercritical fluid (SCF) drying. Kistler prepared aerogels from a number of gelataneous masses and was the first to note their unusually low density and occasional transparency. He remarked that "almost anything that can be made to form a colloidal mass can also be turned into an aerogel".

Silicon dioxide (silica) aerogel has been the type of aerogel most commonly studied because of its ease of preparation and the availability of the principal reagents. It is probably the only aerogel available commercially (Santocel^R - Monsanto Company). One of the first practical uses found for silica aerogel was as an insulating powder filler in cryogenic storage tanks, Dewar flasks, and the like [2,3].

Recently there has been an increased interest in various types of aerogels for several unrelated purposes. Because the material is so porous (with an open pore capillary system), it has been studied as a catalyst support in petroleum cracking processes [4]. Methods have been developed to impregnate the material with catalytically active metals for use with vapor phase reactions, and the surface area has been extensively studied [4,5]. Animal tissue has been converted to an aerogel structure to aid electron microscopic examination of the samples [6].

The work reported here is an extension of a relatively new pursuit rooted in an investigation by Cantin et al. in 1974 [7]. Silica aerogel was proposed as a Cherenkov radiation detector for use in the target assembly of high-energy particle accelerators. This application was based on the fact that aerogel can be made to possess refractive indices in the range below 1.10, a range previously covered only by compressed cryogenic gases. Extremely transparent aerogel free of internal cracks, faults, or other defects was required.

With the successful production of optical-quality aerogel reported, and the low thermal conductivity known, it became the objective of this investigation to synthesize reproducibly transparent silica aerogel and to laminate it between two panes of flat glass to create an aerogel window of good insulating ability. Several such windows were eventually produced. In addition, considerable physical property data for various aerogels were collected, and an important modification in the chemistry of the silicic acid gelation was developed. This new method offers considerable safety improvements over those of previous workers.

CHEMISTRY OF THE ALCOSOL

To form a silica aerogel, one must first prepare colloidal silica in solution. Kistler prepared silica gel by neutralizing sodium silicate solution ($Na_2SiO_3 \cdot nH_2O$) with concentrated hydrochloric acid, but this method is slow and leaves salt in the gel as an undesirable by-product.

Recently, a better method has been proposed [5,7]. Tetramethyl orthosilicate, an ester of monosilicic acid, is catalytically hydrolyzed in an alcoholic solution as follows:

$$Si(OMe)_4 + 4H_2O \xrightarrow[\text{in MeOH}]{\text{[Base]}} Si(OH)_4 + 4MeOH \qquad (1)$$

Highly reactive monosilicic acid is produced along with four equivalents of methanol. The monosilicic acid rapidly condenses forming polymeric silicon dioxide as the final colloidal product:

$$n \cdot Si(OH)_4 \longrightarrow n \cdot SiO_2 + 4n \cdot H_2O \qquad (2)$$

Reaction 1 must be run in a mutual solvent because tetramethyl orthosilicate and water are completely immiscible. Methanol is a logical choice for the solvent because it is produced as a by-product of the ester hydrolysis. The clear, thin, homogeneous reaction mixture in alcohol is called an alcosol, and the solid, glassy gel which eventually forms is called an alcogel. Iler has presented an excellent discussion of this reaction and many other aspects of colloidal silica chemistry in his monograph [8].

Although the tetramethyl orthosilicate synthesis produces aerogels of excellent quality, the reagent is known to be extremely toxic [9]. For that reason, a synthesis has been devised with tetraethyl orthosilicate, which is far less toxic than its methyl homologue [9,10]. The reaction produces monosilicic acid and ethanol and may be written analogously to the methyl silicate hydrolysis:

$$Si(OEt)_4 + 4H_2O \xrightarrow[\text{in EtOH}]{\text{[catalyst]}} Si(OH)_4 + 4EtOH \qquad (3)$$

$$n \cdot Si(OH)_4 \longrightarrow n \cdot SiO_2 + 2n \cdot H_2O \qquad (4)$$

CHEMICAL CONDITIONS FOR AEROGEL SYNTHESIS

When NH_4OH or CH_3NH_2 was used to catalyze reaction 3 in ethanol, a finely divided white precipitate formed which dried to an opaque white powder. Alkaline catalysts were abandoned when hydrofluoric acid was found to produce alcogel of better quality. A further refinement occurred when a combination of HF and HCL was used in catalytic amounts.

The concentrations of the reagents and catalysts were varied to determine their effects on aerogel structure. The following formulation is the composition of the alcosol which has consistently produced the best (transparent and porous) aerogels:

Alcosol Composition "A"

Volumetric Portions @ 25°C	Mole Fractions	
95.6 ml. EtOH	X_{EtOH}	= 0.4423
66.9 ml. Si(OEt)$_4$	$X_{Si(OEt)}$	= 0.0809
32.4 ml. H$_2$O	X_{H_2O}	= 0.4674
0.50 ml. 37 wt % aq. HCl	X_{HCl}	= 0.0016
1.00 ml. 48 wt % aq. HF	X_{HF}	= 0.0078
Σ = 196.4 ml. Ideal Volume (actual volume = 192.6 ml.)	Σ = 1.00	

It can be seen that just three components (EtOH-Si(OEt)$_4$-H$_2$O) make up more than 99 mole % of the alcosol solution. Hence, a ternary diagram accurately represents the composition and solubility limits of the alcosol. Such a diagram, with experimental phase equilibrium curves at two temperatures of interest, is shown in Figure 1.

The alcosol composition given above is shown as point "A" in Figure 1. The elliptical region around point "A" is the region where generally good aerogel has been obtained. In general, a higher initial silicate concentration in the alcosol results in a more transparent aerogel. Because considerable water must be present to hydrolyze the ethyl silicate completely, not much more than 10% silicate can be used before phase separation of the reaction mixture occurs. Only homogeneous alcosols are capable of producing clear aerogel.

Moving in the direction of decreasing ethyl silicate concentration results in successively whiter and more opaque aerogel. It was also found that using only the stoichiometric proportion of water to ethyl silicate (4:1) produced an aerogel that was very clear but too dense and brittle to be of value. Increasing the initial water to silicate ratio to 6:1 increased the porosity without sacrificing clarity. Further water increases above an 8:1 ratio began to yield a porous but opaque, white material.

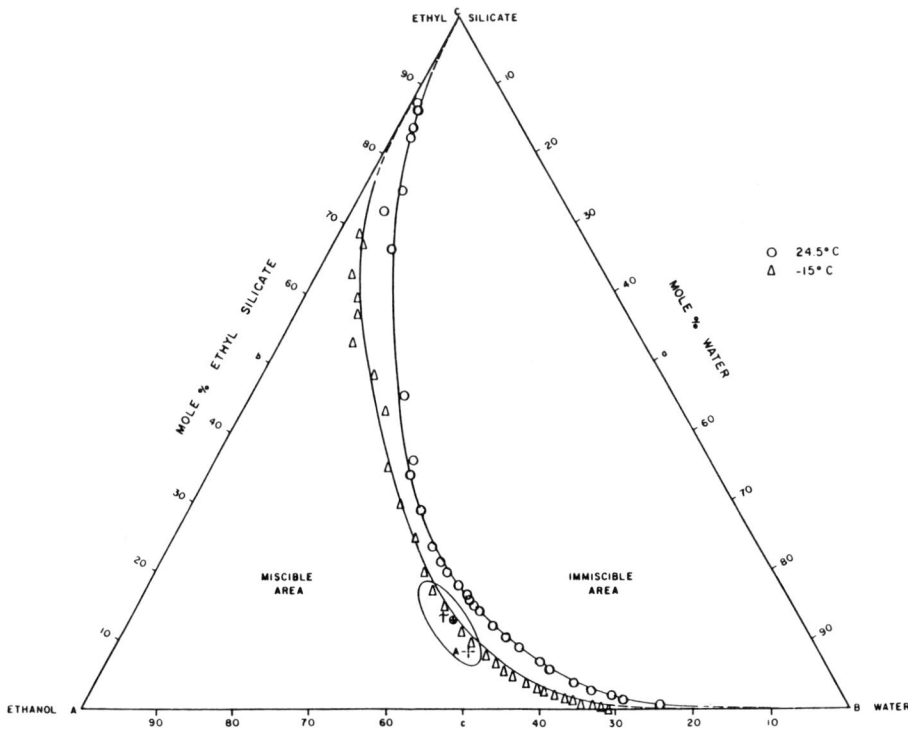

Figure 1. A ternary composition diagram for the tetraethyl silicate-ethanol-water system which shows solubility limits at -15°C and at 24.5°C.

PHYSICAL PROCESSING CONDITIONS

The physical processing conditions of the alcosol are very important from start to finish. It was discovered and repeatedly verified that the alcosols should be mixed at a moderate temperature (ambient) and then cooled quickly and held at rather low temperatures immediately after mixing. Apparently the lowest possible gelation temperature is desirable, but the reaction kinetics slow exponentially making the gelation time rather lengthy. Both the rate of cooling (20°C/hr. was used) and the ultimate temperature (-20°C) influence the transparency of the final product.

The procedure developed for the production of aerogel employing the volumetric proportions given as Composition A is as follows: All of the reagents are liquids at room temperature and they were added (in the order listed) to a 14 cm. Pyrex^R petri dish equipped with magnetic stirring. After mixing for 2-3 minutes, the stirring bar was removed and the dish was covered with a glass lid. The alcosol was placed in a freezer where it cooled to -20°C in about 2-3 hours. After 24 hours at -20°C the solution gelled rather suddenly. At that time a thin layer of cold ethanol was carefully poured onto the exposed gel surface to prevent it from drying. The alcogels were then aged an additional 1-3 days in the freezer before they were removed and allowed to warm to room temperature (12 hours). Within 24 hours after removing the alcogels from the freezer, they were placed in an autoclave and dried supercritically.

SUPERCRITICAL FLUID DRYING

The unit operation of supercritical fluid drying was apparently first proposed by Kistler for the sole purpose of making aerogels [1]. He realized that colloidal gels collapse and densify upon evaporative drying due to the action of the liquid surface tension acting on the walls of a drying pore. The gels would not collapse if a liquid-vapor interface was never allowed to form, but they could not be dried unless the liquid was somehow transformed to a vapor and replaced by dry gas.

There are three ways to convert the liquid into a vapor: evaporation, sublimation, and hypercritical fluid transformation. Evaporation must be avoided because of the surface tension problem. Sublimation, or freeze drying, requires that the liquid first be turned into a frozen solid. Although the liquid → solid → vapor route would avoid the liquid-vapor interface, it would create a discontinuous density transition at the freezing point which might rupture the colloidal network. Furthermore, the mass transfer rate would be very low at the low temperatures involved.

Supercritical removal of the solvent avoids the discontinuous phase transition by effecting a liquid → fluid → gas transformation that never intersects a saturation curve. Figure 2, which is a pressure-volume-temperature diagram for ethanol, is useful in visualizing how this can be done.

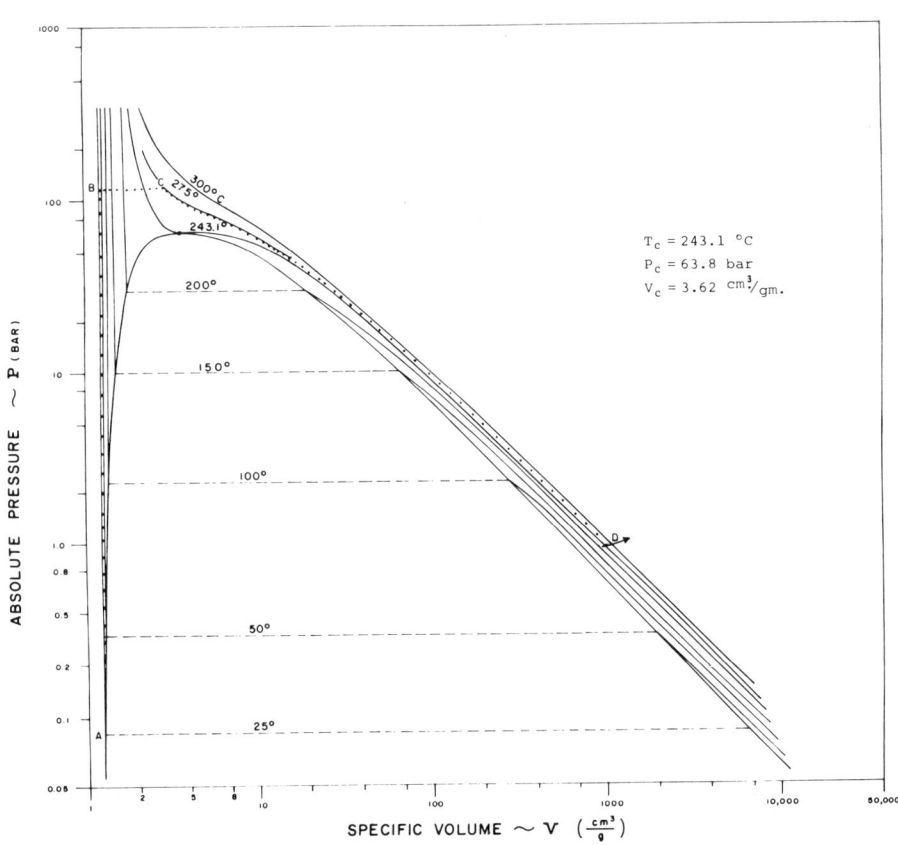

Figure 2. The pressure-volume-temperature diagram for ethanol.

Consider a closed system initially full of liquid ethanol, slightly subcooled. The conditions of the ethanol are shown as point "A" on the diagram. Applying heat to the system raises the liquid temperature and the internal pressure quickly rises. When the system pressure reaches some preselected value above the critical pressure, e.g. point "B", the system may be opened to allow mass to leave in order to keep the pressure constant as the temperature continues to rise. Heating may be continued at constant pressure until the liquid reaches the critical temperature.

Further heating pushes the ethanol into the fluid regime, e.g. point "C".

At point "C", typically at a temperature 10 to 30°C above T_c, a maximum system temperature is established. The escaping ethanol fluid now acts to slowly depressurize the system while the temperature is held at its maximum value. This isothermal depressurization continues until the system pressure equals that of the atmosphere once again. Point "D" has been reached, and there is an equilibrium amount of pure ethanol vapor left in the apparatus.

To remove all remaining traces of potentially condensable vapors, one can use a dry, inert gas such as nitrogen to flush the system. This is the meaning of the arrow passing through point "D". Heating can be discontinued at any time after the nitrogen flush has been started, and the contents return to room temperature, thoroughly dry.

Silica aerogels were produced by placing alcogels in a 20-liter autoclave, filling the autoclave with ethanol, and drying to aerogel by the supercritical route ABCD indicated on Figure 2. Note that there is a small amount of water present. Thus a temperature of 275°C was used to exceed the (unknown) critical temperature of the mixture.

Figure 3 is a diagram of the experimental apparatus. The handpump was used to pressurize the system initially, to compress trapped air in the lines, and to check for leaks. The pressure control valve was a back-pressure regulator with a variable setting adjustment. This was typically set to maintain a system pressure of roughly 120 bar during the heating process. It was bypassed completely when the pressure was vented in the later stages of the run. All tubing was 1/8" or 3/32" type-316 stainless steel fitted with standard high pressure reverse thread couplings.

PHYSICAL PROPERTIES OF SILICA AEROGEL

Two of the most important physical properties associated with the aerogel are its density and porosity. The density can be determined by accurately weighing an irregular sample followed by immersion of the specimen in mercury to determine the bulk volume. The porosity (ε = void volume/total volume) can then be calculated from the known densities of amorphous silicon dioxide (2.2 g/cm^3) and air at room temperature and pressure. Table 1 lists some measured values and the corresponding composition of the alcosol.

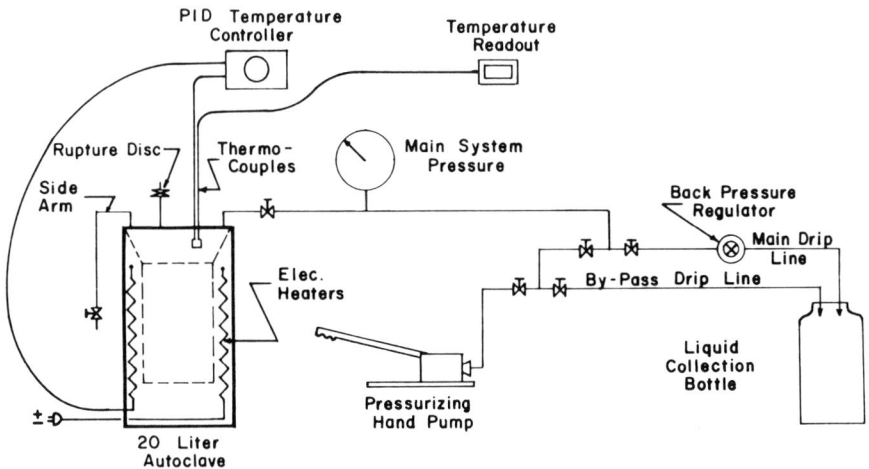

Figure 3. Diagram of the apparatus used for supercritical drying of alcogel.

Table 1. Density and Porosity of Several Aerogel Specimens

Specimen	Mole % Si(OEt)$_4$	Mole % Water	ρ(g/cm^3)	Porosity
13*S	4.2	31.3	0.060-0.065	0.972
13C3	6.1	60.8	0.10-0.11	0.953
20A1	8.0	48.2	0.141	0.937
20B1	8.4	50.2	0.156	0.930

It can be seen that the density increases in direct proportion to the amount of ethyl silicate present in the original alcosol. The denser gels are transparent, and the light material is pure white and opaque. Specimen 20A1 was

453

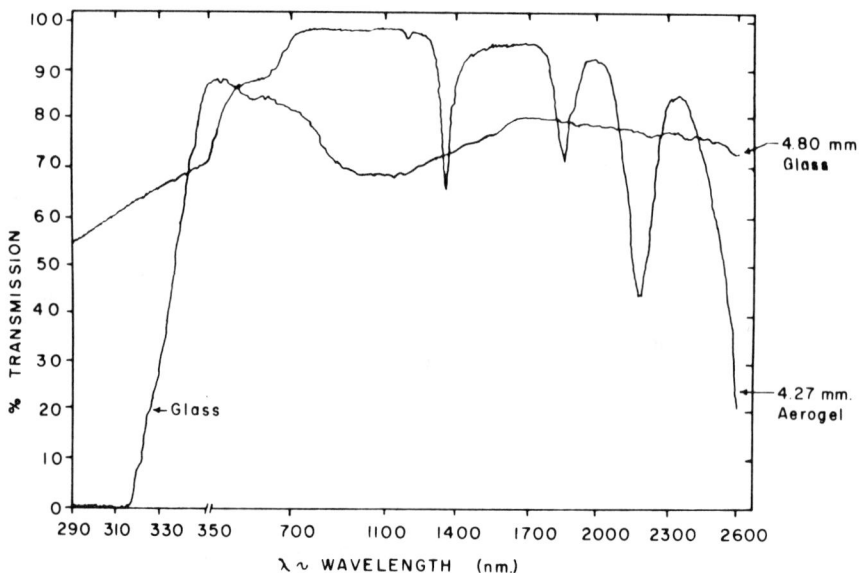

Figure 4. Transmission spectra of aerogel and glass.

probably the most transparent aerogel created over the course of this work.

OPTICAL ABSORPTION

The near infrared-visible-near ultraviolet transmission spectra of an aerogel slab and of commercial float glass are shown in Figure 4. The aerogel was desiccated by heating in an oven @ 270°C for 3 hours. Since both glass and aerogel are amorphous silicon dioxide it seems reasonable that one spectrum would mimic the other, which is indeed the case. Hydroxyl groups (Si-OH and H-OH) are probably responsible for the peaks found in the I.R. spectrum of the aerogel but not in that of glass.

INDEX OF REFRACTION

The refractive index of an aerogel is its most unusual property. No other solid form has an index of refraction lower than 1.3, yet aerogels can be made to cover the range 1.04 to 1.10. The low refractive index of the material creates the visual illusion that the phase boundary is actually disappearing into thin air.

A simple and accurate way of determining the index of refraction of a thin planar sample was used in this investigation. A Michelson interferometer was used to measure the change in path length required to locate bright light fringes before and after the introduction of the aerogel into one of the two light paths. The aerogel sample was desiccated before the measurements were taken. Four readings were recorded resulting in an average value of $n = 1.051 \pm 0.0015$ for the index of refraction. This result is in accord with the index of Cherenkov-radiation-quality aerogel produced for CERN of $n = 1.046$ to 1.054 as reported by Bourdinaud et al. [11]. Their aerogels were prepared by base-catalyzed hydrolysis of tetramethyl orthosilicate.

THERMAL CONDUCTIVITY

The thermal conductivity of the transparent aerogel is of primary interest since the potential merit of aerogel as a window lies in its ability to insulate against the conduction of heat. Two separate thermal conductivity experiments were devised, one steady-state and one unsteady-state measurement. Literature values for comparison can be found in the Chemical Engineers' Handbook [12]. For silica aerogel (250 Å pore diameter) filled with nitrogen at 628 mm Hg, a value of 0.0113 BTU/hr-ft-°F is reported[+].

Steady-state measurements were made of the heat flux and temperature drop across a flat slab of aerogel exposed to the atmosphere. The density of the aerogel sample was 0.119 g/cm^3, its thickness was 10.05 mm, and the cross-sectional area was 25.8 cm^2. The temperatures on the two faces were approximately 99°C and 20°C. Three measurements yielded thermal conductivity values of 0.0275, 0.0308 and 0.0321 BTU/hr-ft-°F, at atmospheric pressure of 746 mm Hg, which may be compared with the value 0.013 BTU/hr-ft-°F for air at 25°C, 1 atm. [13].

A transient heat conduction measurement was also made

[+]Lange [13] reports k = 0.013 BTU/hr-ft-°F @ 248°F. See also reference [14].

in which the temperature of a microthermocouple embedded in the center plane of the aerogel slab was monitored while the surfaces were held at an elevated temperature relative to a uniform initial temperature. Analysis of these data indicated a much lower conductivity, on the order of 0.005 BTU/hr-ft-°F.

One might expect the conductivity of low-density aerogel to be approximately that of the gas filling its pores unless the pore diameters are smaller than the mean-free-path of the gas. Scanning electron micrographs of the aerogel revealed an open structure with an abundance of pores in the 400-600 Å diameter range so that anomalous gas conductivities are possible.

An attempt was made to measure the conductivity of the aerogel at a reduced gas pressure. A steady-state measurement indicated a conductivity value of 0.011 BTU/hr-ft-°F with air at 598 mm Hg in the pores.

On the basis of these observations we concluded that the thermal conductivity of low-density aerogel can be represented approximately as that of the gas filling its pores.

COMPRESSIVE AND TENSILE STRENGTH

Qualitatively, simply pulling and squeezing an aerogel fragment by hand indicates that the material is very weak in tension but moderately strong in compression. It was decided to quantify the strength of the transparent aerogel by subjecting several specimens to loading tests.

Several planar slabs were used for the compressive test and tall necked samples for the tensile test. The analysis was performed on an MTS T5002 Tensile Testing Machine with the applied force vs. linear movement of the grips plotted on an x-y recorder.

At the beginning of the compressive response curve there is a definite linear relationship between stress and strain for which a Hookean modulus can be defined. Cracks in the matrix began to form near the end of this linear region, and became extensive as greater force was applied. When the force was removed the material did not spring back to its original thickness. The average maximum stress in the linear region was $1.1 \times 10^6 \text{N/m}^2$, or 164 psi, with a modulus of $2 \times 10^6 \text{N/m}^2$.

The tensile tests were less accurate due to the very

low forces being measured and because of faults, surface scratches, or other unobservable defects in the neck region. A modulus was not measured, but the observed tensile stress at failure, which ranged from 4000 to 12000 N/m^2, or about 1.0 psi, gives some indication of the low tensile strength of the material.

DIELECTRIC CONSTANT

The dielectric constant (κ) of powdered aerogel was measured in a parallel aluminum plate capacitor. The capacitance was measured with an oscillating bridge operating at 1 MHz. Powder was packed between the plates in increasing amounts in an attempt to verify the linear ($\kappa-1/\kappa+2$) <u>vs.</u> density relationship as predicted by the Clausius-Mossotti equation [15], but no such relationship was found over the small range of densities obtainable. What was found was that the dielectric constant is relatively independent of the packing density in the range $0.14 < \rho < 0.17$ g/cm^3. This value, reported at the average density, is $\kappa = 1.144$.

From electromagnetic theory the dielectric constant is related to the refractive index of the material by the relation $n = \sqrt{\kappa}$. The refractive index of somewhat more transparent aerogel was independently found to be $n = 1.051$, which agrees with $\sqrt{\kappa} = 1.069$ reasonably well. Less transparent material was used in the dielectric test which would have a higher index of refraction.

PRODUCTION OF AN AEROGEL WINDOW

After a recipe (composition, optimum gel temperature, aging time, etc.) leading to suitably transparent aerogel has been developed, it is possible to cast the alcosol in a mold of any desired shape. The chosen geometry must include adequate surface area to allow for mass transfer of the solvent out of the drying solid. One problem noticed experimentally is that alcogels high in ethyl silicate concentration (the most transparent variety) do not support themselves well as tall right vertical cylinders. Such alcogels, while aging, are prone to develop faults or visible cleavage planes near the bottom of the structure, presumably due to the weight of the material above it.

It was originally hoped that large blocks of aerogel could be produced, cut into thin sections, planed or polished flat, and sandwiched between two panes of commercial flat glass thus creating an insulating aerogel window. Several major problems prevented this from being accomplished. The material does not polish at all; every attempt to do so only scratched the surface to the point of opacity. Furthermore, the material is so fragile that breakage and waste would be a significant manufacturing problem.

It was found that the alcogel can be formed *in situ* between two pieces of flat glass and dried from the exposed edges. Removable stainless steel sides were secured to the glass with liquid neoprene, the top plate of glass being held at a fixed spacing above the bottom by several small pins protruding from the metal sides. The mold was precooled to freezer temperature and a freshly mixed batch of alcosol solution was forced into the cavity through a filling port with a syringe. All air bubbles were worked out at this time. The mold remained in the freezer through the gelling and aging process and was then slowly allowed to warm to room temperature.

At this point the sides were carefully removed leaving the top glass plate supported only by the strength of the alcogel, and free to settle as the gel shrinks slightly during the drying process. This "free settling" top glass plate is an essential part of the drying geometry; fixed spacing (such as steel gap spacers at each corner) were found not to work as well.

It was anticipated that there might be a problem evacuating all of the interstitial ethanol through the edges, but this problem did not arise. The drying solid is very permeable to fluid ethanol, and a 48-hour depressurization of a 16 cm. square window (5 mm. thick) was quite sufficient to completely dry the window. A small amount of shrinkage (~ 5-10%) both radially and vertically does occur.

The aerogel forms a tight bond to the glass surfaces. This glass-aerogel bond is actually stronger than the aerogel-aerogel bond in tension; thus the weakest part of the window is located at the centerplane of the aerogel slab. A frame can be cemented around the window edges to protect it from moisture and to strengthen the entire assembly. If desired, the aerogel space can be partially evacuated before sealing. This increases the thermal resistivity of the window, and the aerogel is strong enough in compression to maintain the glass spacing against the pressure of the atmosphere.

CONCLUSIONS

Silica aerogel is a microscopically homogeneous mixture of silicon dioxide and air existing as a solid in a highly expanded porous state. Due to the containment of a large amount of dispersed air, the material displays some bulk properties more characteristic of the gas than of a typical solid. The aerogels produced here ranged in appearance from white to nearly transparent and in texture from soft and spongy to hard, brittle material.

One of our goals was to create an aerogel window with both good optical properties and thermal insulation capability by laminating optically transparent material between two panes of glass. The application of such a window would be numerous; skylights, solar collectors, and architectural windows are several of the more obvious applications. Several aerogel windows of good quality were produced. With a better understanding of the process an even more transparent material could probably be obtained. If smaller average silica particle diameters could be attained, the aerogels would not scatter as much blue light as they presently do. The insulative properties of the aerogel can be adjusted by varying the thickness, pressure, or the gaseous species within the pores.

Apart from window applications, aerogels offer a wide range of peculiar properties to exploit. Because of their low density, they may be of interest in aerospace applications, perhaps in combination with their optical properties. For other applications, the density, index of refraction, and dielectric constant of the material fall in a range not commonly attained by solid materials so they could be used in place of compressed gases. The porosity of the material, in combination with its mechanical strength under tension, might result in an effective gas phase catalyst support or a matrix for diffusional separations.

The practical production of aerogel requires the use of supercritical drying. An engineering challenge is to devise equipment that can perform such a process rapidly and on a large scale.

REFERENCES

1. Kistler, S. S., "Coherent Expanded Aerogels", J. of Phys. Chem., 36, 52 (1932).
2. Probert, S. D., and D. R. Hub, ed., Thermal Insulation, Elsevier Pub. Co. Ltd., 1968, pp. 1-14.
3. Loser, J. B., "Thermophysical Properties of Thermal Insulating Materials", Midwest Research Inst., Kansas City, Mo. (1964).
4. Reis, H. E., "Structure and Sintering Properties of Cracking Catalysts", Adv. in Catalysis, 4, 99 (1952).

5. Teichner, S. J., et al., "Inorganic Oxide Aerogels", Adv. in Coll. and Interfacial Sci., 5, 245 (1976).
6. Lea, P. J., and S. A. Ramjohn, "Investigation of the Substitution of Ethanol with Liquid CO_2 during Critical Point Drying", Microscop. Acta, 83, No. 4, 291 (1980).
7. Cantin, M., M. Casee, L. Koch, R. Jouan, P. Mestreau, and D. Roussel, Nuc. Inst. and Methods, 118, 177 (1974).
8. Iler, R. K., The Colloid Chem. of Silica and Silicates, Cornell Univ. Press, 1955.
9. Sax, N. Irving, Dangerous Properties of Industrial Materials, Van Nostrand Reinhold Co., New York, Third edition, 1968.
10. Cogan, H. D., and C. A. Setterstrom, "Properties of Ethyl Silicate", Chem. Eng. News, 24, No. 18, 2499 (1946).
11. Bourdinaud, J. B., et al., "Use of Silica Aerogel for Cherenkov Radiation Counter", Nuc. Inst. and Methods, 136, 99 (1946).
12. Perry, J. H., and C. A. Chilton, ed., Chemical Engineers' Handbook, 5th ed., McGraw-Hill, 1973, pp. 11-51.
13. Dean, J. A., ed., Lange's Handbook of Chemistry, 10th ed., McGraw-Hill, 1961, p. 874.
14. Lorrain, P., and D. R. Carson, Electromagnetic Fields and Waves, 2nd ed., W. H. Freeman, 1976, p. 116.
15. Kistler, S. S., "The Relation Between Heat Conductivity and Structure in Silica Aerogel", J. Phys. Chem., 39, 79 (1935).

CHAPTER 23

THE ADSORPTION OF PHENOL FROM
DENSE CARBON DIOXIDE ONTO
ACTIVATED CARBON

R. G. Kander
M. E. Paulaitis
  Department of Chemical
    Engineering
  University of Delaware

ABSTRACT

Experimental results are presented for equilibrium loadings of phenol adsorbed onto virgin activated carbon from aqueous solutions and from supercritical carbon dioxide at elevated pressures. Additional results are reported for the cyclic adsorption/regeneration of activated carbon over several cycles. A thermodynamic model is developed for correlating the phase equilibrium data and for evaluating supercritical carbon dioxide as a solvent to regenerate activated carbon saturated with phenol.

INTRODUCTION

Compressed gases or supercritical fluids have several unique physicochemical properties which make them desirable solvents. These properties include: (1) The ability to control solvent density and thus control the solvent power of supercritical fluid (SCF) solvents. (2) The ability to easily produce appreciable changes in SCF solvent densities and thus readily separate the solvent from dissolved solutes. (3) Desirable mass transfer effects, which result in higher separation efficiencies as compared to typical liquid solvents. In many applications, dense CO_2 is the preferred SCF solvent because it is nontoxic, nonflammable, environmentally acceptable, and relatively inexpensive.

Recently, a number of applications utilizing supercritical fluids have been proposed which involve the adsorption of solutes from SCF solvents at elevated pressures. These include SCF chromatography [1] and the regeneration of activated carbon using supercritical CO_2 [2]. One important requirement for developing potential applications such as these is accurate adsorption isotherms for solutes of interest in the presence of a SCF solvent at various temperatures and over a range of fluid densities. At present, however, very little quantitative knowledge about such adsorption isotherms is available. In this paper, an investigation of the adsorption of phenol from supercritical CO_2 onto activated carbon is described. The experimental data and resulting thermodynamic model provide a well-defined, model system which characterize the phase equilibrium behavior involved in adsorption from supercritical fluids at elevated pressures.

EXPERIMENTAL METHODS

Three experiments have been performed: (1) Measurements of equilibrium loadings for phenol onto activated carbon from dilute aqueous solutions over a range of phenol concentrations. (2) Measurements of equilibrium loadings for phenol onto activated carbon from dilute supercritical CO_2 solutions over a range of phenol concentrations and at different temperatures and fluid densities. (3) Measurements of equilibrium loadings as described in (1) above, for phenol onto activated carbon which has been previously regenerated with supercritical CO_2. The activated carbon used in all of these experiments is Amoco GX-31 activated carbon.

I. Aqueous Adsorption Experiments

The equilibrium loading of phenol onto activated carbon from dilute aqueous solution is determined in a batch experiment by contacting an aqueous solution of known phenol concentration with a predetermined amount of activated carbon. After equilibration, the carbon loading is determined by measuring the change in phenol concentration in the aqueous phase.

The experimental procedure involves preparing up to 16 different aqueous solutions with known phenol concentrations ranging from 0 to 10 mmoles/liter. A Perkin-Elmer Model 552 UV-visible spectrophotometer is used to measure these initial concentrations. A known volume of each solution (ca. 200 ml) is then mixed with a predetermined amount of activated carbon (ca. 0.1 gm) in separate 250 ml Erlenmeyer flasks and agitated at constant temperature (New Brunswick Scientific incubator-shaker) until equilibrium is reached (ca. 500 hrs). Equilibrium is obtained when 0.5 ml samples

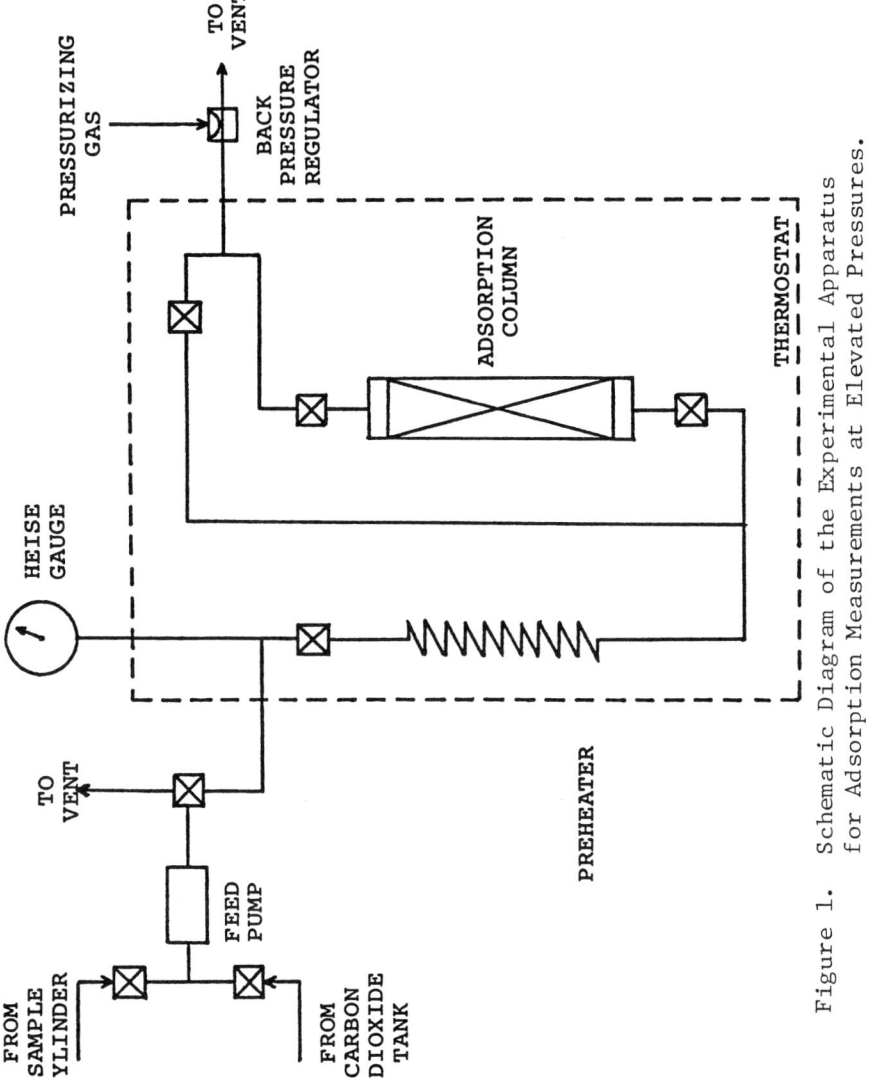

Figure 1. Schematic Diagram of the Experimental Apparatus for Adsorption Measurements at Elevated Pressures.

of the aqueous solution, taken on two consecutive days, show no change in phenol concentration. Since sampling is not initiated until after approximately 400 hrs of contacting, only four or five samples are typically taken prior to equilibration. It has been calculated that this sampling procedure leads to less than a 0.25% change in the overall solution concentration. The equilibrium concentration of phenol in each aqueous solution is then measured using the spectrophotometer, and equilibrium loadings on the activated carbon are determined by simple mass balances.

II. Adsorption from Supercritical CO_2

A schematic diagram of the experimental apparatus for measuring adsorption isotherms from supercritical CO_2 is shown in Figure 1. A mixture of known phenol concentration in liquid carbon dioxide (ca. 1000 psig at ambient temperature) is prepared in a 1000 cc Whitey sample cylinder, which is subsequently cooled in an ice bath. This solution is then charged to the high-pressure liquid feed pump (Milton Roy Co.) where it is compressed, preheated in the constant temperature bath, and passed over the activated carbon bed. Phenol concentrations well below saturation conditions are used, and therefore all of the phenol remains in solution during the initial cooling, compression, and preheating stages. The adsorption column containing the activated carbon bed is constructed of stainless steel high-pressure tubing (6 cm long x 0.2 cm I.D.) with Milton Roy in-line filters placed at each end to retain the carbon bed. The phenol is adsorbed onto the activated carbon and the purified carbon dioxide is then flashed to atmospheric pressure and vented. After the column is saturated, the carbon is removed and weighed to determine the equilibrium loading of phenol.

System pressure is controlled using a back pressure regulator (Grove Co.) and is measured using a Bourdon-type Heise gauge at the inlet to the temperature bath. Fluctuations in pressure due to the reciprocating piston feed pump are typically less than ± 25 psig and flowrates are low enough (less than 5 ml/min) that the pressure drop across the bed is negligible. The temperature within the thermostat is controlled to $\pm 0.3°C$ using a Sargent-Welch Thermonitor and is measured with a calibrated Omega copper-constantan thermocouple.

Saturation of the carbon bed is determined by estimating the time necessary to reach saturation conditions using the following procedure. At the highest fluid-phase concentration of phenol under consideration, a given amount of phenol is passed over the activated carbon bed and the loading is obtained. A larger amount of phenol at the same concentration is then passed over an identical carbon bed and the loading is again determined. This process is continued until the measured loading no longer changes from one run to the next. Thus the minimum amount of phenol needed to saturate the bed is found for the highest phenol concentration in the supercritical fluid. Since the equilibrium loading curve monotonically decreases with decreasing phenol concentration in the fluid phase, this amount of phenol will also saturate the carbon bed at any lower fluid-phase concentrations. The time necessary to saturate the activated carbon bed is thus computed from the known phenol concentration and the flowrate of the fluid phase. All runs are made using twice the time calculated to reach saturation.

Preliminary experiments using pure, supercritical CO_2 were also performed to determine the amount of adsorbed CO_2 present on the activated carbon when measuring carbon loadings. It was found that no significant increase in the weight of the carbon bed could be measured after depressurizing the system at the end of each experiment. Repeated weighings at later times also showed no measurable changes in sample weight. These results suggest that CO_2 adsorption can be neglected when determining the amount of phenol adsorbed on the activated carbon from the measured weight gain of the carbon samples.

III. Cyclic Adsorption/Regeneration Experiments

Both experimental techniques described above are combined in this study to determine the amount of phenol which is irreversibly adsorbed on the carbon bed. Activated carbon samples are first loaded with phenol from aqueous solution as described in the aqueous adsorption experiments. These saturated samples are then regenerated with supercritical CO_2 using the experimental apparatus described in the previous section. This carbon is subsequently re-loaded using an aqueous solution of the same initial phenol concentration, and a second equilibrium carbon loading is obtained. The carbon samples are subjected to three consecutive loading-regeneration cycles with equilibrium loadings measured after each adsorption phase. These loadings for a specific carbon sample can then be compared to show the loss of loading capacity per regeneration cycle due to irreversible phenol adsorption on the activated carbon.

For all the experiments described above, Fisher reagent grade phenol (99.8% pure) is used without further purification. Liquid carbon dioxide ("bone dry") is supplied by Linde. The Amoco GX-31 activated carbon used for the equilibrium measurements is prepared by sieving to a relatively narrow range of particle size (passing through a No. 30 U.S. standard sieve and retained on a No. 40 sieve), washing with distilled water, and drying at 138°C to a constant weight. The carbon samples are then stored in a desiccator to prevent adsorption of moisture from the air. Distilled, de-ionized water (Barnstead Nanopure De-ionizer) is used in preparing all aqueous solutions.

EXPERIMENTAL RESULTS

The adsorption isotherms at 36° and 60°C for phenol onto activated carbon from aqueous solution are presented in Table I. These data were fit to a Toth isotherm [3] of the following form;

$$n = n^o c (b+c^m)^{-1/m} \qquad (1)$$

where n is the carbon loading in mmoles/gm carbon, c is the aqueous phenol concentrations in mmoles/liter, and n^o, b, and m are empirical constants. The parameter n^o represents the

Table I. Experimental Adsorption Isotherms for Phenol on Activated Carbon from Aqueous Solution

T=36°C		T=60°C	
Aqueous Phenol concentration (mmoles/l)	Carbon loading (mmoles/gmC)	Aqueous Phenol concentration (mmoles/l)	Carbon loading (mmoles/gmC)
7.420	5.14	7.725	4.465
5.190	4.70	5.848	4.217
3.070	3.91	.0432	.6597
.948	3.03	.0125	.3208
.481	2.48	.00625	.2000
.213	1.96		
.140	1.71		
.0694	1.35		
.0216	.940		
.0111	.711		

maximum carbon loading at high aqueous phenol concentrations while b and m characterize the curvature of the adsorption isotherm. Table II lists the equation parameters for each isotherm in addition to the average percent errors for the two fits. Figure 2 presents the data from Table I along with the calculated isotherms at the two temperatures. This

Table II. Toth Isotherm Parameters Obtained from Fitting the Experimental Data in Table I

	T=36°C	T=60°C
n^o	21	7
b	.320	.420
m	.1478	.3700
%E	7.0	2.9

$$\%E = \frac{100}{N} \sum_1^N \left\{ \left| \frac{n^{EXP} - n^{CALC}}{n^{EXP}} \right| \right\}$$ where N is the number of data points.

plot shows that carbon loading increases by a factor of approximately 10 to 15 as phenol mole fractions in the aqueous solution increase over three orders of magnitude. The shape of the adsorption isotherms and the order of magnitude of the carbon loadings compare favorably with similar experimental data for different activated carbons [3].

The adsorption isotherms for phenol onto activated carbon from supercritical CO_2 at 36°C/139 atm and 60°C/173 atm are listed in Table III and presented in Figure 3. Specifying pressure in addition to temperature fixes the density of CO_2 and defines a unique loading curve. At 36°C/139 atm the density of CO_2 is 0.787 gm/cc, while at 60°C/173 atm the CO_2 density is 0.658 gm/cc. These two sets of conditions correspond to reduced temperatures/reduced pressures for pure CO_2 of 1.02/1.91 and 1.10/2.38, respectively. Figure 3 shows loading curves that are quite similar to those in Figure 2 for adsorption from aqueous solutions-- i.e., carbon loading again increases with increasing phenol concentration in the SCF phase. In both figures, carbon loading increases with decreasing temperature at fixed phenol concentration in the aqueous or SCF phase. However,

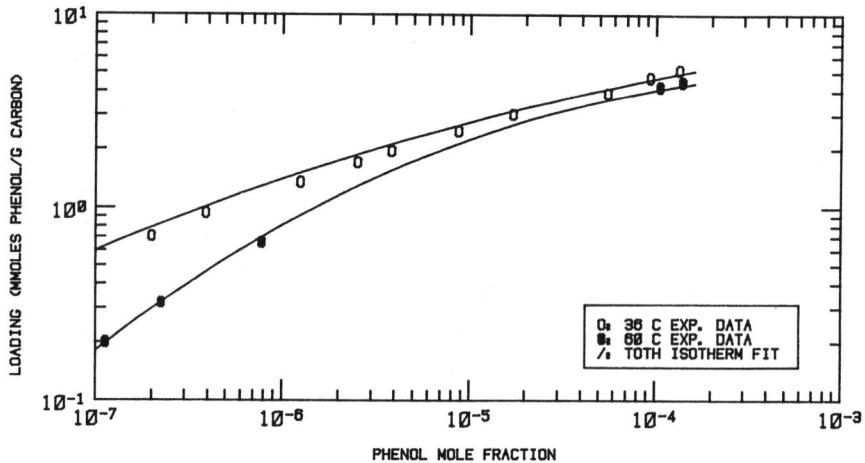

Figure 2. Experimental Adsorption Isotherms for Phenol on Activated Carbon from Aqueous Solution.

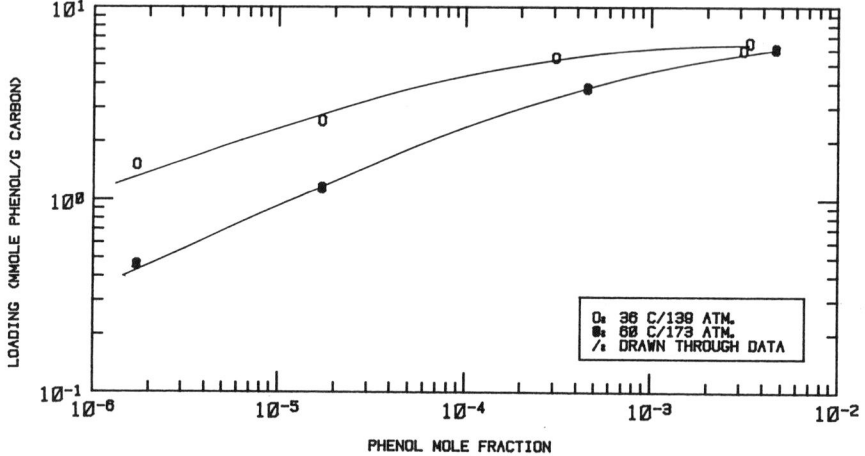

Figure 3. Experimental Adsorption Isotherms for Phenol on Activated Carbon from Supercritical CO_2.

Table III. Experimental Adsorption Isotherms for Phenol on Activated Carbon from Supercritical CO_2

36°C/139 atm		60°C/173 atm	
SCF Phenol concentration (mole frac.)	Carbon Loading (mmoles/gmC)	SCF Phenol concentration (mole frac.)	Carbon Loading (mmoles/gmC)
3.375×10^{-3}	6.50	4.668×10^{-3}	6.04
3.129×10^{-3}	5.94	4.558×10^{-4}	3.78
3.068×10^{-4}	5.45	1.716×10^{-5}	1.149
1.716×10^{-5}	2.561	1.721×10^{-6}	.462
1.721×10^{-6}	1.514		

the increase in carbon loading from 60°C to 36°C is significantly greater for adsorption from supercritical CO_2.

The cyclic adsorption/regeneration results at 36°C, for three regeneration cycles, are presented in Table IV and shown in Figure 4. The data show that carbon loading drops substantially after one regeneration, but does not change appreciably after further adsorption/regeneration cycles. For example, at 10^{-4} mole fraction of phenol in aqueous solution, carbon loading decreases from an initial value of approximately 3.0 mmoles phenol/gm carbon on virgin carbon to nearly 1.4 mmoles phenol/gm carbon after one cycle. Carbon loading after the second cycle decreases only slightly to approximately 1.2 mmoles phenol/gm carbon. These results are consistent with previous observations [4], and indicate that significant reductions in loading capacity of the activated carbon can occur after the first adsorption cycle which are not reversed by regeneration with supercritical CO_2. The discrepancy between the adsorption isotherm for virgin carbon in Figure 4 and the corresponding isotherm at 36°C in Figure 2 is due to the use of carbon as obtained from the supplier for the regeneration experiments--i.e., without preparing the carbon as described in the previous section.

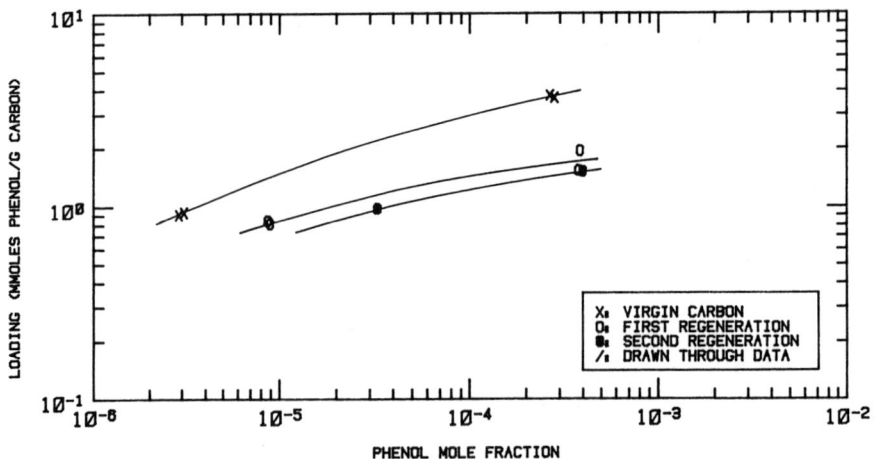

Figure 4. Experimental Adsorption Isotherms for Regenerated Activated Carbon (Three Cycles).

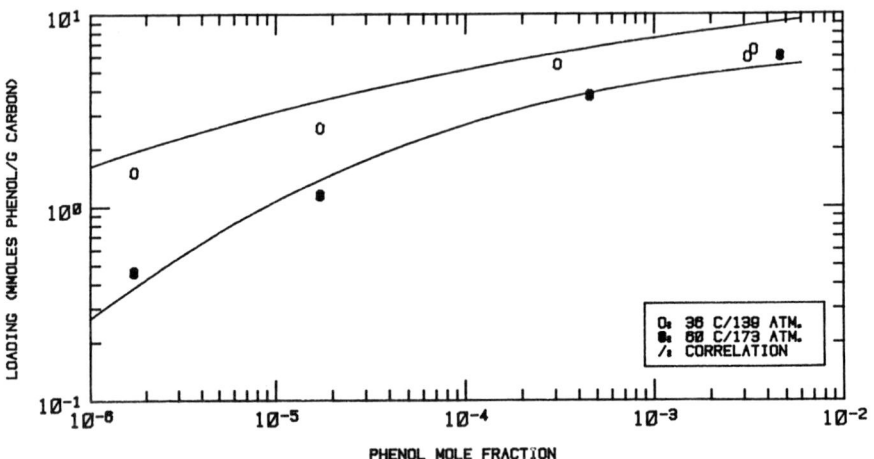

Figure 5. Comparison of Calculated Adsorption Isotherms with the Experimental Data for Phenol from Supercritical CO_2.

Table IV. Experimental Adsorption Isotherms for Phenol on Regenerated Activated Carbon at 36°C

Virgin Carbon:	Phenol conc. (mmoles/l)	Loading (mmoles/gmC)
	6.371	3.65
	6.020	3.76
	.066	.91
	.070	.93

Carbon Regenerated Once:	Phenol conc. (mmoles/l)	Loading (mmoles/gmC)
	8.720	1.95
	8.545	1.54
	.202	.82
	.197	.85

Carbon Regenerated Twice:	Phenol conc. (mmoles/l)	Loading (mmoles/gmC)
	9.017	1.53
	.748	.98

DATA REDUCTION

The experimental carbon loadings for phenol from supercritical CO_2 (Table III and Figure 3) could be correlated using the Toth isotherm expression in Equation (1). However, the value of such a correlation would be restricted to the specific temperatures and CO_2 densities studied. A more useful correlation, which permits reliable interpolation and limited extrapolation of the experimental data, has been developed as described below.

As a first step, the fugacity-loading relationship for phenol on this particular activated carbon is obtained using the experimental results for phenol adsorbed on this carbon from aqueous solution (Table I and Figure 2). The fugacity of phenol (component 2) in aqueous solution at atmospheric pressure and a fixed temperature is given by,

$$f_2 = x_2 \gamma_2 P_2^S \quad (2)$$

where x_2 is the mole fraction of phenol in solution, γ_2 is the corresponding activity coefficient for phenol, and P_2^S is the vapor pressure for pure phenol at the system temperature. Since the experimental mole fractions for phenol are less than 2×10^{-4} (Figure 2), the activity coefficient in Equation (2) can be replaced by its value at infinite dilution, which is evaluated using the UNIFAC group contribution method [5]. Equation (2) can thus be used to compute phenol fugacities associated with the phenol mole fractions in Figure 2, and thereby derive the desired fugacity-loading relationship. Assuming that the fugacity of the adsorbed phenol is independent of the solvent present, this relationship will also apply to phenol adsorbed on this particular carbon from supercritical CO_2.

The next step is to obtain a fugacity-composition relationship for phenol dissolved in supercritical CO_2. This is accomplished using the following expression for fugacities of sparingly-soluble solids in supercritical fluids at a fixed temperature T and pressure P [6]

$$\ln f_2 = \ln x_2 + \ln \gamma_2^\infty(P^o) + \ln f_2^{OL}(P^o) + \int_{P^o}^{P} \frac{\bar{v}_2^\infty dP}{RT} \qquad (3)$$

where x_2 is the mole fraction of phenol in the SCF mixture, γ_2^∞ is the corresponding infinite-dilution activity coefficient at a fixed reference pressure P^o, and f_2^{OL} is the standard-state fugacity for pure phenol at the reference pressure P^o. The last term in Equation (3) is the Poynting correction for pressures P different from the reference pressure, and $\bar{v}_2^\infty$ is the partial molar volume at infinite dilution for phenol in supercritical CO_2. Equation (3) contains two adjustable parameters: γ_2^∞ and the binary interaction parameter used in the mixing rules associated with the Soave-Redlich-Kwong equation of state for evaluating $\bar{v}_2^\infty$ [6]. This latter parameter is assumed to be independent of temperature, while γ_2^∞ is a function of temperature only. Both of these parameters can be obtained from existing experimental data for solid solubilities of phenol in supercritical CO_2 at 36°C and over a range of elevated pressures [7]. The resulting values are 0.13 for the binary interaction parameter and 400 for γ_2^∞ at 36°C and a reference pressure equal to the critical pressure of pure CO_2.

Equating fugacities obtained in these first two steps gives directly the adsorption isotherms for phenol from supercritical CO_2 at 36°C and over a range of SCF densities. No adjustable parameters are required, and thus the experimental isotherm at 36°C and 139 atm in Figure 3 can be predicted using this procedure. To extend this correlation to other temperatures, a single adjustable parameter is used to characterize the temperature dependence of γ_2^∞ in Equation (3)--i.e., the partial molar excess enthalpy for phenol at infinite dilution in supercritical CO_2. This parameter is obtained by fitting the experimental carbon loading for phenol from supercritical CO_2 at 60°C and 173 atm (Table III and Figure 3). The results from this correlation are given in Figure 5 along with the experimental data from Table III. The predicted phenol adsorption at 36°C and 139 atm agrees reasonably well with the experimental data, while the calculated results at 60°C and 173 atm (which includes the one adjustable parameter) show even better agreement with experimental values.

DISCUSSION AND CONCLUSIONS

This thermodynamic correlation can now be used to generate adsorption isotherms for phenol on Amoco GX-31 activated carbon from supercritical CO_2 at selected temperatures and SCF densities. Figures 6 and 7 show two such plots of equilibrium carbon loadings as a function of phenol concentration in the SCF phase at CO_2 densities of 0.80 gm/cc and 0.59 gm/cc, respectively. For each CO_2 density, three adsorption isotherms are given at 35°, 45°, and 55°C. Both figures illustrate that carbon loadings decrease significantly with increasing temperature at constant SCF density. For example, at 10^{-5} mole fraction of phenol in supercritical CO_2 with a density of 0.80 gm/cc (Figure 6), the carbon loading decreases by a factor of nearly 10 when temperature increases from 35° to 55°C. Alternatively, comparing Figures 6 and 7 at a given temperature shows that the effect of solvent density on carbon loading is minimal. For example, at 45°C and a phenol mole fraction in supercritical CO_2 of 10^{-5}, carbon loading changes by about 10% when the CO_2 density is decreased from 0.80 gm/cc to 0.59 gm/cc.

This behavior indicates that the temperature dependence of the adsorption equilibrium dominates any possible solubility effects of the compressible SCF solvent. Comparison of the adsorption isotherms in Figures 6 and 7 with those from aqueous solutions (Figure 2) shows that, although the temperature dependence for the adsorption of phenol from supercritical CO_2 is greater than that from water, carbon loadings are about the same order of magnitude in both cases. Thus it would appear from these results that supercritical CO_2 offers no significant thermodynamic advantage for regenerating activated carbon saturated with strongly-adsorbed

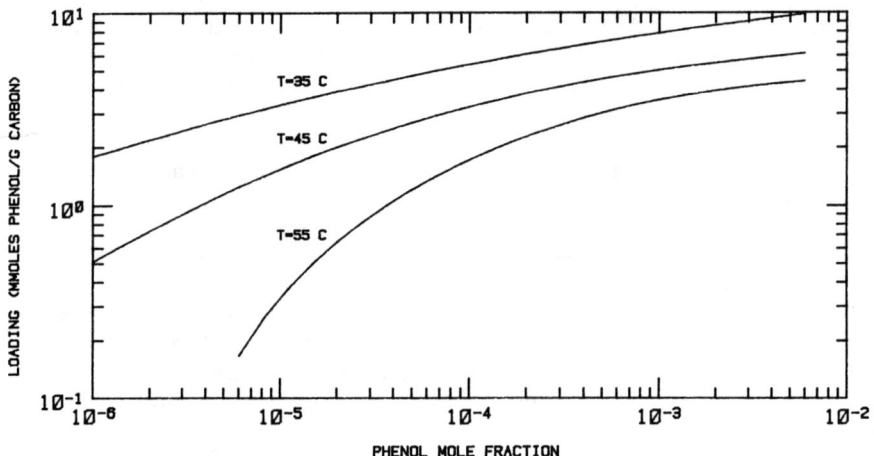

Figure 6. Calculated Adsorption Isotherms for Phenol on Activated Carbon from Supercritical CO_2 at a Density of 0.80 gm/cc.

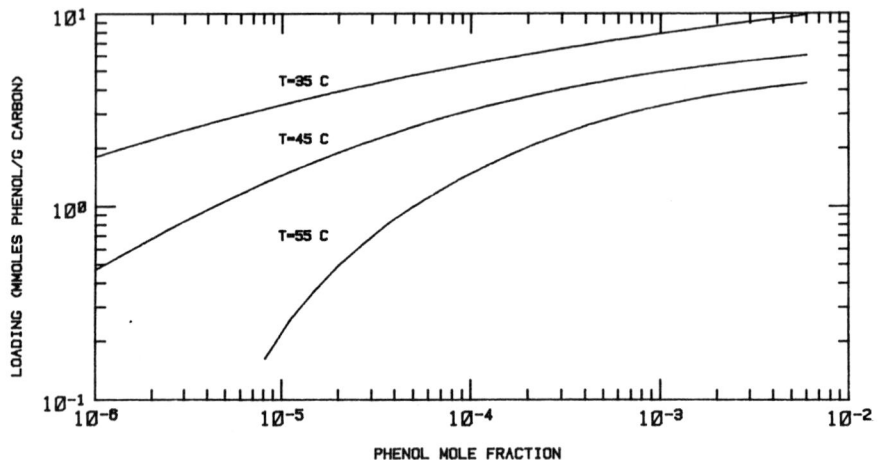

Figure 7. Calculated Adsorption Isotherms for Phenol on Activated Carbon from Supercritical CO_2 at a Density of 0.59 gm/cc.

solutes, such as phenol. Regeneration of activated carbon saturated with organic solutes that do not strongly adsorb, however, may still be a promising alternative to conventional techniques, such as steam regeneration (where the water-solute separation can be costly). In such cases, however, SCF extraction of these solutes from dilute aqueous solution should also be considered as an alternative to the direct regeneration of activated carbon with supercritical fluids. Furthermore, other important SCF solvent properties, such as improved mass transfer, must also be considered in the evaluation of the SCF regeneration process. These properties have been shown to be of primary importance in SCF chromatography [8].

In conclusion, this work describes a method for correlating and reliably extending experimental data for adsorption equilibrium from supercritical fluids. The method can also be used to estimate these isotherms from a limited amount of experimental information--such as carbon loadings from aqueous solution and solubilities in supercritical fluids--on solutes and activated carbons of particular interest. Estimating adsorption equilibria from supercritical CO_2 could be readily accomplished for an appreciable number of solutes, since an extensive amount of experimental data is available on solubilities in supercritical CO_2 [9].

ACKNOWLEDGMENT

The work upon which this publication is based was supported in part by funds provided by the United States Department of the Interior, Office of Water Research and Development Act of 1978, Public Law 95-467.

LITERATURE CITED

1. Klesper, E., Angew. Chem. Int. Ed. Encl. 17, 738 (1978).

2. Modell, M., R. P. de Filippi, and V. Krukonis, "Regeneration of Activated Carbon with Supercritical Carbon Dioxide," Div. Env. Chem., Am. Chem. Soc., Miami, FL, Sept. 1978.

3. Jossens, L., J. M. Prausnitz, W. Fritz, E. U. Schlünder, A. L. Myers, Chem. Eng. Sci. 33, 1097 (1978).

4. Modell, M., R. J. Robey, V. Krukonis, and R. P. de Filippi, "Supercritical Fluid Regeneration of Activated Carbon," 87th Nat. Mtg. AIChE, Boston, MA, Aug. 1979.

5. Magnussen, T., P. Rasmussen, and A. Fredenslund, IEC Proc. Des. Dev. 20, 331 (1980).

6. Mackay, M. E. and M. E. Paulaitis, IEC Fund. 18, 149 (1979).

7. Van Leer, R. A. and M. E. Paulaitis, J. Chem. Eng. Data 25, 257 (1980).

8. Gere, D. R., R. Board, and D. McManigill, Anal. Chem. 54, 736 (1982).

9. Francis, A. W., J. Phys. Chem. 58, 1099 (1958).

CHAPTER 24

ANALYSIS OF DENSE
(SUPERCRITICAL) GAS SYSTEMS

L. G. Randall
  Hewlett-Packard
  Avondale, PA 19311

ABSTRACT

Dense (supercritical) gas solvents are particularly attractive for use with thermally labile and/or involatile substances and have been successfully used as extractive solvents for complex mixtures. Sampling of dense gas systems with on-line or off-line analysis (e.g., gas chromatography, high performance liquid chromatography, mass spectrometry, etc.) will be described in terms of feasibility, sensitivity, and analysis time.

INTRODUCTION

Within the last decade and particularly the last several years there has been a tremendous growth of interest in supercritical or dense gases as solvents, especially for extraction, as evidenced by this symposium and the numerous recent publications (for a current, extensive review, see Ref. 1). Compared to the common separation processes of liquid solvent extraction and distillation, dense gas extraction has several important characteristics:
1. extraction temperatures are usually close to critical temperatures so mild conditions can be used for thermally labile compounds,
2. dense gases can dissolve involatile compounds,
3. compounds can be selectively dissolved by changing the density of the gas since the solvent power of the dense gas is a function of density,
4. three dense gas parameters, density as well as temperature and composition, can be easily varied,
5. essentially complete separation of solvent from solute with high solvent gas recovery can be accomplished by isothermal decompression, isobaric heating or changing

both temperature and pressure,
6. the solutes can be fractionated during the solvent gas/solute separation, and
7. the dense solvent gas viscosity and diffusivity result in more rapid penetration and mass exchange within a matrix (e.g., packed beds, coal, ground tissue).

Because of these appealing properties it is widely acknowledged that supercritical gas solvents will prove to be superior to liquid solvents in many areas such as the food, pharmaceutical, and solid fuel industries. Hence, it seems appropriate to evaluate methods that are presently available for chemical analysis of such high pressure gaseous systems to determine which could most significantly hasten the development of dense (supercritical) gases as solvents.

A few preliminary statements clarifying the term "dense gas" as used in this paper should be made. A dense gas (also referred to as a supercritical fluid, high pressure gas, hyperpressure gas, ultra high pressure gas) is a gas at temperatures and pressures such that its density is comparable to normal liquid densities of the substance. As shown in Figure 1, for a gas at temperatures just above the critical temperature (T_c) of the substance, liquid-like densities are rapidly approached with modest increases in pressure in the range of 0.7 to 2 times the critical pressure (P_c). This author prefers the term "dense gas" to emphasize the fact that in solute/dense gas solutions once the solvent gas is chosen (thus setting the range of the solvent power), the most important parameter governing the solvent power is then the density and not either the pressure or temperature alone.

OVERVIEW

Chemical analysis of dense gas systems is an extremely broad topic because it encompasses a diversity of goals, a large variety of dense gas systems, a multitude of analytical techniques, and many mechanical ways to interface a high pressure system to an analysis scheme. These areas will be briefly outlined with some mention of commonly used sampling and analysis schemes. Then an alternative, powerful approach, Dense Gas Chromatography/Mass Spectrometry (DGC/MS), will be described in greater detail; DGC/MS may be a more versatile, informative technique than any other single method presently used.

Purpose of the Analysis

The most common experiments are solubility studies over a range of temperatures (reduced* temperatures ~ 0.9-1.3,

---

* Reduced parameters are actual values normalized to critical values: reduced pressure = pressure ÷ critical pressure, etc.

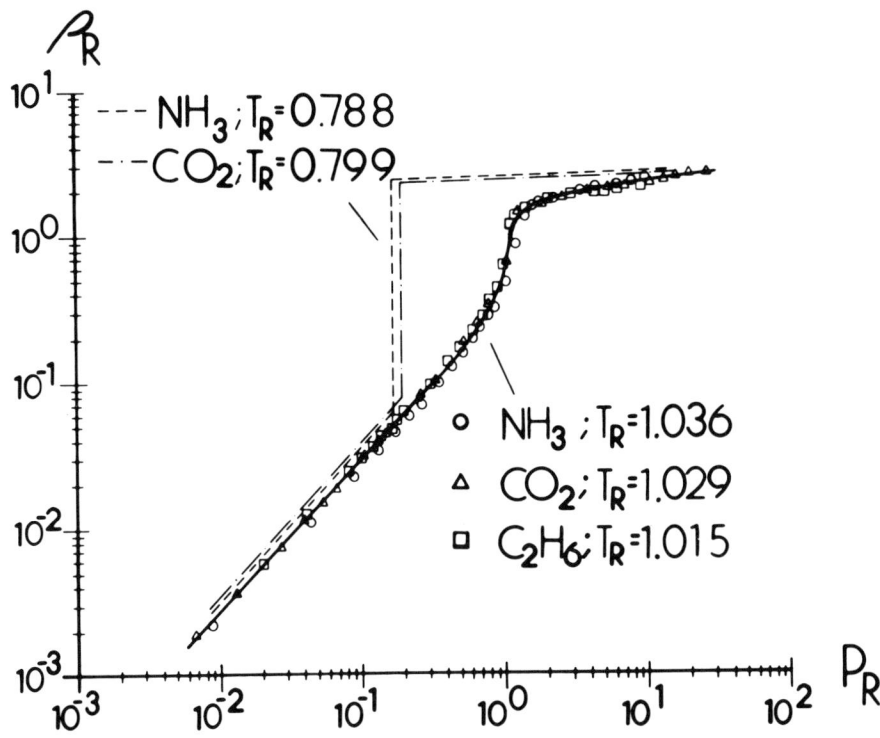

Figure 1. Reduced density (ρ_R) as a function of reduced pressure (P_R) for three gases at reduced temperatures (T_R) close to 1. Reduced isotherms for carbon dioxide ($T_R = 0.799$) and ammonia ($T_R = 0.788$) show density as a function of pressure for subcritical gas and liquid.

T_c ~ 10-300°C) and pressures (reduced pressures of up to 20, and P_c ~ 36-220 atm). Solubilities extend from parts per billion to parts per hundred: e.g., < 1 μg glucose or glycine per normal* liter of carbon dioxide [2] to hundreds of milligrams naphthalene per normal liter carbon dioxide [3].

Information from solubility studies is then applied to mixtures which are often complex matrices such as plant and animal tissues where the component of interest can be a trace

---

* 1 normal liter = 1 liter at 0°C and 1 bar.

or a major component. Matrix characterization for a wide range of solvent gas temperatures and densities where there is interest in all extracted components (trace through major) and where the raw sample size can be grams to kilograms, as in bench top reactors and small pilot plants, is the most ambitious analytical goal.

Still other studies, experimentally very difficult, are concerned with monitoring chemical reactions in the supercritical phase where the solvent gas may be a reactant. Supercritical reactions relevant to extraction processes are already being studied by those interested in supercritical extraction/liquefaction of coal and lignite.

Therefore, the analytical purpose of the proposed experiments--to analyze some component (single knowns to hundreds of unknowns) dissolved by (possibly reactive) dense solvent gases (with solubilities from parts per billion to parts per hundred) from a system (charges of milligrams to kilograms; inert to reactive, complex matrices)--is certainly very broad. The other areas to be discussed (the dense gas system, the analytical method, and the interface between them) are largely defined by the analytical purpose.

Dense Gas System

The variables of dense gas systems which must be considered are
1. number of solutes,
2. chemical range of solutes,
3. identity of solutes,
4. solvent gas critical parameters,
5. solvent gas chemical nature,
6. number of components in the solvent gas (supercritical gas plus entrainers--also referred to as modifiers or azeotrope formers), and
7. support/matrix effects.

Obviously, whatever the analytical purpose is, the component to be extracted must be soluble in the dense solvent gas. The chemical identity of the solvent gas including the presence of an entrainer, the solvent gas density, and the system temperature can all be varied to achieve component miscibility with the solvent gas. When a number of components with a spread in chemical nature and molecular weight are to be extracted, perhaps with selective solvation or selective fractionation upon solute precipitation, the solvent gas will have to be capable of a similar range in its solvent power, requiring adjustment of any or all of the three important parameters, density, composition, and temperature, during the extraction.

The selection of the most appropriate solvent gas for dissolving all desired components then influences the system engineering design (temperature, operating pressure, parameter programming) as well as the analytical method and its design. For example, supercritical toluene at necessarily high temperatures and pressures will probably require more design effort for the entire system than supercritical carbon dioxide at lower temperatures. High temperatures might be avoided while still achieving a desired solvent power if dense solvent mixtures of gas plus entrainer are used, e.g., acetone in carbon dioxide; then it is necessary to determine effective entrainers for particular solute/solvent gas systems and design to exploit the temperature-dependent ternary phase equilibria. In some systems an entrainer in the dense gas interacts with the matrix or support rather than changing the solvent power of the solvent gas. Whatever the entrainer function is, the problems of first preparing the solvent mixture and then minimizing entrainer interference with extract analysis after extraction remain.

Once the solvent gas with specific temperature, density, and composition ranges is chosen, some detection and interface requirements of the analytical scheme are necessarily defined. For example, for ultraviolet (UV) monitoring of a supercritical solution, dual cells would be necessary for toluene (a UV absorber) and for some entrainer/carbon dioxide mixtures but not for carbon dioxide. However, monitoring after cooling and decompressing a supercritical solution to ambient temperature and pressure results in solutes precipitated from an expanded carbon dioxide solution, a liquid solution from a supercritical toluene solution, and either from an entrainer/carbon dioxide solution depending upon entrainer volatility--with each different kind of product entailing different sample handling thereafter.

Analytical Methods and Accompanying Interfaces

A first approach to characterize this part of the analysis problem yields a classification in which there are two choices available for solute detection and quantitation. The first is to monitor the solvent gas/solute solution--which can be done at supercritical conditions, subcritical liquid conditions where the temperature and pressure have been changed to yield a liquid solution, or subcritical gaseous conditions where the pressure has been lowered to yield a vaporized mixture. The second is to separate the solute from the solvent dense gas by a change in temperature and/or pressure and then to measure the solute and solvent independently. However, classification as to solution or isolated solute analysis appears to be less important than the decision whether (and where) the solution changes from supercritical conditions as it travels from the dense gas system to the final detector.

A logic diagram outlining the many possibilities is shown in Figure 2 which distinguishes between those schemes

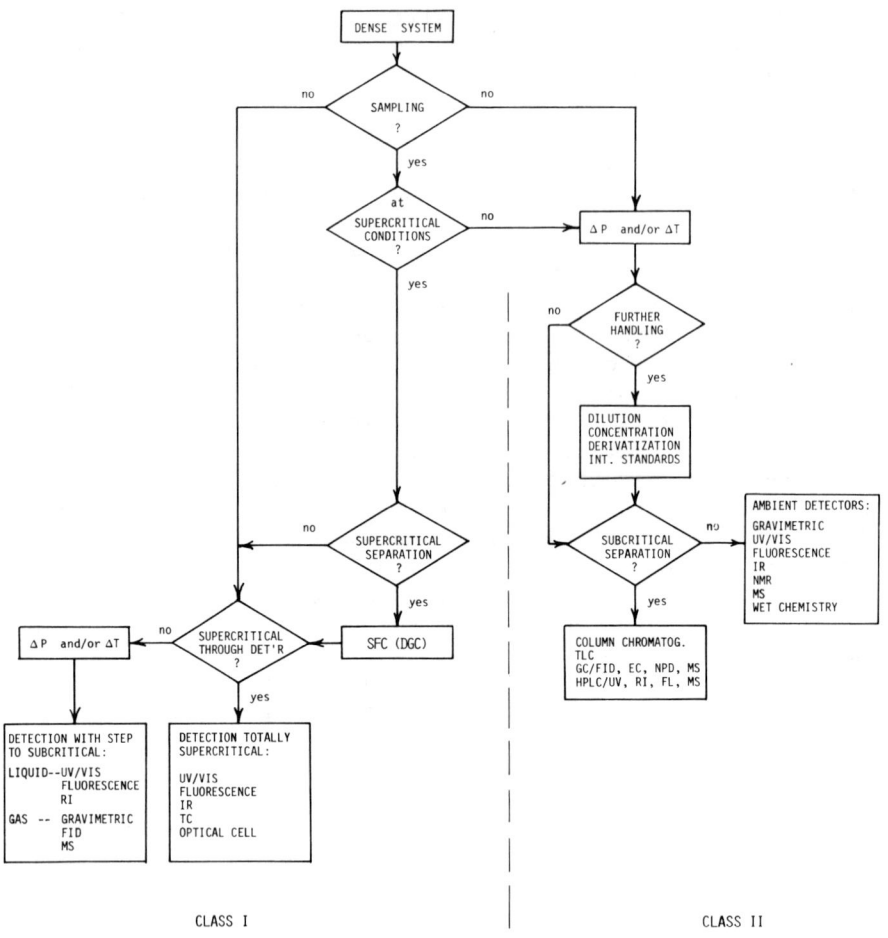

Figure 2. Logic diagram of possible schemes to analyze dense gas systems. (ΔP--change in pressure, ΔT--change in temperature, Int. Standards--internal standards, UV--ultraviolet photometric detection, UV/VIS--ultraviolet/visible photometric detection, RI--refractive index detection, FID--flame ionization detection, MS--mass spectrometric detection, IR--infrared absorption detection, TC--thermal conductivity detection, EC--electron capture detection, NPD--nitrogen-phosphorus detection, FL--fluorescence detection, NMR--nuclear magnetic resonance, TLC--thin layer chromatography, GC--gas chromatography, HPLC--high performance liquid chromatography, SFC--supercritical fluid chromatography, DGC--dense gas chromatography.)

in which supercritical conditions are maintained up to or (partially) through the detector (I) and those in which subcritical conditions are encountered somewhere before the detector usually at or during sampling (II). For example, expansion of a dense gas solution across a pressure reduction valve with solute precipitation directly into a tared vessel for gravimetric measurements belongs to Class I while a similar expansion followed by a liquid solvent rinse of the expansion region and chromatographic quantitation belongs to Class II.

Subcritical Interfaces

The examples just given are simple, particularly compared to most Class I schemes which must separate and detect at high pressures and perhaps high temperatures. Furthermore, Class II schemes which encompass widely used analytical methods have some notable problems also. First, many subcritical analytical techniques may not be well-suited for supercritical extracts. Second, the interfaces* between the supercritical system and the subcritical analytical apparatus presently seem to require much investigator interaction and extract handling. As these are minimized to increase sample throughput and reproducibility, the engineering problems become more difficult.

The interface hardware can be distinguished between on-line and off-line analysis. If an instrument is connected and dedicated to the dense gas system to perform analyses during the extraction, either continuously or at intervals (perhaps with significant interruption of the extraction), it is on-line. In the off-line mode, samples are collected and analyzed asynchronously. Therefore, off-line analysis may be more efficient in terms of uninterrupted extraction time and may allow more analytical techniques to be pursued and optimized per extraction than on-line analysis. However, more sample handling by the investigator is usually required to get the collected sample to the desired instrument in appropriate concentrations and phases.

Both an on-line interface to an instrument and an on-line sample collector for off-line analysis must deal with several questions.

A. Is the subcritical extract a liquid solution of solute plus condensed solvent gas? If so,
   1. is the solute solubility at least as great in the liquid as in the dense gas or will some amount of solute precipitate upon reaching subcritical conditions?

---

* Used here to mean everything between the supercritical system and the final analysis, including the investigator.

2. is the solute concentration in the liquid solution suitable for the chosen detector or is concentration or dilution necessary?
3. are the solution flow rate, temperature, pressure, and desired sampling interval compatible with an on-line instrument?
4. for a liquid with an appreciable vapor pressure at ambient temperature (e.g., CO_2), what is necessary for either a controlled vaporization of the solvent gas or detection at the high pressure?

B. Is the subcritical extract formed by an expansion to a gaseous state? Then,
1. how far is the gaseous solution to be maintained or, alternatively, where should the solute precipitate? When there is to be carefully placed collection of the solute with this type of separation, both temperature control and proper expansion geometry design are necessary so that solute is not lost throughout the expansion region prior to the collection region. If the temperature near the expansion region is too high, solute precipitation can occur causing plugging of the nozzle throat [4]. Expansion geometry determines the placement of standing shock waves which trigger solute condensation and precipitation as the flow passes through them.
2. can the vapor mixture be directly introduced to an on-line instrument or sample collector in terms of volumetric flow rate or is stream splitting necessary?
3. is the precipitated amount of solute suitable for the chosen detection levels or are stream splitting prior to precipitation, longer sampling times or greater volumetric flow rates necessary?
4. is there co-precipitation of an entrainer so that subcritical liquid considerations as outlined above are necessary or is the entrainer sufficiently volatile to be easily evaporated by the expanding gas so that it causes no interference?
5. is it necessary to dissolve the collected precipitate in a liquid solvent for the analysis and with what level of accuracy and reproducibility?

It is not feasible in an overview to discuss any particular interface in depth since so many variations are possible. However, once the dense gas system and the analytical technique are chosen, their interfacing compatibility must be evaluated by addressing the above questions.

Subcritical Analytical Methods

Once a sample of some sort has been produced, many instruments and methods are available for quantitation and/or identification. Table I qualitatively compares some of the more commonly used methods.* Values intermediate between the best and worst cases are presented for sensitivity and speed of analysis which are often inversely proportional. The number of components which can be distinguished with these methods is limited. A gravimetric determination is only useful for one pure component or for a total yield of a multicomponent mixture. The other methods can be used with simple mixtures (2-5 components) of knowns but become practically impossible to apply to mixtures of unknowns. Therefore, some chromatographic method is necessary for component separation prior to detection and quantitation.

A qualitative comparison of the three commonly used analytical chromatographic methods is given in Table II. Column Chromatography, listed in Figure 2, is purposely omitted since it is most commonly used as a fractionation scheme preceding the various spectroscopic or chromatographic analyses.

TLC (Thin Layer Chromatography) can be used off-line as well as directly coupled on-line to a dense gas system as done by Stahl (e.g., Ref. 5) whose on-line method does not force serial sampling/analysis. TLC is sensitive, quantitative, and somewhat compound class specific. It is an inexpensive, fast method for surveys or routine monitoring and is capable of high sample through-put. However, reproducibility is not always high and the technique is difficult to use with complex mixtures (more than 50 components)--particularly unknowns. Furthermore, detection often requires derivatization.

GC (Gas Chromatography) is limited to compounds that have some volatility (vapor pressure > 10 Torr) and thermal stability at GC conditions (up to 450°C) even though GC detectors can range from class specific to universal and are very sensitive. Derivatization can be used to obtain compounds amenable to analysis by GC but, again, is very difficult with unknown mixtures.

HPLC (High Performance Liquid Chromatography) is ideal for involatile and/or thermally labile compounds. Of the commonly used detectors, ultraviolet and fluorescence detection limit the solute class and possibly the mobile phase because of their specificity; refractive index detection, general for solutes, limits gradient solvent programming--an important advantage of HPLC; and mass spectrometric detection, while universal and selective, may require either excessive heating for concentration or volatilization (inappropriate for thermally labile compounds) or eluent splitting which

---

* The discussion presented here purposely ignores any wet chemistry analytical techniques since the possibilities are so vast and also relatively slow for unknown mixtures.

Table I. Some Common Non-Chromatographic Analytical Methods

METHOD	COMPOUND TYPE	SENSITIVITY	SPEED OF ANALYSIS	STRUCTURAL[a] INFORMATION
Gravimetric	Any	mg	minutes	−
UV/VIS (Ultraviolet/Visible Spectrophotometry)	Absorbers	μg	1 min	+
IR (FT) (Infrared-Fourier Transform Spectrophotometry)	Any	200 ng	1 s	++
NMR (Nuclear Magnetic Resonance)	Magnetic Moments	0.1 mg	1 min	++
MS (Mass Spectrometry)	Sufficient Vapor Pressure (>10^{-5} Torr)	ng	1 s	++

[a] Information about the molecular structure of the extracted compound(s) leading to characterization according to chemical class or to identification of the compound: −, none; +, some; ++, extensive.

Table II. Qualitative Comparison of Commonly Used Chromatographic Methods[a]

CHROMATOGRAPHIC METHOD	LEVEL OF QUANTITATION	RANGE OF COMPOUND CLASS	REPRODUC- IBILITY	MOBILE PHASE VERSATILITY	STATIONARY PHASE VERSATILITY	SAMPLE THROUGH-PUT	EXPENSE
TLC	pg–μg	++	–	++	–	fast, parallel	cheap
GC (Most detectors)	ng	vapor pressure > 10 Torr	++	– (temperature is used instead)	++	5 min– 2 h each	moderate to high
HPLC							
Ultraviolet Detection	ng	–	++	+	+	20 min– 2 h each	moderate to high
Fluorescence Detection	ng	–	++	+	+	"	"
Refractive Index Detection	μg	++	++	–	+	"	"
Mass Spectrometer Detection	ng–μg	++	++	+	+	"	"

[a] Definition of symbols for qualitative comparison of the various points: –, poor or limited; +, good or intermediate range; ++, very good or extensive.

decreases sensitivity.

Overall, even though all of the chromatographic methods are useful for specific applications, no one method is a dependably good analytical "match" with dense gas systems--both in terms of possible solute classes and solute identification. Of all the possibilities discussed thus far, HPLC/MS appears to be the best suited.

Supercritical Analysis

The rest of this discussion will concentrate on maintaining supercritical conditions during sampling, separation, and up to or through detection (Class I, Figure 2).

The possible options are continuous flow or sampling at supercritical conditions with monitoring of either
1. the supercritical solution with UV/VIS, fluorescence, and infrared (IR) spectrophotometry and thermal conductivity measurements,
2. the subcritical liquid solution again with UV/VIS, fluorescence and IR spectrophotometry as well as refractive index (RI) or
3. the expanded gas by gravimetric measurements, flame ionization (FID), and mass spectrometry (MS).

As for their use in HPLC, UV/VIS and fluorescence detection are class specific to some degree and are therefore limited in solute universality. Furthermore, with UV/VIS, fluorescence, and IR the dense solvent gas is limited to those gases which do not interfere with solute detection. Refractive index can be added to the detector possibilities [6] if the dense gas solution is condensed to a liquid; however, at supercritical conditions detection by refractive index may be subject to the same limitations for density programming as it is for gradient solvent programming in HPLC. Expansion of the dense gas solution enables gravimetric determination of precipitated solute as discussed before. Carefully controlled expansion of the solution to atmospheric pressure allows flame ionization detection (FID) and expansion to vacuum allows mass spectrometric (MS) detection. The expansion design for FID is the most stringent but it has been done [7-10]. Such detection is certainly more universal for solutes than any of the others heretofore discussed; however, many dense gas systems use organic solvent gases (toluene, pentane, acetone) or organic entrainers in the dense gas (methanol in carbon dioxide) which would make flame ionization detection quite difficult. Mass spectrometric detection is the most selective and most universal of all; selective ion monitoring allows quantitative solute/solvent measurements with minimal interference from the solvent or modified solvent gas. As for the subcritical analysis schemes, some chromatographic separation prior to detection is mandatory for mixtures of any complexity and will now be discussed.

Separations can be performed at supercritical conditions and are referred to as dense gas chromatography (DGC) or supercritical fluid chromatography (SFC); this chromatographic method has been recently reviewed [1,11]. Since a supercritical fluid combines liquid-like densities with gas-like viscosities and diffusivities intermediate between gases and liquids, chromatography with such a mobile phase is a very useful combination of GC and HPLC. About a decade ago interest was high; however, detection problems and the rapid development of HPLC resulted in relatively limited investigation of DGC, especially for complex mixture separation. As indicated above, detection problems have been addressed to some extent by the development of flame ionization detection for supercritical systems and also by the design of a DGC/MS instrument.

A chromatographic separation at the temperatures and pressures actually used for extractions with a variety or combination of detectors--especially the very powerful MS detector--could greatly enhance and accelerate characterization of dense gas systems. Every time a description of a dense gas extraction states that the volatile fraction of the extract was analyzed by GC [12] with the remainder ignored or tediously fractionated by column chromatography [13], perhaps derivatized, and then analyzed by GC or HPLC, there is the question of whether the whole analysis could not have been done much more efficiently with a minimum of sample handling and investigator interaction by DGC/MS or DGC/FID. DGC is a perfect match for the area of dense gas extractions. With current HPLC column technology, capillary column technology [14], and some more effort on detector interfacing, it could become a common indispensable tool for the specialized but rapidly growing area of dense gas systems.

DGC/MS

In 1970 Giddings, Myers and Wahrhaftig [15] proposed a supersonic molecular beam interface between a dense gas chromatograph (DGC) and a mass spectrometer (MS) and such an instrument was subsequently developed at the University of Utah by Wahrhaftig and Randall [4,16]. The instrument can be used with or without the chromatograph to sample high pressure systems (~ 300 atm) quickly or continuously, withdrawing only small volumes if desired.

A mass spectrometer was chosen as the detector because of its advantages of being sensitive, general purpose, and selective while providing extensive structural information-- all well-documented by GC/MS. Indeed, those very important advantages have been the incentive for the tremendous effort expended over the last decade to develop HPLC/MS.

The difficulties inherent in an HPLC/MS coupling led to

the DGC/MS feasibility studies since DGC appeared to be an alternative to HPLC and since the DGC/MS coupling appeared to be easier to achieve.* The latter reason arises from the fact that involatile and/or thermally labile compounds are already in a gaseous state at mild temperatures so that it is not necessary to provide heat of vaporization as in HPLC/MS to volatilize both liquid solvent and solutes. Therefore, an interface between an on-line detecting MS and a dense gas system should operate at the mild temperatures defined by the dense solvent gas and maintain the involatile solute in the vapor state until detection. A supersonic molecular beam interface was chosen since it is well-suited to the problem.

A Supersonic Molecular Beam Interface

A general schematic of a supersonic molecular beam apparatus is shown in Figure 3. A gas at some reservoir

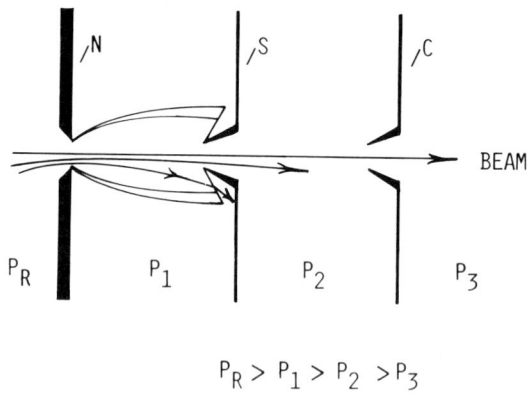

$P_R > P_1 > P_2 > P_3$

Figure 3. Schematic of a supersonic molecular beam apparatus.

pressure, P_R, is expanded through a nozzle, N--the design of which is varied, with converging-diverging, converging, and simple hole being some designs that are commonly used--into a region of lower pressure, P_1, forming a supersonic jet out of which the central core is transmitted thereby forming a

---

* An added benefit would be that a coupling between a high pressure solution of involatile compounds in solvent gas and a mass spectrometer which succeeded in transmitting the solute to the ion source would permit determination of mass spectra of involatile compounds without derivatization.

beam by the skimmer, S, to a lower pressure region, P_2, where the beam is further collimated by the collimator, C, and transmitted to a region of still lower pressure, P_3.

There are several advantages of a supersonic molecular beam interface:
1. a fast expansion which minimizes collisions (and thus clustering and condensation),
2. collisionless transport after molecular flow is established,
3. high to low pressure capability (300 atm to 10^{-3} Torr),
4. possible center-line concentration of heavy species and
5. low temperatures throughout--the temperature being determined by the solvent gas critical temperature.

The factors to be considered in designing the interface, as indicated by prior molecular beam studies, are
1. the expansion structure and process,
2. skimmer interaction with the free jet,
3. beam intensity as a function of nozzle-skimmer separation,
4. mixture component separation,
5. appropriate expansion chamber background pressures, and
6. condensation kinetics.

Only a very brief explanation of each of these will be given since they are discussed in detail in many other references [17-19].

Figure 4 is a schematic of the free jet structure formed by a gas at high pressure expanding into a region of lower but finite pressure through a converging orifice. A shock wave envelope consisting of the barrel shock and a normal shock wave, the Mach disk, is formed. Supersonic flow within

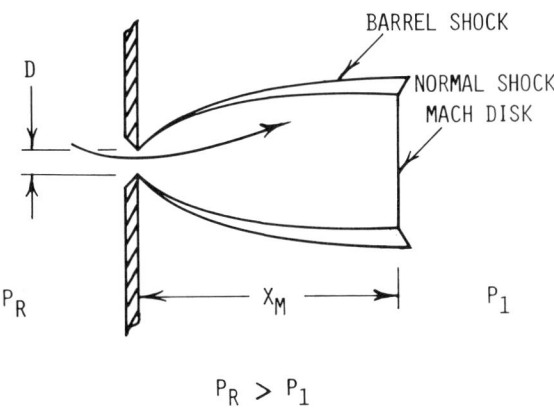

$P_R > P_1$

Figure 4. Free jet shock structure.

the shock structure is continuously expanding along streamlines with a continuously decreasing number of collisions with a possible end result of molecular flow if conditions permit. Passage through the Mach disk shock wave results in collision-dominated, subsonic flow. A molecular beam may be formed by the placement of a properly shaped skimmer within the shock structure as shown in Figure 3.

The design of a "properly shaped skimmer" includes the following points. The outer angle of the skimmer must be small enough to permit shock wave attachment; the inner angle must be large enough to avoid shock formation within the skimmer; the skimmer base to tip distance must be large enough to minimize beam attenuation from molecules rebounding from the skimmer base; and the skimmer lip must be sharp to reduce attenuating viscous layers there.

Furthermore, the axial placement of the skimmer within the jet affects shock wave attachment to the skimmer [17] and also the molecular beam intensity. At large nozzle-skimmer separations typical background scattering attenuates the beam intensity. As the separation decreases, the beam intensity steadily increases until at relatively close nozzle-skimmer separations attenuation due to the increased density between the nozzle and skimmer becomes important and causes a decrease in beam intensity. At very close separations there is an increased but collision-dominated mass flow through the skimmer but metastable species such as clusters and involatile solutes cannot survive as gas phase species to be transmitted to and detected in the mass spectrometer. Hence, there is an optimum nozzle-skimmer separation giving a maximum beam intensity for various species within the beam as shown in Figure 5.

There can be species center-line concentration where the species can be either the heavy or the light component of the expanding mixture or the expansion chamber background gas [20]. The center-line concentration effect arises from competitive mechanisms of background molecule invasion of the semipermeable shock structure and pressure diffusion. Expansion chamber background pressures of $< 10^{-2}$ Torr or an increased pressure ratio results in heavy species center-line concentration--a very desirable effect for the DGC/MS interface.

This center-line concentration is one reason to maintain a low expansion chamber pressure. Another is that the Mach disk placement is determined by the following empirical relationship [17]

$$X_M = 0.67 \, (D) \, (P_R/P_1)^{\frac{1}{2}} \text{ for } 15 < P_R/P_1 < 15,000$$

where X_M is the distance to the Mach disk, D is the nozzle

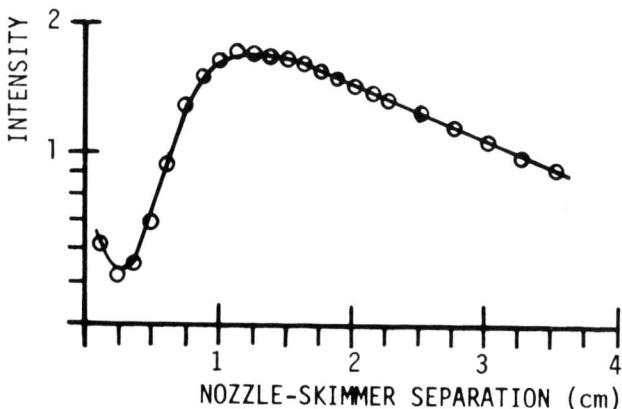

Figure 5. Molecular beam intensity as a function of nozzle-skimmer separation for 8 atm CO_2 and 25 μm nozzle diameter. Intensity (arbitrary units) measured by ion gauge placed on beam center-line downstream of collimator.

throat diameter, P_R is the reservoir pressure, and P_1 is the expansion chamber pressure. An expansion chamber pressure of 10^{-3} Torr compared to 1 atm places the Mach disk about 10^3 times further from the nozzle easing what would be very difficult interface design tolerances at 1 atm with nozzle diameters of 5 to 25 μm. If the first two reasons for low expansion chamber pressures could be ignored, it should be remembered that some kind of differential pumping is mandatory to reach mass spectrometer vacuum ($\sim 10^{-5}$ Torr).

Finally, the most difficult design consideration to evaluate is condensation kinetics. It was feared that the mass spectra of the DGC/MS would be hopelessly complicated by clusters--solvent-solvent, solvent-solute, and solute-solute. In molecular beam work, reservoir conditions and nozzle geometry can be chosen either to enhance clustering of the expanding gas producing clusters of two molecules to millions of molecules or to suppress clustering. An attempt to extrapolate the scaling rules of Hagena and Obert [21] in the manner of Farges et al [22] for the DGC/MS system with a reservoir pressure of 120 atm, a nozzle throat diameter of 12.5 μm, and a reservoir temperature of 305 K for carbon dioxide predicted an average cluster size of 10^4 molecules. In the proposed instrument, even though the reservoir conditions would favor condensation, the very small nozzle throat diameter (with the nozzle design either converging or more simply a hole in a thin plate yielding very fast expansions not yet characterized theoretically [23]) would oppose condensation. Even if the expansion were fast enough to minimize condensation, clusters might still be present in the mass spectra due to an incomplete vaporization of dense gas solution solvated-solute complexes, [(solute)·(solvent)$_n$], during expansion.

Since the proposed interface was to operate at greater pressure ratios ($\sim 10^8$) and with smaller nozzle diameters (5-10 μm) than encountered in prior molecular beam work, its performance was not predictable a priori. Hence, its feasibility had to be demonstrated experimentally.

The DGC/MS Developed at the University of Utah

The DGC/MS instrument has been described in detail recently [16]; only a brief description will be given here. Figure 6 shows a schematic of it. Here the dense gas either

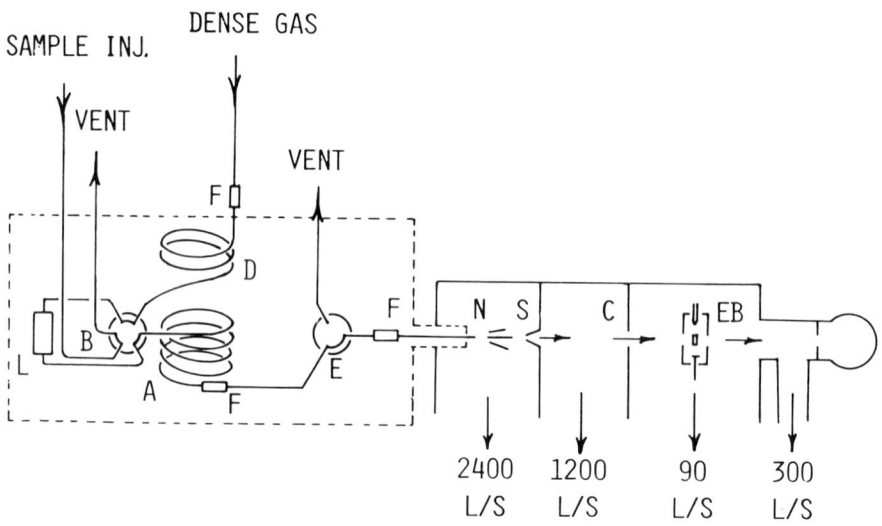

Figure 6. Schematic of the DGC/MS developed at the University of Utah. (Pumping speeds of separate, evacuated chambers in liters/second, L/S, are indicated.)

from a high pressure supply or from a DGC is expanded through a nozzle, N, into a region maintained at a pressure no greater than $\sim 10^{-3}$ Torr where a beam is formed and transmitted to the modified ion source (about 150 mm from skimmer tip to the ionizing electron beam) of a Perkin-Elmer 270 double focussing mass spectrometer (electrostatic sector, 90°/12.7 cm radius; magnetic sector, 60°/15 cm radius). The electron beam, EB, is perpendicular to the molecular beam and ions are accelerated orthogonal to the plane formed by the two beams into the electrostatic sector.

The nozzle, the heart of the interface, is a commercially available pinhole laser-drilled in nickle foil. The foil is mounted in an aligning assembly that can be moved axially with reference to the stationary skimmer to yield a possible nozzle-skimmer separation range of 0 to > 7 cm. The pinhole can be easily changed to permit a variety of nozzle diameters to be used. Figure 7 shows a cross section of the pinhole and its use in the system as a converging nozzle.

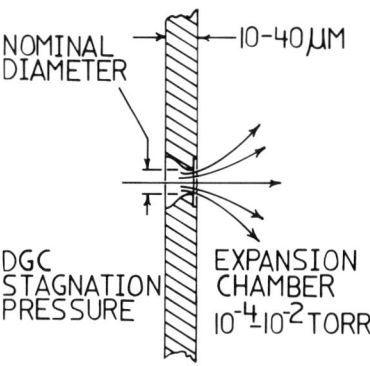

Figure 7. Cross section of a laser-drilled pinhole according to supplier. (Permission by the Ealing Corporation).

The three beam forming components are precisely aligned by an open cage-like structure (Figure 8) that allows fast unobstructed pumping of the nozzle-skimmer region during operation and permits easy removal of the nozzle, skimmer, and collimator when removed as a unit from the vacuum housing. The skimmer design is shown in Figure 9. The collimator is a simple hole in a flat plate.

The DGC components are mounted on a movable frame that is suspended in a thermostatted water bath that is pumped almost to the nozzle orifice--indicated by the dotted line in Figure 6. The major DGC components are shown in Figure 6: a preheat coil, D; an HPLC sampling valve, B (Valco Instruments Co), with a sample loop/chamber, L; a

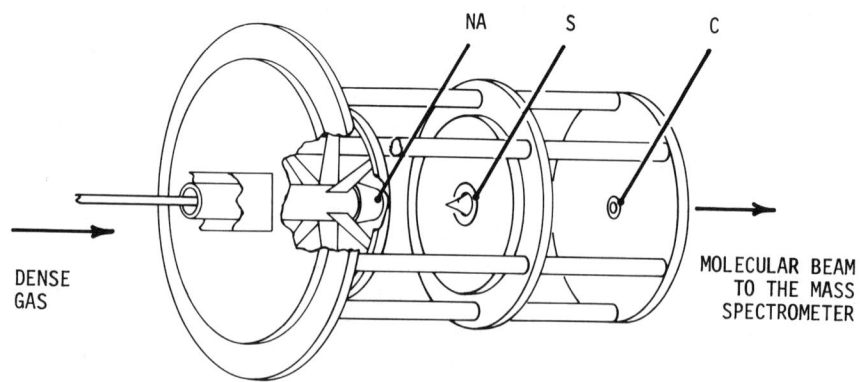

Figure 8. Nozzle assembly (NA), skimmer (S), and collimator (C) mounted in the aligning structure.

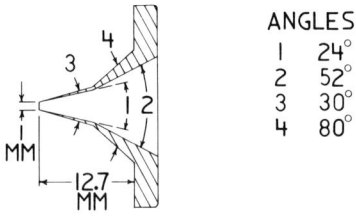

Figure 9. Skimmer design.

chromatographic column, A; a switching valve, E (Valco Instruments Co); and various filters, F.

The initial results with this experimental instrument are quite encouraging. A molecular beam was successfully and routinely formed and various species within it (monomer solvent gas, solvent gas clusters, solutes) could be monitored by the mass spectrometer. Several solvent gases were studied: ethylene, ethane, nitrous oxide, and carbon dioxide. Many test solutes were dissolved and transported to the mass spectrometer: naphthalene, anthracene, 1,3,5-triphenylbenzene, azobenzene, phenanthrene, fluoranthene, pyrene, and caffeine. Anthraquinone was detected in small amounts. A separation of anthracene, fluoranthene, and pyrene with N_2O at 102 atm and 50°C on a Permaphase ODS (DuPont) column was achieved. Solvent gas clusters to as high as 20 molecules (but not the predicted 10^4 molecules/cluster) were observed, but solvent-solute and solute-solute clusters were not observed for DGC/MS operation with reservoir pressures up to 270 atm, temperatures as low as $T_c + 5°$ for T_c = 283 to 310 K (T_c--critical temperature), and nozzle diameters of 5 to 12.5 μm.

CONCLUSIONS

It appears that the widespread development and application of dense gases as extractive solvents may be hindered at this time by the lack of a general analytical method that is truly compatible with the dense gas system in terms of pressure and solvent gas, is generally applicable to many classes of solutes over a wide concentration range while characterizing and perhaps identifying specific solutes, and is fast in terms of on-line capability eliminating investigator handling of the product extract.

The newly developed DGC/MS addresses many of these problems. All of the initial results show that DGC/MS is a feasible analytical method. While such an instrument is rather complicated and expensive and may not develop into a common laboratory tool, it should be very useful on a research scale for many of the very broad goals and dense gas systems outlined here.

REFERENCES

1. L.G. Randall, Sep. Sci. Technol. 17 (1), 1 (1982).
2. E. Stahl and W. Schilz, Chem. Ing. Tech. 50 (7), 535 (1978).
3. R.T. Kurnik, S.J. Holla, and R.C. Reid, J. Chem. Eng. Data 26, 47 (1981).
4. L.G. Randall Frank, Ph.D. Thesis, University of Utah, Salt Lake City (1979).
5. E. Stahl and W. Schilz, Z. Anal. Chem. 280, 99 (1976).
6. W. Asche, Dissertation, Technischen Universität Clausthal, Clausthal (1977).
7. U.S. Patent 3,827,859 (Hag Aktiengesellschaft; O. Vitzthum, P. Hubert, M. Barthels), Filed (1972), Granted (1974).
8. D. Bartmann, Dissertation, Ruhr-Universität Bochum, Bochum (1972).
9. W. Ecknig and H.J. Polster, Chem. Tech. (Leipzig) 31, 89 (1979).
10. J. Künzel, Dissertation, Universität Erlangen-Nürnberg, Erlangen (1979).
11. U. van Wasen, I. Swaid, and G.M. Schneider, Angew. Chem. Int. Ed. Engl. 19 (8), 575 (1980).
12. O.G. Vitzthum, P. Werkhoff, and P. Hubert, J. Food Sci. 40, 911 (1975).
13. K.D. Bartle, A. Calimli, D.W. Jones, R.S. Matthews, A. Olcay, H. Pakdel, and T. Tugrul, Fuel 58, 423 (1979).

14. M. Novotny, S.R. Springston, P.A. Peaden, J.C. Fjeldsted, and M.L. Lee, Anal. Chem. $\underline{53}$ (3), 407A (1981).
15. J.C. Giddings, M.N. Myers, and A.L. Wahrhaftig, Int. J. Mass Spectrom. Ion Phys. $\underline{4}$, 9 (1970).
16. L.G. Randall and A.L. Wahrhaftig, Rev. Sci. Instrum. $\underline{52}$ (9), 1283 (1981).
17. J.B. Anderson, Molecular Beams and Low Density Gasdynamics, edited by P.P. Wegener (Marcel Dekker, New York, 1974), Chap. 1.
18. P. Raghuraman, "Theoretical Study of Molecular Scattering in Supersonic Molecular Beams," Deutsche Luft- und Raumfahrt Forschungsbericht 73-117, Deutsche Forschungs- und Versuchsanstalt für Lüft- und Raumfahrt, Institut für Dynamik verdünnter Gase, Göttingen (1973).
19. R. Campargue and J.P. Breton, Entropie $\underline{42}$, 18 (1971).
20. R. Campargue, J. Chem. Phys. $\underline{52}$, 1795 (1970).
21. O.F. Hagena and W. Obert, J. Chem. Phys. $\underline{56}$, 1793 (1972).
22. J. Farges, M.F. de Feraudy, B. Raoult, and G. Torchet, Proceedings of the 10th International Symposium on Rarefied Gas Dynamics, Progress in Astronautics and Aeronautics Vol. 51, edited by J.L. Potter (American Institute of Aeronautics and Astronautics, New York, 1977), p. 1117.
23. P.P. Wegener, Nonequilibrium Flows, edited by P.P. Wegener (Marcel Dekker, New York, 1969), Chap. 4.

CHAPTER 25

LIQUEFACTION OF BIOMASS
WITH SUPERCRITICAL FLUIDS
IN A HIGH PRESSURE/HIGH
TEMPERATURE FLOW REACTOR

P.Köll, B.Brönstrup,
J.O. Metzger
Universität Oldenburg
Fachb.Chemie, PF 2503
D-2900 Oldenburg,Germany

The dramatic increase of prizes for crude oil since the beginning of the seventies revived the interest, especially in the high industrialized countries, in biomass as renewable energy source and feedstock for chemical industries. This is well-documented in a number of conference reports[1] and monographs[2] on this subject. Mainly different methods of pyrolysis, hydrolysis and hydrogenolysis are described. These methods yield more or less effectively low molecular weight compounds.

The objective of our own work was to disintegrate biomass completely into its single components in an environmentally safe way and to obtain these components in a state, which would allow further chemical treatment and usage. Furthermore we searched for methods, which would allow controlled degradation of biopolymers to low molecular weight compounds of high value.

Firstly, our interest focused on wood and its single components hemicellulose, cellulose and lignin. Secondly on peat, which is very abundant in some parts of the northern hemisphere; thirdly on chitin, which forms the skeleton of crustaceans but is also part of the cell walls of molds and mushrooms and fourthly on sewage sludge, the disposal of which is a severe environmental problem in certain parts of Europe. The mentioned examples of biomass are readily available and are alltogether not used normally as food. Therefore their chemical usage would not directly compete with food production.

In our search for transforming these products in

an efficient way to interesting chemicals, we found, that the special properties of fluids at high temperatures, in our case mainly organic solvents, allowed us to develop an alternative degradation procedure for biomass of different kinds.

The experimental part of this development has been performed in a high pressure/high temperature flow apparatus ("HP-HT-apparatus"), which has been described in detail by us in the literature[3,4] and which is largely made up of commercially available HPLC equipment and a GC furnace to heat the tubular reactor. The only modification necessary is to use a preparative HPLC column (V = 50 ml) as reactor. The biomass sample (1 - 20 g) is transferred into this reactor, which withstands temperatures up to 400°C and pressures up to a few hundred bars.

Thus it is possible to degrade cellulose in supercritical acetone almost quantitatively[3], with anhydrosugars as the main reaction products (see Figure 1). Glucosan $\underline{1}$, 1,6-anhydroglucofuranose $\underline{2}$, dianhydroglucose $\underline{3}$ and levoglucosenone $\underline{4}$ amount to nearly 50%. For comparison the results of typical pyrolyses are shown in Table I[5].

Table I . Products from pyrolysis of cellulose at 300°C (data taken from [5])

	Nitrogen 1 bar	Vacuum
Char	34.2(%)	17.8(%)
Tar	19.1	55.8
Glucosan $\underline{1}$	3.6	28.1
1,6-Anhydrogluco-furanose $\underline{2}$	0.4	5.6

The advantages of the degradation procedure employing supercritical fluids are obvious. The high amount of residue left, called "char", is a characteristic of pyrolytic methods. This is almost totally avoided by our procedure : only 2 % residue was left and 98 % of the cellulose was liquefied. (Reaction conditions : 18 g cellulose, microcrystalline, temperature 250°C raising to 340°C within 10 hours, flow rate 4,5 ml acetone/min.).

The comparatively smooth degradation with supercritical organic solvents is also observed with chitin[3]. Again with acetone as solvent(flow rate

5 ml/min; temperature up from 250°C to 340°C within 7.5 hours, pressure 250 bars) 85 % were liquefied. In the tar obtained, interestingly the amino analogue 5 of glucosan 1 is found in preparatively interesting amounts (see Figure 1). 5 has never been reported as pyrolysis product of chitin and, in fact, this also was our experience in pyrolyzing chitin[6]. The main reaction product in both cases, besides water of course, is acetamide[3,6]. The remaining degradation product consists of a very complex mixture of mainly oxygen- and nitrogen containing heterocycles[7].

With this results in hand, we tried to degrade whole wood. We took birch wood for our investigations[8], because this is a tree, which grows fast even on poor soils and in cold climates and which is not high in value. Table II shows the results we obtained with different solvents under standardized conditions.

Table II. Degradation of birch wood with supercritical fluids (sample weight 3 gram, reaction time 1.0 h, pressure 100 bar, solvent feed rate 1 ml/min). Q gives the ratio of lignin to carbohydrate degradation.

Solvent	Temp. (°C)	Weight loss (%)	Residual lignin (%)	Weight loss of carbohyd. (%)	lignin (%)	Q
Ether	250	20.77	28.02	29.99	---	0
n-Pentane	250	24.29	34.26	38.90	---	0
2-Propanol	250	24.62	17.20	23.39	30.06	1.3
Acetone	250	22.77	19.55	23.72	18.57	0.8
Methanol	250	25.30	13.04	20.25	47.47	2.3
Ethanol	250	21.78	12.59	16.06	46.88	2.9
Ethyl acetate	270	36.76	26.55	42.98	9.44	0.2
2-Butanol	270	31.00	16.52	29.29	38.52	1.3
1-Propanol	270	32.45	11.88	26.92	56.72	2.1
2-Methyl--1-propanol	280	39.61	12.66	35.26	58.78	1.7

Column 3 ("weight loss") gives the percentage of wood degraded. All organic solvents employed dissolve 20-40 % of the birch wood within one hour. Interestingly the carbohydrates and the lignin are

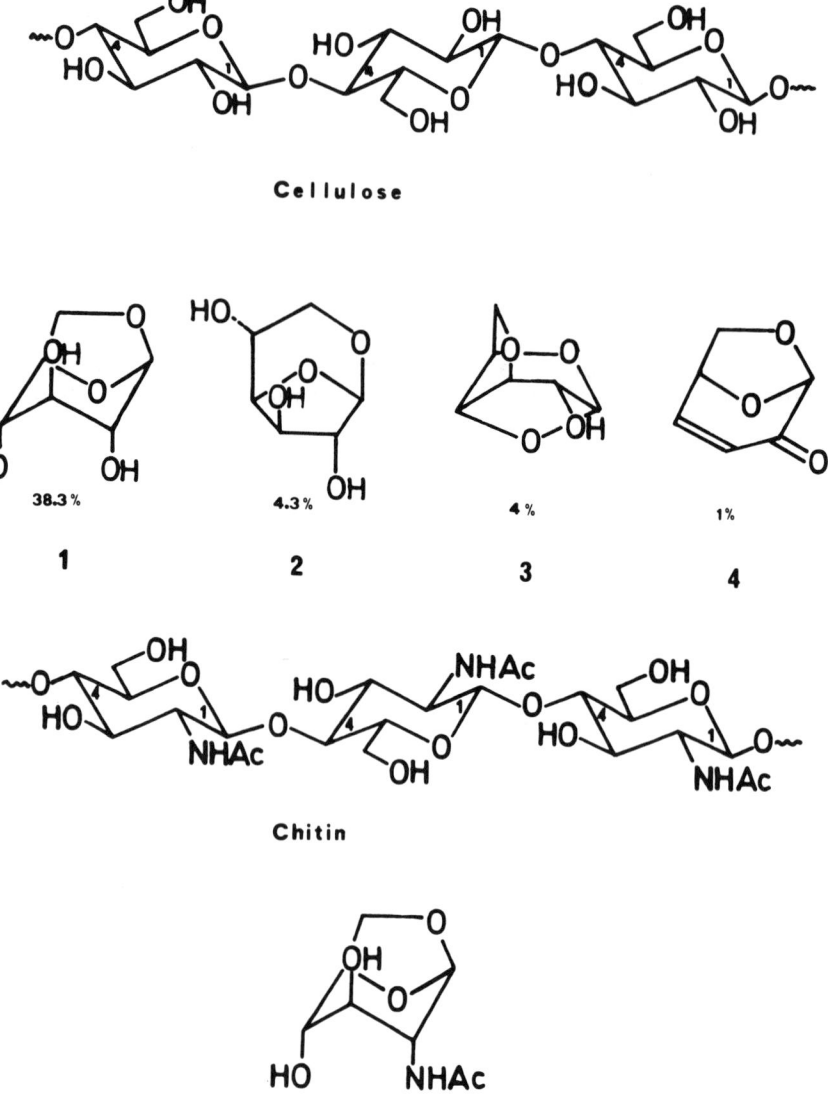

Figure 1. Products of cellulose and chitin degradation with supercritical acetone at 250-340°C.

degraded at different rates, as can be seen from columns 5 and 6. A high ratio Q of lignin to carbohydrate degradation in column 7 shows a high selectivity for lignin. Thus it can be seen that alcohols show strong preference for delignification with minor attack on carbohydrates, while ethers, esters and alkanes on the other hand preferentially attack carbohydrates at 250°C. Methanol shows a high degradation capability; ethanol has a better selectivity.

The selectivity is maintained over the temperature range from 240 - 280°C as was shown with ethanol and propanol-2 as solvents. The amount of degradation increases with temperature. But as can be seen from Figure 2, which shows the temperature influence in case of propanol-2, the ratio of Q decreases beyond 280°C, which can be explained by an increase of the pyrolytic decomposition of carbohydrates. The reaction products form nonhydrolyzable condensation products which are also analyzed as "lignin" and better should be called "humins".

Variation of pressure above critical pressure is of little influence. By increasing same from 100 to 350 bars in the aforementioned experiments the amount of degradation rises only slightly (1 - 2 %). The mass balance shows losses from 0 to 6 % due to formation of gases and low molecular weight compounds which are evaporated with the solvent and therefore escape analysis. This is contradictory to the results obtained by Calimli and Olcay[9] with the supercritical extraction of spruce wood with organic solvents. These outhors performed static experiments in an autoclave and observed high losses due to formation of highly volatile compounds. Our data demonstrate clearly the advantage of a flow apparatus; sensitive compounds are quickly removed from the hot reaction zone, which apparently prevents further decomposition.

Figures 3 and 4 show results obtained by prolonged treatment of birch wood at 250°C with propanol-2 and 94 % ethanol respectively as solvents. In both cases an initial high reaction rate is observed. Further degradation (after 1 hour) proceeds at lower rate. Lignin degradation reaches in Figure 4 almost 80 % of the theoretical level after 4 hours. In principle total degradation of wood can be acchieved with all solvents tested but this requires an inrease of the temperature. Acetone for instance dissolves 92.5 % of wood within 12 hours with a simultaneous temperature increase slowly up from 250 to 340°C.

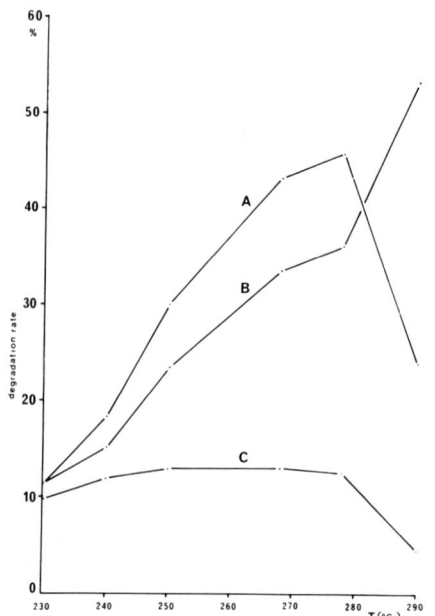

Figure 2. Influence of reaction temperature on degradation of birch wood with propanol-2 (standard conditions: sample weight 3g, solvent feed rate 3 ml/min, reaction time 1 h at 100 bar). Curve A: lignin; B: carbohydrates; C: ratio A:B x 10.

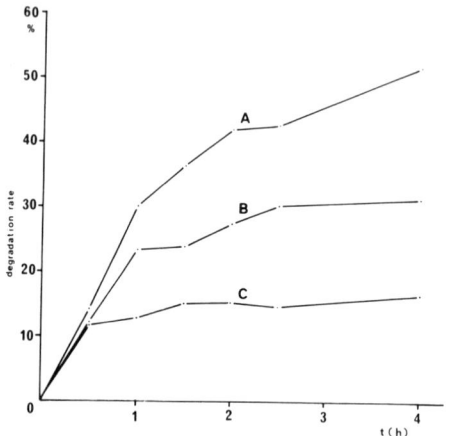

Figure 3. Time dependence of degradation of birch wood with propanol-2 at 250°C (sample weight 3g, solvent feed rate 1 ml/min at 100 bar, reaction time 1 h, Curve A: lignin; B: carbohydrates; C: ratio A:B x 10.

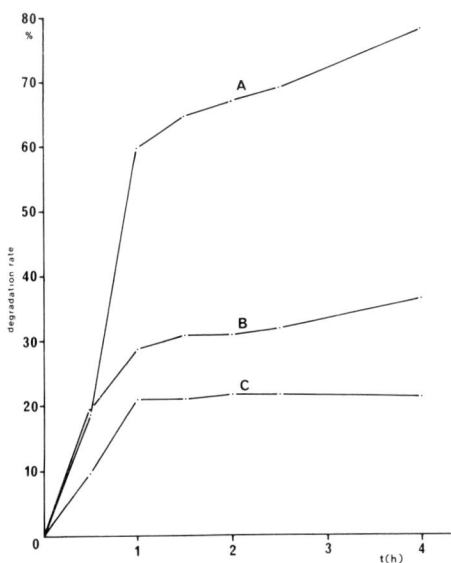

Figure 4. Time dependence of degradation of birch wood with ethanol (6 % water content) at 250°C (sample weight 3g, reaction time 1 h, solvent feed rate 1 ml/min at 100 bar). Curve A: lignin; B: carbohydrates; C: ratio A:B x 10.

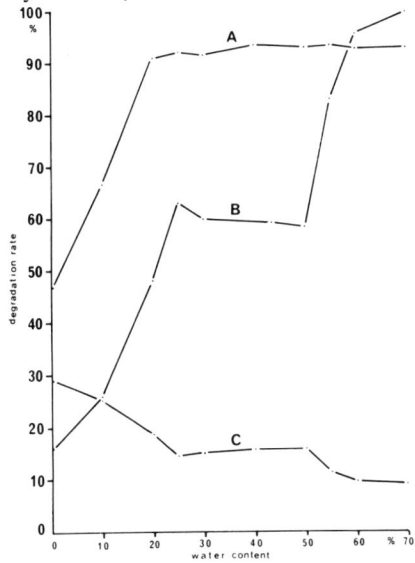

Figure 5. Birch wood degradation in dependence of water content (v/v) of ethanol (sample weight 3g, solvent feed rate 1 ml/min, reaction time 1 h at 100 bar and 250°C). Curve A: lignin, B: carbohydrates; C: ratio A:B x 10.

From Figure 4, in connection with the results given in Table II for absolute ethanol, it can be derived, that a water content of 6 % (v/v) of the solvent significantly increases total degradation. With pure ethanol after 1 hour 16 % of carbohydrates and 47 % of lignin are dissolved (Table II) while with 6 % water content (Figure 4) 28 % and 60 % respectively are degraded. Therefore we tested the properties of different mixtures of ethanol/water systematically. The results are given in Figure 5.

We found - and that corresponds with results obtained by Kleinert already some fifty years ago[10] - that there exists a broad range of mixtures of ethanol/water, where cellulose is left almost unchanged while lignin and hemicelluloses are dissolved on the other hand almost totally. If the water content is raised to 25 % (v/v) (molar ratio ca. 0.5) a dramatic increase of the degradation rates is observed for both lignin and carbohydrates to 92 % and 60 % respectivly. Higher water contents of up to 50 % (v/v) (molar ratio ca. 0.8) do not change the results significantly. The extraction residue amounts to 35 %, which represents the original cellulose content of birch wood. - Can an alternative to preparing pulp for the paper or chemical industries under elimination of problematic inorganic chemicals and the related disposal problems be based on only water and ethanol? It should be mentioned in this context, that chipping or milling of the wood is almost unnecessary because of the high penetrating power of the solvents employed. But the cellulose thus obtained has an average degree of polymerization $\overline{DP}$ of 351 (that's relatively low), a content of alpha-cellulose of 87.2 % (that's low too) and a kappa number of 27.2, which corresponds to lignin left of about 4 % by weight. This is also low for wood pulp. Non hydrolyzable material (by sulfuric acid) amounts also to 4 %. These properties make our cellulose perhaps not so suitable for paper making; the Kleinert procedure[10,11] which works at lower temperatures could be a better alternative. However for chemical and for microbiological purposes the described treatment may have an advantage.

The lignin can be obtained as a fine powder. It has an average molecular weight of about 940, corresponding to 4 to 5 monomeric phenylpropane units. This indicates severe degradation of the original lignin. By gel chromatography (see Figure 6) we

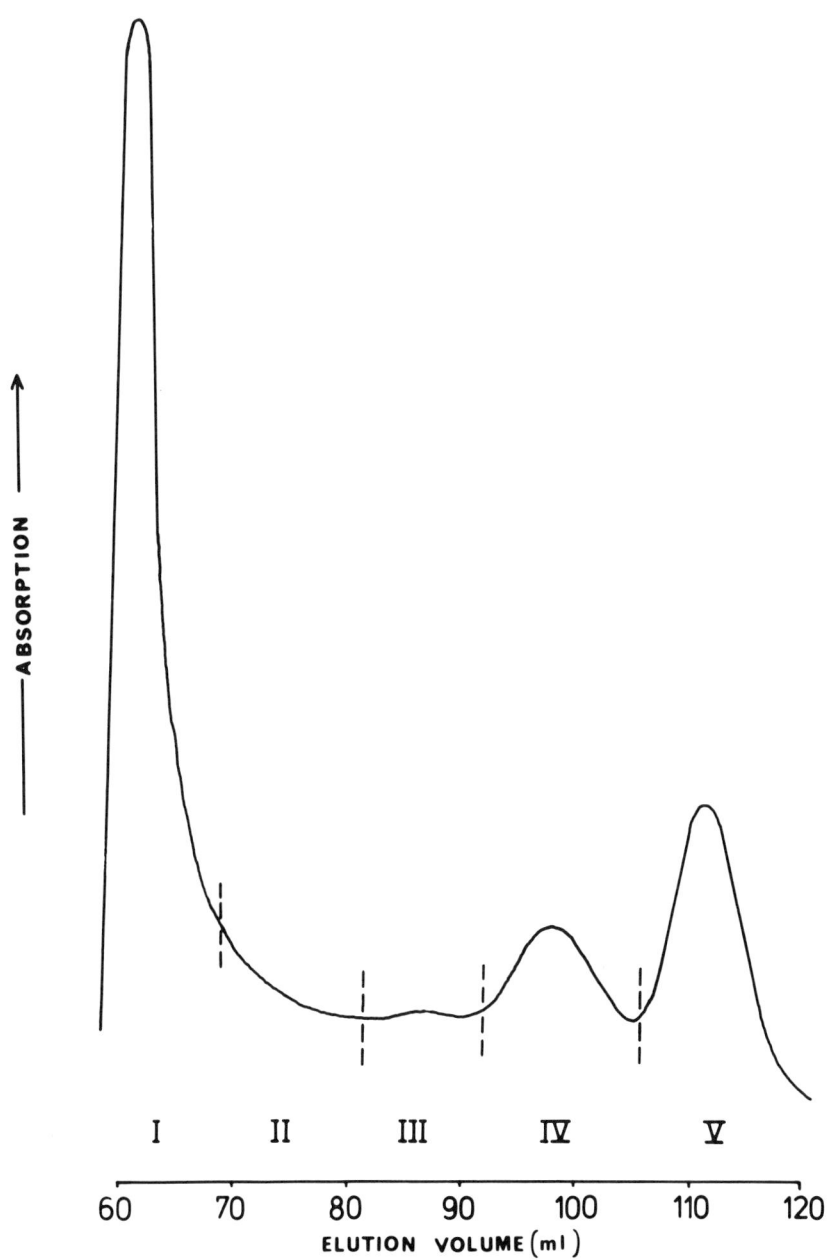

Figure 6. Gel chromatogramm of dissolved lignin (Sephadex LH 20 in DMF). Weight of fractions I-V = I 52.5 %, II 12.6 %, III 5.6 %, IV (dimers) 12.9 % and V (monomers) 16.4 %.

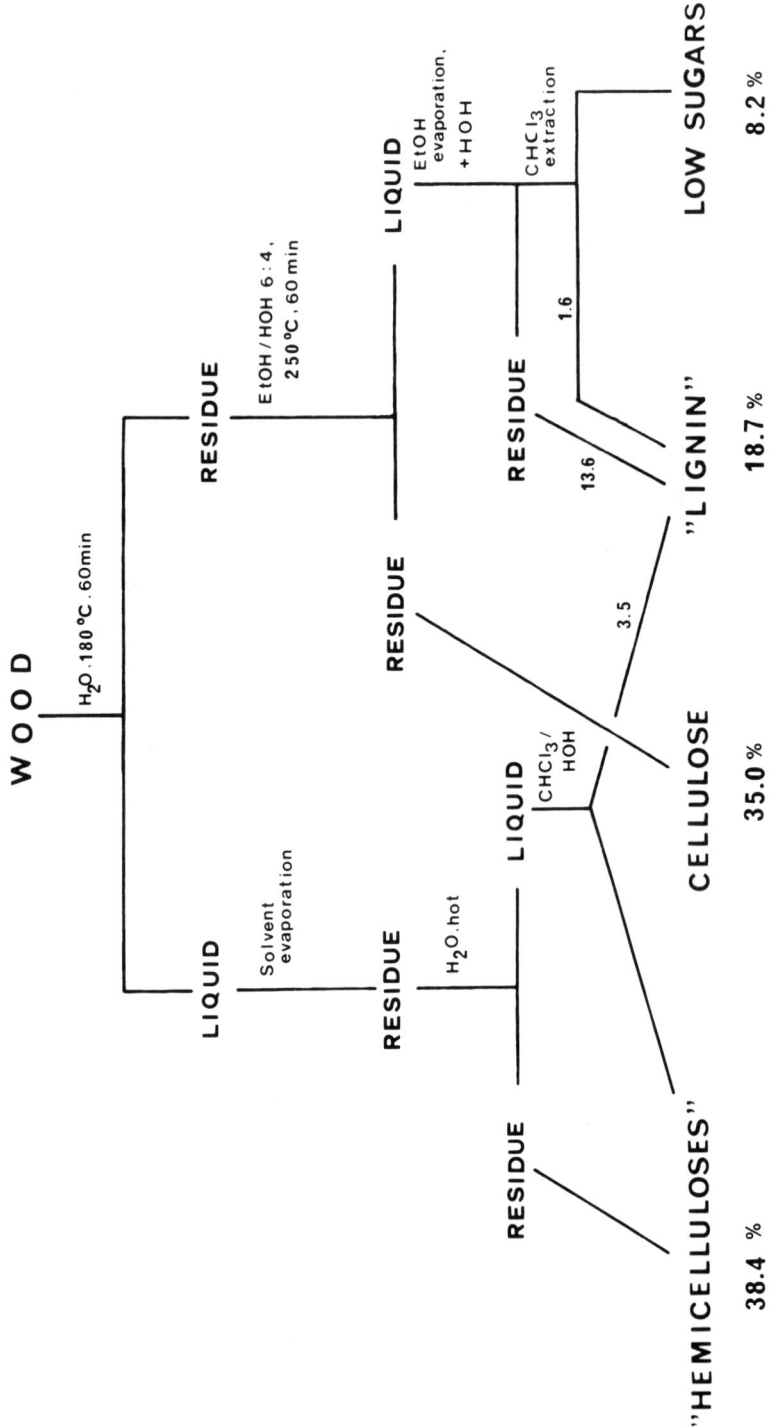

Figure 7. Mass balance of birch wood degradation with water/ethanol mixtures including prehydrolysis with water.

could show that monomeric and dimeric phenols are already present in the extract amounting to almost 30 %. Vacuum destillation of this lignin, which in itself is an interesting product gives more than 50 % yield of mostly monomeric phenols and some dimers.

Hemicelluloses are easily separated from lignin if they are simultaneously extracted from wood, but it is also possible to extract separately with more or less degradation all components from wood. This is shown in Figure 7. Beginning with water alone at moderate temperatures of about $180^{\circ}C$ only hemicelluloses are extracted ("prehydrolysis"); by subsequent use of a water/ethanol mixture at temperatures of around $250^{\circ}C$ lignin is extracted and cellulose is left unchanged. The mass balance approaches 100 % and demonstrates that only a small amount of gaseous products is formed.

From Figure 5 follows that ethanol/water mixtures with more than 60 % water content attack cellulose too. Thus it is possible to liquefy birch wood quantitatively within 1 hour at $250^{\circ}C$. The optimum seems to be in the range of a 70 % water content. Pure water is a bad degradation medium and imposes experimental difficulties because of poor solubility of reaction products in water. It should be mentioned that in most cases the experiments with ethanol/water mixtures have been performed at temperatures below the critical temperatures of these mixtures.

Cellulose degradation has been optimized with respect to low molecular weight compounds[12]. Figure 8 shows the time dependence of the conversion of microcrystalline cellulose ("Avicel") at $250^{\circ}C$ with a 4:6 (v/v) ethanol/water mixture. After 2 hours conversion reaches 95 %. Glucose analyzes for 31 %. Besides it are found 13.5 % ethyl glucosides, 7 % anhydrosugars and 10 % hydroxymethylfurfural (HMF).

From Figure 9 the influence of temperature and the variation of water content on conversion rates and combined yields of monomeric sugars can be derived. The analyses of these products are summarized in Figure 10. By further variation of solvent flow rate the results of Figure 11 can be reached. At $260^{\circ}C$ cellulose is hydrolyzed to a degree of 96 % within 1 hour with a 3:7 (v/v) ethanol/water mixture. Combined monosaccharides (including HMF) analyze for about 75 % of theoretical value. Glucose alone are 56 % of theoretical. This opens new ways

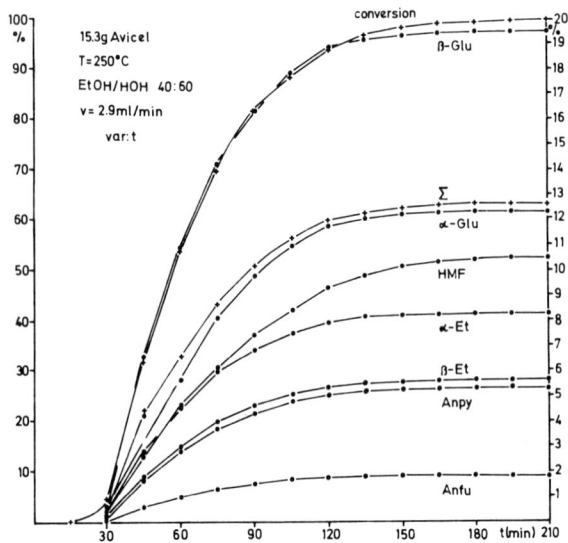

Figure 8. Time dependence of cellulose degradation with an ethanol/water mixture (4:6 v/v). Left-hand scale: conversion and sum of analyzed products (Σ). Right-hand scale: yields (based on charged cellulose) of β-glucose (β-Glu), α-glucose (α-Glu), hydroxymethylfurfural (HMF), ethyl α-glucoside (α-Et), ethyl β-glucoside (β-Et), glucosan (Anpy) and 1,6-anhydroglucofuranose (Anfu).

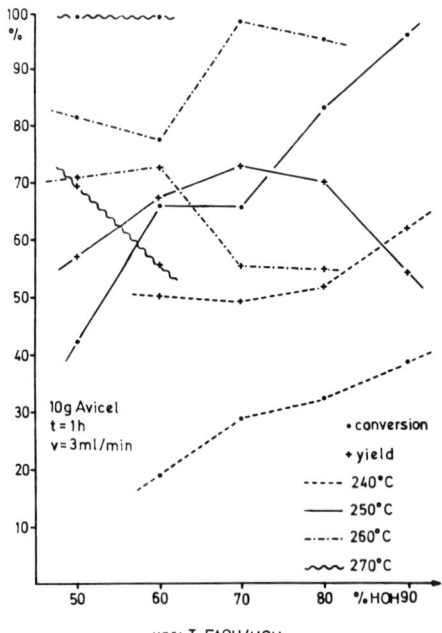

Figure 9.

Influence of water content on degradation of cellulose with ethanol/water mixtures at different temperatures. Conversion curves show amount [%] of degraded cellulose. Yield curves give sum of analyzed products (glucose, ethyl glucosides, anhydroglucoses and HMF) related to amount of degraded cellulose.

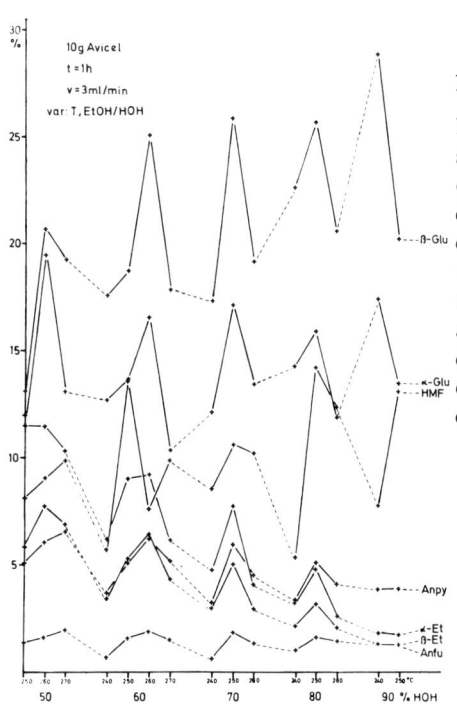

Figure 10.

Influence of temperature on sugar yields in degradation of cellulose with four different ethanol/water mixtures. Yields of β-glucose (β-Glu), α-glucose (α-Glu), hydroxymethylfurfural (HMF), glucosan (Anpy), 1,6-anhydroglucofuranose (Anfu), ethyl α-glucoside (α-Et) and ethyl β-glucoside (β-Et) based on amount of degraded cellulose.

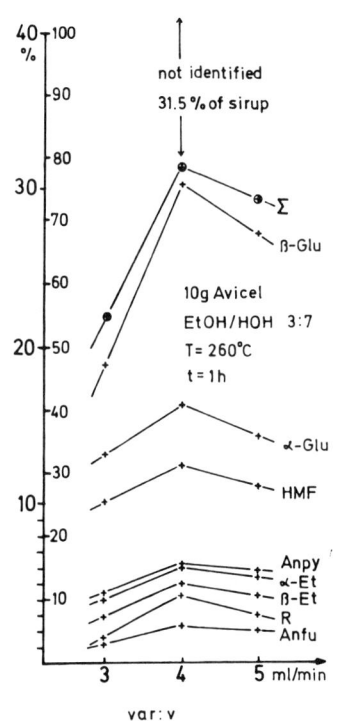

Figure 11.

Influence of solvent feed rate on sugar yields in degradation of cellulose near optimum conditions. Outer left scale gives yields (based on charged cellulose) of β-glucose (β-Glu), α-glucose (α-Glu), hydroxymethylfurfural (HMF), glucosan (Anpy), 1,6-anhydroglucofuranose (Anfu), ethyl α-glucoside (α-Et) and ethyl β-glucoside (β-Et). R = amount of residual cellulose. Inner left scale gives sum (Σ) of analyzed products. Total yield of degradation products obtained as sirup exceeds 100%.

for wood saccharification.

The same ethanol/water mixture was used for peat degradation[13]. Table III gives the results for white peat (low degraded, experiments No. 1 - 4) and black peat (high degraded, experiment No. 5). While almost 90 % of white peat can be dissolved, only 30 % of black peat is degraded under comparable conditions.

Table III. Degradation of peat with an ethanol/water mixture 3:7 (v/v). Experiments No. 1-4: white peat. No. 5: black peat. "Extr". = Extract weight %.

Experiment No.	Reaction Conditions				Results	
	Temp. ($^\circ$C)	Flow Rate (ml/min)	Press. (bar)	Time (min)	Weight loss (%)	Extr. (%)
1	265	1 - 2	150	30	68,0	not d.
2	265	7	150	60	79,8	75,6
3	265	8	150	120	89,9	85,1
4	265 275	7	150	30 30	84,3	80,7
5	275 285	6 - 7	150	30 30	29,5	14,9

Higher weight losses (up to more than 60 %) are also attainable for black peat with solvents of higher ethanol content. The losses due to formation of gases are considerably higher for black peat than for white peat. In neither case the degradation products are liquids (or tars) but solids, which are partly water soluble (up to 55 %).

Contrary to this results with peat, municipal sewage sludge treated in the same manner yields again an oil which has an interestingly high content of long chain fatty acids[14]. Advantage of this treatment is the fact, that it is not necessary to dry the sewage sludge intensively before liquefaction, which is necessary precondition in other processes and which is expensive.

CONCLUSIONS

Organic solvents in the supercritical state such as alkanes, ethers, esters and alcohols ("supercritical fluids") are suitable for biomass disintegration and biopolymer dissolution and degradation in a temperature range of 240 - 340°C. Good properties for the mentioned purposes show also ethanol/water mixtures which are employed as liquids at subcritical conditions. From this it may be concluded, that a high density, an optimum high temperature and good solution properties of fluids are prerequisite for biomass transformation irrespective whether it are supercritical or subcritical fluids. Thus wood can be totally liquefied. On the other hand it is possible to prepare cellulose by dissolving hemicelluloses and lignin only. Obtained lignin gives high yields of monomeric and dimeric phenols by vacuum destillation. High glucose yields are obtained by cellulose degradation with ethanol/water mixtures (3:7 v/v) thus opening new ways for wood saccharification.

LITERATURE

1. Shafizadeh, F., K.V. Sarkanen and A. Tillmann, Eds. Thermal Uses and Properties of Carbohydrates and Lignins (New York: Academic Press, 1976); Tillmann, D.A., K.V. Sarkanen and L.L. Anderson, Eds. Fuels and Energy from Renewable Resources (New York: Academic Press, 1977); Jones, J.L., and S.B. Radding, Eds. Solid Wastes and Residues, Conversion by Advanced Thermal Processes (Washington, D.C.: ACS Symp. Series 76, 1978); Tomlinson, M., Eds. Chemistry for Energy (Washington, D.C.: ACS Symp. Series 90, 1979); St.-Pierre, L.E., and G.R. Brown, Eds. Future Sources of Organic Raw Materials, CHEMRAWN I (Oxford: Pergamon Press, 1980); Jones, J.L., and S.B. Radding, Eds. Thermal Conversion of Solid Wastes and Biomass (Washington, D.C.: ACS Symp. Series 130, 1980); Klass, D.L., Eds. Biomass as a Nonfossil Fuel Source (Washington, D.C.: ACS Symp. Series 144, 1981).

2. The National Research Council. "Renewable Resources for Industrial Materials", (National Academy of Sciences, Washington, D.C.: 1976); Anderson, L.L., and D.A. Tillmann, Eds. Fuels from Waste (New York: Academic Press, 1977); Tillmann, D.A., Wood as an Energy Resource (New York: Academic Press, 1978); Bungay, H.R., Energy, the Biomass Options (New York: J. Wiley a. Sons, 1981); Rider, D.K., Energy: Hydrocarbon Fuels and Chemical Resources (New York: J. Wiley a. Sons, 1981); Sittig, M., Organic and Polymer Waste Reclaiming Encyclopedia (Park Ridge,

N.J.: Noyes Data Corporation, 1981).

3. Köll, P., and J.O. Metzger, Angew. Chem. 90: 802 (1978); Angew. Chem. Int. Ed. Engl. 17: 754 (1978).

4. Metzger, J.O., J. Hartmanns, D. Malwitz and P. Köll in "Chemical Engineering at Supercritical Fluid Conditions". This book.

5. Shafizadeh, F., and Y.L. Fu, Carbohydrate Res. 29: 113 (1973).

6. Köll, P., and J.O. Metzger, Z. Lebensm. Unters. Forsch. 169: 111 (1979).

7. Metzger, J.O., D. Malwitz, and P. Köll, unpublished results (1980).

8. Köll, P., B. Brönstrup and J.O. Metzger, Holzforschung 33: 112 (1978).

9. Calimli, A., and A. Olcay, Holzforschung 32: 7 (1978).

10. Kleinert, Th. N., and K. Tayenthal, U.S.Pat.No. 1.856.567 (1932); Kleinert, Th. N., Cellulosechemie 18: 114 (1940); Das Papier 21: 653 (1967); U.S.Pat.No. 3.585.104 (1971); Tappi 57: 99 (1974); Tappi 58: 170 (1975).

11. Schweers, W., and D. Meier, Holzforschung 33: 25 (1979) and following articles: Baumeister, H., and E. Edel, Das Papier 34: V 9 (1980).

12. Brönstrup, B., Thesis, University of Oldenburg Germany (1982).

13. Köll, P., J.O. Metzger, and B. Brönstrup, Telma, in press.

14. Schuller, D., University of Oldenburg, private communication 1980.

Financial support for part of this work was provided by "Forschungsmittel des Landes Niedersachsen".

CHAPTER 26

THERMAL ORGANIC REACTIONS
IN SUPERCRITICAL FLUIDS

J.O. Metzger, J. Hartmanns,
D. Malwitz, P. Köll
Universität Oldenburg,
Fachbereich Chemie, PF 2503
D-2900 Oldenburg, Germany

ABSTRACT

Intermolecular thermal reactions were studied in a flow-type system at pressures of up to 500 bar and temperatures of up to 500°C. Under these conditions alkanes are added to alkenes (e.g., n-alkenes, acrylonitrile, methyl acrylate, methyl vinyl ketone) as well as to 1.3-dienes and alkynes thus allowing functionalization of saturated hydrocarbons. Acrylic compounds such as methyl acrylate are dimerized thermally yielding dimethyl 2-methylene-glutarate. Benzene reacts with some alkynes in a Diels-Alder type reaction. A high pressure-high temperature flow apparatus ("HP-HT" apparatus) to perform these reactions is described.

INTRODUCTION

It is well known fact that organic reactions are accelerated at higher temperatures. This experience is widely applied in laboratory and industrial work, because thermal energy is cheap and everywhere available.
One therefore would expect that organic chemists use specifically all ranges of temperature to perform chemical reactions. Surprisingly there is a remarkable gap: intermolecular reactions are scarcely performed above 300°C. Pyrolysis reactions are dominating in this range of temperatures. Reactions

of this type are of high technical and economical importance e.g., the thermal cracking of hydrocarbons for olefin production. Some other pyrolytic reactions are also well known e.g., the pyrolysis of esters, the retro-Diels-Alder reaction, and the retro-ene reaction [1].

Organic chemists likewise apply this range of temperature to perform preparatively important reactions related with thermal rearrangements of molecules and formation of new bonds. All these reactions occurring in the gas phase are of intramolecular and monomolecular type, such as thermal cyclizations of unsaturated carbonyl compounds and intramolecular ene reactions. Temperatures of up to 600°C are often used in these cases [1]. These examples clearly demonstrate that application of high temperatures favours also preparatively important reactions with the formation of new bonds.

At temperatures above 300°C some processes of technical importance are of the intermolecular reaction type. First the high pressure polymerization of ethylene has to be mentioned which is carried out at pressures of up to 3500 bar and temperatures of up to 350°C. The production of biphenyl from benzene at 800°C is another example.

Fundamental research on this subject was performed in the thirties and fourties of this century. The theoretical concept of pericyclic reactions for the guided search for thermal reactions have been developed mainly in the last two decades [2]. A sentence of HENDRICKSON refers to this fact: "Many presently unknown pericyclic reactions may have been overlooked simply because relatively simple and well-known potential substrates have not been deliberately heated above 300°C. Many others which do not appear to proceed may be pre-empted by competitive reactions of lower activation energy". [3]

The use of organic solvents for supercritical extraction has stimulated the investigation of reactions at elevated temperatures and high pressures [4]. We observed condensation reactions of acetone during the extraction of biomass with supercritical acetone [5]. 2.1 % diacetone alcohol and 0.6 % mesityloxide are obtained from pure acetone formed apparently by a thermally induced aldol condensation (1)

$$2\ CH_3-\underset{O}{\underset{\|}{C}}-CH_3 \xrightarrow[1.5\ min]{270°C,\ 200\ bar} CH_3-\underset{OH}{\underset{|}{\overset{CH_3}{\overset{|}{C}}}}-CH_2-\underset{O}{\underset{\|}{C}}-CH_3 + CH_3-\overset{CH_3}{\overset{|}{C}}=CH-\underset{O}{\underset{\|}{\overset{CH_3}{\overset{|}{C}}}} \quad (1)$$

Nonactivated alkenes show thermal dimerization [6] e.g., cyclohexene is dimerized to 3-cyclohexyl-cyclohexene (2).

$$2 \; \text{C}_6\text{H}_{10} \xrightarrow{400°C, \; 450 \; \text{bar}, \; 10 \; \text{min}} \text{cyclohexyl-cyclohexene} \quad (2)$$

These along with other observations stimulated us to start a systematical investigation of thermal intermolecular reactions at relatively high temperatures and pressures. Hitherto we have investigated addition reactions of C-H bonds to carbon-carbon double and triple bonds. Some preliminary experiments have been performed with some other systems e.g., addition reactions to C=O bonds. The results are promising. Benzaldehyde is reduced to benzylic alcohol, furfural to furfuryl alcohol by methanol (3).

$$\text{PhCHO} \xrightarrow{400°C, \; 200 \; \text{bar}, \; 5 \; \text{min, Methanol}} \text{PhCH}_2\text{OH} \quad (3)$$

EXPERIMENTAL

All experiments were performed in a high-pressure/high-temperature flow apparatus ("HP-HT" apparatus), which enables relatively unproblematical operation at pressures of up to ca. 500 bar and temperatures of up to 600°C. A schematic diagram of the apparatus is shown in Fig. 1.

The sample in reservoir 1 is pumped with a high pressure pump 3 into reactor 6. The reactor is made from commercial high-quality stainless steel tube (o.d. 1.6 mm, i.d. 0.7 mm). The length of the reactor can be changed from a few meters to 50 m and more depending on the reaction under investigation. This gives much flexibility with respect to the reactor size.

We used also fused silica capillaries as reactor (up to 400°C and 200 bar) to exclude possible transition metal catalysis.

The reactor is heated up to 450°C in GC-furnace 12, up to 600°C in tubular furnace respectively. The reaction mixture is quenched in heat exchanger 9. Reaction pressure is maintained with valve 8; pressure reduction is done in two steps

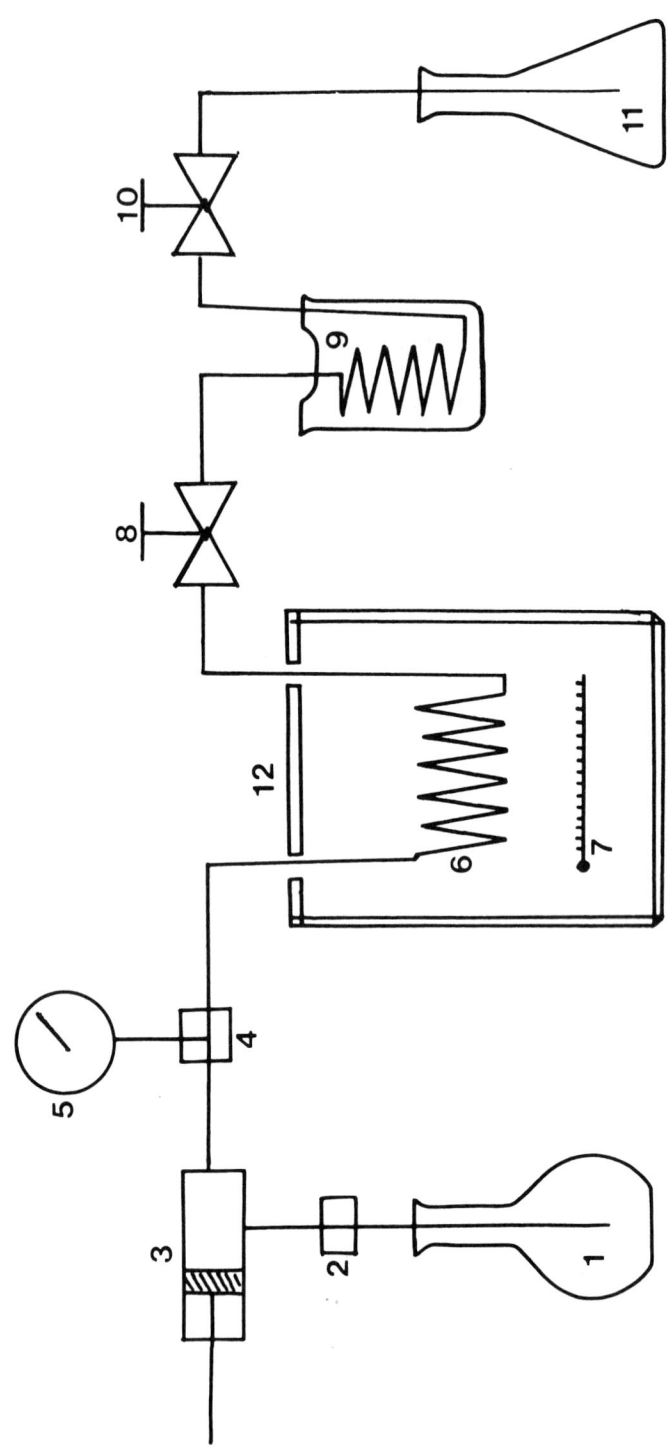

Figure 1. Schematic diagram of the HP-HT apparatus. 1) solvent reservoir; 2) filter; 3) high-pressure pump for up to 500 bar; 4) T-piece; 5) manometer for up to 600 bar; 6) reactor; 7) thermometer; 8) valve; 9) heat exchanger (1.6 mm-diamter capillary); 10) valve; 11) collector; 12) furnace.

with the valves 8 and 10. The same valves are used also to adjust the flow rate.

The HP-HT apparatus is started up as follows. First pure solvent is pumped through the reactor. When the parameters of the reaction (pressure, temperature, residence time) are adjusted, we change from pure solvent to the educt solution with a three-way valve on the low-pressure side. After the passage of four to five reactor volumina we take the first sample for analysis. Assuming that the reaction is already optimized, we can produce on a preparative scale. It is very important - and that is always true for a flow reactor - that for each reaction the parameters temperature, residence time, concentration and pressure are optimized. Due to the rapid adjustment of the reaction parameters the optimization is relatively fast with slightest consumption of reactants.

Analysis: The samples were analyzed qualitatively by GC-MS and quantitatively with internal standard by capillary GC with on-column injection (25 m fused silica capillary OV 101). The products were isolated by distillation or by preparative GC and characterized by NMR, IR, MS.

EXAMPLES

Synthesis of methyl 3-cyclohexyl-2-methyl propionate
10 g (0.1 mol) methyl methacrylate in 336 g (4 mol) cyclohexane were pumped through the HP-HT apparatus at 450°C and 300 bar, with a residence time of 2.5 min. Concentration of the product solution in vacuo gave 13.7 g of crude product. Spinning band distillation yielded 6.7 g = 37 % product (bp 56°C/0.2 torr, n_D^{20} 1.4491).

Synthesis of dimethyl 2-methyleneglutarate
86.1 g (1 mol) methyl acrylate solved in 1000 ml benzene were pumped through the HP-HT apparatus at 400°C, 200 bar, with a residence time of 5 min. Concentration of the product solution in vacuo yielded 73.5 g of crude product. Destillation with a 30 cm - Vigreux column yielded a) dimethyl 2-methyleneglutarate 25g = 29 % (bp 102-104°C/15 torr.) b) Trimethyl 1-hexane-2,4,6-tricarboxylate (trimer) 13.7 g = 17 %, bp 134-138°C/0.2 torr.

RESULTS

1. Thermal Addition of Alkanes to Alkenes - the "Ane Reaction".

We were able to demonstrate that alkanes are adding generally to alkenes at temperatures above 300°C, 200 - 500 bar and reaction times of 1 - 10 min [7].

Table I shows the results of the reaction of cyclohexane with a representative selection of activated, deactivated and normal alkenes.

Table I. Addition of Cyclohexane to Selected Alkenes

$$\bigcirc + CH_2=CHR \longrightarrow \bigcirc-CH_2-CH_2-R \qquad (4)$$

Alkene	Alkane/Alkene	T[°C]	Residence time [min]	Yield [mol %]
Acrylonitrile	100	400	3.5	60
Methyl vinyl ketone	20	400	3	17
Vinyl acetate	20	450	0.5	19
Heptene-1	100	450	4	30

The addition of cyclohexane to methyl acrylate (5) has been investigated more in detail.

$$\bigcirc + CH_2=CH-COOCH_3 \longrightarrow \bigcirc-CH_2-CH_2-COOCH_3 \quad \underline{\underline{1}}$$

$$+ \bigcirc-\underset{\underset{\underline{\underline{2}}}{CH_3}}{\overset{}{CH}}-COOCH_3 \qquad (5)$$

Figure 2a shows the yield of the anti-Markownikow addition product $\underline{\underline{1}}$ increasing from 5 % at 250°C up to 45 % at 450°C.

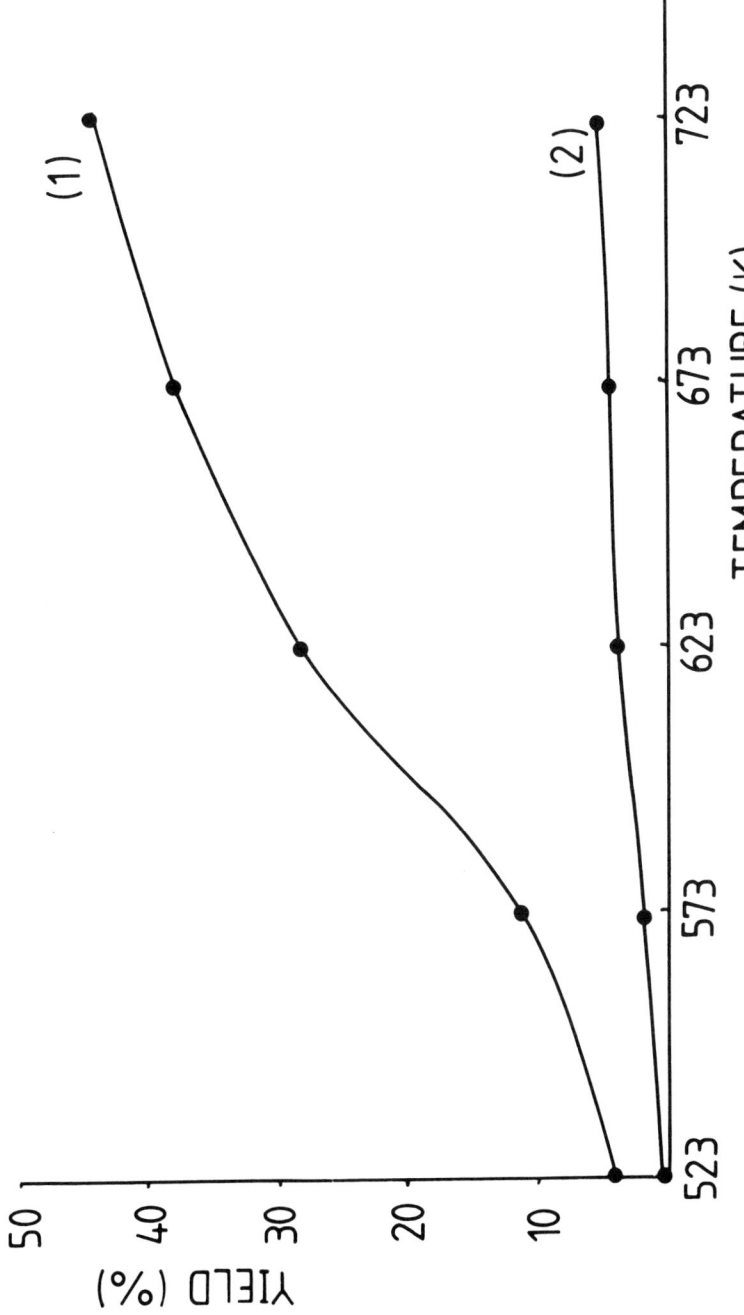

Figure 2a. Addition of cyclohexane to methyl acrylate. Effect of temperature on yield of methyl 3-cyclohexyl propionate 1 and methyl 2-cyclohexyl propionate 2. Reaction conditions: cyclohexane/methyl acrylate 25/1; residence time 10 min; pressure 450 bar; yield based on methyl acrylate charged.

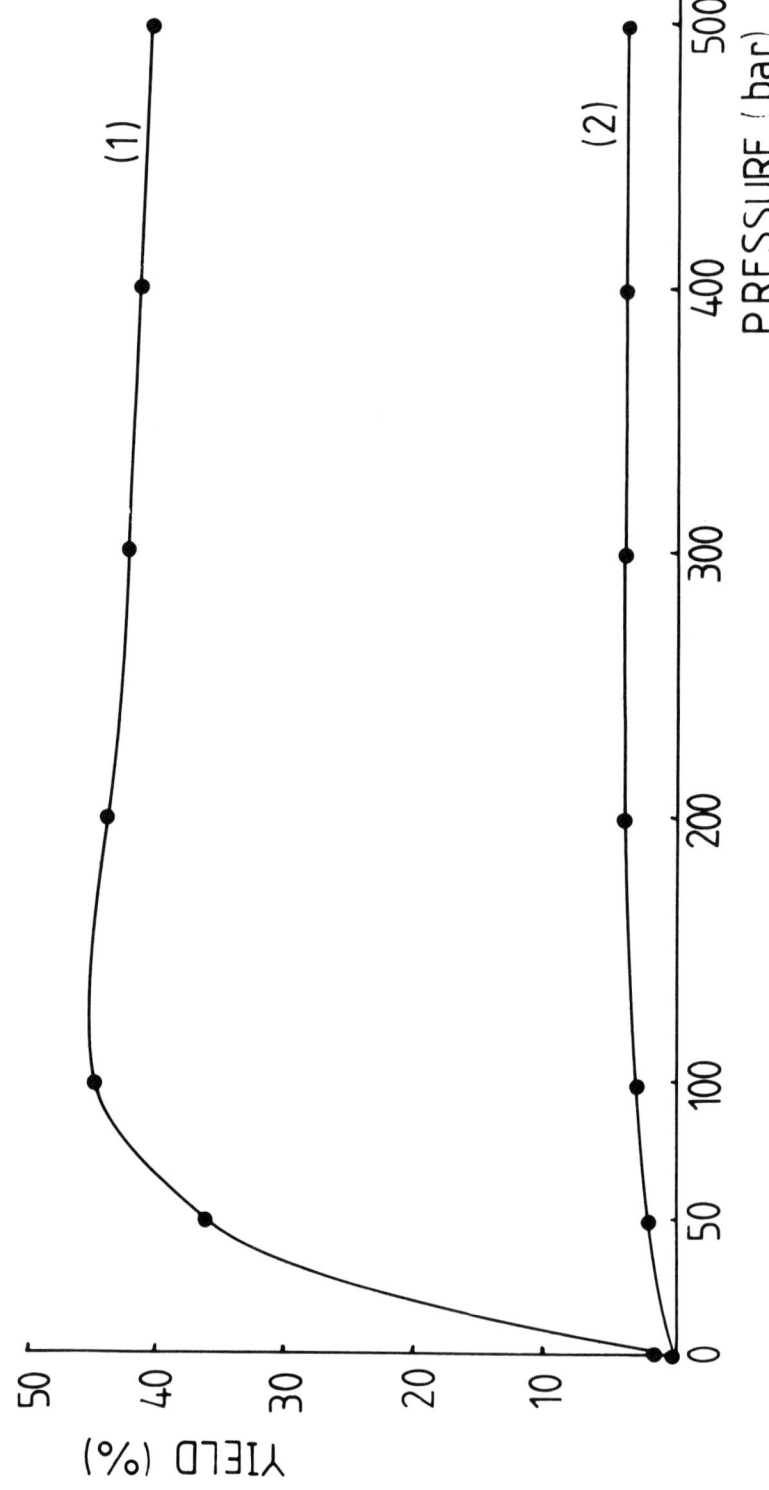

Figure 2b. Effect of pressure on yield of 1 and 2. Reaction condition: cyclohexane/methylacrylate 25/1; residence time 10 min; temperature 723 K.

The Markownikow product 2 is formed in minor amounts.
The influence of pressure shows Figure 2b. The yield is increasing significantly with pressure up to 100 bar. The critical data of cyclohexane are: p_c=40 bar, T_c=280°C. The reaction mixture is supercritical. The essential influence of pressure can be explained by the increase of the density of the reactants allowing an intermolecular reaction with an appropriate rate.

Influence of residence time can be deduced from Figure 2c. At 450°C the reaction is complete after 3 min. At longer reaction times yield is reduced by consecutive reactions.

This example demonstrates unequivocally a remarkable advantage of reactions at elevated temperatures: they proceed extremely fast. On the other hand a rapid withdrawal of products from the reaction zone is necessary to avoid side reactions. This is possible only in a flow reactor.

Figure 2d shows the large effect of the concentration of either reactants. Apparently the concentration of cyclohexane is an important factor in the rate equation. High concentrations of methyl acrylate are favouring the oligomerization of acrylic ester (see section 5).

Table II. Addition of Toluene to Selected Alkenes.

$$\text{Ph-CH}_3 + CH_2=CHR \longrightarrow \text{Ph-CH}_2\text{-CH}_2\text{-CH}_2\text{-R} \tag{6}$$

Alkene	Toluene/Alkene	T[°C]	Residence time [min]	Yield [mol %]
Acrylonitrile	20	400	7	19.5
Methyl vinyl ketone	20	400	3	15
Vinyl acetate	20	380	2	4
Cyclohexene	50	410	10	2

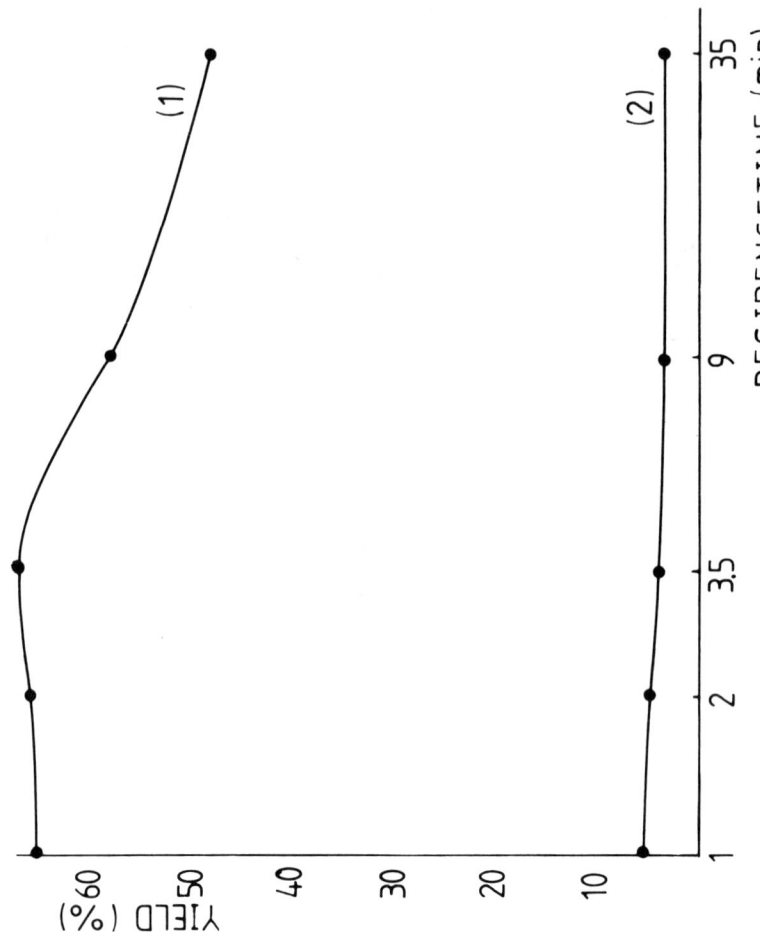

Figure 2c. Effect of residence time on yield of <u>1</u> and <u>2</u>. Reaction conditions: cyclohexane/methyl acrylate 50/1; pressure 200 bar; temperature 723 K.

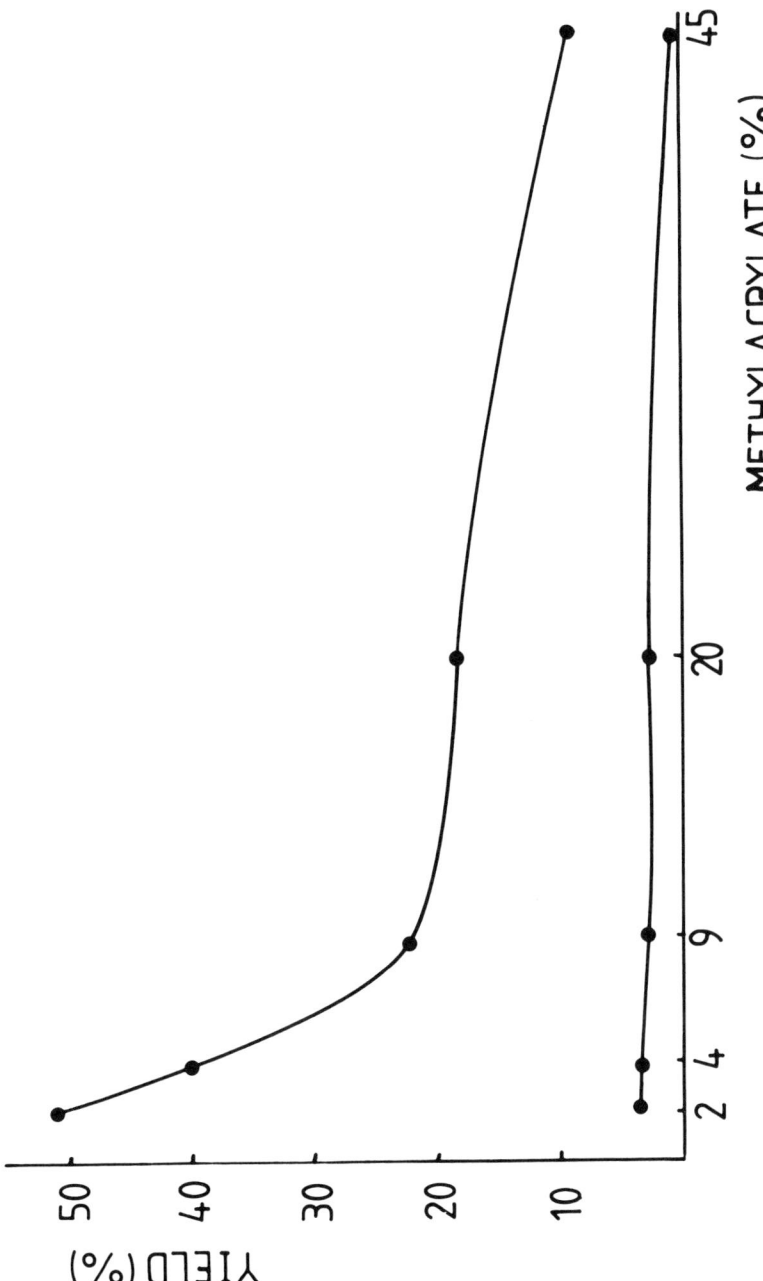

Figure 2d. Effect of concentration of methyl acrylate in cyclohexane on yield of 1 and 2. Reaction conditions: residence time 8 min; pressure 300 bar; temperature 723 K.

Besides cyclohexane we used other cycloalkanes such as cyclopentane and cycloheptane in order to obtain a simple product pattern. This is possible and observed also with alkylaromatics like toluene, xylene, cumene and analogous compounds (see Table II). A mixture of isomeric substitution products is obtained with n-alkanes. Our research with regard to alkane components is in progress. It may be mentioned that to our knowledge benzene is the only inert solvent for alkenes at elevated temperatures.

Currently we are trying to clarify the mechanism of this interesting reaction. Until now we can exclude catalysis of the transition metals of the reactor. The results obtained are identical in fused silica or stainless steel reactors (Figure 3). Addition of radical scavengers (hydroquinone, 2,6-di-tert.-butyl-4-methyl-phenol) did not influence the yield of main reaction product.

In the literature thermal addition of alkanes and alkylaromatics to alkenes has been described only for some special cases [8 - 12]. ALDER [10] compared this reaction with the Diels-Alder reaction and the ene reaction. We therefore proposed the term "ane-reaction" [7].

2. Thermal Addition of Alkanes to 1.3-Dienes.

Alkanes are adding thermally also to 1.3-dienes. Cyclohexane and 1.3-cyclohexadiene (400°C, 200 bar, 4 min, 1 mol % cyclohexadiene in cyclohexane) are yielding two isomeric cyclohexylcyclohexenes (12 % and 7 % respectively based on cyclohexadiene). The detailed structure analysis of these adducts is in progress. Comparable results are obtained with 1.3-butadiene and methyl 2.4-pentadienoate respectively as diene component in the reaction with cyclohexane. An addition product is not obtained with dimethyl 1.3-cyclohexadiene-1.4-dicarboxylate, but the dehydrogenation product dimethyl terephthalate is isolated.

3. Thermal Addition of Alkanes to Alkynes.

Alkanes are adding in a remarkably uniform reaction to alkynes to form alkenes [13]. For instance cyclohexane saturated with acetylene at 0.5 bar and pumped through the "HP-HT" apparatus yields 0.2 % (based on cyclohexane charged) vinylcyclohexane, a commercially interesting monomer.

$$\bigcirc + H-C\equiv C-H \xrightarrow{400°C, 400 \text{ bar}, 2 \text{ min}} \bigcirc-CH=CH_2 \qquad (7)$$

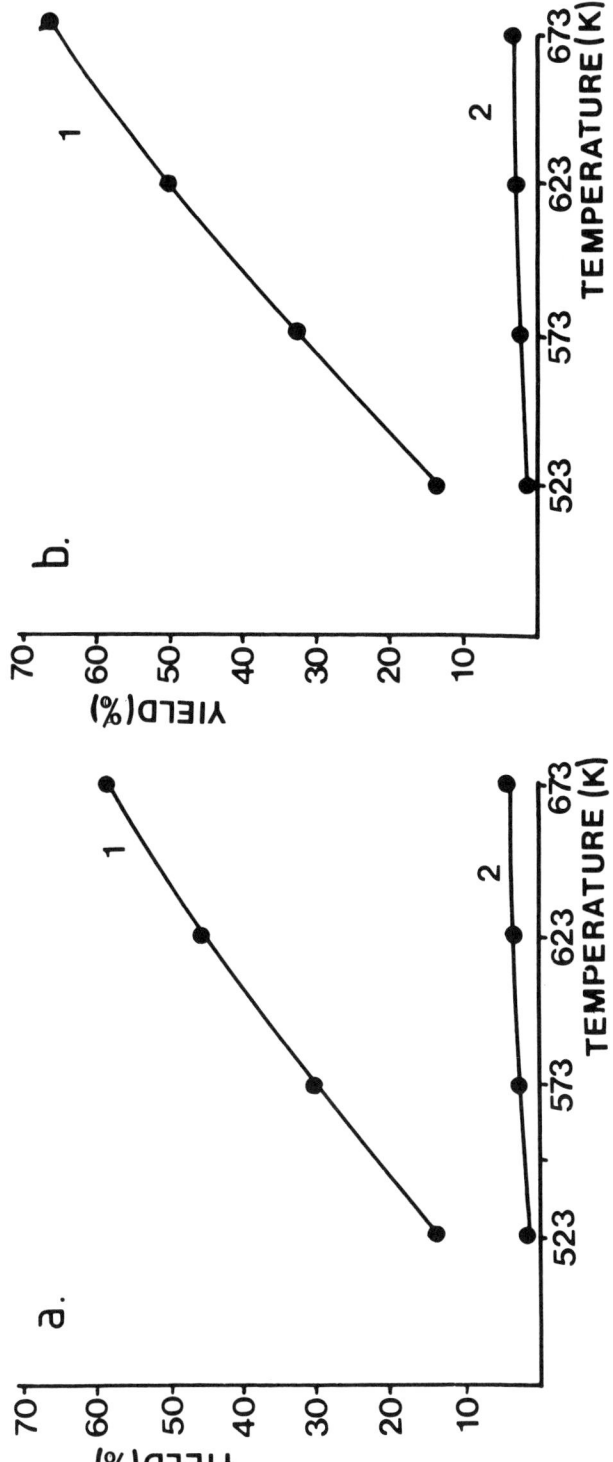

Figure 3. Comparison of the addition of cyclohexane to methyl acrylate in a) stainless steel reactor; b) fused silica reactor. Other reaction conditions are identical: residence time 1 min; pressure 200 bar; cyclohexane/methyl acrylate 100/1. 1:methyl 3-cyclohexyl propionate; 2:methyl 2-cyclohexyl propionate.

Recently KARPF and DREIDING published an intramolecular analogon of this reaction [14].

4. Reaction of Alkynes with Benzene.

In contrast to alkenes alkynes do react with benzene. A patent describes the synthesis of styrene from benzene and acetylene [15]. We could reproduce this result using our technique. Analogous experiments with methyl propiolate yielded methyl benzoate amongst other products. Dimethyl acetylenedicarboxylate gave dimethyl phthalate (8).

Apparently benzene is reacting as diene with an alkyne as dienophile to a barrelene derivative [16] followed by a retro-Diels-Alder-reaction. By GC/MS we could detect styrene in the reaction mixture, which should be formed from acetylene and benzene as mentioned above.

$$\text{C}_6\text{H}_6 + \underset{\text{COOCH}_3}{\overset{\text{COOCH}_3}{\text{C}\equiv\text{C}}} \xrightarrow[\text{10 min}]{300°C, 200 \text{ bar}} \text{[barrelene-COOCH}_3\text{]} \longrightarrow \text{HC}\equiv\text{CH} + \text{C}_6\text{H}_4(\text{COOCH}_3)_2 \quad (8)$$

5. Thermal Dimerization of Acrylic Compounds.

There has been much interest in the catalytic dimerization of acrylic compounds yielding 2-methyleneglutaro derivates, which represent trifunctional compounds and useful intermediates in organic synthesis [18]. Thermal dimerization has not been much investigated. DANUSSO [19] was able to isolate dimethyl 2-methylene-glutarate in connection with experiments of thermal polymerization of methyl acrylate at temperatures of 300°C to 400°C. No yield data were given.

We have carried out the thermal dimerization and oligomerization of methyl acrylate in the HP-HT apparatus (9).

This reaction corresponds to an addition of a vinylic C-H bond to an alkene.

Our results are compiled in Figures 7 and 8 respectively.

Apparently dimerization occurs as the first step of the thermal polymerization of methyl acrylate. At temperatures below 340°C higher oligomers are favoured. With temperature increasing the formation of higher oligomers is reduced in

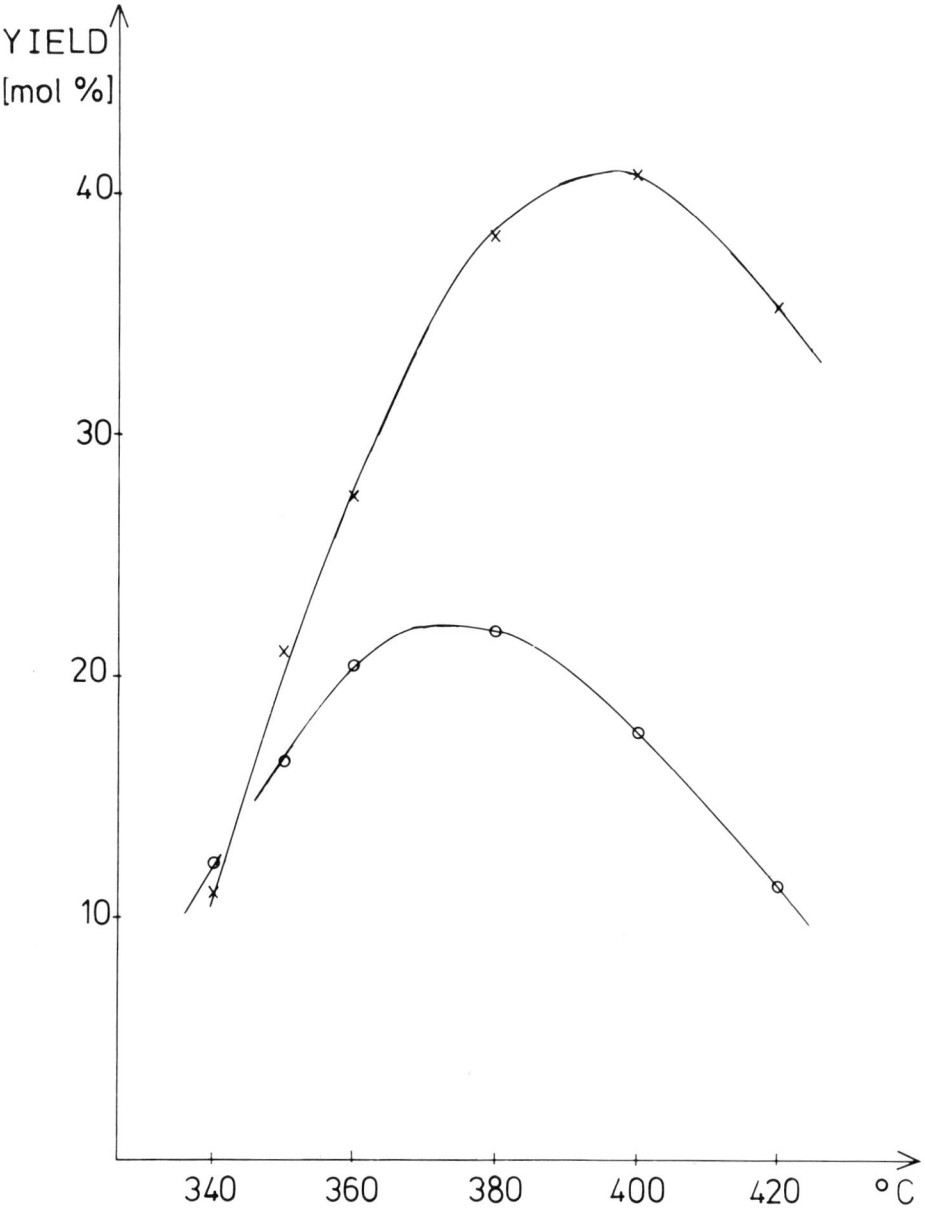

Figure 4. Dimerization of methyl acrylate. Effect of temperature on yield of dimer (xxx) and trimer (ooo). Reaction conditions: 1 mol/l methyl acrylate in benzene; residence time 5 min; pressure 200 bar.

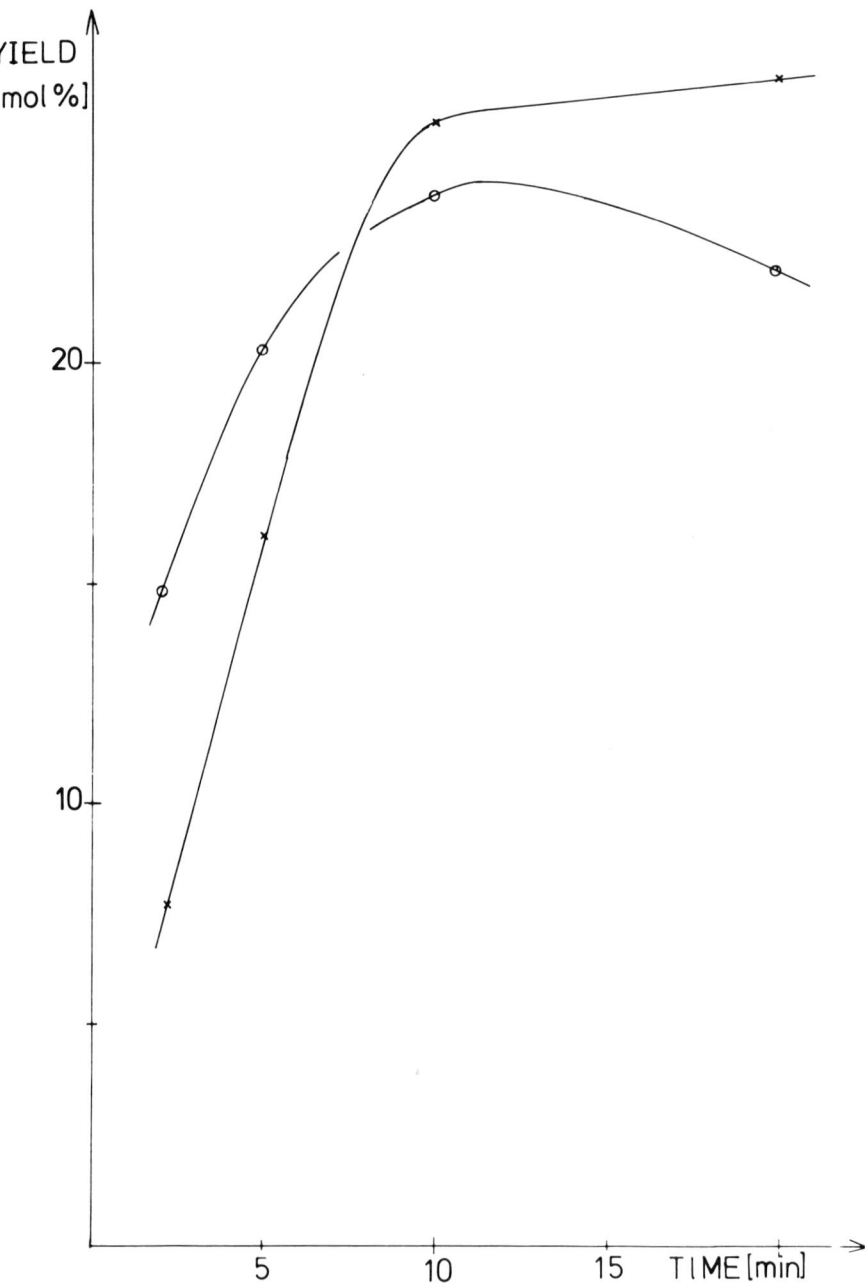

Figure 5. Dimerization of methyl acrylate. Effect of residence time on yield of dimer (x x x) and trimer (o o o). Reaction conditions: 2 mol/l methyl acrylate in benzene; temperature 350°C; pressure 200 bar.

favour of the dimer. Depolymerization reactions are dominating above 400°C.

Methyl vinyl ketone and acrylonitrile are able to dimerize analogously (10,11).

$$2\ CH_2=CH-COOCH_3 \longrightarrow \underset{\underset{COOCH_3}{|}}{CH_2=C}-CH_2-\underset{\underset{COOCH_3}{|}}{CH_2}$$

$$+\ \underset{\underset{COOCH_3}{|}}{CH_2=C}-CH_2-\underset{\underset{COOCH_3}{|}}{CH}-CH_2-\underset{\underset{COOCH_3}{|}}{CH_2} \qquad (9)$$

$$2\ \underset{\underset{\underset{CH_3}{|}}{C=O}}{CH_2=CH} \xrightarrow[5\ min]{380°C,\ 200\ bar} \underset{\underset{\underset{CH_3}{|}}{C=O}}{CH_2=C}-CH_2-CH_2-\underset{\underset{CH_3}{}}{C=O} \qquad (10)$$

$$2\ \underset{\underset{CN}{|}}{CH_2=CH} \xrightarrow[10\ min]{400°C,\ 200\ bar} \underset{\underset{CN}{|}}{CH_2=C}-CH_2-CH_2-CN \qquad (11)$$

CONCLUSIONS

Experimental work of explorative nature has shown that intermolecular thermal reactions of hydrocarbons can proceed with high yields of components which have great potential for the synthetic organic chemist and industrial application. These C-C bond forming reactions proceed at temperatures of 300°C and up which is the range for thermal cracking of hydrocarbons. The high yield of addition products is enhanced by the density of the reactants.

ACKNOWLEDGMENTS

The authors wish to thank Mr. F. Bangert and Mr. D. Neemeyer for their skillful assistance in carrying out the experiments. Thanks also to Mr. W. Schwarting for performing the GC/MS analyses and to Mrs. M. Rundshagen for performing the NMR and IR spectra. Thanks also to Mrs. R. Raphael for her skillful typing of the manuscript. We are grateful to Dr. D. Geere, Hewlett-Packard, Avondale, for providing of the fused-silica-capillaries.

LITERATURE

1. Brown, R.F.C., Pyrolytic Methods in Organic Chemistry (New York:Academic Press, 1980).

2. Woodward, R.B., and R. Hoffmann, The Conservation of Orbital Symmetry (Weinheim: Verlag Chemie GmbH, 1970).

3. Hendrickson, J.B., Angew. Chem. 86: 71 (1974); Angew. Chem. Int. Ed. Engl. 13: 47 (1974).

4. Kershaw, J.R., S. Afr. J. Chem. 31: 15 (1978).

5. Köll, P., and J.O. Metzger, Angew. Chem. 90: 802 (1978); Angew. Chem. Int. Ed. Engl. 17: 754 (1978).

6. Metzger, J.O., and P. Köll, Angew. Chem. 91: 74 (1979); Angew. Chem. Int. Ed. Engl. 18: 70 (1979).

7. Metzger, J.O., J. Hartmanns, and P. Köll, Tetrahedron Lett. 22: 1891 (1981).

8. I.G. Farbenindustrie (J. Binapfl), Deutsches Reichspatent 607.380 (1935).

9. Frey, F.E., and H.J. Hepp, Ind. Eng. Chem. 28: 1439 (1936).

10. Alder, K., and O. Wolff, Liebigs Ann. Chem. 576: 182 (1952).

11. Oltay, E., J.M.L. Penninger, and H. Maatman, Angew. Chem. 84: 947 (1972); Angew. Chem. Int. Ed. Engl. 11: 918 (1972).

12. Nazarova, N.M., and L.Kh. Freidlĭn, Doklady, Akad. Nauk S.S.S.R. 137: 1125 (1961); C.A. 55: 20991i (1961).

13. Metzger, J.O., and P. Köll, Angew. Chem. 91: 75 (1979); Angew. Chem. Int. Ed. Engl. 18: 71 (1979).

14. Dreiding, A.S., and M. Karpf, Helv. Chim. Acta, 62: 852 (1979).

15. Universal Oil Prod. Co., A.P. 2377074 (1942); C.A. 39: 4093 (1945).

16. Ciganek, E., Tetrahedron Lett. 3321 (1967).

17. Hidai, M., and A. Misono, Aspects of Homogenous Catalysis, 2: 159 (1974).

18. Danusso, F., D. Sianesi, and R. Sciaky, Chim. Industria (Milano) 38: 293 (1956).

INDEX

acetamide 501
acetic acid/n-butane/water 189
acetone 41,386,391,416,500-503,516
  biomass extraction 516
  cellulose degradation 500-502
  coal extraction 386,391
  wood degradation 501-503
  wood extraction 416
acetone/carbon dioxide 35
acetonitrile/benzene/n-hexane/water 192
acrylic dimerization 528-531
acrylonitrile dimerization 531
aerogel 445-459
  dielectric constant 457
  index of refraction 455
  optical absorption 454
  physical properties 452-453
  strength 456-457
  synthesis 447-448
  thermal conductivity 455-456
  windows 457-459
alcogel 447-453
alcohol/carbon dioxide/water 101-109
alcohol/water 113-135
alcosol 446-453
  composition 448
alkane addition to alkynes 526
alkane addition to 1.3-dienes 526
alkynes 528
  reaction with benzene 528
ammonia 479
ammonia/nitrogen 26

ammonium sulfate/benzene/ethanol 190
ammonium sulfate/benzene/ethanol/water 192
analysis of gas systems 477, 497
anhydrosugars 509
anthracene/ethylene 143
argon 360
argon/helium 24
argon/krypton 13
argon/neon 24
asphaltenes 432
  as extraction products 432
azeotropes 13-14,19-20,65,107-108,130,132,200,210,298,302,306,348
  double 20
  heterogeneous 19,302,306
  homogeneous 19
  negative 14

benzaldehyde 517
  reduction to benzylic alcohol 517
benzene 55,265,299,369-371,388,391,528
  coal extraction 388,391
  reaction with alkynes 528
benzene/acetonitrile/n-hexane/water 192
benzene/ammonium sulfate/ethanol 190
benzene/ammonium sulfate/ethanol/water 192
benzene/ethylene 37,45-47,50,56-57
benzene/water 298-310
binary systems 3-28,31-79,199-218,221-242,249-253,263-317,345-355,377-393

biomass 499-513,516
  extraction 516
  liquefaction 499-513
biphenyl 139-157
  solubilities in carbon dioxide 139-157
biphenyl/carbon dioxide 139-157
biphenyl/ethylene 143
boiling liquid expanding vapor explosion (BLEVE) 95
n-butane 216
butane/octane 199-200,207, 210-218
butanol 65
n-butanol 55
n-butanol/carbon dioxide 37, 45,48,52,54,56
n-butanol/carbon dioxide/n-octane 64-70
n-butanol/ethane 18
n-butanol/n-octane 66,68-69
1-butanol 386
  coal extraction 386
2-butanol 501
  wood degradation 501

carbon, activated 461-475
  adsorption of phenol 461-475
carbon dioxide 42,49,55,58, 76-77,79,87-89,93,95,101-109,140,265,271,323-324, 327-328,369-371,396,420-421,461-475,479
  density 420-421
  dielectric constant 420-421
  fugacity 271
  liquid 42,95
  solubility 58,101-109
  solubility in ethanol 101-109
carbon dioxide fire extinguishers 95,97-99
  explosions 95,97-99
  ultrasonic screening 99
carbon dioxide/biphenyl 139-157
carbon dioxide/n-butanol 37, 45,48,52,54,56
carbon dioxide/n-butanol/n-octane 64-70
carbon dioxide/n-decane 349, 351-352
carbon dioxide/diphenylamine 24
carbon dioxide/n-dotriacontane 56
carbon dioxide/ethanol/water 107-109
carbon dioxide/helium 26
carbon dioxide/n-hexadecane 37,269-380
carbon dioxide/2-hexanol 14
carbon dioxide/hydrogen 24-25
carbon dioxide/isobutane 266, 268
carbon dioxide/isopropanol/water 107-109
carbon dioxide/methane 210,252
carbon dioxide/mono-olein 45, 49,53,56,59
carbon dioxide/naphthalene 70-75
carbon dioxide/nitrobenzene 18
carbon dioxide/n-octane 14,37, 45,48,52,56
carbon dioxide/2-octanol 14
carbon dioxide/oleic acid 45, 49,53,56-57,59
carbon dioxide/water 16,253-256,396,401
  coal extraction 401
carbon monoxide/hydrogen 24
carbon tetrachloride 77-78
carboxylic acids 57
Carnahan-Starling equation 225
Carnahan-Starling-van der Waals equation 222,225
cellulose 499-500,502,506,509-513
  degradation products 509-511
  pyrolysis products 500
char 500
chemical blasting for chemical disposal 99
chitin degradation 501-502
coal 377-393,395-406,409-416
  chemical analysis 381
  extraction 377-393,395-406, 409-416

hydrogenation residues 419-433
  pore size distribution 390
  See also lignite
coal-derived oil deashing 324
coal extraction 377-393,395-406,409-416
  toluene 381-384,386
coal hydrogenation residues 419-433
  extraction 419-433
coal tar fractions 160-161,165-170
  measuring solubilities 160-161,165-170
  solubility in methane 169-170
computer simulation of multicomponent vaporization 165-166
conformal solution theory 226
corresponding states model 341-355,359-373
  chain molecules 359
critical solvent deashing process 436
cubic equations of state 263-317,323-336
cyclohexane 520-525
  addition to alkanes 520-521
cyclohexane/methane 143
cyclohexane/methylacrylate 521-522,524-525,527
cyclohexene dimerization to 3-cyclohexyl-cyclohexene 517

cis-decalin/n-hexane 355
decane 386,388
  coal extraction 386,388
n-decane/carbon dioxide 349,351-352
n-decane/methane 349-351
dehydroabietic acids 77
dense gas chromatography (DGC) 462,489-490,494-495
dense gas chromatography/flame ionization detection (DGC/FID) 489
dense gas chromatography/mass spectrometry (DGC/MS) 489-497
depolymerization 531
depressurization 81-99
  isentropic 89
dioxane 412-416
  coal extraction 412-416
2,2-dimethylbutane/2,3-dimethylbutane/methane 193
dimethyl 2-methyleneglutarate 519
diphenylamine/carbon dioxide 24
diphenylethane 58
diphenylmethane 58
n-dotriacontane 55,72
n-dotriacontane/carbon dioxide 72-73

eicosane 368
empirical universal equation 367-373
ethane 55,232
ethane/n-butanol 18
ethane/ethanol 18,113-135
ethane/ethanol/water 113-135
ethane/n-heptane 13
ethane/iso-octane 45-46,50,56
ethane/iso-octane/toluene 59-65
ethane/methanol 16
ethane/naphthalene 24
ethane/n-propanol 18
ethane/toluene 61
ethanol 101-109,388,391,451,501,503,505-506
  coal extraction 388,391
  loadings in carbon dioxide 106
  solubility of carbon dioxide 101-109
  wood degradation 501,503,505-506
ethanol/ammonium sulfate/benzene 190
ethanol/ammonium sulfate/benzene/water 192
ethanol/carbon dioxide/water 107-109
ethanol/ethane 18,113,135

ethanol/ethane/water 113-135
ethanol/tetraethyl silicate/
   water 447-449
ethanol/water 130-131,506,
   508-512
   cellulose degradation 511-
      512
   peat degradation 512
   separation by ethane 130-
      131
   wood degradation 506,508,
      511
ether 501
   wood degradation 501
ethyl acetate 501
   wood degradation 501
ethylbenzene 58
ethyl glucosides 509
ethylene 41,55,265,290,327
ethylene/anthracene 143
ethylene/benzene 37,45-47,50,
   56-57
ethylene/biphenyl 143
ethylene/helium 26
ethylene/hexachlorethane 24
ethylene/iso-octane 45-46,50,
   56
ethylene/iso-octane/toluene
   59-65
ethylene/naphthalene 140-150,
   240-241,290-298,333-334,
   349-353
ethylene/n-propanol 54
ethylene/shale oil/toluene 441
ethylene/toluene 45-46,50,56-57
explosions 95,97-99

field variables 5-7,202
fire extinguishers
   See carbon dioxide fire
      extinguishers
flame ionization detection
   (FID) 488
Flory-Huggins model 345,353
fluorescence detection 487-
   488
fugacities 271,312-313,324-
   335
   carbon dioxide 271
furfural 517

reduction to furfuryl alcohol 517
furfuryl alcohol 517

gamma ray transmission for measuring densities 177
gas chromatography (GC) 485
gas-gas demixing 245-260
gas-liquid chromatography (GLC)
   37,41
gas-liquid demixing 252
generalized density variables
   5-7
Gibbs phase rule 3,188
glucosan 1 500-501
glucose 509
glycerides 57,77-78
glycerol 58
glycerol dioleate 77
glycerol mono-oleate 77
glycerol trioleate 77
gravimetric analysis 486,488
Griffiths theory 186,191-195

helium 10
helium/argon 24
helium/carbon dioxide 26
helium ethylene 26
^{3}helium/^{4}helium 185,194,200,
   210
helium/hydrogen 26
helium/methane 26
Helmholtz energy 342-345,353
hemicellulose 499,506,409,513
n-heptane 386,388
   coal extraction 386,388
n-heptane/ethane 13
heteroazeotropes 302,306
hexachlorethane/ethylene 24
n-hexadecane 41,55.58,265,269,
   368
n-hexadecane/carbon dioxide
   37,269-280
hexane 55,386,388,396,400-406
   coal extraction 396,400-406
n-hexane/acetonitrile/benzene/
   water 192
n-hexane/cis-decalin 355
n-hexane/methane 37,45,47,51,
   56-57

n-hexane/n-tridecane 354
1-hexanol 386,388,391
  coal extraction 386,388,391
2-hexanol/carbon dioxide 14
high-performance liquid chromatography (HPLC) 485,487-490,500
high-performance liquid chromatography/mass spectrometry (HPLC/MS) 489-490
hole theory of liquids 246-247
humins 503
hydrofluoric acid 447
  alcogel production 447
hydrogen 10
hydrogen/carbon dioxide 24-25
hydrogen/helium 26
hydrogen/methane 24,238
hydrogen sulfide 328
hydrogen sulfide/methane 16
hydroxymethylfurfural 509-511

infrared/Fourier transform spectrophotometry [IR(FT)] 486
infrared spectrophotometry 486,488
interfacial tension 178
  measurement 178
isobutane/carbon dioxide 266, 268
iso-octane 55
iso-octane/n-butanol/ethanol/water 192
iso-octane/ethane 45-46,50,56
iso-octane/ethane/toluene 59-65
iso-octane/ethylene 45-46,50,56
iso-octane/ethylene/toluene 59-65
iso-octane/toluene 59-62
isopimaric acids 77
isopropanol 101-104,107-109
  loadings in carbon dioxide 101-104,107-109
isopropanol/carbon dioxide/water 107-109

krypton 360
krypton/argon 13
krypton/methane 13

Landau-Griffiths exponents 195
Landau theory 191
lattice models 245-260
Lee-Kesler equation 222
Lennard-Jones model 227-231
Lennard-Jones molecules 227
Lennard-Jones potential 385
Leung-Griffiths model 200-204, 213,218
lignin 415,499,501,503-509,513
  extraction 415
lignite 395-406,409-416,426-433
  chemical analysis 395
  extraction 395-406,409-416
  extraction of hydrogenation residues 426-433
  liquefaction 395-406
liquefied petroelum gas (LPG) 95
local compositions model 330, 333,335
Lummus antisolvent deashing process 436

Mach disk 491-493
macromolecular systems 257-259
Mansoori-Leland approximation 227
Markownikow addition products 520,523
mass spectrometry (MS) 486-489
mean field approximation 224-225,229-230,242
mean field lattice gas model 245-260
mercury injection 31-35
methane 55,159,163,165,169-170, 212,232,346,369-371
methane/carbon dioxide 210,252
methane/cyclohexane 143
methane/n-decane 349-351
methane/2,2-dimethylbutane/ 2,3-dimethylbutane 193
methane/helium 26

methane/n-hexane 37,45,47, 51,56-57
methane/hydrogen 24,238
methane/hydrogen sulfide 16
methane/krypton 13
methane/methanol 238-239
methane/naphthalene 151,223
methane/nitrogen 199-203,207-212
methane/n-octane 143
methane/propane 346-347
methane/water 159,252-253,255
methanol 231-232,236-239,386, 388,391,396,399-406,447, 501,503,517
  coal extraction 386,388, 391,399-406
  reduction agent 517
  solubility 236-239
  wood degradation 501,503
methanol/ethane 16
methanol/methane 238-239
methanol/toluene 391-393
  coal extraction 391-393
methanol/water 404-406
methylacrylate/cyclohexane 521-522,524-525-527
methyl 3-cyclohexyl-2-methyl propionate synthesis 519
methyl palmitate 77
2-methyl-1-propanol 501
  wood degradation 501
methyl vinyl ketone dimerization 531
mixing rules 54,56,323-336, 472
  geometric mean 54
Moldover-Gallagher model 200-208,218
molecular beam 490-496
monatomic fluids 360
mono-olein 55,57
mono-olein/carbon dioxide 45, 49,53,56,59
monosaccharides 509

naphthalene 55,140-150,265, 269,282,288,290,327,335, 388
  fugacity 282,335
  solubility in ethylene 140-150

naphthalene/carbon dioxide 70-75,148-151,280-291,331-332, 334
naphthalene/ethane 24
naphthalene/ethylene 140-150, 240-241,290-298,333-334, 349,353
naphthalene/methane 151,223
near-critical conditions 199-218
neon 10
neon/argon 24
neophytadiene 77
nicotine 77
nitrobenzene/carbon dioxide 18
nitrogen 33,35,41,91-92,212, 328,388,500
  adsorption 388
  carrier gas 35
  cellulose pyrolysis 500
  liquid 33
nitrogen/ammonia 26
nitrogen/methane 199-203,207-212
nitrogen/oxygen 13
nonrandom two liquid model 330,333,335
nuclear magnetic resonance (NMR) spectroscopy 403,486
nucleation 82-83,89,93,95
  homogeneous 82-83,89,93,95

octane 55,216,231-232,235-238, 369-371
  solubility 235-238
n-octane/n-butanol 66,68-69
octane/n-butanol/carbon dioxide 64-70
octane/carbon dioxide 14,37,45, 48,52,56
n-octane/methane 143
octane/propane 218
1-octanol 386,388,391
  coal extraction 386,388,391
2-octanol/carbon dioxide 14
oil deashing 440
oil hydrogenation residues, extraction 419-433
oil sand extraction 419-433
oil shale
  See shale oil
oleic acid 55,57,59

oleic acid/carbon dioxide 45,49,53,56-57,59

packed-bed equilibrium cell 160,162
Padé approximant 229
Padé equation 221,230,239-242
  quantum corrections 239-240
paraffinic solvent 436
paraffin production 424
peat degradation 499,512
Peng-Robinson equation 85-86, 113-135,139,152,222-223, 225,242,263-317
pentane 41,324,386,391,426-433,501
  coal extraction 386,391
  hydrogenation residues extraction 426-433
  wood degradation 501
perturbation theory 28,221-242,342
perturbed-hard-chain equation 359-373
petroleum reservoir fluids 173-180
phase rule 3,5-9
phenol 461-475,509
  adsorption from carbon dioxide onto activated carbon 461-475
  fugacity 471-472
photon correlation spectroscopy 178-179
Pitzer acentric factor 360
Plöcker one fluid model 353-354
polyethylene/n-heptane 258-259
polyethylene/n-hexane 258
polyethylene/n-octane 258
polymerization 257
polymer scaling theories 361
Poynting correction 325
propane 324,423-433
  deasphalting 324
  hydrogenation residues extraction 426-433
  oil sand residues extraction 423-427
propane/methane 346-347

propane/octane 218
propane/perfluorocyclobutane 348-349
propane/shale oil/toluene 441
n-propanol/ethane 18
n-propanol/ethylene 54
1-propanol 386,391,501
  coal extraction 386,391
  wood degradation 501
2-propanol 501,503-504
  wood degradation 501,503-504
pyridine 386
  coal extraction 386

quaternary systems 192

Raoult's law 13
Redlich-Kwong equation 45-76, 222-223,225,239-242
refractive index detection 487-488
resynthesis reactions 77
rigid lattice model

sapphire 175-178
  window 178
SASOL 162
separator bottoms extraction 430-432
sewage sludge degradation 499, 512
shale oil 435-443
shale oil/entrainer/supercritical gas 437-443
shale oil/ethylene/toluene 441
silica aerogel
  See aerogel
skimmer 492-496
Soave-Redlich-Kwong equation 324-335,472
  modified form 324-335
Soave's correlation 72
sparged cells 163
sparged gas-liquid contactors 160,162
spinodal states 81-94,99,247-250,252,257,264,272,282,314
static cell 31-37,46
statistical mechanics 28
supercritical fluid chromatography (SFC)

541

See dense gas chromatography
supercritical fluid (SCF) drying 445,450-452,459

tar sands 435
 mineral content 435
ternary systems 59-79,101-109,113-135,189-193,435-443,447-449
tetraethyl orthosilicate 447
tetraethyl silicate/ethanol/water 447-449
tetrahydrofurane 416
 wood extraction 416
tetralin 386
 coal extraction 386
thermal compressor 33
thermal conductivity test 488
thin layer chromatography (TLC) 485,487
toluene 55,324,386,391,412,414,416,424-433
 coal extraction 386,391,412,414,416
 deashing 324
 hydrogenation residues extraction 426-433
 oil sand extraction 424,427
toluene/ethane 61
toluene/ethane/iso-octane 59-65
toluene/ethylene 45-45,50,56-57
toluene/ethylene/iso-octane 59-65
toluene/ethylene/shale oil 441
toluene/iso-octane 59-62
toluene/propane/shale oil 441
Toth isotherm 466-467,471
tricritical points 185-195
n-tridecane/n-hexane 354
2,2,5-trimethyl hexane 369-371
triolein 58

ultrasonic screening of fire extinguishers 99
ultraviolet monitoring 481,487
ultraviolet/visible spectrophotometry (UV/VIS) 486,488

universal equation 367-373

vacuum distillation residue extraction 428-433
van der Waals equation 10,200,224-225,230,264,324,326-330,333,360
 modified form 324,326-330,333
van der Waals one fluid model 226-227,230,234,238,242,344,348,353-354
van der Waals surface tension theory 195
Van Laar equation 330,333,335
vinyl chloride 95
virial equation 384
virial expansion 222
viscosity measurement 177-178
visual cell system 175-180
 safety features 179

wall cooling 96
 effect on depressurization 96
water 265,267,369-371,399,401
 coal extraction 399,401
water/acetonitrile/benzene/n-hexane 192
water/ammonium sulfate/benzene/ethanol 192
water/carbon dioxide 16,253-256,396,401
 coal extraction 401
water/carbon dioxide/ethanol 107-109
water/carbon dioxide/isopropanol 107-109
water/ethane/ethanol 113-135
water/ethanol 130-131,506,508-412
 cellulose degradation 511-512
 peat degradation 512
 separation by ethane 130-131
 wood degradation 506,508-511
water/ethanol/tetraethylsilicate 447-449
water/methane 159,252-253,255
windows 178,457-459
 aerogel 457-459

sapphire 178
wood 410-411,415-416,499-513
  degradation 499-513
  extraction 410-411,415-416
  liquefaction 513

saccharification 512-513

xenon 360
xylene 386,391
  coal extraction 386,391

DATE DUE